THE CHEMISTRY
OF POLYPEPTIDES

Essays in Honor of Dr. Leonidas Zervas

THE CHEMISTRY
OF POLYPEPTIDES

Essays in Honor of Dr. Leonidas Zervas

Edited by P. G. Katsoyannis

Chairman, Department of Biochemistry
Mount Sinai School of Medicine
City University of New York
New York, New York

PLENUM PRESS • NEW YORK - LONDON

Library of Congress Cataloging in Publication Data

Main entry under title:

The Chemistry of polypeptides.

Includes bibliographies.
CONTENTS: Katsoyannis, P. G. The scientific work of Leōnidas Zervas.–Bodan-
szky, M. and Klausner, Y. S. Active esters and the strategy of peptide synthesis –
Young, G. T. The facilitation of peptide synthesis by the use of picolyl esters. [etc.]
1. Polypeptides–Addresses, essays, lectures. 2. Zerbas, Leōnidas Theodorou, 1902–
I. Zerbas, Leōnidas Theodorou, 1902–   II. Katsoyannis, P. G., ed.
QP552.P4C47   574.1'9245   73-79423
ISBN 0-306-30730-8

QP552
P4
C47

© 1973 Plenum Press, New York
A Division of Plenum Publishing Corporation
227 West 17th Street, New York, N.Y. 10011

United Kingdom edition published by Plenum Press, London
A Division of Plenum Publishing Company, Ltd.
Davis House (4th Floor), 8 Scrubs Lane, Harlesden, London, NW10 6SE, England

Printed in the United States of America

CONTRIBUTORS TO THIS VOLUME

H. C. Beyerman

Laboratory of Organic Chemistry
Technische Hogeschool
Delft, The Netherlands

Miklos Bodanszky

Department of Chemistry
Case Western Reserve University
Cleveland, Ohio

M. Brenner

Institut für Organische Chemie
Universität Basel
Basel, Switzerland

Evangelos Bricas

Equipe de Recherche No. 15, C.N.R.S. and
Laboratoire des Peptides
Institut de Biochimie
Université Paris-Sud
Orsay, France

L. C. Craig

Rockefeller University
New York, New York

E. W. B. de Leer

Laboratory of Organic Chemistry
Technische Hogeschool
Delft, The Netherlands

P. Fankhauser

Institut für Organische Chemie
Universität Basel
Basel, Switzerland

M. Fridkin

Departments of Organic Chemistry and Biophysics
The Weizmann Institute of Science
Rehovot, Israel

Joseph S. Fruton

Kline Biology Tower
Yale University
New Haven, Connecticut

Horst Hanson

Physiologisch-Chemisches Institut
Martin-Luther-Universität
Halle-Wittenberg, GDR

Ralph Hirschmann

Merck Sharp & Dohme Research Laboratories
Division of Merck & Co., Inc.
West Point, Pennsylvania

J. Hirt

Laboratory of Organic Chemistry
Technische Hogeschool
Delft, The Netherlands

E. Katchalski

Departments of Organic Chemistry and Biophysics
The Weizmann Institute of Science
Rehovot, Israel

Panayotis G. Katsoyannis

Mount Sinai School of Medicine
of the City University of New York
New York, New York

Yakir S. Klausner

Department of Chemistry
Case Western Reserve University
Cleveland, Ohio

Kálmán Medzihradszky

Institute of Organic Chemistry
Eötvös L. University
Budapest, Hungary

R. B. Merrifield

Rockefeller University
New York, New York

Yu. A. Ovchinnikov

Shemyakin Institute for Chemistry of
Natural Products
USSR Academy of Sciences
Moscow, USSR

A. Patchornik

Departments of Organic Chemistry and Biophysics
The Weizmann Institute of Science
Rehovot, Israel

Iphigenia Photaki

Laboratory of Organic Chemistry
University of Athens
Athens, Greece

Josef Rudinger

Institut für Molekularbiologie und Biophysik
Eidgenössische Technische Hochschule
Zürich, Switzerland

Irving L. Schwartz

Department of Physiology and Biophysics
Mount Sinai School of Medicine of the
City University of New York and
Medical Research Center
Brookhaven National Laboratory
New York, New York

Daniel F. Veber

Merck Sharp & Dohme Research Laboratories
Division of Merck & Co., Inc.
West Point, Pennsylvania

Erich Wünsch

Max-Planck-Institut für Biochemie
Abteilung für Peptidchemie
München, Germany

G. T. Young

The Dyson Perrins Laboratory
Oxford University
Oxford, England

PREFACE

Leonidas Zervas, Emeritus Professor of organic chemistry at the University of Athens, and past president of the National Academy of Greece, celebrated his seventieth birthday this past year. For almost fifty years Zervas devoted his scientific skills and perception to the advancement of chemistry, particularly in the field of peptides and proteins. Indeed, his efforts, along with those of his teacher, co-worker, and friend, Max Bergmann, laid the foundations for a new era in the chemistry of peptides and proteins.

Many of his colleagues and former students felt that it would be most appropriate at this time to honor him with a commemorative volume. They have contributed to this volume chapters describing some of their work and reviewing the advancements in particular areas of polypeptide chemistry. They dedicate this volume to Leonidas Zervas as an expression of their esteem and appreciation for the role he has played for the past half-century in the field of peptides and proteins.

July, 1973   *P. G. Katsoyannis*

CONTENTS

Chapter 10
Survey of the Synthetic Work in the Field of the Bacterial Cell Wall Peptides 205

Evangelos Bricas

Chapter 11
Intracellular Proteolysis and Its Demonstration with Synthetic and Natural Peptides and Proteins as Substrates 237

Horst Hanson

CHAPTER 1

THE SCIENTIFIC WORK OF LEONIDAS ZERVAS

Panayotis G. Katsoyannis

Mount Sinai School of Medicine of the City University of New York
New York, N.Y.

Leonidas Zervas was born in Megalopolis, Greece, in 1902. He graduated from high school in the town of Kalamata in 1918 and attended the School of Natural Sciences, Chemistry Section, University of Athens, for 2 years. In 1921, he interrupted his studies in Athens and moved to Berlin, where he attended the Philosophical Faculty (Chemistry Section), the University of Berlin, and received his Ph.D. degree in chemistry in 1926.

Upon completion of his studies in Berlin, he went to Dresden to the Kaiser-Wilhelm Institut, of the Kaiser-Wilhelm Gesellschaft, known today as the Max Planck Gesellschaft, where he carried out his thesis work. The late Max Bergmann was then director of the Institute. Zervas worked with Bergmann as a research associate (1926–1929), and later as Head of the Organic Chemistry Section (1929–1934). During the latter 5 years, he served also as the Deputy Director of the Institute. In 1933, notwithstanding the protests of the Kaiser-Wilhelm Gesellschaft, Bergmann suffered the fate of many other distinguished scientists and was forced to resign. Zervas, persuaded by Bergmann and the Kaiser-Wilhelm Gesellschaft, remained as Deputy Director of the Institute to finish up the work in progress. He resigned that post in 1934 and joined Bergmann at the Rockefeller Institute of Medical Research in New York, the present-day Rockefeller University. In 1937, Zervas was named Professor of Organic Chemistry and Biochemistry at the University of Thessaloniki, Greece. In 1939, he was appointed Professor of Organic Chemistry at the University of Athens. He held this chair until his retirement in 1968. He was elected a member of the National Academy of Greece (Academy of Athens) in 1956 and served as its President in 1969–1970.

Zervas was the recipient of many honors from distinguished scientific societies and universities. Among these may be mentioned his honorary membership in the American Society of Biological Chemists. The University of Basel, on the occasion of its Fifth Centennial Celebration, conferred on Zervas an honorary doctorate degree. He is among the founders and served for 10 years as a Vice-President of the Royal Hellenic Research Foundation.

The scientific work of Zervas is concerned with problems of pure organic chemistry and with chemical problems closely related to biochemistry. Specifically, this work deals with chemical aspects of carbohydrates, with methods of phosphorylation of hydroxy and amino compounds, and with the chemistry of amino acids, peptides, and proteins. Many of his publications on amino acids and peptides during the period from 1926 to 1936 represented a common effort with his teacher, coworker, and friend, Max Bergmann. In a very real sense, the present volume is also a tribute to that great scientist. In this chapter I will try to present a brief survey of the scientific contributions of Zervas and, regretfully, omit the relevant contributions of other investigators.

Throughout his scientific career, Zervas has shown a particular interest in problems related to carbohydrates and amino sugars. In this interest, he follows the tradition that Bergmann has inherited from Emil Fischer. Zervas, in his first published work in this area, reported (in 1930) on the synthesis and the determination of the stereochemical configuration of sturacitol [1] and polygallitol [2] (Fig. 1). Sturacitol was synthesized by catalytic hydrogenation of tetra-O-acetyl-2-hydroxy-D-glucal. Its conversion, through a series of intermediates, to D-fructose established sturacitol as 1,5-anhydro-D-mannitol. Similarly, the conversion of tetra-O-acetyl-α-D-glucosyl bromide via catalytic hydrogenation and deacetylation to polygallitol proved the latter compound to be 1,5-anhydro-D-sorbitol.

When it became known that in natural products D-fructose is present in the furanoid configuration, intensive efforts were directed toward the preparation of such compounds. The synthesis of several crystalline deriva-

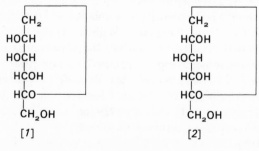

Fig. 1

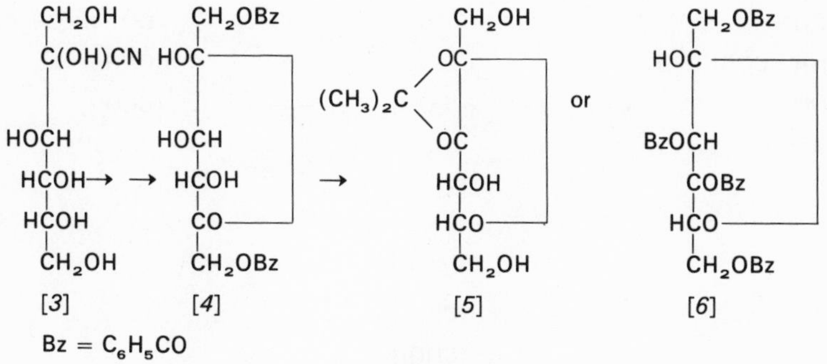

Fig. 2

tives of fructofuranose was accomplished. For these syntheses, the cyano-hydrin of D-fructose [3] served as the starting material (Fig. 2). Partial benzoylation in the 1,6-positions followed by removal of hydrogen cyanide led to 1,6-di-O-benzoyl-D-fructose [4], which in turn was converted to 2,3-isopropylidene-D-fructofuranose [5] or 1,3,4,6-tetra-O-benzoyl-D-fructo-furanose [6]. Both these products proved very useful in further synthetic processes.

Of particular interest are the studies carried out in the early 1930s related to D-glucosamine. Isolation of chitobiose upon acetolysis of chitin led to the elucidation of the general structure of this polysaccharide. The finding that chitobiose is N-acetyl-β-D-glucosaminyl-(1 → 4)-N-acetyl-D-glucosamine implied that chitin has the general structure of cellulose, the only difference being the replacement of the hydroxyl group at C-2 in cellulose by the N-acetyl moiety in chitin. The stereochemical configuration of D-glucosa-mine was established as that of 2-amino-D-glucose by taking advantage of the stereospecificity of peptidases in the hydrolysis of peptide bonds involving L-α-amino acids. In these studies, D-glucosamine was oxidized to D-glucosaminic acid [7], which in turn was coupled with glycine to yield glycylglucosaminic acid [8] or glucosaminylglycine [9] (Fig. 3). Both the latter dipeptides were not hydrolyzed by proteolytic enzymes. The corresponding dipeptides of the D-mannosaminic acid [10], obtained by epi-merization of D-glucosaminic acid, were readily hydrolyzed by peptidases. Tetra-O-acetyl-β-D-glucosamine [12], which proved to be a very useful intermediate in the synthesis of glycopeptides, was prepared in the following manner (Fig. 4): The amino group of D-glucosamine was initially protected by conversion to a Schiff's base, the hydroxyl groups were acetylated [11], and this product was partially deblocked with 1 equivalent of strong acid. Benzoylation of compound [12] followed by deacetylation produced N-benzoyl-D-glucosamine [13], which in turn, upon interaction with acetone, was converted to an oxazolidone derivative [14]. The oxazolidone was an

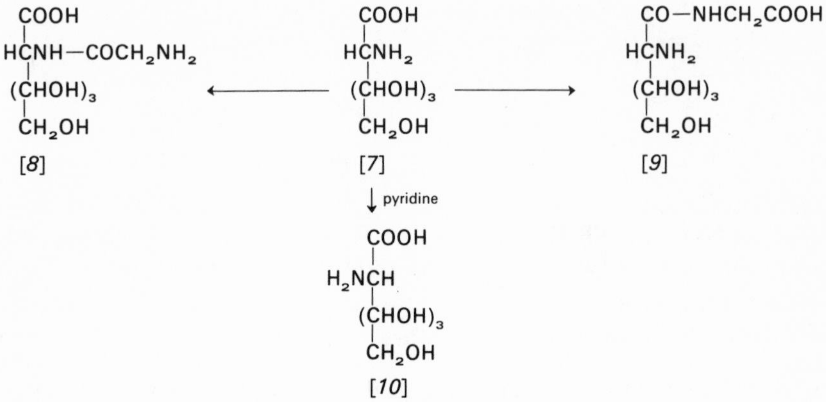

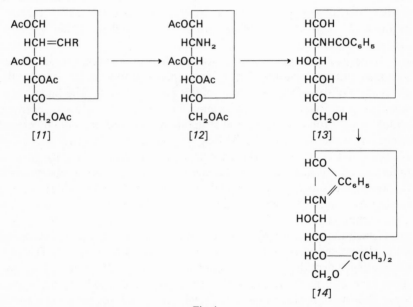

COOH
|
HCNH—COCH₂NH₂
|
(CHOH)₃
|
CH₂OH
[8]

COOH
|
HCNH₂
|
(CHOH)₃
|
CH₂OH
[7]

CO—NHCH₂COOH
|
HCNH₂
|
(CHOH)₃
|
CH₂OH
[9]

↓ pyridine

COOH
|
H₂NCH
|
(CHOH)₃
|
CH₂OH
[10]

Fig. 3

AcOCH
|
HCH=CHR
|
AcOCH
|
HCOAc
|
HCO—
|
CH₂OAc
[11]

AcOCH
|
HCNH₂
|
AcOCH
|
HCOAc
|
HCO—
|
CH₂OAc
[12]

HCOH
|
HCNHCOC₆H₅
|
HOCH
|
HCOH
|
HCO—
|
CH₂OH
[13]

HCO
\
| CC₆H₅
//
HCN
|
HOCH
|
HCO—
|
HCO——C(CH₃)₂
| /
CH₂O
[14]

Fig. 4

important intermediate in the synthesis of D-glucosaminofuranosides. The acetylated Schiff's base [11] served also as the starting material for the synthesis of D-glucosaminopyranosides.

It had been known for some time that a benzyl or substituted benzyl group attached to an oxygen atom as in alcohols, to an amino nitrogen or to a sulfur atom as in thioethers, is removed by hydrogenolysis as toluene or the

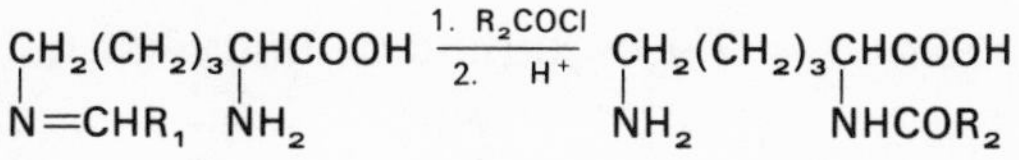

$$CH_2(CH_2)_3CHCOOH \xrightarrow[2.\ H^+]{1.\ R_2COCl} CH_2(CH_2)_3CHCOOH$$

$$N{=}CHR_1 \quad NH_2 \qquad\qquad NH_2 \qquad NHCOR_2$$

$$[15] \qquad\qquad\qquad\qquad [16]$$

Fig. 5

corresponding substituted derivative. In 1930, Bergmann and Zervas began applying this interesting reaction in a variety of synthetic processes. The synthetic sequence generally involved the preparation of intermediates whose functional groups were appropriately protected, i.e., by benzylation. The protecting groups were removed by catalytic hydrogenation in the final step of the synthesis. Originally, this sequence was used for protection of hydroxyl groups of carbohydrates through the formation of cyclic acetals with benzaldehyde. Such protected intermediates were subsequently employed for the synthesis of a variety of carbohydrate derivatives. Of more significance, however, was the extension of this reaction to the now famous "carbobenzoxy method" so widely used today in the synthesis of peptides. This will be discussed subsequently.

Zervas, in his doctoral thesis, referred to Schiff's bases of amino acids with aromatic aldehydes. The basic amino acids give rise to monoderivatives. Thus lysine with aromatic aldehydes produces a Schiff's base with the N^ε group only [15], and the free α-amino function can then be involved in the synthesis of N^α-acyllysine derivatives [16] (Fig. 5). Such an approach has indeed been used by Zervas in recent years in the synthesis of lysine-containing peptides. The Schiff's base of arginine proved to be a very insoluble compound even in dilute alkali, and this property was extensively employed for the removal of arginine from protein hydrolysates.

Because of its structure, arginine possesses many unusual properties, a fact that did not escape the scrutiny of Bergmann and Zervas. One of their early findings was the realization that the free ester of arginine [17] rapidly undergoes disproportionation to give rise to racemic α,δ-guanido valeric acid anhydride [19] and ornithine [20], presumably through an arginyl-arginine [18] intermediate (Fig. 6). It was further shown that with arginine derivatives with an unprotected guanidino group, rapid racemization takes place when the α-amino and carboxyl groups participate in some kind of ring formation. Thus racemization of an L-phenylalanyl-L-arginine dioxopiperazine derivative was found to take place rapidly (half-life approximately 30 min). Racemization of cyclic derivatives, e.g., hydantoins or dioxopiperazines, of other amino acids proceeds slowly even in the presence of dilute alkali (half-life more than 15 hr).

Of particular importance was the study of the racemization of optically active N-acylamino acids under the action of catalytic amounts of acid

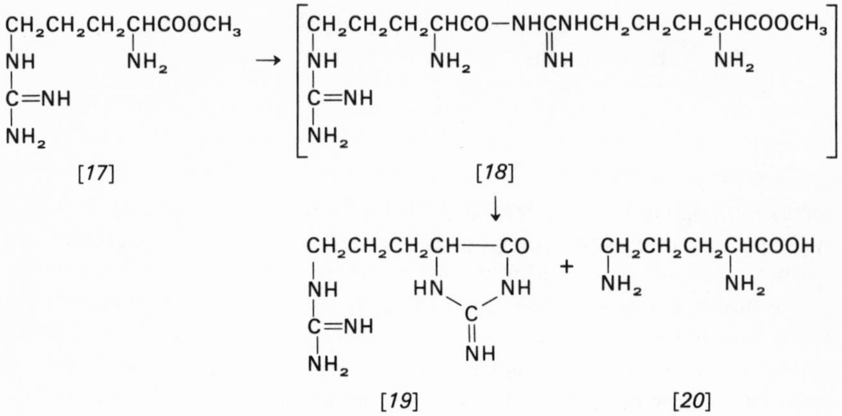

$$CH_2CH_2CH_2CHCOOCH_3$$
$$| \qquad\qquad\qquad |$$
$$NH \qquad\qquad\quad NH_2$$
$$|$$
$$C=NH$$
$$|$$
$$NH_2$$

[17]

$\rightarrow$

$$\left[CH_2CH_2CH_2CHCO-NHCNHCH_2CH_2CH_2CHCOOCH_3 \right.$$
$$\left. \quad NH \qquad NH_2 \qquad NH \qquad\qquad\qquad NH_2 \right.$$
$$\left. \quad C=NH \right.$$
$$\left. \quad NH_2 \right]$$

[18]

$\downarrow$

$$CH_2CH_2CH_2CH——CO$$
$$|\qquad\qquad\quad HN\quad NH$$
$$NH\qquad\qquad\qquad C$$
$$|\qquad\qquad\qquad\quad \|$$
$$C=NH\qquad\qquad NH$$
$$|$$
$$NH_2$$

[19]

$+$

$$CH_2CH_2CH_2CHCOOH$$
$$| \qquad\qquad\qquad |$$
$$NH_2 \qquad\qquad NH_2$$

[20]

Fig. 6

$$RCHCOOH \underset{+H_2O}{\overset{-H_2O}{\rightleftarrows}}$$
$$|$$
$$NHCOCH_3$$

$$RCH——CO \rightleftharpoons RC===COH$$
$$\quad N\quad O \qquad\qquad N\quad O$$
$$\quad\quad C \qquad\qquad\qquad C$$
$$\quad\quad | \qquad\qquad\qquad |$$
$$\quad\quad CH_3 \qquad\qquad\quad CH_3$$

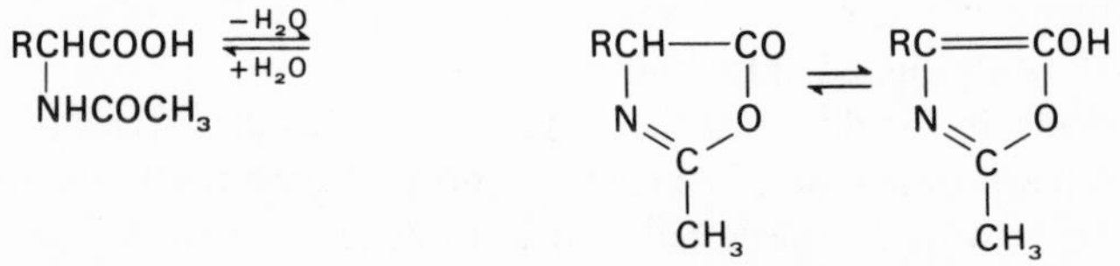

Fig. 7

anhydrides such as acetic anhydride. Under such conditions, racemization occurs rapidly, presumably through the formation and subsequent enolization of an oxazolone (Fig. 7). Excess of acetic anhydride leads to the quantitative formation of racemic oxazolones. N-Acylpeptides behave in exactly the same fashion. In this case, however, racemization is not confined only to the C-terminal amino acid residue. Thus interaction of acetic anhydride with N-acetyl-L-phenylalanyl-L-tyrosine leads not only to complete racemization of the tyrosine residue but also to partial racemization of the penultimate phenylalanine residue. Apparently, this occurs because the oxazolone [21] formed undergoes further intramolecular tautomerizations [22] and [23] (Fig. 8). The assumption that an oxazolone intermediate is formed was based on the fact that N-acyl derivatives of amino acids containing a secondary amino group (which therefore cannot form an oxazolone) usually are not racemized under the abovementioned conditions. It is well known that occasionally racemization also takes place during peptide bond formation. In recent years, intensive studies by several investigators, and particularly by Young, on this critical matter have indicated that in many instances this racemization also proceeds via oxazolone formation. It should be pointed out, however, that other investigators have reported cases where racemization without oxazolone formation may take place by simple proton abstraction from the asymmetrical α-carbon atom.

Racemic oxazolones, obtained from amino acids with excess of acetic anhydrides [24] or readily synthesized unsaturated oxazolones [25] (Fig. 9), on interaction with amino acids produced dipeptides. In spite of the obvious disadvantages, this approach was used by Bergmann and Zervas for the synthesis of peptides on several occasions in those early years. Application of the oxazolone method of synthesis was also extended to histidine peptides with most interesting results. Interaction of acetic anhydride with histidine was shown to yield the N^{im}-acetyloxazolone [26], which, upon reaction with an amino acid, gave a mixture of N^{α}-acetyl-DL-histidylamino acid and N-acetylamino acid (Fig. 10). The formation of N-acetylamino acid occurred through the transfer of the N^{im}-acetyl moiety of the histidine derivative to the α-amino group of the amino acid. This was the first time that an amide linkage was formed via an imidazole intermediate.

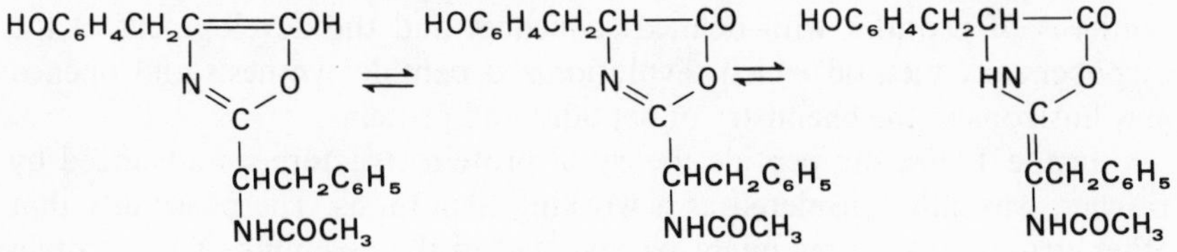

Fig. 8

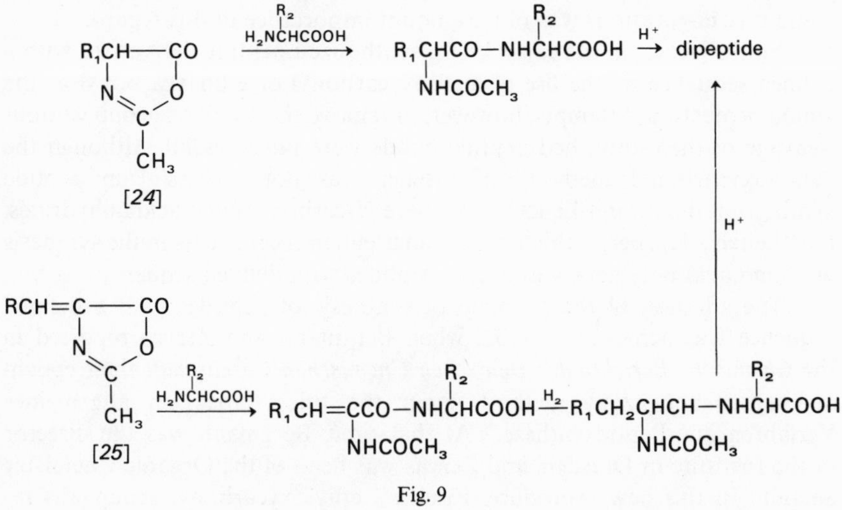

Fig. 9

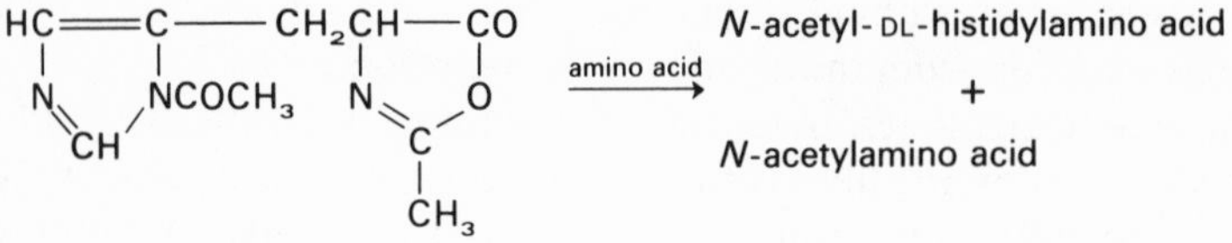

HC════C─────CH$_2$CH───CO   *N*-acetyl- DL-histidylamino acid
  │   │   │   │   amino acid
  N╲ ╱NCOCH$_3$ N═╲ ╱O   ────────→   +
   ╲CH╱   C
                        │   *N*-acetylamino acid
                       CH$_3$

[*26*]

Fig. 10

Subsequent studies have shown that acylpyrimidines, such as benzoyl theobromine and N,N^1-diacyldioxopiperazine, can also serve as acylating reagents.

The work described thus far was concerned with chemical changes common to almost all amino acids and with special chemical reactions characteristic of the side chains of certain amino acids.

The principal scientific achievement of Bergmann and Zervas was the synthesis of peptides with defined sequences and the development of the carbobenzoxy method which revolutionized peptide synthesis and opened new horizons in the chemistry of peptides and proteins.

In the 1930s, the peptide theory of protein structure, as advanced by Fischer, was still considered as a working hypothesis. The possibility that other structural features might be involved in the assembly of the protein molecule was not excluded. The elucidation of the specificity of proteolytic enzymes and the study of their action on proteins and synthetic peptides were central to the resolution of this problem. The availability of synthetic peptides containing all the naturally occurring amino acids of both the L and D configurations was of paramount importance in this regard.

Emil Fischer, in the early 1900s, synthesized peptide derivatives with a defined sequence by the use of methoxycarbonyl or ethoxycarbonyl as the amino protectors. Attempts, however, to remove the blocking group without cleavage of the established peptide bonds were unsuccessful. Although the "alkyloxycarbonyl method" of Fischer was not successful in peptide synthesis, it did permit Leuchs to prepare *N*-carboxyamino acid anhydrides, the "Leuchs–Körper," which proved so useful in recent years in the synthesis of amino acid polymers, and of polypeptides with defined sequences.

The solution of the problem of synthesis of peptides with a defined sequence was achieved in 1932, when Bergmann and Zervas reported in the *Chemische Berichte der Deutschen Chemischen Gesellschaft* their epoch-making carbobenzoxy method under the title "Über ein allgemeines Verfahren zur Peptidsynthese." At that time, Bergmann was the director of the Institute in Dresden, and Zervas was head of the Organic Chemistry Section. In this new procedure, Fischer's ethyloxycarbonyl group was replaced by the benzyloxycarbonyl (carbobenzoxy) moiety as the amino protector, which could be readily removed by catalytic hydrogenation under very mild conditions. The first publication dealt with the synthesis of a variety of N-carbobenzoxyamino acids [27], their conversion to acid chlorides, [28] or azides, and their subsequent coupling with other amino acids (Fig. 11). The N-carbobenzoxydipeptides [29] thus produced on catalytic hydrogenation were converted to the corresponding N-carbamic acids [30], which spontaneously decompose to produce free peptides [31] and carbon dioxide.

In this same publication, it was also reported that N-carbobenzoxyamino acids, unlike other N-acylamino acids, were not racemized upon interaction with acetic anhydride or during their conversion to acid chlorides. The latter derivatives, however, are unstable and, on exposure to heat, eliminate benzyl chloride with formation of N-carboxy-α-amino acid anhydrides. This last reaction is a method of synthesis of "Leuchs-Körper." The first report of Bergmann and Zervas was followed by some 30 publications by the same authors, or by Zervas alone, describing applications of the carbobenzoxy method to various amino acids and amino sugars, and to the problem of protection of the carboxyl group and the secondary functional groups of amino acids during peptide synthesis.

Some of the first applications of the carbobenzoxy method dealt with the synthesis of glycopeptides of D-glucosamine [32], the preparation of esters of amino acids with glucose [33], and some 25 years later the synthesis of the D-glucosylamide of L-aspartic acid [34], as shown in Fig. 12.

Extension of the carbobenzoxy method to glutamic and aspartic acids permitted the blocking of one of the carboxyl groups of these acids by esterification with benzyl alcohol, allowing the other carboxyl group free

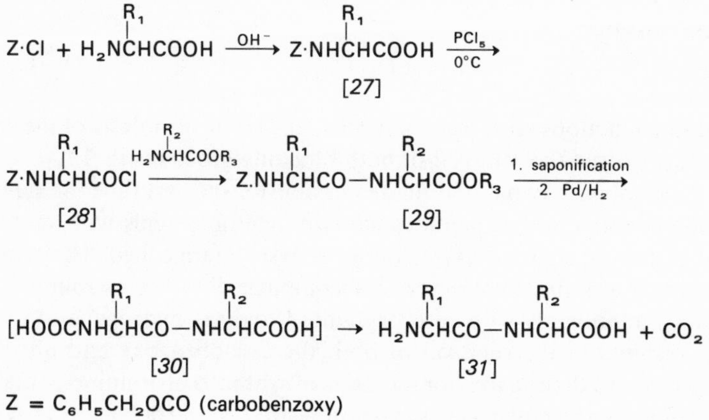

$$Z\cdot Cl + H_2N\overset{R_1}{\underset{|}{C}}HCOOH \xrightarrow{OH^-} Z\cdot NH\overset{R_1}{\underset{|}{C}}HCOOH \xrightarrow[0^\circ C]{PCl_5}$$

[27]

$$Z\cdot NH\overset{R_1}{\underset{|}{C}}HCOCl \xrightarrow{H_2N\overset{R_2}{\underset{|}{C}}HCOOR_3} Z.NH\overset{R_1}{\underset{|}{C}}HCO-NH\overset{R_2}{\underset{|}{C}}HCOOR_3 \xrightarrow[2.\ Pd/H_2]{1.\ saponification}$$

[28]   [29]

$$[HOOCNH\overset{R_1}{\underset{|}{C}}HCO-NH\overset{R_2}{\underset{|}{C}}HCOOH] \rightarrow H_2N\overset{R_1}{\underset{|}{C}}HCO-NH\overset{R_2}{\underset{|}{C}}HCOOH + CO_2$$

[30]   [31]

$Z = C_6H_5CH_2OCO$ (carbobenzoxy)

Fig. 11

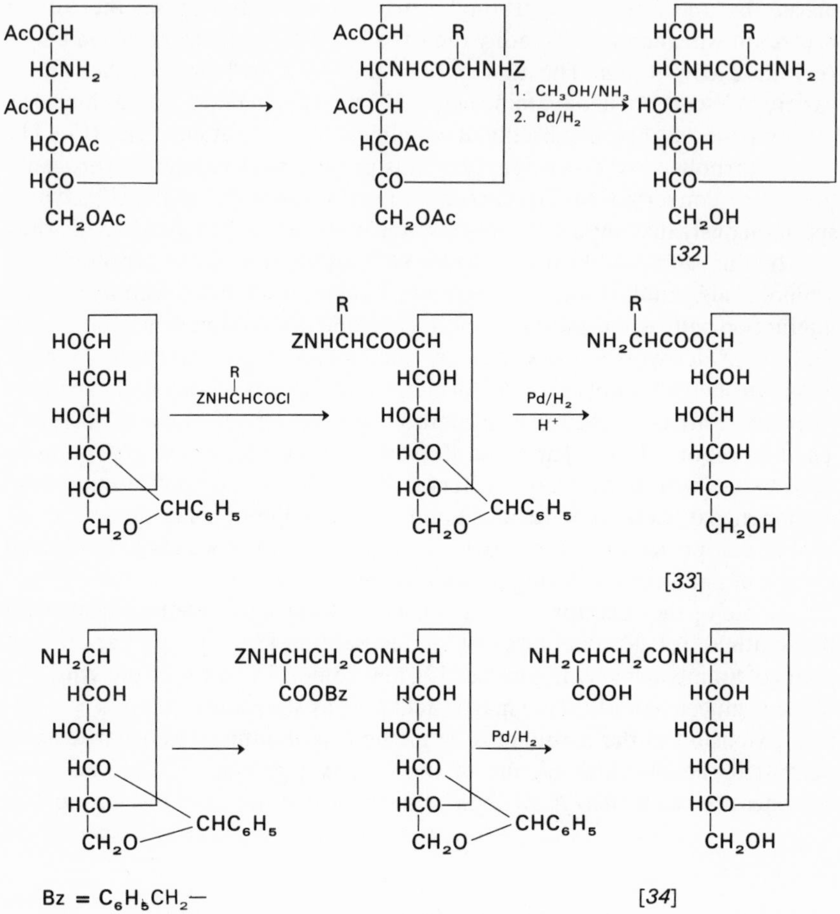

AcOCH
HCNH₂
AcOCH
HCOAc
HCO—
CH₂OAc

⟶

AcOCH   R
HCNHCOCHNHZ
AcOCH
HCOAc
CO—
CH₂OAc

$\xrightarrow[\text{2. Pd/H}_2]{\text{1. CH}_3\text{OH/NH}_3}$

HCOH   R
HCNHCOCHNH₂
HOCH
HCOH
HCO—
CH₂OH

[*32*]

HOCH
HCOH
HOCH
HCO
HCO—CHC₆H₅
CH₂O

$\xrightarrow[]{\text{ZNHCHCOCl}}$ (R)

ZNHCHCOOCH   R
HCOH
HOCH
HCO
HCO—CHC₆H₅
CH₂O

$\xrightarrow[\text{H}^+]{\text{Pd/H}_2}$

NH₂CHCOOCH   R
HCOH
HOCH
HCOH
HCO—
CH₂OH

[*33*]

NH₂CH
HCOH
HOCH
HCO
HCO—CHC₆H₅
CH₂O

⟶

ZNHCHCH₂CONHCH
COOBz   HCOH
HOCH
HCO
HCO—CHC₆H₅
CH₂O

$\xrightarrow[]{\text{Pd/H}_2}$

NH₂CHCH₂CONHCH
COOH   HCOH
HOCH
HCOH
HCO—
CH₂OH

Bz = C₆H₅CH₂—   [*34*]

Fig. 12

for coupling reactions with other amino acids. Hydrogenolysis of the resulting product caused the removal of both blocking groups (Fig. 13).

The carbobenzoxy method made possible for the first time the synthesis of proline peptides and of peptides containing amino acids with secondary functional groups, such as lysylhistidine or lysyglutamic acid. The synthesis of *C*-terminal arginine peptides was accomplished by the coupling of carbobenzoxyamino acids with nitroarginine. Catalytic hydrogenation of the product resulted in the removal of both the carbobenzoxy and nitro protecting groups. In their search for methods of synthesis of arginine-containing peptides, Zervas and coworkers have synthesized N^α-carbobenzoxy-, N^α,N^ω-dicarbobenzoxy-, and $N^\alpha,N^\omega,N^{\omega^1}$-tricarbobenzoxyarginine. It was then found that coupling of N^α,N^ω-dicarbobenzoxyarginine with amino acids results in the formation of the desired protected dipeptide along with N^α-N^ω-dicarbobenzoxyanhydroarginine as a byproduct. Such a lactame derivative, however, is not formed when $N^\alpha,N^\omega,N^{\omega^1}$-tricarbobenzoxyarginine is the starting material.

The great variety of oligopeptides, prepared by the carbobenzoxy method, were subsequently employed as substrates for studying the specificity of proteolytic enzymes. The first enzymatic studies were carried out in the Institute in Dresden in collaboration with Schleich and were continued at the Rockefeller Institute in New York. In this laboratory, Bergmann and Zervas were joined by Fruton. Zervas departed for Greece in 1935, and the synthetic and enzymatic studies were continued by Bergmann and Fruton until Bergmann's untimely death in 1944. Dr. Fruton covers the enzymatic studies elsewhere in the present volume. It should be emphasized here that, with the synthesis of suitable substrates for all the proteolytic enzymes, the specificity of the enzymes was elucidated and the peptide theory of protein structure was firmly established.

Parallel to the carbobenzoxy procedure for peptide synthesis, a method for the stepwise degradation of peptides was developed which also utilized the hydrogenolytic removal of the protecting benzyl group (Fig. 14). In this procedure, an N-acylpeptide azide [35] is subjected to Curtius degradation in the presence of benzyl alcohol to give a benzylurethan derivative [36]. Catalytic hydrogenation and heating of the latter product yield an N-acylpeptide amide [37] with one amino acid residue less than the starting peptide derivative, and an aldehyde [38] characteristic of the C-terminal amino acid residue. The resulting N-acylpeptide amide [37], on heating with hydrazine, is converted to the corresponding hydrazide, which could be

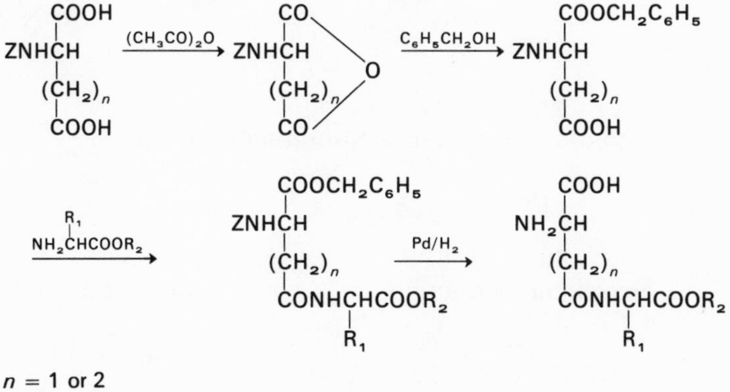

$n = 1$ or 2

Fig. 13

further degraded by the same series of reactions. This stepwise method for peptide degradation is now only of historical interest but demonstrates the expert chemical ingenuity of its discoverers.

The original carbobenzoxy procedure could not be applied for the synthesis of sulfur-containing peptides because the presence of sulfur inhibited the hydrogenolysis of the carbobenzoxy group. However, the introduction of the sodium in liquid ammonia method by du Vigneaud and the discovery of various acidolytic procedures by other investigators have made possible the use of the carbobenzoxy group in the synthesis of cysteine- or cystine-containing peptides and opened the way for the synthesis of the pituitary hormones, and finally of insulin. The chemistry of cystine-containing peptides, and particularly of the asymmetrical ones, presents special problems. The contribution of Zervas and his coworkers in this particular area is discussed by Dr. Photaki elsewhere in this volume.

Even today, some 40 years after its introduction in peptide synthesis, the carbobenzoxy moiety is the most frequently used group as an amino protector. There are instances, however, where its removal is a difficult task. Such is the case with carbobenzoxylated long polypeptide chains, which possess highly unfavorable solubility properties and which are very insoluble even in the solvents usually used for the decarbobenzoxylation step. Removal of the carbobenzoxy group is also problematic, particularly during chain elongation when carboxyl groups and secondary functions of the constituent amino acid residues are protected with blocking moieties that are labile under conditions used for decarbobenzoxylation. In such instances, the use of amino protectors which can be removed under very mild conditions such as trityl (triphenylmethyl) and arylsulfenyl becomes obligatory.

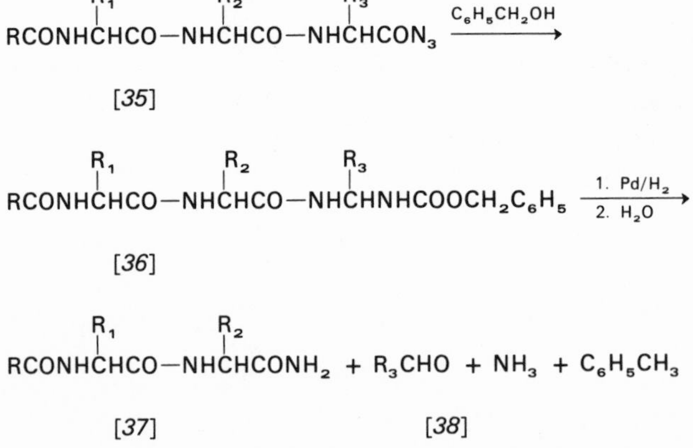

$$\overset{R_1}{\underset{|}{RCONHCHCO}}-\overset{R_2}{\underset{|}{NHCHCO}}-\overset{R_3}{\underset{|}{NHCHCON_3}} \xrightarrow{C_6H_5CH_2OH}$$

[*35*]

$$\overset{R_1}{\underset{|}{RCONHCHCO}}-\overset{R_2}{\underset{|}{NHCHCO}}-\overset{R_3}{\underset{|}{NHCHNHCOOCH_2C_6H_5}} \xrightarrow[\text{2. } H_2O]{\text{1. Pd/H}_2}$$

[*36*]

$$\overset{R_1}{\underset{|}{RCONHCHCO}}-\overset{R_2}{\underset{|}{NHCHCONH_2}} + R_3CHO + NH_3 + C_6H_5CH_3$$

[*37*]   [*38*]

Fig. 14

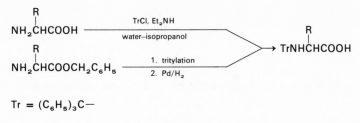

Fig. 15

The use of the trityl moiety as an amino protector, and its sensitivity in acid media, has been known for many years. However, large-scale use of this group in peptide synthesis became possible when Zervas and his coworkers developed efficient methods for the preparation of optically pure N-tritylamino acids. This was accomplished by direct tritylation of amino acids in aqueous solution in the presence of diethylamine or indirectly by selective hydrogenolysis of the benzyl group from the readily prepared N-tritylamino acid benzyl esters (Fig. 15). The steric hindrance of the trityl group in coupling N-tritylamino acids was overcome in large measure by the use of carbodiimides or mixed anhydrides of diphenylphosphoric acid, for peptide bond formation. Detritylation of peptides was readily accomplished by alcoholysis in the presence of 1 equivalent of a strong acid.

Steric hindrance by the trityl group in the tritylamino acids did not permit, however, the use of all the existing peptide bond forming reagents with these protected amino acids. To overcome this problem, it was considered desirable to set apart the massive trityl group from the crucial region of the molecule by the use of the tritylsulfenyl moiety as the amino protector. It was indeed found that N-tritylsulfenylamino acids could participate in coupling reactions, although in relatively low yields. Removal of the N-tritylsulfenyl group from the protected peptides was accomplished upon interaction with 2 equivalents of hydrogen chloride. Further search for amino protectors with more favorable properties led to the development of yet another method of protection; the newest of all, the so-called NPS method.

In this method, which is now widely used in peptide synthesis, o-nitrophenylsulfenyl chloride (NPS chloride) interacts with amino acid esters, or amino acids in alkaline medium, to produce in good yield NPS-amino acids.

Coupling of NPS-amino acids with amino acid, or peptide esters, is effected with any of the known coupling methods in very good yields. Removal of the NPS group from the coupling product is accomplished under very mild acidic conditions (Fig. 16).

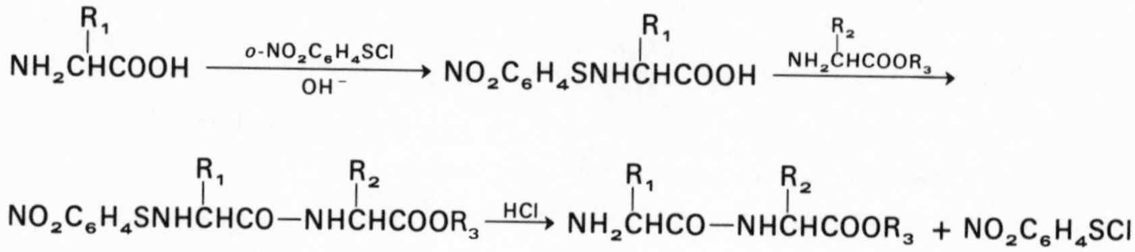

$$\underset{\underset{NH_2CHCOOH}{\overset{R_1}{|}}}{} \xrightarrow[OH^-]{o\text{-}NO_2C_6H_4SCl} \underset{\underset{NO_2C_6H_4SNHCHCOOH}{\overset{R_1}{|}}}{} \xrightarrow{\overset{R_2}{\underset{NH_2CHCOOR_3}{|}}}$$

$$\underset{\overset{R_1}{|}\quad\overset{R_2}{|}}{NO_2C_6H_4SNHCHCO-NHCHCOOR_3} \xrightarrow{HCl} \underset{\overset{R_1}{|}\quad\overset{R_2}{|}}{NH_2CHCO-NHCHCOOR_3} + NO_2C_6H_4SCl$$

Fig. 16

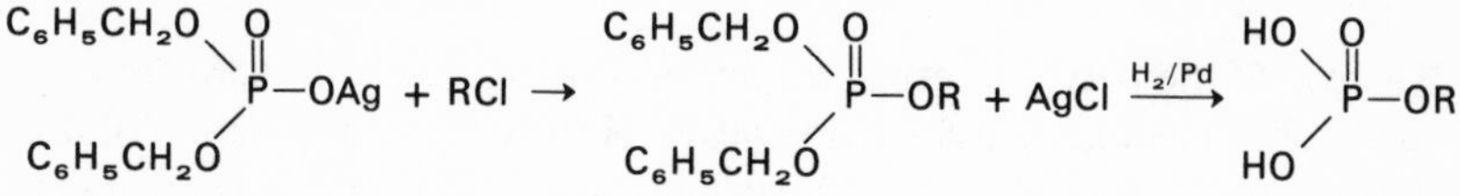

[39]   [40]   [41]   [42]

Fig. 17

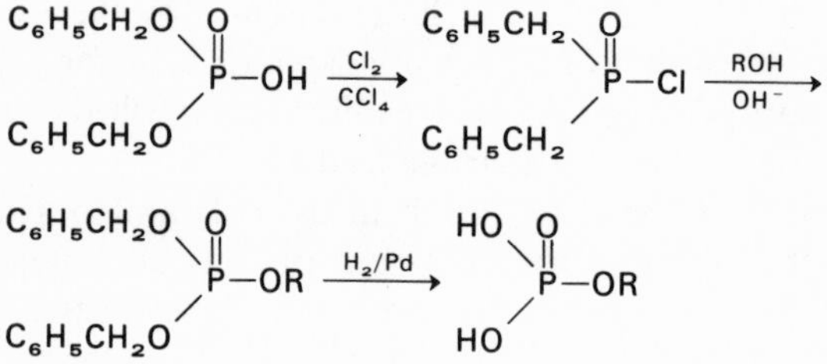

Fig. 18

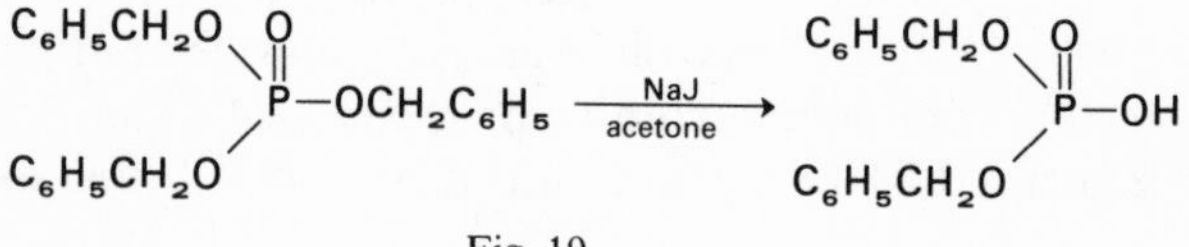

Fig. 19

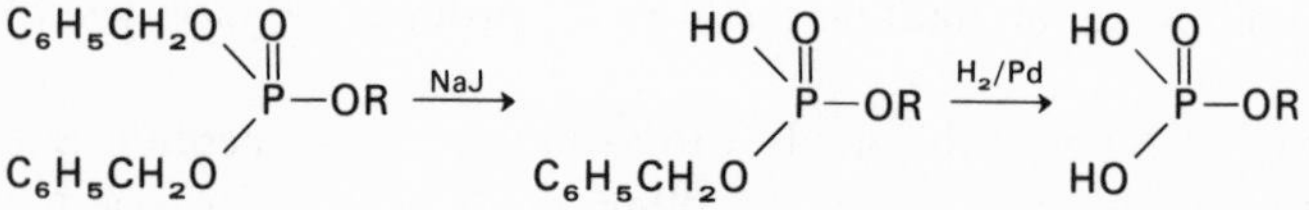

[43]   [44]   [45]

Fig. 20

The contribution of Zervas to peptide chemistry encompasses not only the development of methods for amino protection but also the development of methods for the blocking of the carboxyl function of amino acids. The method of carboxyl group protection by conversion to a benzyl ester, which can be readily deblocked by hydrogenolysis, as was mentioned earlier, is still widely used in peptide synthesis. The spectrum of carboxyl protectors has been further increased by the addition of the diphenyl-methyl moiety, which can be removed either by acidolysis or by hydrogenoly-sis. Finally, in cases where deblocking of the carboxyl group by acid or alkali is prohibited, the use of the phenacyl moiety was the solution. The acid-stable phenacyl ester group can be readily cleaved by sodium thiophenoxide and, though less effectively, by catalytic hydrogenation.

Since the early years of his scientific career, Zervas has shown particular interest in methods of synthesis and the study of the properties of phos-phorylated amino and hydroxy compounds. His first effort in this area was in 1931, when he undertook the synthesis of dibenzylphosphorochloridate in an attempt to synthesize phosphorarginine, which has just been found in nature. This early work, however, was interrupted because the carbobenzoxy method was simultaneously developed, and its exciting potential for peptide synthesis dominated the scene in the Institute, as it did for many years to come. Soon after he returned to Greece, Zervas returned to the problem of phosphorylation. In a preliminary publication in 1939 under the title "Über eine neue Phosphorylierungsmethode," he states that the problems associated with the synthesis of the unstable dibenzylphosphorochloridate have been put aside for an alternate procedure of phosphorylation. In the new approach, the silver salt of the dibenzylphosphoric acid [39] interacts with a halide of the compound to be phosphorylated [40] (Fig. 17). The resulting triester phosphate [41], upon calalytic hydrogenation, is con-verted to the desired phosphorylated compound [42].

Soon after this first publication, World War II broke out. With the foreign occupation of Greece during World War II, and the anomalous conditions in the aftermath, all research efforts came to a halt. In this period, Sir Alexander Todd, continuing work along similar lines, prepared in a simple and ingenious manner dibenzylphosphochloridate, which he used for various syntheses. This reagent proved, as Zervas had envisioned, eminently suitable for the phosphorylation of hydroxy and amino com-pounds (Fig. 18) and was utilized by Todd in his epoch-making syntheses of nucleotides.

Resumption of research activities at the University of Athens led to the development of a simple method for the synthesis of phosphoric acid di-benzylesters. This method involved the anionic cleavage of tribenzyl-phosphates by NaJ dissolved in acetone (Fig. 19). Similarly, dibenzylalkyl-phosphates [43] with NaJ are converted to the corresponding mono-

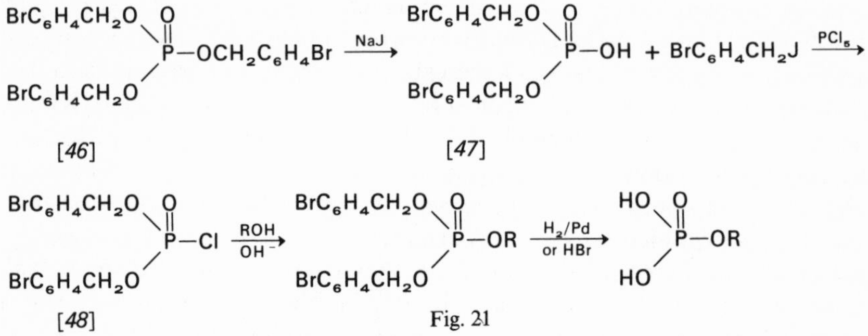

$$
\underset{\underset{\text{BrC}_6\text{H}_4\text{CH}_2\text{O}}{\overset{\text{BrC}_6\text{H}_4\text{CH}_2\text{O}}{}}{\text{P}}\overset{\text{O}}{\underset{}{\|}}\text{—OCH}_2\text{C}_6\text{H}_4\text{Br} \xrightarrow{\text{NaJ}} \underset{\underset{\text{BrC}_6\text{H}_4\text{CH}_2\text{O}}{\overset{\text{BrC}_6\text{H}_4\text{CH}_2\text{O}}{}}{\text{P}}\overset{\text{O}}{\underset{}{\|}}\text{—OH} + \text{BrC}_6\text{H}_4\text{CH}_2\text{J} \xrightarrow{\text{PCl}_5}
$$

[46]   [47]

$$
\underset{\underset{\text{BrC}_6\text{H}_4\text{CH}_2\text{O}}{\overset{\text{BrC}_6\text{H}_4\text{CH}_2\text{O}}{}}{\text{P}}\overset{\text{O}}{\underset{}{\|}}\text{—Cl} \xrightarrow[\text{OH}^-]{\text{ROH}} \underset{\underset{\text{BrC}_6\text{H}_4\text{CH}_2\text{O}}{\overset{\text{BrC}_6\text{H}_4\text{CH}_2\text{O}}{}}{\text{P}}\overset{\text{O}}{\underset{}{\|}}\text{—OR} \xrightarrow[\text{or HBr}]{\text{H}_2/\text{Pd}} \underset{\underset{\text{HO}}{\overset{\text{HO}}{}}}{\text{P}}\overset{\text{O}}{\underset{}{\|}}\text{—OR}
$$

[48]   Fig. 21

benzylalkylphosphates [44], which can be used either for further synthetic processes or can be hydrogenolyzed to the alkylphosphates [45] (Fig. 20). In a similar series of reactions, and under controlled conditions, tetra-benzylpyrophosphates can be converted to tri- or *sym*-dibenzyl derivatives. *p*-Substituted (bromo nitro- or iodo-) tribenzylphosphates [46] are converted to phosphoric diesters [47], which are transformed to crystalline acid chlorides [48] (Fig. 21). The latter compounds can be used as phosphorylating agents. Removal of the *p*-substituted benzyl groups is achieved by hydrogenolysis or upon treatment with a dilute solution of HBr in anhydrous organic solvents.

Interaction of the aforementioned *p*-substituted dibenzylphospho-chloridates with amino acid esters and catalytic hydrogenation of the phosphorylated product in alkaline medium led to the synthesis of *N*-phosphorylated amino acids. The latter compounds are extremely sensitive in acid medium and are completely hydrolyzed even at pH 4.0 (Fig. 22).

The *N*-dibenzylphosphorylamino acids are stable compounds and can be used as intermediates in peptide synthesis. Removal of the amino protector in the final synthetic step is accomplished either by catalytic hydrogenation or by HBr in anhydrous organic solvents.

These few pages of enumeration of Zervas's work are hardly adequate to describe and do justice to the man and the scientist as I know him. It was

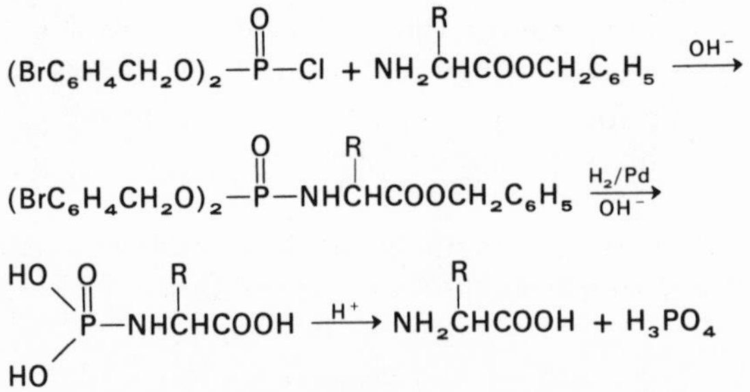

$$
(\text{BrC}_6\text{H}_4\text{CH}_2\text{O})_2\text{—}\overset{\overset{\text{O}}{\|}}{\text{P}}\text{—Cl} + \text{NH}_2\overset{\overset{\text{R}}{|}}{\text{C}}\text{HCOOCH}_2\text{C}_6\text{H}_5 \xrightarrow{\text{OH}^-}
$$

$$
(\text{BrC}_6\text{H}_4\text{CH}_2\text{O})_2\text{—}\overset{\overset{\text{O}}{\|}}{\text{P}}\text{—NH}\overset{\overset{\text{R}}{|}}{\text{C}}\text{HCOOCH}_2\text{C}_6\text{H}_5 \xrightarrow[\text{OH}^-]{\text{H}_2/\text{Pd}}
$$

$$
\underset{\underset{\text{HO}}{\overset{\text{HO}}{}}}{\text{P}}\overset{\text{O}}{\underset{}{\|}}\text{—NH}\overset{\overset{\text{R}}{|}}{\text{C}}\text{HCOOH} \xrightarrow{\text{H}^+} \text{NH}_2\overset{\overset{\text{R}}{|}}{\text{C}}\text{HCOOH} + \text{H}_3\text{PO}_4
$$

Fig. 22

through his personal efforts and devotion that a level of chemical research equal to that of the more scientifically aware and advanced countries was achieved in Greece. I had the privilege of working in his laboratory as an undergraduate and as a graduate student. As a teacher, he rightfully demanded much from his students, but equally showed a kind and generous concern for their progress. I welcome this opportunity to express, for myself and all his other students, our gratitude to Professor Zervas for his encouragement and guidance in the development of our scientific interests and careers.

REFERENCES

The following is a list, in chronological order, of publications by Zervas and coworkers:

Bergmann, M., Ensslin, H., and Zervas, L., 1925, Über die Aldehydverbindungen der Aminosäuren, *Berichte* **58**:1034.

Bergmann, M., and Zervas, L., 1926, Über die Aldehydverbindungen der Aminosäuren und ihre präparative Verwendung, *Z. Physiol. Chem. Hoppe-Seyler's* **152**:282.

Zervas, L., 1926, Über die Aldehydverbindungen der Aminosäuren, Dissertation, Berlin.

Bergmann, M., and Zervas, L., 1927, Synthese des Glycocyamins aus Arginin und Glykokoll. Ein Beitrag zur Kreatinfrage, *Z. Physiol. Chem. Hoppe-Seyler's* **172**:277.

Bergmann, M., and Zervas, L., 1928, Synthese des Kreatins aus Sarkosin und Arginin, *Z. Physiol. Chem. Hoppe-Seyler's* **173**:80.

Bergmann, M., and Zervas, L., 1928, Zur Kenntnis des Histidins. Peptidbildung durch Acylwanderung, *Z. Physiol. Chem. Hoppe-Seyler's* **175**:145.

Bergmann, M., and Zervas, L., 1928, Notiz über Synthese von DL-Histidylglycin, *Z. Physiol. Chem. Hoppe-Seyler's* **175**:154.

Zervas, L., and Bergmann, M., 1928, Das sog. Arginyl-arginin von E. Fischer ein Bisguanido-*n*-valeriansäure-anhydrid, *Berichte* **61**:1195.

Bergmann, M., and Zervas, L., 1929, Über katalytische Racemisierung von Aminosäuren und Peptiden, *Biochem. Z.* **203**:280.

Bergmann, M., Zervas, L., and Köster, H., 1929, Autoracemisierung argininhaltiger Aminoäureanhydride, *Berichte* **62**:1901.

Bergmann, M., Zervas, L., and du Vigneaud, V., 1929, Synthese arginin-haltiger Peptide, *Berichte* **62**:1905.

Bergmann, M., du Vigneaud, V., and Zervas, L., 1929, Acylwanderung und Spaltungsvorgänge bei acylierten Dioxopiperazinen, *Berichte* **62**:1909.

Zervas, L., 1930, Synthese des Styracits, *Berichte* **63**:1689.

Bergmann, M., Zervas, L., and Silberqueit, E., 1931, Über die Biose des Chitins, *Naturwissenschaften* **19**:20.

Bergmann, M., and Zervas, L., 1931, Synthesen mit Glucosamin, *Berichte* **64**:974.

Bergmann, M., and Zervas, L., 1931, Neue Dismutationsprodukte der Zucker, *Berichte* **64**:1434.

Bergmann, M., and Zervas, L., 1931, Ergäzung zu unserer Mitteilung: Neue Dismutationsproduckte der Zucker, *Berichte* **64**:2032.

Zervas, L., 1931, Über Benzyliden-glucose und ihre Verwendung zu Synthesen. I-Benzoylglucose, *Berichte* **64**:2289.

Bergmann, M., Zervas, L., and Lebrecht, F., 1931, Über die Dehydrierung von Aminosäuren und einen Übergang zur Pyrrol-reihe, *Berichte* **64**:2315.

Bergmann, M., and Zervas, L., 1931, Über das Arcain, *Z. Physiol. Chem. Hoppe-Seyler's* **201**:208.

Bergmann, M., Zervas, L., and Silberqueit, E., 1931, Über Glucosaminsäure und ihre Desaminierung, *Berichte* **64**:2428.

Bergmann, M., Zervas, L., and Silberqueit, E., 1931, Über Chitin und Chitobiose, *Berichte* **64**:2436.

Bergmann, M., and Zervas, L., 1932, Über ein allgemeines Verfahren der Peptidsynthese, *Berichte* **65**:1192.

Bergmann, M., and Zervas, L., 1932, Über die Synthese von Glucopeptiden des Glucosamins, *Berichte* **65**:1201.

Bergmann, M., Zervas, L., Schleich, H., and Leinert, F., 1932, Über proteolytische Fermente, Verhalten von Prolinpeptiden, *Z. Physiol. Chem. Hoppe-Seyler's* **212**:72.

Bergmann, M., Zervas, L., and Greenstein, J. P., 1932, Synthese von Peptiden des D-Lysins, *Berichte* **65**:1962.

Bergmann, M., Zervas, L., and Schleich, H., 1932, Über proteolytische Enzyme. Bindungsart des Prolins in der Gelatine, *Berichte* **65**:1747.

Bergmann, M., Zervas, L., and Salzmann, L., 1933, Synthese von L-Glutamin, *Berichte* **66**:1288.

Zervas, L., 1933, Synthese von D-Glucuronsäure, *Berichte* **66**:1326.

Bergmann, M., and Zervas, L., 1933, Über Isoglutamin, *Z. Physiol. Chem. Hoppe-Seyler's* **221**:51.

Zervas, L., and Sessler, P., 1933, Über eine neue Acetonfructose, *Berichte* **66**:1698.

Bergmann, M., and Zervas, L., 1934, Über proteolytische Enzyme III. Mitt. Über die Wirkungsweise und Spezifität von Dipeptidase, *Z. Physiol. Chem. Hoppe-Seyler's* **224**:II.

Bergmann, M., and Zervas, L., 1933, Eiweisstoffe, *in "Handbuch der Pflanzenanalyse, herausgegeben von Professor Klein,"* Vol. II, p. 229.

Bergmann, M., Zervas, L., Salzmann, L., and Schleich, H., 1934, Über Dipeptide mit vorwiegend Sauren Eigenschaften und ihr fermentatives Verhalten, *Z. Physiol. Chem. Hoppe-Seyler's* **224**:17.

Bergmann, M., Zervas, L., Rinke, H., and Schleich, H., 1934, Synthese von Dipeptiden des Lysins und ihr Verhalten gegen proteolytische Fermente, *Z. Physiol. Chem. Hoppe-Seyler's* **224**:26.

Bergmann, M., Zervas, L., Rinke, H., and Schleich, H., 1934, Über Dipeptide von epimeren Glucosaminsäuren und ihr Verhalten gegen Dipeptidase. Konfiguration des D-Glucosamins, *Z. Physiol. Chem. Hoppe-Seyler's* **224**:33.

Bergmann, M., Zervas, L., and Rinke, H., 1934, Neues Verfahren zur Synthese von Peptiden des Arginins, *Z. Physiol. Chem. Hoppe-Seyler's* **224**:40.

Bergmann, M., Zervas, L., and Schleich, H., 1934, Über proteolytische Enzyme IV. Mitt. Spezifität und Wirkungsweise der sogen. Carboxy-polypeptidase, *Z. Physiol. Chem. Hoppe-Seyler's* **224**:52.

Bergmann, M., Zervas, L., and Overhoff, J., 1934, Notiz über synthetische Zucker-Aminosäureverbindungen, *Z. Physiol. Chem. Hoppe-Seyler's* **224**:56.

Bergmann, M., Zervas, L., and Engler, J., 1935, Über Isoglucal und Protoglucal, *Annalen* **508**:25.

Bergmann, M., Zervas, L., Fruton, J. S., Schneider, F., and Schleich, H., 1935, On proteolytic enzymes V. On the specificity of dipeptidase, *J. Biol. Chem.* **109**:325.

Bergmann, M., Zervas, L., and Fruton, J. S., 1935, On proteolytic enzymes VI. On the specificity of papain, *J. Biol. Chem.* **111**:224.

Bergmann, M., Zervas, L., and Ross, W. F., 1935, On proteolytic enzymes VII. The synthesis of peptides of L-lysine and their behavior with papain, *J. Biol. Chem.* **244**:III.

Bergmann, M., Zervas, L., and Schneider, F., 1936, A method for the stepwise degradation of polypeptides, *J. Biol. Chem.* 113:341.

Bergmann, M., and Zervas, L., 1936, On proteolytic enzymes IX. Inactivation of papain, *J. Biol. Chem.* 114:711.

Bergmann, M., Zervas, L., and Fruton, J. S., 1936, On proteolytic enzymes XI. The specificity of the enzyme papain peptidase I, *J. Biol. Chem.* 115:III.

Zervas, L., 1939, Über eine neue Phosphorylierungsmethode. I-Glucosylphosphat, *Naturwissenschaften* 27:317.

Zervas, L., and Papidimitriou, E., 1940, Über die Konstitution des Styracits. Umwandlung von Aldosen in Ketosen, *Berichte* 73:174.

Zervas, L., and Panagopoulos, K., 1945, Enzymological studies on l-phosphoro-D-glucose, *Chim. Chron.* 10:1.

Zervas, L., and Dilaris, I., 1955, Dealkylation and debenzylation of triesters of phosphoric acid. Phosphorylation of hydroxy and amino compounds, *J. Am. Chem. Soc.* 77:5354.

Zervas, L., and Katsoyannis, P., 1955, *N*-Phosphoroamino acids and peptides, *J. Am. Chem. Soc.* 77:5357.

Zervas, L., 1955, Phosphorylation et tritylation des acides amines. Nouvelles methodes de la synthese peptidique, *in* Résumés des Communications XIVe Congrés International de Chimie pure et appliquée, Zürich, p. 224.

Zervas, L., and Zioudrou, Chr., 1956, Catalytic reduction of acetobromo-sugars, *J. Chem. Soc.*, 214.

Zervas, L., and Theodoropoulos, 1956, Trityl amino acids and peptides. A new method of peptide synthesis, *J. Am. Chem. Soc.* 78:1359.

Zervas, L., and Dilaris, I., 1956, Entalkylierung und Entbenzylierung von neutralen Pyrophosphorsäureester. Pyrophosphorylierung von Oxy- und Aminoverbindungen, *Berichte* 89:925.

Zervas, L., 1956, Phosphorylation and tritylation of amino acids, *Chim. Chron.* 21:3.

Zervas, L., Winitz, M., and Greenstein, J. P., 1956, The percarbobenzoxylation of L-arginine, *Arch. Biochem. Biophys.* 65:No. 2.

Zervas, L., Winitz, M., and Greenstein, J. P., 1957, Studies on arginine peptides I. Intermediates in the synthesis of *N*-terminal and *C*-terminal arginine peptides, *J. Org. Chem.* 22:1515.

Zervas, L., Winitz, M., and Greenstein, J. P., 1958, A synthesis of L-arginyl-L-arginine, *Arch. Biochem. Biophys.* 75:No. I.

Stelakatos, G. C., Theodoropoulos, D. M., and Zervas, L., 1959, On the trityl method for peptide synthesis, *J. Am. Chem. Soc.* 81:2884.

Konstas, S., Photaki, I., and Zervas, L., 1959, Uberführung von D-Glucosamin in Oxazolon und Oxazolinderivate, *Berichte* 92:1288.

Zervas, L., Benoiton, L., Weiss, E., Winitz, M., and Greenstein, J. P., 1959. Preparation and disulfide interchange reactions of unsymmetrical open-chain derivatives of cysteine, *J. Am. Chem. Soc.* 81:1729.

Zervas, L., Otani, T., Winitz, M., and Greenstein, J. P., 1959, Studies on arginine peptides II. Synthesis of L-arginyl-L-arginine and other *N*-terminal arginine dipeptides, *J. Am. Chem. Soc.* 81:2878.

Zervas, L., and Konstas, S., 1960, Über Glucosaminide, *Berichte* 93:435.

Zervas, L., and Photaki, I., 1960, Über Cystein- und Cystinpeptide, *Chimia* 14:375.

Zervas, L., Winitz, M., and Greenstein, J. P., 1961, Studies on arginine peptides III. On the structure of tricarbobenzoxy-L-arginine, *J. Am. Chem. Soc.* 83:3300.

Bezas, B., and Zervas, L., 1961, On peptides of L-lysine, *J. Am. Chem. Soc.* 83:719.

Coutsogeorgopoulos, Ch., and Zervas, L., 1961, On β-D-glucosylamides of L-amino acids and nicotinic acid, *J. Am. Chem. Soc.* 83:1885.

Cosmatos, A., Photaki, I., and Zervas, L., 1961, Peptidsynthesen über N-phosphorylamino-säure-phosphorsäureanhydride, *Berichte* **94**:2644.

Zervas, L., 1962, Über Cystein- und Cystinpeptide, *Coll. Czech. Chem. Commun.* **27**:2242.

Zervas, L., and Photaki, I., 1962, On cysteine and cystine peptides I. New S-protecting groups for cysteine, *J. Am. Chem. Soc.* **84**:3887.

Zervas, L., Photaki, I., and Ghelis, N., 1963, On cysteine and cystine peptides II. S-A-Cyl-cysteines in peptide synthesis, *J. Am. Chem. Soc.* **85**:1937.

Zervas, L., Photaki, I., Cosmatos, A., and Ghelis, N., 1963, On cysteine and cystine peptides, *in* "Peptides: Proceedings of the Fifth European Peptide Symposium" (G. T. Young, ed.), p. 27, Pergamon Press, Oxford.

Gazis, E., Bezas, B., Stelakatos, G. S., and Zervas, L., 1963, On the protection of α-amino and carboxyl groups for peptide synthesis, *in* "Peptides: Proceedings of the Fifth European Peptide Symposium" (G. T. Young, ed.), p. 17, Pergamon Press, Oxford.

Zervas, L., Borovas, D., and Gazis, E., 1963, New methods in peptide synthesis I. Trimethyl-sufenyl and o-nitrophenylsulfenyl groups as N-protecting groups, *J. Am. Chem. Soc.* **85**:3660.

Zervas, L., Photaki, I., Cosmatos, A., and Bororas, D., 1965, On cysteine and cystine peptides III. Synthesis of fragment of insulin containing the intrachain disulfide bridge, *J. Am. Chem. Soc.* **87**:4922.

Stelakatos, G. C., Paganou, A., and Zervas, L., 1965, New methods in peptide synthesis. Part III. Protection of carboxyl group, *J. Chem. Soc.* (*C*) **1965**:1191.

Cosmatos, A., Photaki, I., and Zervas, L., 1966, The synthesis of an oxytocin-type fragment of insulin, *in* "Proceedings of the Sixth European Peptide Symposium, Athens, 1963" (L. Zervas, ed.), p. 301, Pergamon Press, Oxford.

Gazis, E., Borovas, D., Hamalidis, Ch., Stelakatos, G. S., and Zervas, L., 1966, New methods in peptides synthesis, *in* "Proceedings of the Sixth European Peptide Symposium, Athens, 1963" (L. Zervas, ed.), Pergamon Press, Oxford.

Zervas, L., and Hamalidis, Ch., 1965, New Methods in Peptide Synthesis II. Further Example of the Use of the o-Nitrophenylsulfenyl Groups for the Protection of Amino-Groups, *J. Am. Chem. Soc.* **87**:99.

Zervas, L., Cosmatos, A., and Diamandis, P., 1965, Diphenylmethylester der Phosphorsaure, *Experientia* **21**:5.

Zervas, L., Photaki, I., Yovanidis, C., Taylor, J., Phocas, I., and Bardakos, V., 1967, Some problems concerning amino, carboxyl, and side-chain protection of peptides, *in* "Peptides: Proceedings of the Eighth European Symposium" (H. C. Beyerman, A. Van De Linde, and W. Maassen Van Den Brink, eds.), p. 28, North-Holland, Amsterdam.

Phocas, I., Yovanidis, C., Photaki, I., and Zervas, L., 1967, New methods in peptide synthesis. Part IV. N–S transfer of N-o-nitrophenylsulfenyl groups in cystein peptides, *J. Chem. Soc.* (*C*) **1967**:1506.

Taylor-Papadimitriou, J., Yovanidis, C., Paganou, A., and Zervas, L., 1967, New methods in peptide synthesis. V. On α- and γ-diphenylmethyl and phenacyl esters of L-glutamic-acid, *J. Chem. Soc.* (*C*) **1967**:1831.

Photaki, I., Phocas, I., Taylor-Papadimitriou, J., and Zervas, L., 1968, On cysteine and cystine peptides *in* "Peptides" (V. Bricas, ed.), p. 201, North-Holland, Amsterdam.

Zervas, L., Photaki, I., and Phocas, I., 1968, Notiz über S-trityl-L-cystein, *Chem. Ber.* **101**:3332.

Zervas, L., 1970, Über die N-trityl- und N-o-nitrophenylsulfenyl-Methode zur Peptidsynthese, *Z. Naturforsch.* **15b**:322.

Photaki, I., Taylor-Papadimitriou, J., Sakarellos, C., Mazarakis, P., and Zervas, L., 1970, On cysteine-cystine peptides. Part V. S-Trityl and S-diphenylmethyl cysteine and cystine peptides, *J. Chem. Soc.* (*C*) **1970**:2683.

CHAPTER 2

ACTIVE ESTERS AND THE STRATEGY OF PEPTIDE SYNTHESIS

Miklos Bodanszky and Yakir S. Klausner

Department of Chemistry
Case Western Reserve University
Cleveland, Ohio

I. INTRODUCTION

After the epoch-making discovery of the still unsurpassed carbobenzoxy groups by Bergmann and Zervas (1932), the first major breakthrough in peptide synthesis came in the early 1950s with the introduction of mixed anhydrides by Wieland and Bernhard (1951), Boissonnas (1951), and Vaughan and Osato (1952). The impetus provided by the availability of an easily removable aminoprotecting group and efficient carboxyl activation culminated in the synthesis of oxytocin by du Vigneaud and his coworkers (du Vigneaud *et al.*, 1953, 1954; Katsoyannis and du Vigneaud, 1954; Ressler and du Vigneaud, 1954; Swan and du Vigneaud, 1954). For even more ambitious endeavors, however, carboxyl activation in the form of mixed anhydrides seemed to be not entirely satisfactory. Unsymmetrical anhydrides yield—at least in principle—two acylation products, the desired peptide (A) and an acyl derivative (B) of the amino component (Fig. 1). With the proper choice of the "activating" acid in the mixed anhydride, the formation of the undesired byproduct B can be kept at a minimum, but it is unlikely that the relative electrophilicities of the two carbonyl carbons could be so different

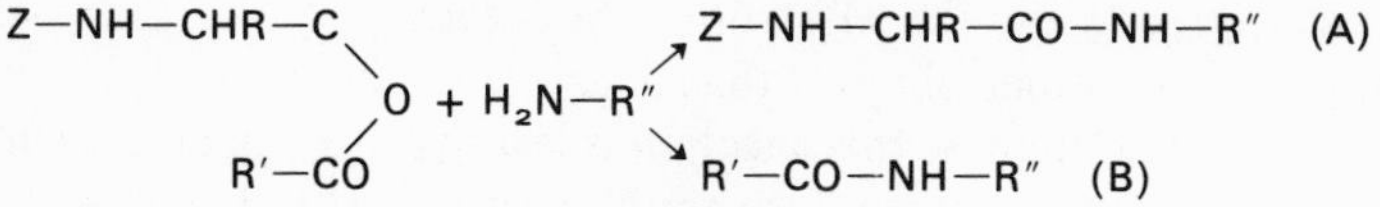

$$Z-NH-CHR-C \diagdown \qquad \diagup Z-NH-CHR-CO-NH-R'' \quad (A)$$
$$O + H_2N-R''$$
$$R'-CO \diagup \qquad \diagdown R'-CO-NH-R'' \quad (B)$$

Fig. 1

$$Z-NH-CHR-CO-OR' + H_2NR'' \rightarrow Z-NH-CHR-CO-NH-R'' + R'-OH$$

Fig. 2

that the main product, peptide A, would be completely uncontaminated by some small amount of the byproduct B. [A combination of electron release and of steric hindrance, such as in pivaloyl mixed anhydrides (Zaoral, 1962), might approach the ideal of a single acylation product.] In the preparation of smaller peptides, recrystallization of A is usually sufficient for the removal of B, but in compounds containing several amino acid residues, the differences in the properties of A and B are small, and more elaborate methods of purification may become necessary. The obvious desire for activated derivatives of acylamino acids (or peptides) that would yield a single acylation product led to the development of activated and active esters. In principle, aminolysis of esters results in the formation of only the desired amide and of the alcohol corresponding to the leaving group (Fig. 2). While the aminolysis of simple esters such as methyl esters could be applied (Fischer, 1902) for the preparation of peptides, the reaction rates, at least at room temperature, are insufficient for practical synthesis. Moreover, the *protection* of the carboxyl function of the amino component in the form of methyl or ethyl esters requires that the aminolysis of the *activated* carboxyl derivative should proceed orders of magnitude faster than the aminolysis of the protecting group; otherwise, a complex mixture of products has to be expected.

In his investigation of mixed anhydrides, Wieland included the "anhydrides" of acylamino acids with thiophenol (Wieland *et al.*, 1951). These thiophenyl esters proved to be useful acylating agents, but, perhaps because their thiol ester character rather than their aryl ester nature was emphasized, no immediate development of the aryl ester principle followed. About the same time, Schwyzer *et al.* (1955a) embarked on a systematic examination of substituents that can "activate" methyl esters. The nitril group turned out to be most effective in this respect; and the cyanomethyl esters of protected amino acids (Schwyzer *et al.*, 1955b,c) were successfully applied in the synthesis of peptides.

The fundamental studies of Gordon *et al.* (1948, 1949) on the ammonolysis esters directed the attention of one of the authors of this chapter (M.B.) to the extremely high reactivity of vinyl and especially of phenyl esters. Since this reactivity rests on the electron-withdrawing effect of the π-electron system, it became rather obvious that negative substituents of the aromatic ring should further increase this effect, thus leading to an enhancement of the electrophilic character of the carbonyl carbon. The increased electrophilicity facilitates the formation of a—presumably tetrahedral—inter-

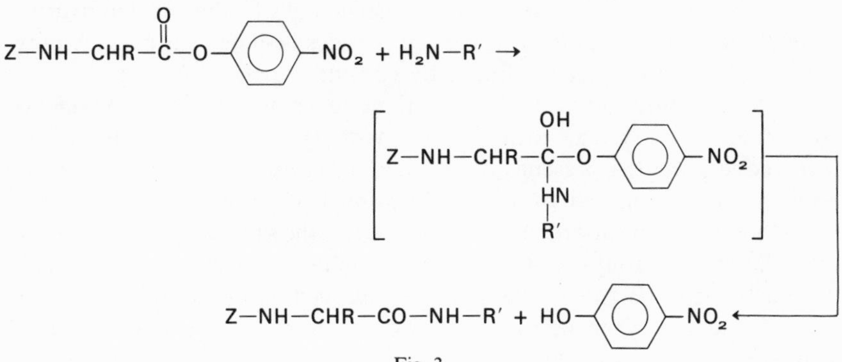

$$Z-NH-CHR-\overset{\overset{\displaystyle O}{\|}}{C}-O-\langle\bigcirc\rangle-NO_2 + H_2N-R' \rightarrow$$

$$\left[Z-NH-CHR-\overset{\overset{\displaystyle OH}{|}}{\underset{\underset{\displaystyle R'}{|}}{C}}-O-\langle\bigcirc\rangle-NO_2 \right]$$

$$Z-NH-CHR-CO-NH-R' + HO-\langle\bigcirc\rangle-NO_2 \leftarrow$$

Fig. 3

mediate with the nucleophile (the amino component) (Fig. 3). Elimination of the leaving group (nitrophenolate) from this intermediate results in the desired amide. At least in principle, a single acylation product is formed. These reactions can be carried out in reasonable time even under mild conditions. Similar considerations prompted the group of Kenner (Farrington et al., 1955) to propose p-nitrothiophenyl esters as reactive intermediates. These and the activated methyl esters of Schwyzer found a somewhat limited application in other laboratories; the nitrophenyl ester method (Bodanszky, 1955) was better received and is still one of the most widely used coupling procedures (Jones, 1970).

II. THE DEVELOPMENT OF ACTIVE ESTERS

Activation of methyl esters with electron-withdrawing substituents lost much of its appeal after aryl esters found their way into the praxis of peptide synthesis. The ethynyl group plays the role of the negative substituent in propargyl esters (Bodanszky, 1957), but the value of this type of activated ester remained unexplored. For the preparation of cyanomethyl esters, bromoacetonitril rather than chloroacetonitril was proposed by Taschner et al. (1965c): the higher reactivity of the bromo derivative allows esterification at room temperature, and thus the racemization previously observed, mainly with N-benzyloxycarbonyl-S-benzylcysteine (Iselin et al., 1955), can be avoided. Even this improvement could not overcome the disadvantage of only moderate reactivity of cyanomethyl esters. The rate of the bimolecular reaction between an active ester and the nucleophile (the amino component) is highly dependent on the concentration of the reactants. In the case of larger peptides, more activation is needed to compensate for the necessarily lower concentrations. Such higher activation was found when a phenol or a substituted hydroxylamine is the "alcohol" part of the ester. Since these alcohols can also be described as weak acids, the question of why their acyl derivatives should not be looked upon as mixed anhydrides is quite legitimate. A pragmatic differentiation can be found by considering the reactivity of active esters toward different nucleophiles. Active esters readily react when attacked by an amino group but are rather inert to hydroxyl groups unless the reaction is forced, e.g., by the presence of an ester-exchange catalyst. In fact, active esters are often purified by recrystallization from boiling alcohol, a procedure inconceivable for mixed anhydrides. For the same reason a protection of the hydroxyl groups in serine and threonine side-chains is necessary if "overactivated" (Brenner, 1967) derivatives such as mixed anhydrides are applied, but not when active esters, e.g., nitrophenyl esters, are used for acylation.

In Table I, an attempt is made to summarize the types of aryl esters that gained some significance in peptide synthesis or are promising in this respect. Their enumeration is incomplete: the literature is too abundant for an exhaustive survey. Also, the fundamental investigations of Pless and

Table I. Substituted Phenols Proposed as Alcohol Components of Active Esters of Acylamino Acids

Phenol	Reference	Phenol	Reference
HS—⬡	Wieland *et al.* (1951)	HSe—⬡	Farrington *et al.* (1957)
HO—⬡—NO$_2$	Bodanszky (1955)	SeH (on naphthalene)	Jakubke (1965c)
HO—⬡ (NO$_2$)	Bodanszky (1955)		
HO—⬡ (O$_2$N)	Bodanszky (1955)	HO—⬡—CN	Schwyzer *et al.* (1960)
HO—⬡—NO$_2$ (O$_2$N)	Bodanszky (1955)	HO—⬡—SCH$_3$ (with O and O on the S)	Johnson and Jacobs (1968a,b), Johnson and Ruettinger (1970), Schwyzer and Sieber (1958)
HS—⬡—NO$_2$	Farrington *et al* (1955)		

Table I (continued)

Phenol	Reference	Phenol	Reference
	Kuprýszewski (1961)		Jakubke (1965a), Jakubke and Voigt (1966)
	Stewart (1961)		Jakubke (1965b)
	Barth and Losse (1964)		Pless and Boissonnas (1963)
	Kaczmarek et al. (1967)		Kuprýszewski and Formela (1963)
	Kuprýszewski and Muzalewski (1967)		Kisfaludy et al. (1970), Kovacs et al. (1967)
	Kuprýszewski et al. (1968)		Taschner et al. (1965a,b)
	Mitin and Nadezhdina (1968)		Lloyd and Young (1968, 1971)
	Losse et al. (1965)		Dutta and Morley (1971)
	Jones and Young (1968)		

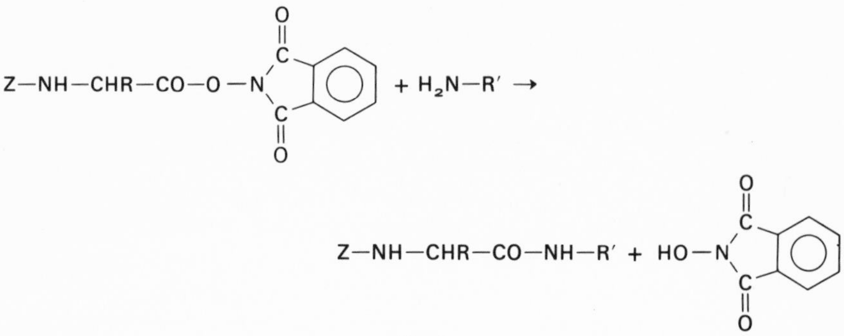

$$Z-NH-CHR-CO-O-N\overset{\displaystyle O}{\underset{\displaystyle O}{}} + H_2N-R' \rightarrow$$

$$Z-NH-CHR-CO-NH-R' + HO-N\overset{\displaystyle O}{\underset{\displaystyle O}{}}$$

Fig. 4

Table II. Hydroxylamine Derivatives Proposed as "Alcohol" Components of Active Esters of Acylamino Acids

Hydroxylamine derivative	Reference	Hydroxylamine derivative	Reference
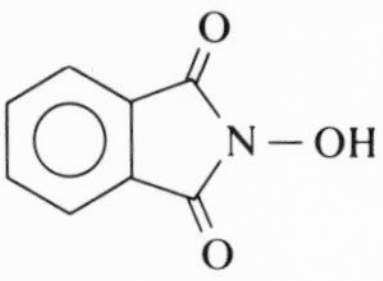 (phthalimide N−OH)	Nefkens and Tesser (1961), Nefkens et al. (1962)	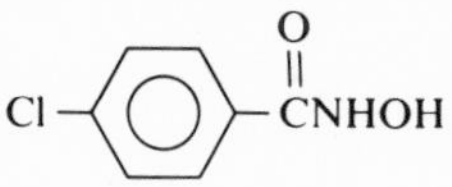 Cl−(C₆H₄)−CNHOH	Govindachari et al. (1966)
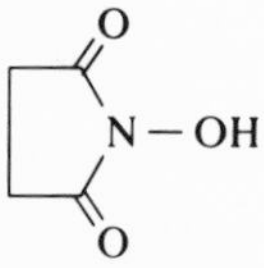 (succinimide N−OH)	Anderson et al. (1963, 1964)	$CH_3-\underset{CH_3}{\overset{CH_3}{C}}-CNHOH$	Rajappa et al. (1967)
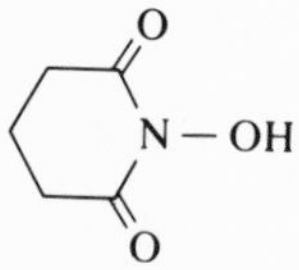 (glutarimide N−OH)	Jeschkeit (1968)	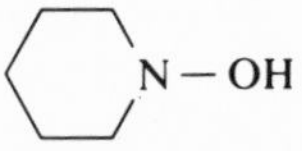 (piperidine N−OH)	Beaumont et al. (1965a,b), Handford et al. (1965)
$C_2H_5OCNHOH$	Jeschkeit (1969)	$\underset{R}{\overset{R}{C}}=N-OH$	Bittner et al. (1965), Losse et al. (1964), Fujino and Nishimura (1969)
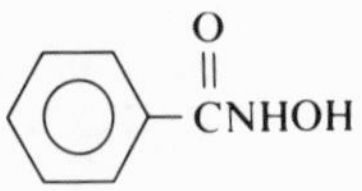 (phenyl CNHOH)	Lubiewska-Nakonieczna et al. (1970), Taschner et al. (1967)	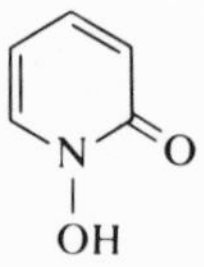 (2-pyridone N−OH)	Paquette (1965), Sarantakis et al. (1968), Taylor et al. (1970)

Boissonnas (1963) on the relationship between the dissociation constants of phenols and the reactivities of the corresponding esters in aminolysis, a study that included a large number of substituted phenols, is represented only by the one proposed as optimal: 2,4,5-trichlorophenol.

A novel idea in the development of active esters was suggested by Nefkens and Tesser (1961, 1962), who recommended the application of N-hydroxyphthalimide instead of the usual phenols as the alcohol component of active esters (Fig. 4). These active intermediates are closer to mixed anhydrides than most substituted aryl esters: their alcohol component is a hydroxamic acid. Nevertheless, the new acylating agents are sufficiently selective to warrant designation as active esters. Also, while further development first led to the quite analogous (and quite popular) esters of N-hydroxysuccinimide (Anderson et al., 1963, 1964), later studies focused more on the enhancement in reactivity caused by the N-atom next to the hydroxyl group and extended the investigations to a large variety of hydroxylamine derivatives, many of which are not hydroxamic acids. Table II sums up the more significant members of this group.

The high reactivity of vinyl esters, already mentioned briefly in this chapter, was explored (Weygand and Steglich, 1961) as a practical approach to peptide bond formation only to a limited extent. Closely related enol esters are the reactive intermediates formed by the addition of acylamino acids to several so-called coupling reagents. These intermediates (Table III) are enol esters or analogous compounds with an electron-withdrawing π-electron system and could be regarded as active esters. In a similar sense, the oxazolinones (azlactones) derived from acylamino acids also belong to this category: their reactivity as acylating agents rests on the same principle (Fig. 5).

A recent trend in peptide synthesis is the use of "additives" with coupling reagent, particularly with dicyclohexylcarbodiimide (DCC). The earlier proposals, simultaneously and independently made by Rothe and Kunitz (1957) and by Elliott and Russel (1957), to prepare p-nitrophenyl ester from protected amino acids, p-nitrophenol, and DCC and to use the resulting product without isolation is a procedure somewhat similar to the one introduced by Wünsch and Drees (1966) and Weygand et al. (1966b): the application of DCC in the presence of N-hydroxysuccinimide. There is, however, a notable difference in the choice of additives. While p-nitrophenol does not prevent racemization (Kovacs et al., 1969; Zimmerman and Anderson, (1967), N-hydroxysuccinimide is quite efficient in this respect. Even more effective additives, 1-hydroxybenzotriazole and its derivatives, were recommended by König and Geiger (1970). Perhaps a few words of caution need to be added here. These additives are indeed useful and allow a very convenient "Eintopf-Verfahren," a one-pot procedure. But, while racemization

Table III. Reactive Intermediates Formed from Acylamino Acids and Coupling Reagents

Coupling reagent	Intermediate	Reference
Ethoxyacetylene	$$\underset{\underset{\displaystyle CH_2=\overset{\displaystyle }{C}-OC_2H_5}{\mid}}{\underset{O}{\overset{R\quad O}{\underset{\mid\quad\parallel}{ZHNCHC}}}}$$	Arens (1955), Panneman *et al.* (1959)
Dicyclohexylcarbodiimide	ZHNCHC (R, O) — O — NH–C=N (cyclohexyl groups)	Sheehan and Hess (1955)
Diphenylketene-*p*-tolylimine	CCHNHZ (O, R) — O — CH–C=N—CH$_3$ (diphenyl)	Stevens and Munk (1958)
Dimethylamino-*t*-butyl-acetylene	ZHNCHC (R, O) — O — t—C$_6$H$_9$—CH=C—N(CH$_3$)$_2$	Buyle and Viehe (1964)
N-Ethyl-5-phenylisoxazolium-3'-sulfonate (Woodward's reagent K)	SO$_3$H (ring) O—CCHNHZ (O, R); C=CHCNHCH$_3$ (O)	Woodward *et al.* (1961, 1966)
N-Ethylbenzisoxazolium fluoroborate	OCCHNHZ (O) (ring); CNHC$_2$H$_5$ (O)	Kemp (1967), Kemp and Woodward (1965)
2-Ethyl-7-hydroxybenzisoxazolium fluoroborate	CNHC$_2$H$_5$ (O) (ring); OH; O=CCHNHZ (O, R)	Kemp and Chien (1967)

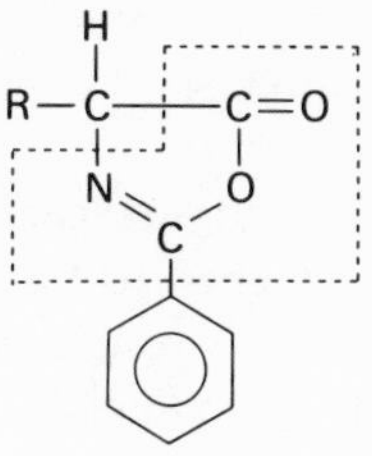

Fig. 5

might be greatly reduced, it is not necessarily entirely eliminated in each instance. The optical purity of the product should depend on the relative rates of the reactions of the active intermediate (O-acyl isourea derivative) with the amino component, with the additive, and with itself (azlactone formation). These rates obviously depend on the concentration of the reactants and probably also on their nature. It was interesting to note (Bodanszky et al., 1972) that in the rather hindered model system, acetyl-L-isoleucylglycine ethyl ester (Bodanszky and Conklin, 1967), N-hydroxy-succinimide is a more efficient suppressor of the racemization caused by DCC than the perhaps too bulky 1-hydroxybenzotriazole that was found superior in other model systems (Weygand et al., 1963a,b; 1966a). These observations suggest that the applicability of additives for individual problems of fragment condensation may need to be reexamined from case to case.

III. ACTIVE ESTERS IN STEPWISE SYNTHESES

An invaluable property of the benzyloxycarbonyl group, its ability to prevent racemization of activated derivatives of protected amino acids, was already recognized by Bermann and Zervas (1932) at the time when they introduced this group. The protection against racemization provided by urethan-type amino-protecting groups is probably the most reliable measure found so far in combatting this central problem. Slow changes with time in the values of specific rotations of benzyloxycarbonyl amino acid p-nitrophenyl esters, when solutions of these are exposed to bases (Bodanszky and Birk-himer, 1960), demonstrate that the protection is not absolute, but sufficient for all practical purposes. Notable exceptions are the derivatives of S-benzyl-cysteine and O-benzylserine, in which base-catalyzed racemization may take place through mechanisms (Kovacs et al., 1968) other than the usual one, which involves an oxazolinone (azlactone) intermediate. Since the urethan-type protecting groups prevent azlactone formation (Bodanszky and Ondetti, 1966), but not, e.g., the direct abstraction of the proton from the α-carbon atom (Kovacs et al., 1971) or the reversible elimination–addition of benzylmercaptane (Maclaren *et al.*, 1958), in these cases other measures, such as the use of hindered amines (Bodanszky and Bodanszky, 1967), become necessary.

The benzyloxycarbonyl, *tert*-butyloxycarbonyl groups, etc., can prevent racemization only if they are attached to the amino group of an activated *amino acid*, but are ineffective in this respect if used for the protection of a *peptide*. This well-documented fact (Vaughan, 1952; Vaughan, and Eichler, 1953) explains why the stepwise strategy starting with the *N*-terminal residue has not been practical so far. Building of long chains through condensation of larger fragments is possible if glycine or proline is the *C*-terminal residue of the carboxyl components or if procedures free from racemization are applied. The general belief that the azide method satisfies this requirement was badly shaken in recent years (Anderson *et al.*, 1966; Kemp *et al.*, 1970; Sieber *et al.*, 1971), and the expectation that the racemization can be completely excluded by the use of proper additives in DCC-mediated couplings still remains to be confirmed. Earlier claims that some coupling reagents, e.g., isoxazolium salts (Woodward *et al.*, 1961, 1966), yield optically pure products could not be justified in later experiments (Kemp *et al.*, 1970; Woodward and Woodman, 1969). For all these reasons, a reliable approach to long chains is the stepwise strategy (Bodanszky and du Vigneaud, 1959*a,b*; cf. also Bodanszky, 1960) in which synthesis starts with the *C*-terminal residue and proceeds through stepwise lengthening of the chain. By the incorporation of a single amino acid in each step, full advantage is taken of the racemization-preventing ability of the urethan-type protecting groups.

Unless a single product is formed—and in high yield—in each step of such a synthetic approach, purification of the intermediates, losses of valuable materials will render stepwise synthesis impractical. The expected second products formed in mixed anhydride reactions (cf. above) leave it questionable whether or not their stepwise application is justified, even though some attempts (Sarges and Witkop, 1965; Tilak, 1970) were made in this direction. The first proposal for the stepwise strategy was made (Bodanszky and du Vigneaud, 1959*a,b*) after the use of active esters was well established. The real significance of active esters indeed lies in chain-lengthening. During the preparation of di- and tripeptides, diketopiperazine formation often competes with the desired reaction, and therefore higher activation than found in the commonly used active esters is desirable. From the tripeptide stage on, however, it becomes more advantageous to apply active esters exclusively for the addition of single residues. Not only is the problem of racemization avoided, but also the difference in the properties of active esters and those of the product, a protected peptide, is usually so great that separation of starting material and product is relatively simple. Often, washing the peptide with common organic solvents such as ethanol, ethyl acetate, or chloroform is sufficient for the removal of unreacted active ester. When necessary, excess active ester can be removed via reaction with an amine (Löw and Kisfaludy, 1965) such as

$$H_2N-CH_2CH_2-N\begin{array}{c} \diagup CH_3 \\ \diagdown CH_3 \end{array}$$

followed by extraction of the resulting amide

$$Z-NH-CHR-CO-NH-CH_2-CH_2-N\begin{array}{c} \diagup CH_3 \\ \diagdown CH_3 \end{array}$$

with dilute acid.

Because of the ease of separation from the product, active esters of protected amino acids can be applied in excess. This excess assures that all the amino component is acylated. Complete reaction of the latter is important both for high yields and for simple isolation techniques: no unreacted amine has to be removed. A not unimportant aspect (Bodanszky, 1971) of the use of active esters in excess is the suppression of intramolecular side-reactions in the peptide. As the chain is lengthened, the excess of acylating agent can be gradually increased, with the consquence that the reaction between the two reactants becomes pseudo-unimolecular. The rate, therefore, can remain more or less the same in each step.

The necessity to synthesize a long chain singlehandedly, the reduced possibility of teamwork, is an obvious disadvantage of the stepwise strategy. To some extent, this drawback is offset by the relative stability of active esters. Unlike chlorides, azides, or mixed anhydrides, most active esters of protected amino acids can be stored, and hence they can be prepared in larger amounts. The purity and optical purity of the preparations can be checked, and samples of the same material can be used from time to time. In fact, many such active esters became commercially available, thus permitting a kind of cooperation between chemists.

Stepwise application of active esters was first demonstrated on the example of oxytocin (Bodanszky and du Vigneaud, 1959a,b; Bodanszky, 1960). Subsequently, the same approach was used for the preparation of the vasopressins (Bodanszky et al., 1960, 1964) and of many analogues of the pituitary hormones (Schröder and Lübke, 1965). A similar synthesis of bradykinin (Nicolaides and DeWald, 1961) still involved the same chain length : nine amino acids. In an outstanding achievement of peptide chemistry, the synthesis of porcine adrenocorticotropin by Schwyzer and Sieber (1966), a pentadecapeptide was assembled stepwise with nitrophenyl esters as acylating agents. An analogous combination of stepwise chain-building with fragment condensation is now generally used, as shown, e.g., in the syntheses of insulin (Katsoyannis *et al.*, 1964; Kung *et al.*, 1965; Meienhofer *et al.*, 1963), ribonuclease A (Hirschmann *et al.*, 1969), and ribonuclease T_1 (Yanaihara *et al.*, 1969; Beacham *et al.*, 1971). The longest chain built so far, entirely stepwise, with active esters as acylating agents through isolated intermediates is porcine secretin (Bodanszky and Williams, 1967; Bodanszky *et al.*, 1967), consisting of 27 amino acid residues. There was no indication in this work that a limit had been approached.

IV. ACTIVE ESTERS IN SOLID-PHASE PEPTIDE SYNTHESIS

While several polymer-bound active esters were proposed (Fridkin *et al.*, 1965, 1968; Wieland and Birr, 1966) as acylating agents, this kind of synthetic approach has found only limited application so far. More important is the use of active esters in syntheses where the amino component is attached to an insoluble polymeric support. In his first full paper on solid-phase peptide synthesis (SPPS), Merrifield (1963) expressed the view that active esters would be ideal in the new technique. However, he found the *p*-nitrophenyl esters of protected amino acids, probably because of an unfortunate choice of solvent, not sufficiently reactive and therefore proposed the general use of DCC for coupling. Subsequently, it was shown (Bodanszky and Sheehan, 1964) that active esters, especially in dimethylformamide solution, can be successfully applied in SPPS. In the introduction of asparagine and glutamine residues, partial dehydration of the carboxamide group to a nitril (Gish *et al.*, 1956; Ressler, 1956) is caused by DCC, a strong argument against the use of this coupling reagent. These two residues are indeed activated in the form of *p*-nitrophenyl esters, purified (Bodanszky *et al.*, 1963) from the nitril formed during esterification, and then used for acylation.

Active esters offer several advantages in SPPS. Since complete acylation of the amino component is imperative, excess acylating agent has to be used. Whereas in DCC couplings the excess protected amino acid is lost in the form of *N*-acyl-dicyclohexylurea, the excess active esters can be recovered. Furthermore, the leaving group appears, e.g., as *p*-nitrophenol in the filtrate, and thereby the monitoring of the acylation reaction by ultraviolet absorption becomes a simple and practical possibility. More significant than these technical details is the gain in tactical freedom when active esters are used instead of the overactivating DCC. Several functional groups, e.g., side-chain hydroxyls and carboxyls, can be left without protection if moderately reactive esters are the acylating agents, but protection of the same groups is absolutely necessary when protected amino acids and DCC are used *in excess*. Protection of side-chain carboxyl groups in the form of esters precludes the removal of the completed chain from the resin by ammonolysis or hydrazinolysis (Bodanszky and Sheehan, 1966) and also by ester exchange (Beyerman *et al.*, 1968). Thus the entire plan of a synthesis is a function of the method of acylation.

These advantages led to the exclusive use of active esters in a few syntheses (Hörnle, 1967; Weber *et al.*, 1967). The lack of more general acceptance is probably due to the relatively low reaction rates (Rudinger and Gut, 1967; Lübke, 1971) of active esters. In dichloromethane, the most frequently used solvent in SPPS, *p*-nitrophenyl esters react extremely slowly, but even in dimethylformamide the rates are disappointing sometimes. Further disappointment is caused when, for the sake of better rates, highly active compounds such as pentachlorophenyl ester or hydroxysuccinimide esters (Klostermeyer, 1968; Klostermeyer *et al.*, 1967) are chosen. The combined hindrance caused by the side-chains of the activated amino acid and of the *N*-terminal residue of the amino component, by the matrix of the resin, and by the growing peptide chain (Bath, 1970; Hagenmaier, 1970) could be sufficient to render some active esters useless in SPPS. For similar reasons, incomplete acylations were observed (Lübke, 1971) with *p*-nitrophenyl ester of (protected) isoleucine, even when applied in considerable excess and for long reaction time.

The mere fact that some active esters, highly reactive in solution, perform poorly with a resin-bound amino component suggests that a special search is needed for active esters tailored for SPPS. The 2,4,5-trichlorophenyl esters seem to be superior in this respect (Bodanszky and Bath, 1969) to *p*-nitrophenyl esters. Still better performance is expected from *o*-nitrophenyl esters, which are more reactive, even in solution, than the *p*-isomers and are remarkably fast in SPPS (Bodanszky and Bath, 1969). Moreover, the rates observed with *o*-nitrophenyl esters are only slightly dependent on the solvent (Bodanszky and Greenwald, 1972), and therefore the procedures would need not be limited to dimethylformamide. Recent reports on esters of 2-hydroxypyridine (Dutta and Morley, 1971) also offer promise for SPPS.

V. CONCLUSIONS

The choice between powerful condensing agents, such as DCC, and moderately reactive derivatives of acylamino acids, active esters, is a decisive one. In the building of long chains with residues that have functional side-chains, many problems are encountered. The probability of side-reactions is more or less proportional to the degree of activation, and therefore a gain in efficiency requires sacrifices in selectivity. Application of DCC, even with additives, necessitates "global protection" as shown in the synthesis of glucagon (Wünsch, 1967). With moderate carboxyl activation, it becomes possible to plan syntheses with "minimal protection" (Beacham *et al.*, 1971; Hirschmann *et al.*, 1969; Hofmann, 1971; Ondetti *et al.*, 1968; Yanaihara *et al.*, 1969). In principle, only the ε-amino group of lysine and —SH groups of cysteine residues need to be protected. The difficulties encountered in the purification of longer chains carrying numerous protecting groups suggest that, in the synthesis of large molecules, minimal protection is to be preferred, and hence further application of active esters, both in solution and in SPPS, can be expected in the future.

REFERENCES

Anderson, G. W., Zimmerman, J. E., and Callahan, F. M., 1963, *N*-Hydroxysuccinimide esters in peptide synthesis, *J. Am. Chem. Soc.* **85**:3039.

Anderson, G. W., Zimmerman, J. E., and Callahan, F. M., 1964, The use of Esters of *N*-hydroxysuccinimide in peptide synthesis, *J. Am. Chem. Soc.* **86**:1839.

Anderson, G. W., Zimmerman, J. E., and Callahan, F. M., 1966, Racemization control in the synthesis of peptides by the mixed carbonic–carboxylic anhydride method, *J. Am. Chem. Soc.* **88**:1338.

Arens, J. F., 1955, The chemistry of acetylenic ethers. XIII. Acetylenic ethers as reagents for the preparation of amides, *Rec. Trav. Chim.* **74**:769.

Barth, A., and Losse, G., 1964, Darstellung *N*-geschützter Aminosäure-4-(phenylazo)-phenylester und ihre Verwendung als farbige carboxylaktivierte Verbindungen zur Peptidsynthese, *Z. Naturforsch.* **19b**:264.

Bath, R. J., 1970, Problems of steric hindrance in peptide synthesis, Ph.D. dissertation, Case Western Reserve University, Cleveland, Ohio.

Beacham, J., Dupuis, G., Finn, F. M., Storey, H. T., Yanaihara, C., Yanaihara, N., and Hofmann, K., 1971, Studies on polypeptides. XLIX.

Fragment condensations with peptide derivatives related to the primary structure of ribonuclease T_1, *J. Am. Chem. Soc.*, **93**:5526.

Beaumont, S. M., Handford, B. O., and Young, G. T., 1965a, Some Observations concerning racemisation during coupling reactions, *Acta Chim. Acad. Sci. Hung.* **44**:37.

Beaumont, S. M., Handford, B. O., Jones, J. H., and Young, G. T., 1965b, The use of esters of *NN*-dialkylhydroxylamines in peptide synthesis and as selective acylating agents, *Chem. Commun.* **1965**:53.

Bergmann, M., and Zervas, L., 1932, Über ein allgemeines Verfahren der Peptid-Synthese, *Chem. Ber.* **65**:1192.

Beyerman, H. C., Hindriks, H., and De Leer, E. W. B., 1968, Alcoholysis of Merrifield-type peptide-polymer bonds, *Chem. Commun.* **1968**:1668.

Bittner, S., Knobler, Y., and Frankel, M., 1965, Active α-aminoacyl-*O*-derivatives of *N*-substituted hydrosylamine. *O* to *N* migration and a method for peptide synthesis, *Tetrahedron Letters* **1965**:95.

Bodanszky, M., 1955, Synthesis of peptides by aminolysis of nitrophenyl esters, *Nature* **175**:685.

Bodanszky, M., 1957, Aminolysis of propargyl esters: A new method in peptide synthesis, *Chem. Ind.* **1957**:524.

Bodanszky, M., 1960, Stepwise synthesis of peptides by the nitrophenyl-ester method, *Ann. N.Y. Acad. Sci.* **88**:655.

Bodanszky, M., 1971, The principle of excess in the synthesis of secretin, *in* "Prebiotic and Biochemical Evolution" (A. P. Kimball and J. Oró, eds.), pp. 216–222, North-Holland, Amsterdam.

Bodanszky, M., and Bath, R. J., 1969, Active esters and resins in peptide synthesis: The role of steric hindrance, *Chem. Commun.* **1969**:1259.

Bodanszky, M., and Birkhimer, C. A., 1960. Racemization studies with the aid of active esters, *Chimia* **14**:368.

Bodanszky, M., and Bodanszky, A., 1967, Racemization in peptide synthesis. Mechanism— Specific models, *Chem. Commun.* **1967**:591.

Bodanszky, M., and Conklin, L. E., 1967, A simple method for the study of racemization in peptide synthesis, *Chem. Commun.* **1967**:773.

Bodanszky, M., and du Vigneaud, V., 1959*a*, Synthesis of oxytocin by the nitrophenyl ester method, *Nature* **183**:1324.

Bodanszky, M., and du Vigneaud, V., 1959*b*, A method of synthesis of long peptide chains using a synthesis of oxytocin as an example, *J. Am. Chem. Soc.* **81**:5688.

Bodanszky, M., and Greenwald, S., 1972, unpublished.

Bodanszky, M., and Ondetti, M. A., 1966, "Peptide Synthesis," Interscience, New York.

Bodanszky, M., and Sheehan, J. T., 1964, Active esters and resins in peptide synthesis, *Chem. Ind.* **1964**:1423.

Bodanszky, M., and Sheehan, J. T., 1966, Active esters and resins in peptide synthesis, *Chem. Ind.* **1966**:1597.

Bodanszky, M., and Williams, N. J., 1967, Synthesis of secretin. I. The protected tetradecapeptide corresponding to sequence 14–27, *J. Am. Chem. Soc.* **89**:685.

Bodanszky, M., Meienhofer, J., and du Vigneaud, V., 1960, Synthesis of lysine-vasopressin by the nitrophenyl ester method, *J. Am. Chem. Soc.* **82**:3195.

Bodanszky, M., Denning, G. S., Jr., and du Vigneaud, V., 1963, *p*-Nitrophenyl carbobenzoxy-L-asparaginate, *Biochem. Prep.* **10**:122.

Bodanszky, M., Ondetti, M. A., Birkhimer, C. A., and Thomas, P. L., 1964, Synthesis of arginine-containing peptides through their ornithine analogs. Synthesis of arginine vasopressin, arginine vasotocin and L-histidyl-L-phenylalanyl-L-arginyl-L-tryptophylglycine, *J. Am. Chem. Soc.* **86**:4452.

Bodanszky, M., Ondetti, M. A., Levine, S. D., and Williams, N. J., 1967, Synthesis of secretin. II. The stepwise approach, *J. Am. Chem. Soc.* **89**:6753.

Bodanszky, M., Birns, M. T., and Greenwald, S., 1972, unpublished.

Boissonnas, R. A., 1951, Une nouvelle méthode de synthese peptidique, *Helv. Chim. Acta* **34**:874.

Brenner, M., 1967, Critical evaluation of coupling methods, *in* "Peptides: Proceedings of the Eighth European Peptide Symposium" (H. C. Beyerman, A. Van De Linde, and W. Maasen Van Den Brink, eds.), pp. 1–7, North-Holland, Amsterdam.

Buyle, R., and Vieche, G. H., 1964, Peptidsynthesen mit Inaminen, *Angew. Chem.* **76**:572.

Dutta, A. S., and Morley, J. S., 1971, Polypeptides. Part XII. The preparation of 2-pyridyl esters and their use in peptide synthesis, *J. Chem. Soc.* (*C*) **1971**:2896.

du Vigneaud, V., Ressler, C., Swan, J. M., Roberts, C. W., Katsoyannis, P. G., and Gordon, S., 1953, The synthesis of an octapeptide amide with the hormonal activity of oxytocin, *J. Am. Chem. Soc.* **75**:4879.

du Vigneaud, V., Ressler, C., Swan, J. M., Roberts, C. W., and Katsoyannis, P. G., 1954, The synthesis of oxytocin, *J. Am. Chem. Soc.* **76**:3115.

Elliott, D. F., and Russel, D. W., 1957, Peptide synthesis employing *p*-nitrophenyl esters prepared with the aid of *N,N'*-dicyclohexylcarbodiimide, *Biochem. J.* **66**:49P.

Farrington, J. A., Kenner, G. W., and Turner, J. N., 1955, The preparation of *p*-nitrophenyl thiol esters and their application to peptide synthesis, *Chem. Ind.* **1955**:601.

Farrington, J. A., Hextall, P. J., Kenner, G. W., and Turner, J. M., 1957, Peptides. Part VII. The preparation and use of *p*-nitrophenyl thiol esters, *J. Chem. Soc.* **1957**:1407.

Fischer, E., 1902, Über einige Derivate des Glykocolls, Alanins und Leucins, *Chem. Ber.* **35**:1095.

Fridkin, M., Patchornik, A., and Katchalski, E., 1965, A synthesis of cyclic peptides utilizing high molecular weight carriers, *J. Am. Chem. Soc.* **87**:4646.

Fridkin, M., Patchornik, A., and Katchalski, E., 1968, Use of polymers as chemical reagents. II. Synthesis of bradykinin, *J. Am. Chem. Soc.* **90**:2953.

Fujino, M., and Nishimura, O., 1969, Use of esters of simple ketoximes in peptide synthesis, *Chem. Pharm. Bull.* **17**:1937.

Gish, D. T., Katsoyannis, P. G., Hess, G. P., and Stedman, R. J., 1956, Unexpected formation of anhydro compounds in the synthesis of asparaginyl and glutaminyl peptides, *J. Am. Chem. Soc.* **78**:5954.

Gordon, M., Miller, J. G., and Day, A. R., 1948, Effect of structure on reactivity. I Ammonolysis of esters with special reference to the electron release effects of alkyl and aryl groups, *J. Am. Chem. Soc.* **70**:1946.

Gordon, M., Miller, J. G., and Day, A. R., 1949, Effect of structure on reactivity. II. Influence of solvents on ammonolysis of esters, *J. Am. Chem. Soc.* **71**:1245.

Govindachari, T. R., Nagarajan, K., and Rajappa, S., 1966, Synthesis of *O*-acylbenzohydroxamic acids and their use in peptide synthesis, *Tetrahedron* **22**:3367.

Hagenmaier, H., 1970, The influence of the chain length on the coupling reaction in solid phase peptide synthesis, *Tetrahedron Letters* **1970**:283.

Hanford, B. O., Jones, J. H., Young, G. T., and Johnson, T. F. N., 1965, Amino-acids and peptides. Part XXIV. The use of esters of 1-hydroxypiperidine and of other *NN*-dialkyl-hydroxylamines in peptide synthesis and as selective acylating agents, *J. Chem. Soc.* **1965**:6814.

Hirschmann, R., Nutt, R. F., Veber, D. F., Vitali, R. A., Varga, S. L., Jacob, T. A., Holley, F. W., and Denkewalter, R. G., 1969, Studies on the total synthesis of an enzyme. V. The preparation of enzymatically active material, *J. Am. Chem. Soc.* **91**:507.

Hofmann, K., 1971, Synthetic studies with ribonuclease T₁ using a minimum side chain protection, *in* "Peptides—1969: Proceedings of the Tenth European Peptide Symposium" (E. Scoffone, ed.), p. 130, North-Holland, Amsterdam.

Hörnle, S., 1967, Synthese der Rinder-Insulin-A-Kette nach der Merrifield-Methode unter ausschliesslicher Verwendung von *tert.*-Butyloxycarbonylaminosäure-*p*-nitrophenylestern, *Z. Physiol. Chem.* **348**:1355.

Iselin, B., Feurer, M., and Schwyzer, R., 1955, Über aktivierte Ester. V. Verwendung der Cyanmethylester-Methode zur Herstellung von (*N*-Carbobenzoxy-*S*-benzyl-L-cysteinyl)-L-tyrosyl-L-isoleucin auf verschiedenen Wegen, *Helv. Chim. Acta* **38**:1508.

Jakubke, H. D., 1965a, Darstellung *N*-geschützter Aminosäureester des 8-Hydroxychinolins und deren Verwendung als reaktive Zwischenprodukte für die Peptidsynthese, *Z. Naturforsch.* **20b**:273.

Jakubke, H. D., 1965b, Über aktivierte Ester; Peptidsynthese mit *N*-Acylaminosäure-8-thiochinolyl Estern, *Z. Chem.* **5**:453.

Jakubke, H. D., 1965c, Aminoacyl-Derivate des 2-Selenonaphtols und ihre Verwendung zur Peptidsynthese, *Annalen* **682**:244.

Jakubke, H. D., and Voigt, A., 1966, Untersuchungen über die peptid-chemische Verwendbarkeit von Acylaminosäure-chinolyl-(8)-estern, *Chem. Ber.* **99**:2419.

Jeschkeit, H., 1968, Peptidsynthesen mit N-Hydroxyglutarimidestern, *Z. Chem.* **8**:20.

Jeschkeit, H., 1969, Peptidsynthesen mit N-Hydroxycarbamaten, *Z. Chem.* **9**:266.

Johnson, B. J., and Jacobs, P. M., 1968*a*, A new carboxy-protecting group for peptide synthesis and its direct conversion into an activated ester suitable for peptide formation: 4-(Methylthio)phenyl and 4(methylsulphonyl)phenyl esters, *Chem. Commun.* **1968**:73.

Johnson, B. J., and Jacobs, P. M., 1968*b*, The 4-(methylsulfonyl)phenyl activated ester. Susceptibility to racemization, *J. Org. Chem.* **33**:4524.

Johnson, B. J., and Ruettinger, T. A., 1970, Further studies on N-acylamino acid esters of 4-(methylthio)phenol, *J. Org. Chem.* **35**:255.

Jones, J. H., 1970, Peptide synthesis, *in* "Amino-acids, Peptides, and Proteins," Vol. 2 (G. T. Young, ed.), pp. 143–191, The Chemical Society, London.

Jones, J. H., and Young, G. T., 1968, Amino-acids and peptides. Part XXVIII. Anchimeric acceleration of aminolysis of esters. The use of mono-esters of catechol in peptide synthesis, *J. Chem. Soc. (C)* **1968**:436.

Kaczmarek, M., Kupryszewski, G., and Wajcht, J., 1967, N-Acylated *p*-propionylphenyl esters and their application to the synthesis of peptides, *Zeszyty Nauk., Mat. Fiz. Chem. Wyzsza Ssk. Pedagog. Gdansku* **7**:143; *Chem. Abst.* **70**:47840j, 1969.

Katsoyannis, P. G., and du Vigneaud, V., 1954, The synthesis of *p*-toluenesulfonyl-L-isoleucyl-L-glutaminyl-L-aspargine and related peptides, *J. Am. Chem. Soc.* **76**:3113.

Katsoyannis, P. G., Fukuda, K., Tometsko, A., Suzuki, K., and Tilak, M., 1964, Insulin peptides. X. The synthesis of the B-chain of insulin and its combination with natural or synthetic A-chain to generate insulin activity, *J. Am. Chem. Soc.* **86**:930.

Kemp, D. S., 1967, The N-ethylbenzisoxazolium cation. II. The mechanism of reactions of the cation with nucleophiles, *Tetrahedron* **23**:2001.

Kemp, D. S., and Chien, S. W., 1967, A new peptide coupling reagent, *J. Am. Chem. Soc.* **89**:2743.

Kemp, D. S., and Woodward, R. B., 1965, The N-ethylbenzisoxazolium cation. I. Preparation and reactions with nucleophilic species, *Tetrahedron* **21**:3019.

Kemp, D. S., Wang, S. W., Busby, G., III, and Hugel, G., 1970, Microanalysis by successive isotope dilution. A new assay for racemic content, *J. Am. Chem. Soc.* **92**:1043.

Kisfaludy, L., Roberts, J. E., Johnson, R. H., Mayers, G. L., and Kovacs, J., 1970, Synthesis of N-carbobenzoxyamino acid and peptide pentafluorophenyl esters as intermediates in peptide synthesis, *J. Org. Chem.* **35**:3563.

Klostermeyer, H., 1968, Synthese von Gramicidin S mit Hilfe der Merrifield Methode, *Chem. Ber.* **101**:2823.

Klostermeyer, H., Halstrøm, J., Kusch, P., Föhles, J., and Lunkenheimer, W., 1967, Experiences with the solid phase method of peptide synthesis, *in* "Peptides: Proceedings of the Eighth European Peptide Symposium" (H. C. Beyerman, A. Van De Linde, and W. Maasen Van Den Brink, eds.), p. 113, North-Holland, Amsterdam.

König, W., and Geiger, R., 1970, Eine neue Methode zur Synthese von Peptiden: Aktivierung der Carboxylgruppe mit Dicyclohexylcarbodiimid unter Zusatz von 1-Hydroxy-benzotriazolen, *Chem. Ber.* **103**:788.

Kovacs, J., Kisfaludy, L., and Ceprini, M. Q., 1967, On the optical purity of peptide active esters prepared by N,N'-dicyclohexylcarbodiimide and complexes of N,N'-dicyclohexylcarbodiimide-pentachlorophenol and N,N'-dicyclohexylcarbodiimide-pentafluorophenol, *J. Am. Chem. Soc.* **89**:183.

Kovacs, J., Mayers, G. L., Johnson, R. H., and Ghatak, U. R., 1968, Racemization studies in peptide chemistry. Re-investigation of the "β-elimination-readdition" mechanism of N-benzyloxycarbonyl-S-benzylcysteine derivatives, *Chem. Commun.* **1968**:1066.

Kovacs, J., Kisfaludy, L., Ceprini, M. Q., and Johnson, R. H., 1969, Investigations on the stereospecificity of peptide active phenyl ester formation and coupling, *Tetrahedron* **25**:2555.

Kovacs, J., Cortegiano, H., Cover, R. E., and Mayers, G. L., 1971, Isoracemization of N-carbobenzoxyamino acid ester derivatives, *J. Am. Chem. Soc.* **93**:1541.

Kung, Y. T., Du, Y. C., Huang, W. T., Chen, C. C., Ke, L. T., Hu, S. C., Jiang, R. Q., Chu, S. Q., Niu, C. I., Hsu, J. Z., Chang, W. C., Chen, L. L., Li, H. S., Wang, Y., Loh, T. P., Chi, A. H., Li, C. H., Shi, P. T., Yieh, Y. H., Tang, K. L., and Hsing, C. Y., 1965, Total synthesis of crystalline bovine insulin, *Sci. Sinica (Peking)* **14**:1710.

Kuprýszewski, G., 1961, Amino acid chlorophenyl esters. II. Synthesis of peptides by aminolysis of active N-protected amino acid 2,4,6-trichlorophenyl esters, *Roczniki Chem.* **35**:595; *Chem. Abst.* **57**:27121i, 1961.

Kuprýszewski, G., and Formela, M., 1963, Amino acid chlorophenyl esters. VII. Reaction of phthaloylglycine chlorophenyl esters with benzylamine, *Roczniki Chem.* **37**:161; *Chem. Abst.* **59**:11651e, 1963.

Kuprýszewski, G., and Muzalewski, F., 1967, N-Acylated p-sulfamoylphenyl esters and their application for the synthesis of peptides, *Zeszyty Nauk., Mat. Fiz. Chem. Wyzsza Szk. Pedagog. Gdansku* **7**:159; *Chem. Abst.* **70**:47839r, 1969.

Kuprýszewski, G., Muzalewski, F., and Przybylski, J., 1968, The application of active N-acylated amino acid [N-(2-pyrimidyl)sulfonamido] phenyl esters for the synthesis of peptide bonds, *Roczniki Chem.* **42**:1009; *Chem. Abst.* **70**:4582j, 1968.

Lloyd, K., and Young, G. T., 1968, The use of acylamino-acid esters of 2-mercaptopyridine in peptide synthesis, *Chem. Commun.* **1968**:1400.

Lloyd, K., and Young, G. T., 1971, Amino-acids and peptides. Part XXXIV. Anchimerically assisted coupling reactions: The use of 2-pyridyl thiol esters, *J. Chem. Soc.* (*C*) **1971**:2890.

Losse, G., Barth, A., and Schatz, K., 1964, N-Geschützte Aminoacyl-oxime als neue Carboxyl-aktivierte Verbindungen zur Peptid-Synthese, *Annalen* **677**:185.

Losse, G., Hoffmann, K. H., and Hetzer, G., 1965, Peptidsynthesen mit O-[Cbo-Aminoacyl]-Oximen und O-[Cbo-Aminoacyl]-Pyrazolon-Enolen, *Annalen* **684**:236.

Löw, M., and Kisfaludy, L., 1965, Some observations with N-hydroxysuccinimide esters, *Acta Chim. Acad. Sci. Hung.* **44**:61.

Lubiewska-Nakonieczna, L., Rzeszotarska, B., and Taschner, E., 1970, Aminosäuren und Peptide. XV. Synthese aktiver Benzhydroxamester N-geschützter Aminosäuren und Peptide und ihre Anwendung zur Darstellung optisch einheitlicher Peptide, *Annalen* **741**:157.

Lübke, K., 1971, Solid phase synthesis of [Arg^8] vasopressin, in "Peptides—1969: Proceedings of the Tenth European Peptide Symposium" (E. Scoffone, ed.), p. 154, North-Holland, Amsterdam.

Maclaren, J. A., Savige, W. E., and Swan, J. M., 1958, Amino acids and peptides. IV. Intermediates for the synthesis of certain cystine-containing peptide sequences of insulin, *Aust. J. Chem.* **11**:345.

Meienhofer, J., Schnabel, E., Bremer, H., Brinkhoff, O., Zabel, R., Sroka, W., Klostermeyer, H., Brandenburg, D., Okuda, T., and Zahn, H., 1963, Synthese der Insulinketten und ihre Kombination zu insulinaktiven Präparaten, *Z. Naturforsch.* **18b**:1120.

Merrifield, R. B., 1963, Solid phase peptide synthesis. I. The synthesis of a tetrapeptide, *J. Am. Chem. Soc.* **85**:2149.

Mitin, Yu. V., and Nadezhdina, L. B., 1968, Esters of N-substituted amino acids and dimethyl-amino phenols, *Zh. Org. Khim.* **4**:1181; *Chem. Abst.* **69**:87427r, 1968.

Nefkens, G. H. L., and Tesser, G. I., 1961, A novel activated ester in peptide synthesis, *J. Am. Chem. Soc.* **83**:1263.

Nefkens, G. H. L., Tesser, G. I., and Nivard, R. J. F., 1962, Synthesis and reactions of esters of N-hydroxyphthalimide and N-protected amino acids, *Rec. Trav. Chim.* **81**:683.

Nicolaides, E. D., and DeWald, H. A., 1961, Studies on the synthesis of polypeptides. Bradykinin, *J. Org. Chem.* **26**:3872.

Ondetti, M. A., Narayanan, V. L., von Saltza, M., Sheehan, J. T., Sabo, E. F., and Bodanszky, M., 1968, The synthesis of secretin. III. The fragment-condensation approach, *J. Am. Chem. Soc.* **90**:4711.

Panneman, H. J., Marx, A. F., and Arens, J. F., 1959, Derivatives of amino acids and peptides. IV. The synthesis of peptides by means of ethoxyacetylene, *Rec. Trav. Chim.* **78**:487.

Paquette, L. A., 1965, Electrophilic additions of acyl and sulfonyl halides to 2-ethoxypyridine 1-oxide. A new class of activated esters and their application to peptide synthesis, *J. Am. Chem. Soc.* **87**:5186.

Pless, J., and Boissonnas, R. A., 1963, Über die Geschwindigkeit der Aminolyse von verschiedenen neuen, aktivierten, *N*-geschützten α-Aminosäurephenylestern, insbesondere 2,4,5-Trichlorphenylestern, *Helv. Chim. Acta* **46**:1609.

Rajappa, S., Nagarajan, K., and Iyer, V. S., 1967, Hydroxamic acids and their derivatives. III. Preparation of esters of pivalohydroxamic acid and their use in peptide synthesis, *Tetrahedron* **23**:4805.

Ressler, C., 1956, Formation of α,γ-diaminobutyric acid from asparagine-containing peptides, *J. Am. Chem. Soc.*, **78**:5956.

Ressler, C., and du Vigneaud, V., 1954, The synthesis of the tetrapeptide amide *S*-benzyl-L-cysteinyl-L-prolyl-L-leucylglycinamide, *J. Am. Chem. Soc.* **76**:3107.

Rothe, M., and Kunitz, F. W., 1957, Cyclische Peptide. II. Synthese cyclischer Oligopeptide der ε-Aminocapronsäure. Konstitutionaufklärung der ringförmigen Bestandteile von Polycaprolactam, *Annalen* **609**:88.

Rudinger, J., and Gut, V., 1967, Discussion, in "Peptides: Proceedings of European Peptide Symposium" (H. C. Beyerman, A. Van De Linde, and W. Maasen Van Den Brink, eds.), p. 89, North-Holland, Amsterdam.

Sarantakis, D., Sutherland, J. K., Tortorella, C., and Tortorella, V., 1968, 2-Fluoropyridine *N*-oxide and its reactions with amino-acid derivatives, *J. Chem. Soc.* (*C*) **1968**:72.

Sarges, R., and Witkop, B., 1965, Gramicidin A. VI. The synthesis of valine- and isoleucine-gramicidin A, *J. Am. Chem. Soc.* **87**:2020.

Schröder, E., and Lübke, K., 1965, "The Peptides," Vol. II, Academic Press, New York.

Schwyzer, R., and Sieber, P., 1958, Verdoppelungsreaktionen beim Ringschluss von Peptiden. II. Cyclo-glycyl-glycyl-DL-phenylalanyl-glycyl-glycyl-DL-phenylalanyl. Vervendung aktivierter Ester zur Synthese macrocyclischer peptide. Molekulargewichtbestimmungen in Dimethylsulfoxyd, *Helv. Chim. Acta* **41**:2190.

Schwyzer, R., and Sieber, P., 1966, Die Totalsynthese des β-Corticotropins (adrenocorticotropes Hormon; ACTH), *Helv. Chim. Acta* **49**:134.

Schwyzer, R., Iselin, B., and Feurer, M., 1955a, Über aktivierte Ester. I. Aktivierte Ester der Hippursäure und ihre Umsetzungen mit Benzylamin, *Helv. Chim. Acta* **38**:69.

Schwyzer, R., Feurer, M., Iselin, B., and Kägi, H., 1955b, Über aktivierte Ester. II. Synthese aktivierter Ester von Aminosäure-Derivaten, *Helv. Chim. Acta* **38**:80.

Schwyzer, R., Feurer, M., and Iselin, B., 1955c, Über aktivierte Ester. III. Umsetzungen aktivierter Ester von Aminosäure-II und Peptid-Derivaten mit Aminen und Aminosäureestern, *Helv. Chim. Acta* **38**:83.

Schwyzer, R., Iselin, B., Rittel, W., and Sieber, P., 1960, Aryl esters through sulfites and application to gramicidin S, *Chem. Abst.* **54**:7579e.

Sheehan, J. C., and Hess, G. P., 1955, A new method of forming peptide bonds, *J. Am. Chem. Soc.* **77**-1067.

Sieber, P., Brugger, M., and Rittel, W., 1971, Remarks concerning racemisation with the azide method, in "Peptides—1969: Proceedings of the Tenth European Peptide Symposium" (E. Scoffone, ed.), pp. 60–61, North-Holland, Amsterdam.

Stevens, C. L., and Munk, M. E., 1958, Nitrogen analogs of ketones. V. Formation of the peptide bond, *J. Am. Chem. Soc.* **80**:4069.

Stewart, F. H. C., 1968, Benzyloxycarbonylamino acid 2,6,-dichloro-4-nitrophenyl esters and the reaction of 2,6-dichloro-4-nitrophenol with *N,N'*-dicyclohexylcarbodiimide, *Austr. J. Chem.* **21**:477.

Swan, J. M., and du Vigneaud, V., 1954, The synthesis of L-glutaminyl-L-asparagine, L-glutamine and L-isoglutamine from *p*-toluenesulfonyl-L-glutamic acid, *J. Am. Chem. Soc.* **76**:3110.

Taschner, E., Rzeszotarska, B., and Lubiewska, L., 1965*a*, Preparation of peptides with 3-pyridyl esters, *Angew. Chem. Internat. Ed.* **4**:594.

Taschner, E., Rzeszotarska, B., and Lubiewska, L., 1965*b*, Aminosäuren und Peptide. XIII. Darstellung aktiver Pyridyl-(3)-ester *N*-geschützter Aminosäuren und Peptide sowie deren Anwendung zur Peptid-Synthese, *Annalen* **690**:177.

Taschner, E., Rzeszotarska, B., and Kuziel, A., 1965*c*, New way for the formation of peptide bonds without racemisation, *Acta Chim. Acad. Sci. Hung.* **44**:67.

Taschner, E., Rzeszotarska, B., and Lubiewska, L., 1967, A new method for the synthesis of benzhydroxamic esters of *N*-protected amino-acids and peptides and their use for the synthesis of peptides without racemisation, *Chem. Ind.* **1967**:402.

Taylor, E. C., Kinzle, F., and McKillop, A., 1970, Thallium in organic synthesis. XII. Improved syntheses of 1-acyloxyl-2-(1H)-pyridone class of active esters, *J. Org. Chem.* **35**:1672.

Tilak, M. A., 1970, New non-solid phase method for quick, quantitative synthesis of analytically pure peptides without intermediate or final purifications. I. Synthesis of a nonapeptide, *Tetrahedron Letters* **1970**:849.

Vaughan, J. R., Jr., 1952, Preliminary investigations on the preparation of optically active peptides using mixed carbonic–carboxylic acid anhydrides, *J. Am. Chem. Soc.* **74**:6137.

Vaughan, J. R., Jr., and Eichler, J. A., 1953, The preparation of optically-active peptides using mixed carbonic–carboxylic acid anhydrides, *J. Am. Chem. Soc.* **75**:5556.

Vaughan, J. R., Jr., and Osato, L., 1952, The preparation of peptides using mixed carbonic–carboxylic acid anhydrides, *J. Am. Chem. Soc.* **74**:676.

Weber, U., Hörnle, S., Grieser, G., Herzog, K. H., and Weitzel, G., 1967, Struktur und Wirkung von Insulin. II. Synthetische A-Ketten mit variierter Sequenz, *Z. Physiol. Chem.* **348**:1715.

Weygand, F., and Steglich, W., 1961, Peptid-Synthesen mit Acylaminosäurevinylestern, *Angew. Chem.* **73**:757.

Weygand, F., Prox, A., Schmidhammer, L., and König, W., 1963*a*, Gaschromatographische Untersuchung der Racemisierung bei Peptidsynthesen, *Angew. Chem.* **75**:282.

Weygand, F., Prox., A., Schmidhammer, L., and König, W., 1963*b*, Gas chromatographic investigation of racemization during peptide synthesis, *Angew. Chem. Internat. Ed.* **2**:183.

Weygand, F., Prox. A., and König, W., 1966*a*, Racemisierung bei Peptidsynthesen, *Chem. Ber.* **99**:1451.

Weygand, F., Hoffmann, D., and Wünsch, E., 1966*b*, Peptidsynthesen mit Dicyclohexylcarbodiimid unter Zusatz von *N*-Hydroxysuccinimid, *Z. Naturforsch.* **21b**:426.

Wieland, Th., and Bernhard, H., 1951, Über Peptid-Synthesen. 3. Mitteilung. Die Verwendung von Anhydriden aus *N*-acylierten Aminosäuren und Derivaten anorganischer Säuren, *Annalen* **572**:190.

Wieland, Th., and Birr, C., 1966, Peptidesynthesen mit am Harz aktivierten Bausteinen, *Angew. Chem.* **78**:303.

Wieland, Th., Schäfer, W., and Bockelmann, E., 1951, Über Peptidsynthesen. V. Über eine bequeme Darstellungsweise von Acylthiophenolen und ihre Verwendung zu Amid- und Peptid-Synthesen, *Annalen* **573**:99.

Woodward, R. B., and Woodman, D. J., 1969, Azlactone formation in the isoxazolium salt method of peptide synthesis, *J. Org. Chem.* **34**:2742.

Woodward, R. B., Olofson, R. A., and Mayer, H., 1961, A new synthesis of peptides, *J. Am. Chem. Soc.* **83**:1010.

Woodward, R. B., Olofson, R. A., and Mayer, H., 1966, A useful synthesis of peptides, *Tetrahedron Suppl.* **8**:321.

Wünsch, E., 1967, Die Totalsynthese des Pankreas-Hormons Glucagon, *Z. Naturforsch.* **22b**:1269.

Wünsch, E., and Drees, F., 1966, Zur Synthese des Glucagons. X. Darstellung der Sequenz 22–29, *Chem. Ber.* **99**:110.

Yanaihara, N., Yanaihara, C., Dupuis, G., Beacham, J., Camble, R., and Hofman, K., 1969, Studies on polypeptides. XLII. Synthesis and characterisation of seven fragments spanning the entire sequence of ribonuclease T_1, *J. Am. Chem. Soc.* **91**:2184.

Zaoral, M., 1962, Amino acids and peptides. XXXVI. Pivaloyl chloride as a reagent in the mixed anhydride synthesis of peptides, *Coll. Czech. Chem. Commun.* **27**:1273.

Zimmerman, J. E., and Anderson, G. W., 1967, The effect of active ester components on racemization in the synthesis of peptides by the dicyclohexylcarbodiimide method, *J. Am. Chem. Soc.* **89**:7151.

CHAPTER 3

THE FACILITATION OF PEPTIDE SYNTHESIS BY THE USE OF PICOLYL ESTERS*

G. T. Young

The Dyson Perrins Laboratory
Oxford University
Oxford, England

I. INTRODUCTION: THE "FUNCTIONAL HANDLE"

It has for long been clear that the routine synthesis of large peptides and proteins requires some facilitation of the methods used so successfully for the synthesis of peptide hormones such as oxytocin, the vasopressins, and their analogues and extended with some difficulty to peptides of 30–40 residues. The solid-phase method of synthesis was the first result of this recognition. Despite its convenience for the synthesis of small peptides, it suffers from the "inborn defect" that one of the reactants is a polymer and reactions are therefore difficult to force to completion; in consequence, the products are mixtures, which in the case of small peptides can usually be separated fairly readily, but for large peptides and proteins purification may be difficult or impossible. (For a valuable critical review of current methods of peptide synthesis, see Wünsch, 1971.) An alternative approach (Camble *et al.*, 1967) is to arrange that the required product is recovered *after* the reaction has been completed in homogeneous solution by means of a functional group acting as a "handle," incorporated in the growing peptide. In this way, one can combine the advantages of homogeneous reactions—the mild conditions, the ability to follow the progress of the reaction and to

---

*Abbreviations follow the Rules of the IUPAC-IUB Commission on Biochemical Nomenclature. Additional abbreviations used here are Cha, β-cyclohexylalanine; Har, homoarginine; Pcp, pentachlorophenyl; Pic, 4-picolyl; Tcp, 2,4,5-trichlorophenyl.

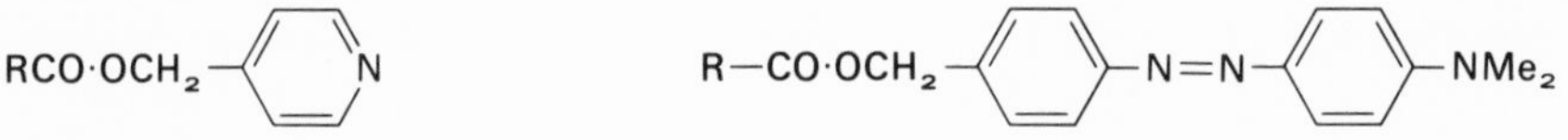

$$RCO \cdot OCH_2 - \bigcirc N \qquad\qquad R-CO \cdot OCH_2 - \bigcirc - N=N - \bigcirc - NMe_2$$

[1]   [2]

Fig. 1

establish its completion, and the ability to characterize and, if necessary, to purify each intermediate—with some of the convenience of the solid-phase method.

The most appropriate position for such a "handle" is clearly as a protection of the C-terminal carboxy group, although in particular cases a side-chain function could be used alternatively or (as will be seen) additionally; an example has very recently appeared (Schafer and Carlsson, 1972) in which the histidine side-chain is used in this way. Since the carboxy component is acidic, or may give rise to acidic contaminants, a basic handle should ensure a simple separation, but other types can clearly be envisaged. On the whole, there appeared to be advantages in using a weakly basic group in order that the product could be liberated as the free base under mild conditions and so transferred to organic solvents. We therefore investigated the use of 4-picolyl esters (Fig. 1 [1]) as carboxy protection (Camble et al., 1967, 1969; Garner et al., 1968); this first choice has proved to be very satisfactory for our purpose and has been used in all subsequent syntheses. Simultaneously, Wieland and Racky (1968) had been examining the analogous use of p-dimethylamino-azobenzyl esters [2], the colored coupled product being adsorbed on Sulfoethyl-Sephadex; the tripeptide L-prolyl-L-alanyl-L-phenylalanine was synthesized in this way. Recently, the "handle" principle has been proposed for use in polynucleotide synthesis (Hata et al., 1971).

II. 4-PICOLYL ESTERS OF α-AMINO ACIDS

4-Picolyl esters proved to be of considerable interest in themselves as new carboxy-protecting groups (Camble et al., 1969). They are prepared by the reaction of a benzyloxycarbonylamino acid tetramethylguanidinium salt (or triethylammonium salt) with 4-picolyl chloride, or (better) by the condensation of the benzyloxycarbonylamino acid with 4-picolyl alcohol by means of dicyclohexylcarbodiimide. In contrast to benzyl esters, they are relatively stable to acid; for example, after 24 hr in 45% (w/v) hydrogen bromide in acetic acid at room temperature, benzyloxycarbonylglycine 4-picolyl ester gave only glycine 4-picolyl ester dihydrobromide, and no glycine could be detected in the product. Presumably, protonation of the pyridyl nitrogen hinders further attack by protons. The benzyloxycarbonyl group can therefore be removed without affecting the ester, and this is the usual route for the preparation of amino acid 4-picolyl esters as their dihydro-bromides. Most of these are reasonably stable when pure, but some are hygroscopic, and in general such salts should be stored in a desiccator and used without undue delay. The acylamino acid 4-picolyl esters and the amino acid 4-picolyl ester salts prepared so far are listed in Tables I and II, respectively. More recently (Maclaren 1972), amino acid 4-picolyl esters have been prepared directly using ethyl acetoacetate to provide temporary pro-tection of the amino group, and this should prove a valuable route.

4-Picolyl esters may be cleaved readily by cold alkali, by hydrogenation in the presence of palladium on charcoal, by sodium in liquid ammonia, and by electrolytic reduction at a mercury cathode under the mild conditions used for the electrolytic removal of the nitrogroup from nitroarginine and its peptides (Scopes *et al.*, 1965). Benzyl esters are not cleaved by electroly-tic reduction, and this property is one of the most interesting of the 4-picolyl group, in connection with its use as a general method of carboxy protection. In work with Dr. Wayne B. Watkins, it has been observed that the 4-picolyl group, but not the benzyloxycarbonyl group, is removed by the catalytic hydrogenation of *N*-benzyloxycarbonyl-*S*-benzyl-L-cysteinylglycine 4-picolyl ester and analogous peptides; the hydrogenation of similar peptides con-taining methionine or *S*-benzylmethylcysteine failed completely, but electro-lytic reduction successfully removed the 4-picolyl group. The main product from the hydrogenolysis of the 4-picolyl residue appears to be 4-methyl-piperidine, which gives a bright violet coloration with ninhydrin, but 4-methylpyridine has also been detected (Schafer, 1970; Fletcher, 1971). Dr. M. Fridkin of the Weizmann Institute of Sciences, Rehovoth, Israel, kindly examined the action of anhydrous hydrogen fluoride on benzyloxycarbonyl-glycyl-L-phenylalanine 4-picolyl ester and found only partial cleavage of the picolyl ester group after 1 hr at room temperature. The esters may be con-verted into amides in the usual way, and into hydrazides for the further coupling of fragments.

III. USE OF 4-PICOLYL ESTERS FOR THE FACILITATION OF PEPTIDE SYNTHESIS

The use of 4-picolyl esters for the facilitation of synthesis was examined first in the synthesis of L-leucyl-L-alanylglycyl-L-valine (Camble *et al.*, 1967, 1969), a simple model used analogously in the early days of solid-phase synthesis (Merrifield, 1963). α-Amino protection was provided by benzyloxy-carbonyl, cleaved by hydrogen bromide in acetic acid. In the first step, benzyloxycarbonylglycine was condensed with L-valine 4-picolyl ester (liberated *in situ* from the dihydrobromide by triethylamine) by means of dicyclohexylcarbodiimide in tetrahydrofuran; the acylating agents were

Table I. 4-Picolyl Esters of Acylamino Acids

Ester[a]	Method of Preparation[b]	Recrystallization solvent	Melting point (°C)	$[\alpha]_D^{20}$	Reference
Z-Ala-OPic	A	Ether	111–112.5	−20.2 (c 1 in Me_2NCHO)	Camble et al. (1969)
Z-Arg(NO$_2$)-OPic	A	Dioxan	151–153	−13.4 (c 1 in Me_2NCHO)	Camble et al. (1969)
Z-Asp(OPic)	c	Ethyl acetate	158–160	−20 (c 0.8 in Me_2NCHO)	Garner and Young (1971)
Boc-Asp(OPic)	c	Ethyl acetate	133–135	−23 (c 1 in Me_2NCHO)	Garner and Young (1971)
Z-Glu(OPic)	c	Ethyl acetate	138–139	−12.3 (c 1.2 in Me_2NCHO)	Garner and Young (1971)
Boc-Glu(OPic)	c	Ethyl acetate	156–158	−16 (c 1 in Me_2NCHO)	Garner and Young (1971)
Z-Gly-OPic	A	Ether	71.5–73.5	—	Camble et al. (1969)
Boc-Thr(Bzl)-OPic	B(i)	Ether–petroleum	71–72	−23 (c 1 in $CHCl_3$)	J. G. Warnke and G. T. Young (unpublished)
Z-Ile-OPic	A	Ether	50.5–52	−10.3 (c 1.1 in Me_2NCHO)	Camble et al. (1969)
Z-Leu-OPic	A	—	Oil	−12.1 (c 2 in EtOAc)	Camble et al. (1969)
Z-Met-OPic	A	—	Oil	−26.3 (c 2 in EtOH)	Camble et al. (1969)
Z-Orn(Pipoc)-OPic	B(i)	—	Oil	+3 (c 1.2 in $CHCl_3$)	T. G. Pinker and G. T. Young (unpublished)
Boc-Orn(Pipoc)-OPic	B(i)	EtOAc–petroleum	118–119	+1 (c 1.2 in $CHCl_3$) (+6.8 at 365 nm)	T. G. Pinker and G. T. Young (unpublished)
Z-Phe-OPic	A	Ether	87.5–89.5	−33.0 (c 1 in Me_2NCHO)	Camble et al. (1969)

Table I (concluded)

Ester[a]	Method of Preparation[b]	Recrystallization solvent	Melting point (°C)	$[\alpha]_D^{20}$	Reference
Z-Pro-OPic	A	—	Oil	−34.7 (c 1 in EtOAc)	Camble et al. (1969)
Boc-Tyr(Bzl)-OPic	B(i)	EtOAc–petroleum	78–80	−18.5 (c 1 in Me$_2$NCHO)	J. G. Warnke and G. T. Young (unpublished)
Z-Val-OPic	A	Ether	65–66	−12.1 (c 1 in Me$_2$NCHO)	Camble et al. (1969)

[a] Pic, 4-picolyl; Pipoc, piperidino-oxycarbonyl. All optically active amino acids are of the L form.
[b] Method A: from 4-picolyl chloride; method B, from 4-picolyl alcohol and dicyclohexylcarbodiimide (i) in dichloromethane, (ii) in pyridine. It is advantageous to add the dicyclohexylcarbodiimide gradually at 0–5°.
[c] By partial saponification of the diester.

Table II. Salts of Amino Acid 4-Picolyl Esters[a]

Salts	Recrystallization Solvent	Melting point (°C)	$[\alpha]_D^{20}$	Reference
Ala-OPic, 2HBr	Aqueous acetone	165–167	+2.0 (c 1 in Me$_2$NCHO)	J. G. Warnke and G. T. Young (unpublished)
Arg(NO$_2$)-OPic, 2HBr	Aqueous acetone	170–175(d.)	−2.6 (c 1.1 in H$_2$O)	Camble et al. (1969)
Asp-(OPic)-OPic, 3HBr, H$_2$O	Aqueous ethanol	161–162;[b] 174–175	+1.2 (c 1 in H$_2$O)	Garner and Young (1971)
Glu(OPic)-OPic, 3HBr, H$_2$O	Ethanol	140	+8.0 (c 1.1 in H$_2$O)	Garner and Young (1971)
Gly-OPic, 2HBr	Aqueous acetone	193–195		Camble et al. (1969)
Leu-OPic, 2HBr	Ethanol–ether	165–167(d.)	−2.5 (c 2 in H$_2$O)	Camble et al. (1969)
Met-OPic, 2HBr	Ethanol	164–165	+3.7[c] (c 1 in H$_2$O)	J. A. Maclaren (1972)
Orn(Pipoc)-OPic, 2HBr, 3H$_2$O	Aqueous acetone	—	−0.2 (c 0.9 in H$_2$O) (+4° at 365 nm)	T. G. Pinker and G. T. Young (unpublished)
Phe-OPic, 2HBr	Ethanol	182–185(d.)	−11.1 (c 2 in H$_2$O)	Camble et al. (1969)
Pro-OPic, 2HBr	Ethanol	158–162	−27.1 (c 1 in H$_2$O)	Camble et al. (1969)
Val-OPic, 2HBr	Aqueous acetone	197.5–199(d.)	+1.7 (c 1 in H$_2$O)	Camble et al. (1969)

[a] Prepared by the action of hydrogen bromide in acetic acid on the benzyloxycarbonyl derivative. All optically active amino acids are of the L form.
[b] The higher melting point is observed with a higher rate of heating.
[c] The rotation given for this compound by Camble et al. (1969) is erroneous.

used in 2.5-M proportions with respect to the amino component, and the coupling reaction was continued until thin layer chromatography could detect no aminocomponent. After filtration from dicyclohexylurea and triethylamine hydrobromide, the product was adsorbed on moist Sulfoethyl-Sephadex C-25 (H^+ form); the resin was washed thoroughly with tetra-hydrofuran to remove nonbasic contaminants (including the excess of acylating agents), and the product was eluted by means of triethylamine in aqueous tetrahydrofuran. Evaporation of the eluate gave analytically pure benzyloxycarbonylglycyl-L-valine 4-picolyl ester in 92% yield. The benzyl-oxycarbonyl group was cleaved by means of hydrogen bromide in acetic acid, the process being repeated until the protected tetrapeptide 4-picolyl ester was obtained in 61% overall yield from L-valine 4-picolyl ester dihydro-bromide. Hydrogenation of the whole product gave tetrapeptide, which after precipitation from ethanol by ether but with no further purification, was pure by elemental analysis, amino acid analysis, and thin layer chroma-tography; no dipeptide or tripeptide could be detected in the product, showing that each coupling reaction had proceeded to completion. A second synthesis using p-nitrophenyl esters for coupling (4-M proportions) gave an overall yield of protected tetrapeptide 4-picolyl ester of 65%.

The most satisfactory conclusion from this preliminary examination was of the effectiveness of the separation procedure. It was shown at this time that, as expected, the H^+ form of the ion exchanger decomposes t-butoxycarbonyl groups, but this difficulty was overcome by saturation of the ion exchanger with 3-bromopyridine (Camble et al., 1969; Garner et al., 1968); this very weak base (pK_a 2.8) is displaced by the picolyl ester.

Clearly, the acidic phase by which the weakly basic product is separated need not be a solid phase, and in the next synthesis (Garner and Young, 1969, 1971; Garner et al., 1971), that of Val[5]-angiotensin II, the product from each coupling reaction was extracted from an organic solvent into aqueous citric acid (chosen in order to avoid cleavage of t-butoxycarbonyl groups), except in the last stage; the benzyloxycarbonyl-octapeptide 4-picolyl ester was insoluble in aqueous citric acid, and it was therefore separated on Sulfoethyl-Sephadex saturated with 3-bromopyridine. The synthesis is out-lined in Fig. 2; the overall yield of protected octapeptide 4-picolyl ester from the C-terminal component was 38%. The pleasing feature of this synthesis was that the crude protected intermediates, without further purification, had satisfactory elemental analyses and were pure as judged by thin layer chromatography in several solvents. In all cases, the trifluoroacetate salts obtained after the removal of t-butoxycarbonyl groups by means of tri-fluoroacetic acid were also analytically pure.

It was recognized that as the peptide increases in size the effectiveness of the picolyl group in separations would become weaker, and that it might

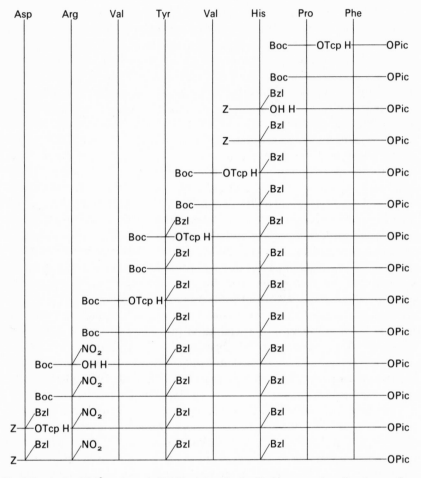

Fig. 2. Synthesis of Val^5-angiotensin II. Dicyclohexylcarbodiimide was used to effect the coupling of protected histidine and arginine.

prove important to introduce additional picolyl groups as the chain lengthens. The angiotensin case seemed a useful example here, and in a further synthesis (Garner *et al.*, 1971; Garner and Young, 1971) the terminal aspartic acid residue was introduced as α-2,4,5-trichlorophenyl β-4-picolyl benzyloxycarbonyl-L-aspartate. In contrast to the β-benzyl ester, the protected octapeptide (Fig. 3) so formed was readily soluble in aqueous citric acid. It should be noted that at the stage at which this additional picolyl group is added, the simple separation procedure may not be adequate to remove the excess of acylating agent.

Z-Asp-Arg-Val-Tyr-Val-His-Pro-Phe-OPic
　　OPic NO₂　　Bzl　　　Bzl

Fig. 3

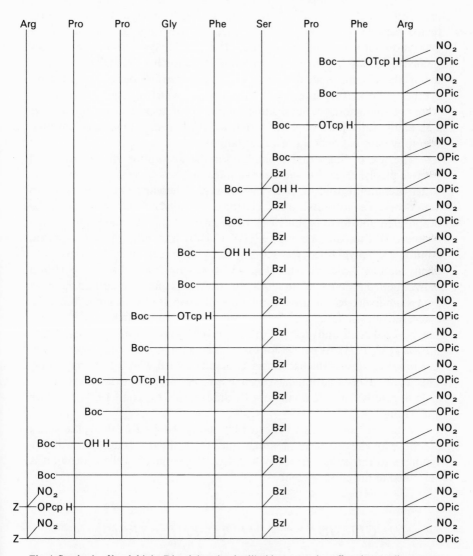

Fig. 4. Synthesis of bradykinin. Dicyclohexylcarbodiimide was used to effect the coupling except when active esters are indicated.

Boc—Ser—Cys—Gly—Phe—OPic
         | |
        Bzl   Btm

Fig. 5

The procedure was next tested on a synthesis of bradykinin (Garner *et al.*, 1971; Schafter *et al.*, 1971), using the steps shown in Fig. 4. The protected peptide picolyl ester was isolated at each stage except the last by extraction into aqueous citric acid; the benzyloxycarbonyl-nonapeptide picolyl ester was isolated on a column of Sulfoethyl-Sephadex C-25 (H^+ form), the overall yield from the *C*-terminal residue being 42%. In this synthesis, the intermediate trifluoroacetate salts, which often are hydroscopic, were not characterized, but again each intermediate protected peptide ester was analytically pure after the usual separation, and thin layer chromatography detected only traces of contaminants.

The use of Sulfoethyl-Sephadex for the absorption of the protected peptide picolyl esters has the disadvantage that it is necessary to have some water present (in order to swell the resin) and therefore one is restricted to water-miscible solvents; after elution, the solvent and base must be evaporated and there is some risk of damage during this stage. During the synthesis (Burton and Young, 1971) of a highly lipophilic peptide fragment, *t*-butoxycarbonyl-*O*-benzyl-L-seryl-*S*-benzylthiomethyl-L-cysteinylglycyl-L-phenylalanine 4-picolyl ester (Fig. 5), it was observed that the partition between ethyl acetate and aqueous citric acid was unfavorable, and aqueous potassium hydrogen sulfate was no more effective. It was then found that the macroreticular sulfonic acid resin Amberlyst-15 is advantageous, because it can be used with anhydrous organic solvents (Burton and Young, 1971; Burton *et al.*, 1971). As before, for use with acid-labile protecting groups such as *t*-butoxycarbonyl the exchanger is first saturated with 3-bromopyridine. This method of isolation has been used very largely in later syntheses, including those of nine analogues of bradykinin (Fletcher, 1971; Fletcher and Young, 1972).

Table III lists syntheses effected by the picolyl ester method; the yields are those for the protected peptides, since the method is concerned with the synthesis to this stage. In nearly all cases, the removal of the protection is described in the cited reference.

IV. CONCLUSIONS FROM RECENT EXPERIENCE

It may be helpful if now the essential steps in the procedure are reviewed in the light of our experience in these syntheses.

Table III. Protected Peptides Synthesized by the Picolyl Ester Method

Protected peptide	Coupling methods	Isolation procedure	Overall yield (%)[a]	Reference
Z-Leu-Ala-Gly-Val-OPic	DCCI	SE-Sephadex	61	Camble et al. (1969)
Z-Leu-Ala-Gly-Val-OPic	Active ester	SE-Sephadex	65	Camble et al. (1969)
Z-Asp(Bzl)-Arg(NO$_2$)-Val-Tyr(Bzl)-Val-His(Bzl)-Pro-Phe-OPic Protected [Val^5]-angiotensin II	Mainly active ester	Citric acid, SE-Sephadex	38	Garner and Young (1971)
Z-asp (OPic)-Arg(NO$_2$)-Val-Tyr(Bzl)-Val-His(Bzl)-Pro-Phe-OPic Protected [Val^5]-angiotensin II	Mainly active ester	Citric acid	38	Garner and Young (1971)
Z-Arg(NO$_2$)-Pro-Pro-Gly-Phe-Ser(Bzl)-Pro-Phe-Arg(NO$_2$)OPic Protected bradykinin	Mainly active ester	Citric acid, SE-Sephadex	42	Schafer et al. (1971)
Boc-Arg(NO$_2$)-Pro-Pro-Gly-Phe-Ser(Bzl)-Pro-Phe-Arg(NO$_2$)-OPic	DCCI/HOBt	Citric acid, Amberlyst-15	13[b]	Fletcher and Young (1972)
Z-Arg(NO$_2$)-Pro-Pro-Gly-Cha-Ser(Bzl)-Pro-Phe-Arg(NO$_2$)-OPic Protected [Cha^5]-bradykinin	DCCI/HOBt Active ester	Citric acid, Amberlyst-15	37	Fletcher and Young (1972)
Z-Arg(NO$_2$)-Pro-Pro-Gly-Phe-Ser(Bzl)-Pro-Cha-Arg(NO$_2$)-OPic Protected [Cha^8]-bradykinin	DCCI/HOBt	Amberlyst-15	42	Fletcher and Young (1972)
Z-Arg(NO$_2$)-Pro-Pro-Gly-Cha-Ser(Bzl)-Pro-Cha-Arg(NO$_2$)-OPic Protected [Cha^{5,8}]-bradykinin	DCCI/HOBt	Amberlyst-15	33	Fletcher and Young (1972)
Z-Har(NO$_2$)-Pro-Pro-Gly-Phe-Ser(Bzl)-Pro-Phe-Arg(NO$_2$)-OPic Protected [Har^1]-bradykinin	DCCI/HOBt	Citric acid, Amberlyst-15	16[b]	Fletcher and Young (1972)
Z-Har(NO$_2$)-Pro-Pro-Gly-Cha-Ser(Bzl)-Pro-Phe-Arg(NO$_2$)-OPic Protected [Har^1, Cha^5]-bradykinin	DCCI/HOBt	Citric acid, Amberlyst-15	37	Fletcher and Young (1972)
Z-Val-Pro-Pro-Gly-Phe-Ser(Bzl)-Pro-Phe-Arg(NO$_2$)-OPic Protected [Val^1]-bradykinin	DCCI/HOBt	Citric acid, Amberlyst-15	14[b]	Fletcher and Young (1972)
Z-Val-Pro-Pro-Gly-Phe-Thr(Bzl)-Pro-Phe-Arg(NO$_2$)-OPic Protected [Val^1, Thr^6]-bradykinin	DCCI/HOBt	Citric acid, Amberlyst-15	23	Fletcher and Young (1972)
Z-Leu-Pro-Pro-Gly-Phe-Thr(Bzl)-Pro-Phe-Arg(NO$_2$)-OPic Protected [Leu^1, Thr^6]-bradykinin	DCCI/HOBt	Citric acid, Amberlyst-15	23	Fletcher and Young (1972)
Boc-Ser(Bzl)-Cys(Btm)-Gly-Phe-OPic	DCCI/HOBt	Amberlyst-15	76.5	Burton and Young (1971)

[a]Calculated on the C-terminal amino acid 4-picolyl ester hydrobromide.
[b]In these cases, loss was incurred when the common intermediate protected heptapeptide, Boc-Gly-Phe-Ser(Bzl)-Pro-Phe-Arg(NO$_2$)-OPic, crystallized on the Amberlyst column.

A. Removal of the α-Amino Protection: "Deprotection Step"

t-Butoxycarbonyl has normally been used for α-amino protection, and removal by trifluoroacetic acid presents no problems peculiar to this procedure. Hydrogen chloride in dioxane was satisfactorily used for deprotection in some early syntheses (Schafer et al., 1971), but in other cases the reaction has been incomplete because the hydrochloride of the protected peptide 4-picolyl ester coprecipitated with the hydrochloride of the amino compound (unpublished work with Dr. M. Löw). The choice of solvent is of course critical here. This point should therefore be watched if acids other than trifluoroacetic acid are used for deprotection.

B. Liberation of the Amino Component from Its Salt Before Coupling

It is important that the amino component shall be liberated completely, since any unchanged salt will prevent completion of the acylation step and may follow the protected peptide 4-picolyl ester in the separation. It is true that fractional elution of the Amberlyst column will usually separate amino component from the t-butoxycarbonyl derivative, only the latter being eluted by pyridine (unpublished work with P. P. Nash), but clearly it is preferable to acylate completely.

We have found the following procedures to be most satisfactory in our work:

1. When the coupling reaction is to be carried out in dimethylformamide, the salt of the amino component is dissolved in this solvent, excess (approximately 4 equivalents) of triethylamine is added, and after 5–15 min the excess of triethylamine is removed at room temperature on a rotary evaporator. The evaporation should not be prolonged unduly, as triethylamine may be lost from the triethylammonium trifluoroacetate, regenerating the salt of the amino component.

2. When the coupling reaction is to be carried out in other solvents (e.g., tetrahydrofuran), it is convenient to dissolve or suspend the salt in chloroform and then to add an excess of triethylamine; when solution is complete, the solution is evaporated to dryness and the residual amino component is taken up in the required solvent.

3. For coupling reactions involving amino acid esters or dipeptide esters in which dioxopiperazine formation is an important side-reaction, the salt is dissolved in the coupling solvent and the acylating agent and lastly a slight excess of the hindered amine, ethyldiso-propylamine, are added. The dioxopiperazine is of course readily removed by our separation procedure [unless it has a basic side-

chain, e.g., when using C-terminal aspartic acid as its di-4-picolyl ester; in the only such case we have encountered, little or no dioxopiperazine was formed and no difficulty was experienced (unpublished work with P. P. Nash)]. Its coproduct, 4-picolyl alcohol, can, however, be a source of impurity when dicyclohexylcarbodiimide is the condensing agent, as some 4-picolyl ester of the acylamino acid can be formed; this will accompany the desired produce in the isolation step and if not removed will result in the building up of a shortened peptide chain. Such contaminants have so far been observed only in small amounts and can usually be removed by virtue of their greater solubility in organic solvents such as ether or light petroleum.

C. Coupling Reactions

In principle, of course, the picolyl ester procedure can be used with any method of coupling which does not give rise to basic contaminants. As we have seen, active esters were used in a number of earlier syntheses, but completion of hindered reactions (e.g., the coupling of proline-2 to proline-3 in bradykinin; Schafer et al., 1971) was unacceptably slow. The use of dicyclohexylcarbodiimide alone has been found to cause extensive trifluoroacetylation when such hindered reactions are effected, as in these cases, in the presence of trifluoroacetate anion, but the addition of 1-hydroxybenzotriazole reduces the extent of this side-reaction to a very low figure (Fletcher et al., 1972), and this procedure (with preformation of the ester) has been used satisfactorily in all recent syntheses. The coupling, with tetrahydrofuran or dimethylformamide as solvent, is continued until thin layer chromatography reveals no amino component (sensitivity 0.1–0.2%), usually within 1 hr of the addition of the amino component.

D. Isolation of the Product

The citric acid procedure is convenient and usually gives high recoveries, but its application is limited to those cases in which the protected peptide (1) partitions satisfactorily between aqueous citric acid and a water-immiscible solvent, and (2) can be extracted from the basified aqueous layer into an organic solvent. The satisfaction of these conditions will of course depend on the amino acids present and the type of protection used; they are most likely to be met up to the tetrapeptide stage. It is necessary to make trials to see if this method of isolation is appropriate, and this is best done by removing the solvent from the reaction mixture, adding 0.7 M citric acid to the residue, and extracting with solvents of increasing polarity, starting with ether. By thin layer chromatography, it is easily seen whether the contaminants are being removed and whether any product is being lost. In the bradykinin series, a mixture of equal volumes of ether and ethyl acetate proved to be generally satisfactory. The aqueous layer is then brought to pH 8.5 and solvents of increasing power are used to extract the product; the effectiveness of the purification is checked by thin layer chromatography of the extracts. If suitable solvents cannot be found, the Amberlyst procedure is used.

The isolation on Amberlyst-15 (saturated with 3-bromopyridine) has the advantage of being applicable in our experience in every case, and although the recoveries may on average be a little lower this method is commonly chosen because of its general applicability. The exchange of the picolyl ester for the 3-bromopyridine on the resin appears to be rather slow, and although it is convenient to use columns of resin the procedure is essentially batchwise. The details will not be repeated here, but some important points resulting from recent experience will be mentioned. The preferred solvent for application to the column is ethyl acetate, but dichloromethane and (purified) dimethylformamide have been used; the coupling solvent is therefore (when necessary) evaporated and replaced. It is helpful at this stage to wash the solution with aqueous sodium hydrogen carbonate, to remove much of the 1-hydroxybenzotriazole, and with water to remove salts which reduce the capacity of the column. If the product is not soluble in ethyl acetate or dichloromethane, the washing may be carried out on the residue from the evaporation; the final residue is dried at 1-mm pressure and then taken up in dimethylformamide for application to the column. The eluate must be recycled until no product can be detected by thin layer chromatography in a sample of eluate concentrated to $\frac{1}{10}$ vol and loaded heavily (approximately 1 mg). It is inadvisable to leave the product on the resin longer than is necessary. Nonbasic coproducts and byproducts are then washed off by means of ethyl acetate or by dimethylformamide; it is essential that the dimethylformamide shall be free from dimethylamine, which will elute the product, and it should be freshly distilled from phthalic acid. The absence of product in these washes is again monitored by thin layer chromatography of a concentrated sample. The product is then eluted by a solution of pyridine in pure dimethylformamide (25% v/v). The eluate containing product is evaporated to dryness at 1-mm pressure, and the residue is triturated with ether, or precipitated from an organic solvent by ether or petrol, to remove traces of 3-bromopyridine. 3-Bromopyridine is soluble in water and so may be removed in this way from water-insoluble protected peptides. The purity of the final product is confirmed at this stage by thin layer chromatography at heavy loading. The yield, from one protected intermediate to the next (i.e., including the deprotection step), is usually 85–90%, and the product is normally pure by elemental and amino acid analysis, no contaminants, or traces only, being detectable by thin layer chromatography at approximately

1 mg loading. Needless to say, to obtain consistently high yields requires skillful and careful working.

The syntheses achieved so far establish the use of this procedure for the preparation of pure peptides of 8–9 residues in high overall yield. Such protected peptides can clearly be used as fragments for the synthesis of larger peptides, and syntheses illustrating this route, as well as those continuing the stepwise elongation of the chain, are in progress. The picolyl ester method has the advantage over conventional synthesis (in solution) of the simple and effective isolation of pure product, which not only saves time but provides high yields at each step. In comparison with solid-phase synthesis, the advantages are in the milder conditions required for reactions in solution, in the evidence of purity and the characterization available at each step, and in the resulting cumulative evidence for the structure of the final product; the protected product normally needs no further purification, although this may of course be required after removal of the protecting groups. The overriding importance of the purity of peptides used in pharmacological work has been emphasized recently (Vane, 1971; Bangham et al., 1971). As regards scale, the Amberlyst column separation has been used on a 10 mM scale (e.g., batches of 7 g of a protected tetrapeptide) (Burton and Young, 1971), and there should be little difficulty in handling larger quantities. The procedure is no doubt more laborious and time consuming than is solid-phase synthesis, but given the required protected amino acids, and provided that the synthesis is not delayed by the analytical procedures, small peptides can be prepared quite quickly; for example, protected [1-L-homoarginine]-bradykinin was synthesized in 16 days (Fletcher, 1971; Fletcher and Young, 1972).

REFERENCES

Bangham, D. R., Calam, D. H., Parsons, J. A., and Robinson, C. J., 1971, Purity of synthetic peptide preparations, *Nature* 232:631.

Burton, J., and Young, G. T., 1971, The use of amberlyst ionexchanger in the picolyl ester method of peptide synthesis, *Israel J. Chem.* 9:201.

Burton, J., Fletcher, G. A., and Young, G. T., 1971, The use of a macroreticular ion-exchange resin in the isolation stage of the picolyl ester method of peptide synthesis, *Chem. Commun.* 1971:1057.

Camble, R., Garner, R., and Young, G. T., 1967, Novel facilitation of peptide synthesis, *Nature* 217:247.

Camble, R., Garner, R., and Young, G. T., 1969, Amino-acids and peptides. Part XXX. Facilitation of peptide synthesis by the use of 4-picolyl esters for carboxy-group protection, *J. Chem. Soc. (C)* 1969:1911.

Fletcher, G. A., 1971, Synthesis of peptide hormones, D. Phil. thesis, Oxford.

Fletcher, G. A., and Young, G. T., 1972, Amino-acids and peptides. Part XXXVI. The synthesis of analogues of bradykinin by the picolyl ester method, *J. Chem. Soc.* (Perkin I) 1972: p. 1867.

Fletcher, G. A., Löw, M., and Young, G. T., 1972, Amino-acids and peptides. Part XXXVII. Trifluoroacetylation during the coupling of *t*-butoxycarbonylamino-acids with peptide esters in the presence of trifluoroacetate anion, *J. Chem. Soc.* (Perkin I) **1973**: 1162.

Garner, R., and Young, G. T., 1969, Facilitation of peptide synthesis by the use of 4-picolyl esters: Synthesis of Val[5]-Angiotensin-II, *Nature* **222**: 177.

Garner, R., and Young, G. T., 1971, Amino-acids and peptides. Part XXXIII. Synthesis of Val[5]-angiotensin-II by the picolyl ester method, *J. Chem. Soc.* (*C*) **1971**: 50.

Garner, R., Schafer, D. J., Watkins, W. B., and Young, G. T., 1968, A new repetitive method of peptide synthesis, *in* "Peptides—1968: Proceedings of the Ninth European Peptide Symposium, Orsay, 1968" (E. Bricas, ed.), p. 145, North-Holland, Amsterdam.

Garner, R., Schafer, D. J., and Young, G. T., 1971, The use of 4-picolyl esters to facilitate peptide synthesis, *in* "Peptides—1969, Proceedings of the Tenth European Peptide Symposium, Abano Terme, 1969" (E. Scoffone, ed.), p. 102, North-Holland, Amsterdam.

Hata, T., Tajima, K., and Mukaiyama, T., 1971, A simple protecting group protection—purification "handle" for polynucleotide synthesis. I, *J. Am. Chem. Soc.* **93**: 4928.

Maclaren, J. A., 1972, A convenient preparative method for esters of amino-acids, *Austral. J. Chem.*, **25**: 1293.

Merrifield, R. B., 1963, Solid phase peptide synthesis. I. The synthesis of a tetrapeptide, *J. Am. Chem. Soc.*, **85**: 2149.

Schafer, D. J., 1970, Studies in peptide synthesis, D. Phil. thesis, Oxford.

Schafer, D. J., and Carlsson, L., 1972, Simplified synthesis of histidine peptides, *J. Chem. Soc. Chem. Commun.* **1972**: 276.

Schafer, D. J., Young, G. T., Elliott, D. F., and Wade, R., 1971, Amino-acids and peptides. Part XXXII. A simplified synthesis of bradykinin by the use of the picolyl ester method. *J. Chem. Soc.* (*C*) **1971**: 46.

Scopes, P. M., Walshaw, K. B., Welford, M., and Young, G. T., 1965, Amino-acids and peptides. Part XXI. Removal of the nitro-group from nitroarginine and peptides of nitroarginine by electrolytic reduction, *J. Chem. Soc.* **1965**: 782.

Vane, J. R., 1971, Purity and stability of synthetic peptides such as angiotensins and kinins, *Nature* **230**: 382.

Wieland, Th., and Racky, W., 1968, Peptidsynthese mit Farbigen am ionenaustacher fixierbaren Benzylestern, *Chimia* **22**: 375.

Wünsch, E., 1971, Synthesis of naturally occurring polypeptides, problems of current research, *Angew. Chem. Internat. Ed.* **10**: 786.

CHAPTER 4

ON CYSTEINE AND CYSTINE PEPTIDES

Iphigenia Photaki

Laboratory of Organic Chemistry
University of Athens
Athens, Greece

I. INTRODUCTION

Of the amino acids found in natural peptides and proteins, cysteine and cystine fall into a special group. The specific properties of the sulfhydryl and corresponding disulfide groups become associated with the peptide chain, creating special problems, both chemical and stereochemical, characteristic of these amino acids. This chapter is mainly concerned with current work on cysteine-cystine peptides being carried out at the Laboratory of Organic Chemistry at the University of Athens, and for this reason does not include a complete bibliography of the subject. From the extensive list of references, only those dealing with the same subject or those considered useful for an understanding of the problem have been selected.

The inclusion of cystine in the peptide allows the formation of symmetrical or unsymmetrical peptides, with cross-linking of chains through $-S-S-$ bridges. Furthermore, cyclic peptides can be formed by intrachain disulfide formation when more than one cysteine group is present in the peptide chain (Fig. 1).

$$S-CH_2-\underset{\underset{NHR_1}{|}}{\overset{\overset{NHR_1}{|}}{C}}HCOR_2$$
$$S-CH_2-\underset{\underset{NHR_1}{|}}{C}HCOR_2$$

$$S-CH_2-\underset{\underset{NHR_1}{|}}{\overset{\overset{NHR_1}{|}}{C}}HCOR_2$$
$$S-CH_2-\underset{\underset{NHR_3}{|}}{C}HCOR_4$$

"Symmetrical" cystine peptide   "Unsymmetrical" cystine peptide

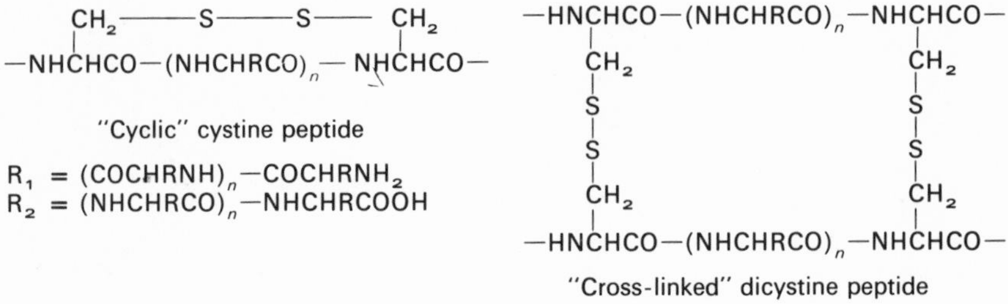

$$
\begin{array}{cccccc}
\text{CH}_2 & \!\!\!-\!\!\! & \text{S} & \!\!\!-\!\!\! & \text{S} & \!\!\!-\!\!\! & \text{CH}_2 \\
| & & & & & | \\
-\text{NHCHCO} & \!\!\!-(\text{NHCHRCO})_n-\!\!\! & \text{NHCHCO}- \\
\end{array}
$$

"Cyclic" cystine peptide

R_1 = (COCHRNH)$_n$—COCHRNH$_2$
R_2 = (NHCHRCO)$_n$—NHCHRCOOH

—HNCHCO—(NHCHRCO)$_n$—NHCHCO—
    | |
   CH$_2$ CH$_2$
   | |
   S   S
   | |
   S   S
   | |
   CH$_2$ CH$_2$
   | |
—HNCHCO—(NHCHRCO)$_n$—NHCHCO—

"Cross-linked" dicystine peptide

Fig. 1. Cystine peptides.

II. S-PROTECTION OF CYSTEINE DURING PEPTIDE SYNTHESIS: TRANSFORMATION OF S-PROTECTED CYSTEINE PEPTIDES TO CYSTINE PEPTIDES

A. S-Benzylcysteines

Cystine is usually incorporated into a peptide chain as S-protected cysteine residues which are unblocked at the end of the synthesis and oxidized to the disulfide form.

It is a well-known fact that the first naturally occurring cysteine peptide to be synthesized was glutathione. The discovery of its structure was followed by its chemical synthesis by Harington and Mead (1935) and by du Vigneaud and Miller (1936). These syntheses represent the first application of the car-bobenzoxy method of Bergmann and Zervas (1932) to the synthesis of a biologically active peptide. In this case, a modification of the original carbo-benzoxy method was used, since the presence of cysteine-cystine in the peptide did not allow removal of the benzyloxycarbonyl (Z) group by catalytic hydrogenolysis due to poisoning of the catalyst. Harington removed the carbobenzoxy group using phosphonium iodide, a reaction now considered as acid catalyzed (Ben-Ishai and Berger, 1952; Albertson and McKay, 1953).

A more important step was made, however, in the application of the carbobenzoxy method to cysteine peptides when Sifferd and du Vigneaud (1935) used S-benzylcysteine [1]. This amino acid can be easily incorporated in a peptide chain and at the end the S-protecting group as well as the N-Z group is removed in one operation by the action of sodium in liquid ammonia (Na/liq. NH$_3$) (Fig. 2).

By using the benzyloxycarbonyl group for N-protection and the benzyl group for S-protection, together with new coupling methods, the problems involved in synthesizing such sulfur peptide hormones as oxytocin, vaso-pressin (du Vigneaud et al., 1953, 1957, 1958), and even insulin could be overcome (Katsoyannis et al., 1963; Meienhofer et al., 1963; Kung et al.,

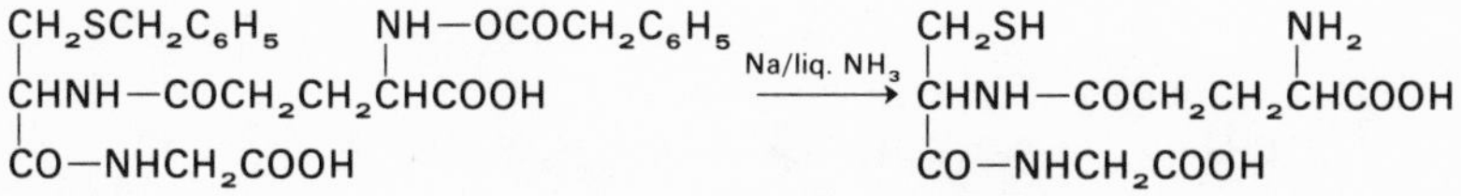

$$\begin{array}{ccc} \underset{|}{CH_2SCH_2C_6H_5} & \underset{|}{NH-OCOCH_2C_6H_5} & \xrightarrow{\text{Na/liq. NH}_3} & \underset{|}{CH_2SH} & \underset{|}{NH_2} \\ \underset{|}{CHNH-COCH_2CH_2CHCOOH} & & & \underset{|}{CHNH-COCH_2CH_2CHCOOH} \\ CO-NHCH_2COOH & & & CO-NHCH_2COOH \end{array}$$

Fig. 2. Synthesis of glutathione by du Vigneaud.

1965). It is beyond the scope of this chapter to review the synthesis of oxytocin, vasopressin, and insulin. Even today, S-benzylcysteine constitutes the most widely used S-protected cysteine, since this protecting group is practically unaffected by hydrogen bromide, which is now used in place of catalytic hydrogenolysis to remove the N-Z group in cases where the latter method is not feasible (Ben-Ishai and Berger, 1952; Anderson et al., 1952; Boissonnas and Preitner, 1953). An example of a synthesis in which the N-Z group is removed by hydrogen bromide in acetic acid (HBr/AcOH) and the S-benzyl group by Na/liq. NH_3 is the synthesis of L-cysteinyl-L-cysteine by Izumiya and Greenstein (1954) (Fig. 3).

However, in recent years it has been noted that cleavage with Na/liq. NH_3 can be accompanied by side-reactions such as desulfurization and rupture of peptide bonds, particularly those involving the amino side of proline. According to recent findings (Katsoyannis et al., 1971), cleavage of the chain at the amino-terminal side of the proline residue is completely avoided by carrying out the sodium–liquid ammonia reaction in the presence of a large excess of sodium amide.

A modified benzyl group i.e., the p-nitrobenzyl group [2] has been introduced (Berse et al., 1957) as an S-protecting group able to be removed by catalytic hydrogenolysis (Fig. 4). However, even if its removal under such mild conditions were complete, something which is difficult to achieve (Ondetti and Bodanszky, 1962; Hiskey and Tucker, 1962), the simultaneous hydrogenolytic cleavage of an —S—S— bridge already existing in the molecule cannot be excluded, since cystine, its esters, and some cystine products can be catalytically reduced to the corresponding SH derivatives (Bergmann and Michalis, 1930; Theodoropoulos, 1953; Zervas and Theodoropoulos, 1955; Zervas, 1956).

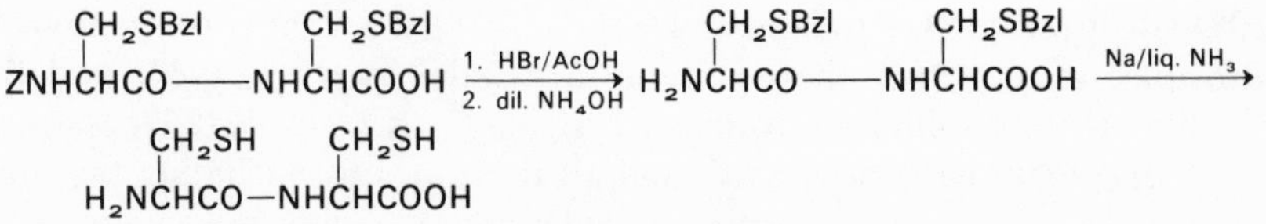

$$\begin{array}{cccc} \underset{|}{CH_2SBzl} & \underset{|}{CH_2SBzl} & & \underset{|}{CH_2SBzl} & \underset{|}{CH_2SBzl} \\ ZNHCHCO\text{——}NHCHCOOH & \xrightarrow[\text{2. dil. NH}_4\text{OH}]{\text{1. HBr/AcOH}} & H_2NCHCO\text{——}NHCHCOOH & \xrightarrow{\text{Na/liq. NH}_3} \\ \end{array}$$

$$\begin{array}{cc} \underset{|}{CH_2SH} & \underset{|}{CH_2SH} \\ H_2NCHCO-NHCHCOOH \end{array}$$

Fig. 3. Synthesis of L-cysteinyl-L-cysteine.

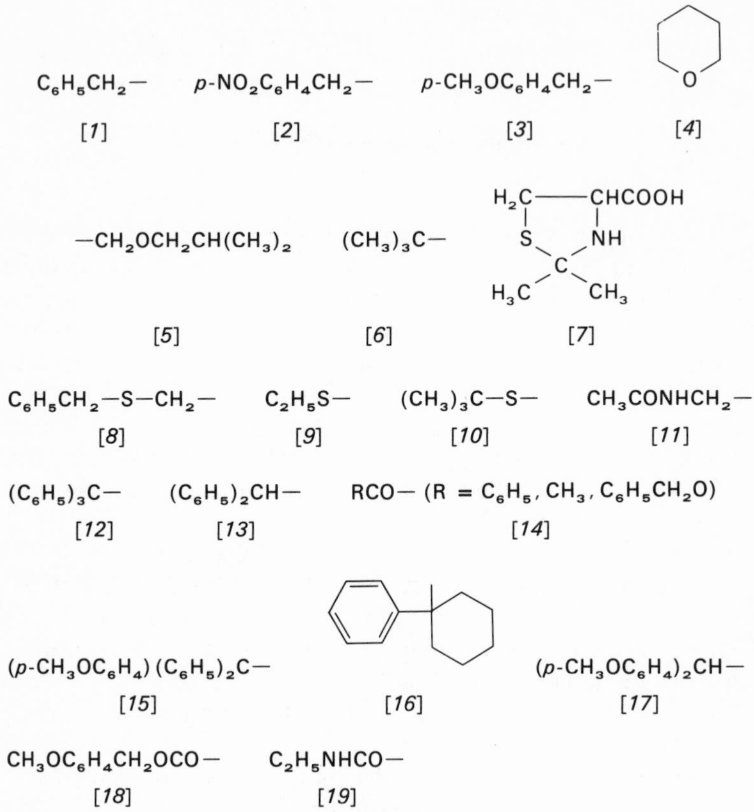

$C_6H_5CH_2-$ p-$NO_2C_6H_4CH_2-$ p-$CH_3OC_6H_4CH_2-$

[1]   [2]   [3]   [4]

$-CH_2OCH_2CH(CH_3)_2$ $(CH_3)_3C-$

[5]   [6]   [7]

$C_6H_5CH_2-S-CH_2-$ C_2H_5S- $(CH_3)_3C-S-$ $CH_3CONHCH_2-$

[8]   [9]   [10]   [11]

$(C_6H_5)_3C-$ $(C_6H_5)_2CH-$ $RCO-$ $(R = C_6H_5, CH_3, C_6H_5CH_2O)$

[12]   [13]   [14]

$(p$-$CH_3OC_6H_4)(C_6H_5)_2C-$ $(p$-$CH_3OC_6H_4)_2CH-$

[15]   [16]   [17]

$CH_3OC_6H_4CH_2OCO-$ C_2H_5NHCO-

[18]   [19]

Fig. 4. *S*-Protecting groups.

Another derivative of the benzyl group, i.e., the *p*-methoxybenzyl group [3] which can be removed by reduction with Na/liq. NH_3 or by acidolytic cleavage by anhydrous hydrogen fluoride, has been used in the synthesis of oxytocin (Sakakibara *et al.*, 1967).

B. Various S-Protecting Groups for Cysteine Not Including S-Aralkyl and S-Acyl Groups

In recent years, many other *S*-protecting groups (Fig. 4) such as the tetrahydropyranyl [4] (Holland and Cohen, 1958), the isobutyloxymethyl [5] (Brownlee *et al.*, 1964), the *tert*-butyl [6] (Callahan *et al.*, 1963), and the 2,2'-dimethylthiazolidine derivative of cysteine [7] (King *et al.*, 1957; Sheehan and Yang, 1958) have been used, and all these groups, including the thiazolidine derivative, can be split off under more or less mild conditions. The *S*-isobutyloxymethyl group, for instance, can be removed either by

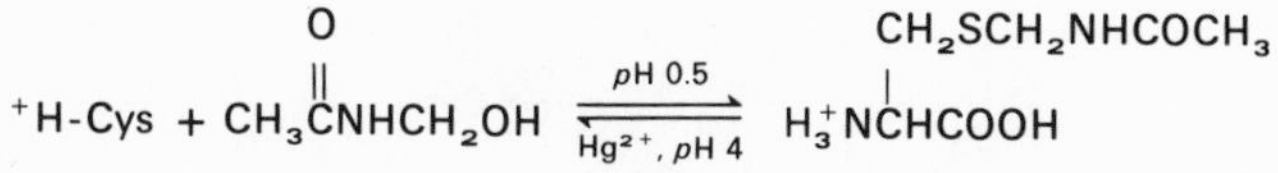

$$^+H\text{-Cys} + CH_3\overset{\displaystyle O}{\overset{\|}{C}}NHCH_2OH \xrightarrow[Hg^{2+}, pH\,4]{pH\,0.5} H_3^+N\overset{\displaystyle CH_2SCH_2NHCOCH_3}{\overset{|}{C}}HCOOH$$

Fig. 5. S-Acetamidomethyl-L-cysteine.

2 N HBr/AcOH (Pimlott and Young, 1958) or by the action of thiocyanogen or sulfenylthiocyanate (Hiskey and Sparrow, 1970). However, these groups have not as yet found a wide application.

The S-benzylthiomethyl group [8] introduced by Pimlott and Young (1958) as an S-protecting group can be removed at room temperature by means of mercuric acetate in formic acid (Brownlee et al., 1964). This method, although reported to present difficulties to some research workers (Katsoyannis, 1961), has been successfully used in a synthesis of glutathione and homoglutathione (Camble et al., 1968).

The ethylmercapto group [9] has also been proposed as an S-protecting group; it can be removed by treatment with thiophenol or thioglycollic acid at 45°C, and it is stable under acidic conditions. This group has been used in a synthesis of glutathione and oxytocin (Inukai et al., 1967) and also in the synthesis of an insulin A chain (Weber, 1969).

Finally, the tert-butylmercapto group [10] may be mentioned, which can be removed by sulfitolysis (Wünsch and Spangenberg, 1971). However, to our knowledge, it has not yet been used in cysteine peptide synthesis.

A very promising S-protecting group seems to be the S-acetamidomethyl [11] group (Veber et al., 1968). This group is stable under acidic and alkaline conditions but is cleaved by mercury ions at pH 4 and has been used in a ribonuclease synthesis (Hirschmann et al., 1969) (Fig. 5).

Cysteine peptides can also be formed by reduction of symmetrical cystine peptides. The resulting SH-peptides can be converted to a suitable S-protected derivative and used further for the preparation of longer cystine peptides. This principle was applied in the first synthesis of oxytocin (Ressler and du Vigneaud, 1954) and also in a synthesis of the B chain of insulin (Zahn and Schmidt, 1970).

C. S-Trityl (Tri) Cysteine

A variety of other S-protecting groups, i.e., trityl [12], diphenylmethyl [13], and acyl [14] groups, will be more extensively described, since they include the contribution of our laboratory in this field. The main feature of these groups is that they can be removed selectively by procedures which do not attack sensitive groups already existing in the molecule.

Tritylation of cysteine (Amiard et al., 1956) according to the method of Zervas (1955) and Zervas and Theodoropoulos (1956) and of its methyl

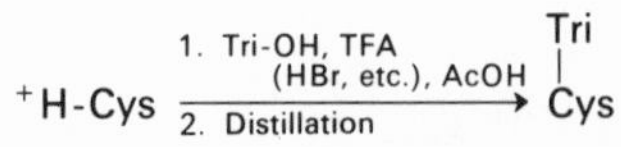

1. Tri-OH, TFA
(HBr, etc.), AcOH
$^+$H-Cys $\xrightarrow{\qquad\qquad\qquad}$ Tri
2. Distillation |
Cys

Fig. 6. Preparation of S-trityl-L-cysteine.

ester (Theodoropoulos, 1953; Zervas and Theodoropoulos, 1956) yields the corresponding N,S-ditritylderivatives. By heating the filtrate of N,S-ditrityl-cysteine with aqueous acetic acid, which results in detritylation of the N-trityl groups (Hillmann-Elies et al., 1953; Theodoropoulos, 1953; Amiard et al., 1956; Zervas and Theodoropoulos, 1956), Velluz isolated S-tritylcysteine with mp 202–203°C, $[\alpha]_D + 19°$ (c 2, 0.1 N NaOH). S-Trityl-L-cysteine is prepared (Fig. 6) in good yield by selective tritylation of cysteine protected at the amino group by protonation, i.e., by the action of trityl-chloride on cysteine salts (hydrochloride, p-toluenesulfonate) dissolved in dimethylformamide (Zervas and Photaki, 1960, 1962) or from cysteine hydrochloride and tritylcarbinol in acetic acid in the presence of 1.1 moles of boron trifluoride ethereate (Hiskey and Adams, 1965a). S-Tritylation of cysteine can be accomplished more easily through the interaction of trityl-carbinol and cysteine or its salts dissolved in trifluoracetic acid (Photaki et al., 1970) and more easily still when the solvent is acetic acid (Zervas–Photaki; Zervas et al., unpublished data). S-Trityl-L-cysteine prepared in our laboratory by one of the above methods or by selective N-detritylation of isolated, pure N,S-ditritylcysteine shows even after repeated recrystal-lizations, mp 181–182°C, $[\alpha]_D + 16.2°$ (c 2, 0.1 N NaOH). The S-trityl-L-cysteine thus prepared is analytically, optically, and chromatographically pure. Several amino- or carboxyl-protected derivatives of this key inter-mediate are also known in pure condition, i.e., N-benzyloxycarbonyl, N-formyl (Zervas and Photaki, 1962), N-o-nitrophenylsulfenyl (Nps) (Zervas et al., 1963c), N-tert-butyloxycarbonyl (Zervas et al., 1965; Zahn et al., 1969), N-benzhydryl and N-phthaloyl (Hiskey et al., 1967), as well as methyl (Zervas and Photaki, 1962), diphenylmethyl (Hiskey and Adams, 1965b), and phenacyl esters (Zervas et al., unpublished data).

The methods used in selective removal of the S-Tri group are of especial interest. Velluz et al. (1956), who first worked on this problem, report two examples of splitting of the S-Tri group: (1) from S-tritylglutathione in chloroform solution by bubbling through dry HCl gas for a few minutes and (2) from bis-S,S'-trityloxytoceine dissolved in a mixture of $CH_2Cl_2 : 10$ N HCl (5:1). In these cases, the yield of the final product is very low. As a matter of fact, Velluz's reported yield in the step of detritylation is 15% in the case of glutathione, whereas no yield of oxytocin is reported even as biological activity in a bioassay. In our experience (Zervas and Photaki, 1962), the extent of S-detritylation of S-Tri-L-cysteine and of numerous

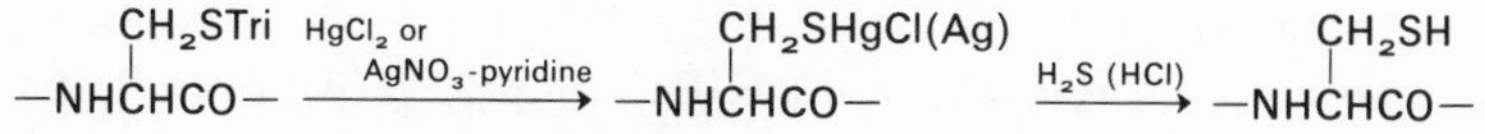

$$CH_2STri \xrightarrow[\text{AgNO}_3\text{-pyridine}]{\text{HgCl}_2 \text{ or}} CH_2SHgCl(Ag) \xrightarrow{\text{H}_2\text{S (HCl)}} CH_2SH$$
$$-NHCHCO- \qquad\qquad -NHCHCO- \qquad\qquad -NHCHCO-$$

Fig. 7. Removal of S-trityl group by Hg^{2+} or Ag^+.

derivatives of it never exceeds 40–50% even after more than 30 min bubbling of hydrogen chloride in a chloroform solution. An even lower percentage of cleavage was observed using a mixture of CH_2Cl_2 and 10 N HCl (5:1). The low degree of S-detritylation using Velluz's conditions is understandable bearing in mind the experimental findings of Hanson and Law (1965) and of our laboratory (Photaki et al., 1970).

In the light of results obtained by the examination of numerous S-trityl peptides (Photaki et al., 1970), certain suggestions can be made regarding the splitting off of the S-Tri group. The removal of this group by Hg^{2+} or Ag^+ (Zervas and Photaki, 1962), followed by the action of hydrogen sulfide on the thiolate thus formed, is to be recommended. In some cases, the removal of silver can be done by treatment with hydrochloric acid (Fig. 7).

Regarding the acidic conditions for splitting off the S-Tri group, one must keep in mind that TFA and HBr, and to a lesser extent HCl, in AcOH remove the S-Tri group by an equilibrium reaction. Therefore, in order to use TFA or HBr (HCl) in AcOH as cleavage reagents the reaction conditions must be carefully controlled. In order to avoid retritylation, these reagents should not be removed by distillation, but water must be added before proceeding to either titration (oxidation) or isolation of the S-deblocked cysteine derivative. If this is not done, evaporation of the above solvent reactants results in reformation of the S-Tri derivatives (cf. p. 64). Considering this, the acidic conditions for the splitting off of the S-Tri group can be summarized as follows:

1. Trifluoracetic acid is a good cleavage reagent only for S-tritylcysteine and for peptides bearing an S-tritylcysteine residue at the amino end. For all other peptides, the optimal conditions for cleavage need to be determined. In the cases so far studied, the percentage cleavage did not exceed approximately 60%. For preparative purposes, the percentage cleavage can be increased by isolation of the product and repetition of the reaction with TFA.

2. The removal of S-trityl groups by TFA is almost fully suppressed in the presence of triphenylcarbinol or triphenylmethyl chloride. This fact can be used to great advantage in cases where a polypeptide bears other side-chain protecting groups [e.g., N-t-butyloxycarbonyl (Boc), t-butyl esters (OBu^t), diphenylmethyl ester (ODpm), etc.] quantitatively removable by TFA.

3. Taking the precautions already mentioned, HBr/AcOH is the reagent of choice for removal of the S-Tri group in all peptides. Triphenylcarbinol inhibits the removal of the S-Tri group, whereas triphenylmethyl chloride has no appreciable effect (Hanson and Law, 1965).
4. Hydrogen chloride in acetic acid removes the S-Tri group to a lesser extent; cleavage is also suppressed in the presence of triphenylcarbinol.

When the trityl group contains *p*-methoxy substituents [*15*], as in the *p*-methoxyphenyl-diphenylmethyl-L-cysteine (mp 148–150°C) (Zervas *et al.*, unpublished data), the removal of the trityl group under acidic conditions is achieved much more readily. For more details of these procedures, the reader is referred to the original literature—work by Zervas and Photaki in the years 1961–1970.

Special oxidative procedures which easily cause splitting off of the trityl group simultaneously cause the formation of disulfides. This is true of the thiocyanogen method (Hiskey and Tucker, 1962) and of the more simple method of Kamber and Rittel (1968), who used iodine in methanol. This reaction probably occurs through a sulfenyl iodide (or thiocyanate) intermediate because of the ease of formation of the stable carbonium ion (Fig. 8).

Oxidative splitting of the S-Tri group is especially useful in the preparation of cyclic cystine peptides and will be discussed below.

The incorporation of the *S*-trityl protected cysteine into a peptide chain does not present difficulties. All the known *N*-protected *S*-tritylcysteines with the exception of the N-Tri derivative are easily coupled with other amino acids either by the mixed anhydride method using alkylcarbonic acids or by dicyclohexycarbodiimide. As in the case of other *N*-tritylamino acids, *N,S*-ditritylcysteine cannot be coupled by mixed anhydrides because of stereochemical inhibition. This inhibition can be overcome by using either carbodiimide (Amiard *et al.*, 1956) or a mixed anhydride with diphenyl phosphoric acid (Cosmatos *et al.*, 1961) as coupling agent. Many peptide derivatives of *N*-protected S-Tri cysteine have been prepared, and are reported in the literature.

When, however, the peptide chain to be lengthened contains groups such as OBu^t or ODpm, then the use of N-Tri or N-Nps cysteine is recommended

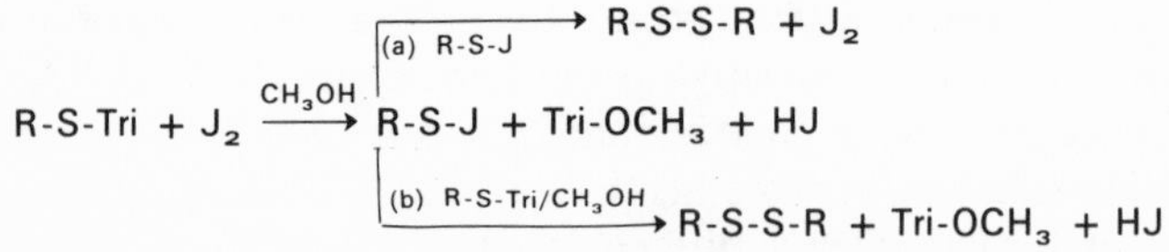

$$R\text{-S-Tri} + J_2 \xrightarrow{CH_3OH} \begin{cases} \text{(a) R-S-J} \rightarrow R\text{-S-S-R} + J_2 \\ R\text{-S-J} + Tri\text{-OCH}_3 + HJ \\ \text{(b) R-S-Tri/CH}_3\text{OH} \rightarrow R\text{-S-S-R} + Tri\text{-OCH}_3 + HJ \end{cases}$$

Fig. 8. Oxidative removal of *S*-trityl by iodine.

because these groups can be removed under conditions which in practical terms do not affect either the above groups or the S-Tri group. If required, when other acid-labile groups are absent from the molecule, the N-Z group could be used in synthesis of S-Tri peptides, since after the splitting with HBr/AcOH the SH group could be retritylated during the evaporation of the HBr/AcOH (Hirotsu *et al.*, 1967), especially if triphenylcarbinol is added (cf. p. 66).

Recently, Köning *et al.* (1968) have studied different *S*-protecting groups such as the phenyl-cyclohexyl group [*16*] which also give stabilized carbonium ions on acidolysis. This group can be introduced by the action on cysteine hydrochloride of the corresponding carbinol using boron trifluoride etherate as a catalyst and can be removed by TFA.

D. S-Diphenylmethyl (Dpm) Cysteine

The use of the Dpm group as a new *S*-protecting group for cysteine was introduced at the same time as the S-Tri group (Zervas and Photaki, 1960, 1962; Hiskey and Adams, 1965*a*). S-Dpm-cysteine, originally prepared (Zervas and Photaki, 1962) by heating a solution of diphenylmethyl chloride and cysteine salts in dimethylformamide solution, can be prepared in better yield by several methods, e.g., by the reaction of cysteine *p*-toluenesulfonate in HBr/AcOH solution with diphenylmethyl chloride (Hanson and Law; 1965); by heating a solution of cysteine hydrochloride and benzhydrol in acetic acid in the presence of 1.1 equivalent of boron trifluoride ethereate followed by evaporation of the solvent (Hiskey and Adams, 1965*a*); at room temperature from cysteine hydrochloride and benzhydrol in 1 N HBr/AcOH followed by distillation or even in TFA at room temperature without distillation (Photaki *et al.*, 1970). The simplest method is by warming a solution of cysteine *p*-toluenesulfonate or hydrochloride in acetic acid with benzhydrol for a few hours (Zervas–Photaki; Zervas *et al.*, unpublished data). For the incorporation of S-Dpm-L-cysteine into a peptide chain, several *N*-protected derivatives can be used, e.g., the N-Z or the *N*-formyl (Zervas and Photaki, 1962), or the N-Boc (Zahn *et al.*, 1969).

Boiling trifluoracetic acid (15–30 min) in the presence of phenol is an excellent cleavage reagent for the S-Dpm group both from the S-Dpm-cysteine and S-Dpm-cysteinylpeptides, and from peptides bearing an S-Dpm-cysteine residue at positions other than the *N*-terminal (Zervas and Photaki, 1962; Photaki *et al.*, 1968*b*, 1970). Cleavage occurs even at room temperature (20–30°C, 16 hr), especially in dilute solutions, and the efficiency of splitting of the *S*-Dpm group is as high as 90% (Photaki *et al.*, 1970). Hydrogen bromide splits off to some extent the S-Dpm group i.e., 45–50% at 50–55°C or 10% at room temperature. The *S*-bis-*p*-methoxyphenylmethyl group [*17*] introduced by Hanson and Law (1965) as a cysteine *S*-protecting group, is cleaved with TFA appreciably faster than the S-Dpm (Hanson and Law; 1965; Photaki *et al.*, 1970). According to Hiskey and Tucker (1962), the S-Dpm group can be removed to a great extent, but less easily than the S-Tri, by oxidation with thiocyanogen or a sulfenylthiocyanate. This is perhaps unexpected, because iodine splits off the S-Dpm group only to a small extent (Kamber and Rittel, 1968).

The best method to split off S-Tri groups from a peptide chain bearing S-Tri and S-Dpm groups is treatment with Hg^{2+} or Ag^+ (Zervas *et al.*, 1965; Hiskey *et al.*, 1970*b*), which removes only S-Tri groups. As yet, no working method has been found for removing S-Dpm selectively without attacking at the same time the S-Tri groups.

Bearing in mind the conditions for removal of the S-Dpm group, it is clear that *N*-protecting groups can be removed selectively without affecting the S-Dpm group, while with the right choice of *N*- and carboxyl-protecting groups the S-Dpm itself can be selectively removed.

E. S-Acyl Groups

The S-acyl groups, in particular the *S*-benzoyl, form a third series of *S*-protecting groups which are used in peptide synthesis (Zervas *et al.*, 1962, 1963*a,b*). As expected, the *S*-acyl groups are especially susceptible to alkali, e.g., the *S*-benzoyl (Bz) group is completely removed by 0.2 N NaOH (15 min) and partially (50%) by 2 N ammonia (15 min). However, it is preferable to remove *S*-acyl groups by methanolysis in the presence of sodium methoxide in an atmosphere of hydrogen (Fig. 9), since this reaction proceeds rapidly and almost quantitatively without causing β-elimination or racemization.

The absence of β-elimination during this methanolysis has been confirmed by Hiskey *et al.* (1970*a*). Further, the *N*-decarbobenzoxylating agents TFA and HBr/AcOH do not attack *S*-benzoylcysteine to any significant extent and thus it can be incorporated into a peptide chain as the N-Z derivative (Zervas *et al.*, 1963*a,b*). For lengthening the chain, other *N*-protected S-Bz cysteines such as the N-Nps (Zervas and Hamalidis, 1965), *N*-formyl (Zervas and Photaki, 1962), or N-Boc (Photaki, 1966) can be used, the N-Nps always being used when other sensitive groups (OBu^t, ODpm, etc.) are present in the molecule. *S*-Acetyl (Ac) cysteine is practically resistant to TFA but is cleaved to a great extent by HBr/AcOH, so that other *N*-protecting groups must be used in place of the benzyloxycarbonyl in this case.

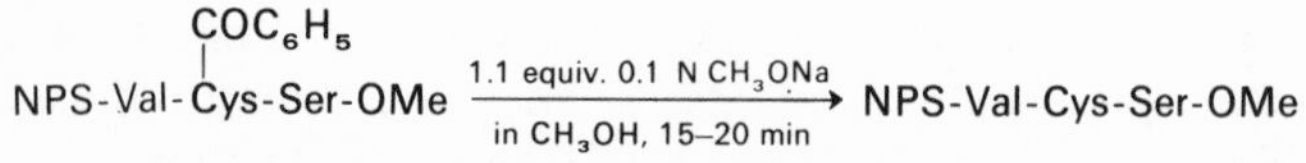

$$\text{NPS-Val-}\underset{\underset{\text{Cys}}{|}}{\overset{\overset{\text{COC}_6\text{H}_5}{|}}{}}\text{-Ser-OMe} \xrightarrow[\text{in CH}_3\text{OH, 15–20 min}]{\text{1.1 equiv. 0.1 N CH}_3\text{ONa}} \text{NPS-Val-Cys-Ser-OMe}$$

Fig. 9. Removal of *S*-benzoyl group by methanolysis.

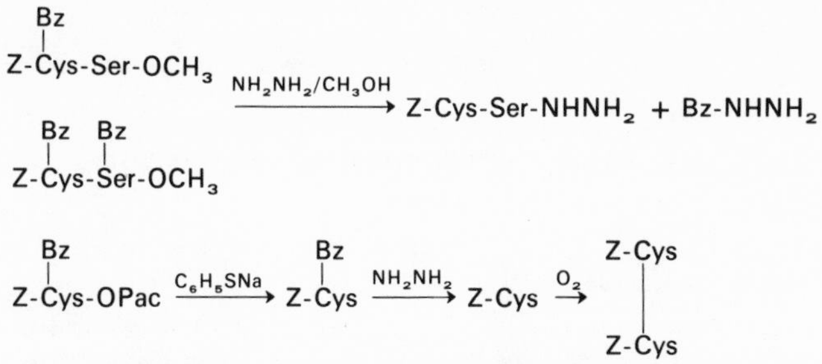

$$
\begin{array}{l}
\overset{\displaystyle Bz}{\underset{\displaystyle |}{}} \\
\text{Z-Cys-Ser-OCH}_3 \\
\overset{\displaystyle Bz \quad Bz}{\underset{\displaystyle | \quad |}{}} \\
\text{Z-Cys-Ser-OCH}_3
\end{array}
\xrightarrow{\text{NH}_2\text{NH}_2/\text{CH}_3\text{OH}}
\text{Z-Cys-Ser-NHNH}_2 \;+\; \text{Bz-NHNH}_2
$$

$$
\overset{Bz}{\underset{|}{\text{Z-Cys-OPac}}}
\xrightarrow{\text{C}_6\text{H}_5\text{SNa}}
\overset{Bz}{\underset{|}{\text{Z-Cys}}}
\xrightarrow{\text{NH}_2\text{NH}_2}
\text{Z-Cys}
\xrightarrow{\text{O}_2}
\begin{array}{l}
\text{Z-Cys} \\
| \\
\text{Z-Cys}
\end{array}
$$

Fig. 10. Removal of S-benzoyl group by hydrazinolysis.

Recently, we have been using hydrazine in dilute methanolic solution for deblocking of acyl groups (Fig. 10). The S-acyl group is rapidly removed as acylhydrazide without causing β-elimination (Photaki; Zervas and Sakarellos, unpublished data). Of course, in the case of a peptide ester the corresponding hydrazide is obtained after this treatment. If the formation of the hydrazide is undesirable, the phenacyl (Pac) group is used as carboxyl-protecting group (Stelakatos et al., 1966). In this case, the phenacyl group is first removed by thiophenolate, and subsequently the S- and, if present, the O-benzoyl groups are removed by hydrazine (Zervas–Photaki; Zervas and Ferderigos, unpublished data).

Another example of an S-acyl group is the ethylcarbamoyl group [19], which was introduced by Guttmann (1966) and used by him for a synthesis of glutathione and of oxytocin. This group is also stable under acidic and neutral conditions but is readily cleaved by basic reagents without de-sulfurization or racemization.

S-Carbobenzoxy- and S,N-dicarbobenzoxy-cysteine can also be used, and they deserve special consideration. Berger et al. (1956) have prepared N,S-dicarbobenzoxy- and S-Z-L-cysteine as well as S-Z-cysteine carbonic acid anhydride, which they have used only for preparation of cysteine polymers. The S-Z group was removed from the final product using Na/liq. NH$_3$. In our experience, for the synthesis of peptides the alcoholysis of the S-Z group with 0.1 N sodium methoxide in methanol is to be preferred even if it proceeds more slowly than in the case of the corresponding S-benzoyl. Boiling TFA does not differentiate between N- and S-Z groups, both of them being cleaved, while the S-Z group survives treatment with 2 N HBr/AcOH (Fig. 11) to a very great extent (Zervas et al., 1963a,b). Thus S-carbobenzoxy-

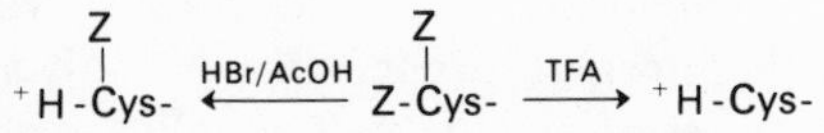

$$
\overset{Z}{\underset{|}{{}^+\text{H-Cys-}}}
\xleftarrow{\text{HBr/AcOH}}
\overset{Z}{\underset{|}{\text{Z-Cys-}}}
\xrightarrow{\text{TFA}}
{}^+\text{H-Cys-}
$$

Fig. 11. Behavior of S-Z group under acidic conditions.

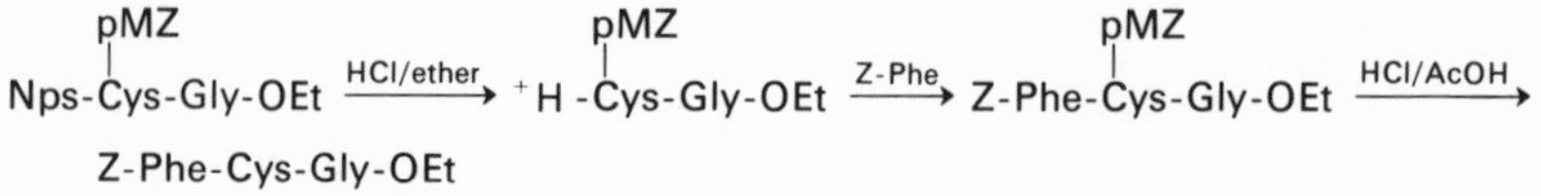

$$\underset{\text{Nps-Cys-Gly-OEt}}{\overset{\text{pMZ}}{|}} \xrightarrow{\text{HCl/ether}} \overset{\text{pMZ}}{\underset{+\text{H -Cys-Gly-OEt}}{|}} \xrightarrow{\text{Z-Phe}} \overset{\text{pMZ}}{\underset{\text{Z-Phe-Cys-Gly-OEt}}{|}} \xrightarrow{\text{HCl/AcOH}}$$

Z-Phe-Cys-Gly-OEt

Fig. 12. Use of S-pMZ-L-cysteine in peptide synthesis.

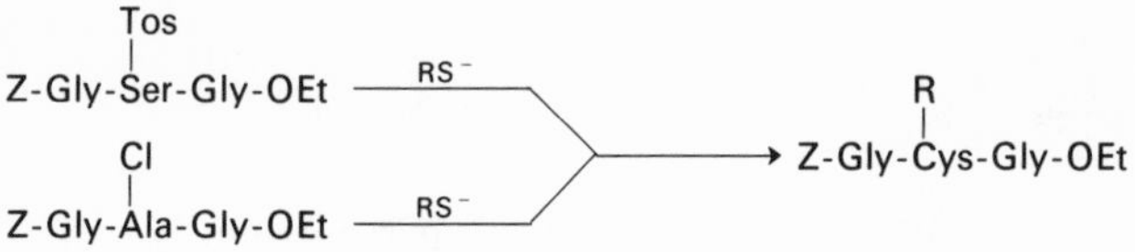

$$\begin{array}{c} \overset{\text{Tos}}{\underset{\text{Z-Gly-Ser-Gly-OEt}}{|}} \xrightarrow{\text{RS}^-} \\ \overset{\text{Cl}}{\underset{\text{Z-Gly-Ala-Gly-OEt}}{|}} \xrightarrow{\text{RS}^-} \end{array} \Bigg\rangle \longrightarrow \overset{R}{\underset{\text{Z-Gly-Cys-Gly-OEt}}{|}}$$

Tos = $CH_3C_6H_4SO_2$, R = C_6H_5CO, CH_3CO

Fig. 13. Transformation of O-tosyl-L-serine peptides and of L-chloroalanine peptides to S-acyl-L-cysteine peptides.

cysteine can also be used for peptide synthesis in conjunction with the carbobenzoxy method for N-protection (Zervas et al., 1963a,b; Sokolovsky et al., 1964). Instead of the N-Z, other derivatives of S-Z-Cys such as N-formyl (Katsoyannis, 1961) or N-Nps (Zervas and Hamalidis, 1965) can be used.

In the last years, the use of p-methoxybenzyloxycarbonyl group (pMZ) [18] as an S-acyl protecting group has been and is still being investigated (Photaki, 1970). S-pMZ-L-cysteine in the form of its N-Nps derivative can be incorporated into a peptide chain in the presence of sensitive groups such as the OBu^t or ODpm (Fig. 12). The deblocking of the S-acyl group can be effected both by alcoholysis and by acidic reagents such as HBr/AcOH or TFA. The latter method of deacylation is more acceptable because the method of alcoholysis will result in transesterification of any other carboxyl esters which might be present in the molecule. In addition, the acidic deblocking would be especially useful for long peptide chains containing other acid-sensitive protecting groups; in this case, the free SH-polypeptide could be produced in one step.

An alternative method of incorporating S-acylcysteines into a peptide chain involves the use of β-chloroalanine as starting material (Photaki and Bardakos, 1966; Wilchek et al., 1966). After incorporation into a peptide chain, the β-chloro group can be transformed to an S-acyl group (Fig. 13). Furthermore, peptides of serine can also be transformed to S-acylcysteine peptides by displacement reactions of their O-tosyl derivatives (Photaki and Bardakos, 1965a,b; Zioudrou et al., 1965).

The possibility of an $S \rightarrow N$ migration should be taken into account when using the S-acylcysteine derivatives for peptide synthesis. In fact, such a migration has been shown to occur when N-Z-glycine is coupled with S-Bz-L-cysteine methyl ester via the carbodiimide method (Zervas et al., 1965). The resulting compound is a protected S-dipeptide ester, i.e., S-(N-

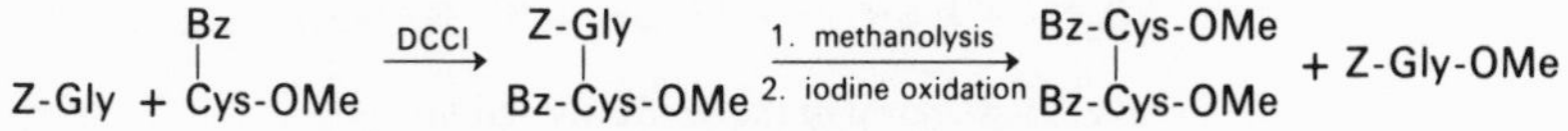

$$\text{Z-Gly} + \underset{\substack{| \\ \text{Cys-OMe}}}{\text{Bz}} \xrightarrow{\text{DCCl}} \underset{\substack{| \\ \text{Bz-Cys-OMe}}}{\text{Z-Gly}} \xrightarrow[\text{2. iodine oxidation}]{\text{1. methanolysis}} \underset{\substack{| \\ \text{Bz-Cys-OMe}}}{\text{Bz-Cys-OMe}} + \text{Z-Gly-OMe}$$

Fig. 14. $S \rightarrow N$ migration of the S-benzoyl group.

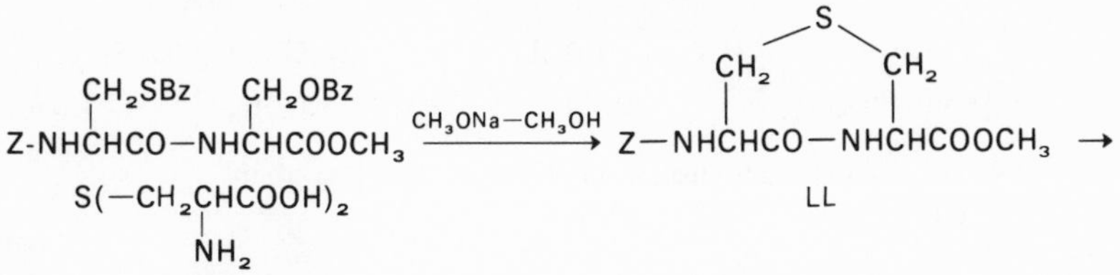

$$\underset{\substack{| \\ \text{S}(-\text{CH}_2\text{CHCOOH})_2 \\ | \\ \text{NH}_2}}{\underset{\substack{\text{CH}_2\text{SBz} \quad \text{CH}_2\text{OBz} \\ | \qquad\qquad | \\ \text{Z-NHCHCO}-\text{NHCHCOOCH}_3}}{}} \xrightarrow{\text{CH}_3\text{ONa}-\text{CH}_3\text{OH}} \underset{\text{LL}}{\underset{\substack{\text{CH}_2 \qquad \text{CH}_2 \\ | \qquad\qquad | \\ \text{Z}-\text{NHCHCO}-\text{NHCHCOOCH}_3}}{\overset{\text{S}}{\overbrace{\qquad\qquad}}}} \rightarrow$$

Fig. 15. Formation of cyclo-L-lanthionyl from cysteinylserine.

benzyloxycarbonyglycyl)-N-benzoyl-L-cysteine methyl ester, as shown by methanolysis. Upon methanolysis, Z-glycine is split off and N-benzoyl-L-cysteine methyl ester is formed and can be isolated as the corresponding cystine derivative (Fig. 14). The results of the methanolysis show unambiguously if such a migration has taken place.

Acyl migration has also been observed by Sokolovsky et al. (1964) and by Hiskey et al. (1966). In our experience, such an $S \rightarrow N$ migration is not observed when an S-acylcysteinyl peptide ester is coupled with N-protected amino acids. In this case, migration is not favored because of the lower basicity of cysteinyl peptide esters; furthermore, to minimize the possibility of an $S \rightarrow N$ migration the following procedure is adopted: to the solution of peptide ester salt the equivalent amount of triethylamine is added and immediately thereafter the N-protected amino acid in 10% excess followed by the addition of carbodiimide (Zervas et al., 1963a). An alternative procedure which is recommended is to add the peptide ester salt to an already prepared mixed anhydride solution of the N-protected amino acid and then, under cooling, add dropwise the equivalent amount of triethylamine (Zervas et al., 1965).

A very interesting case proved to be the methanolysis of N-Z-L-cysteinyl-L-serine methyl ester in which both the cysteine SH and the serine OH groups are protected by benzoylation. In this case, the serine moiety is converted by β-elimination to an α-aminoacrylic acid residue; the simultaneously deblocked SH group adds to the double bond, yielding among other things the methyl ester of N-benzyloxycarbonyl-cyclo-L-lanthionyl methyl ester (Fig. 15) and a dimer product of N-benzyloxycarbonyl-meso-lanthionyl methylester containing a 14-membered ring. Hydrolysis of this product gives L- or meso-lanthionine (Zervas and Ferderigos, 1973). This reaction is reminiscent of the synthesis of lanthionine by the interaction of L-cysteine and α-acylaminoacrylic acid in strong alkaline solution (Schöberl and Wagner, 1956).

$$R-S-SO_3Na + R'SH \longrightarrow R-SH + R'-S-SO_3Na$$

Fig. 16. Removal of the $S-SO_3Na$ group.

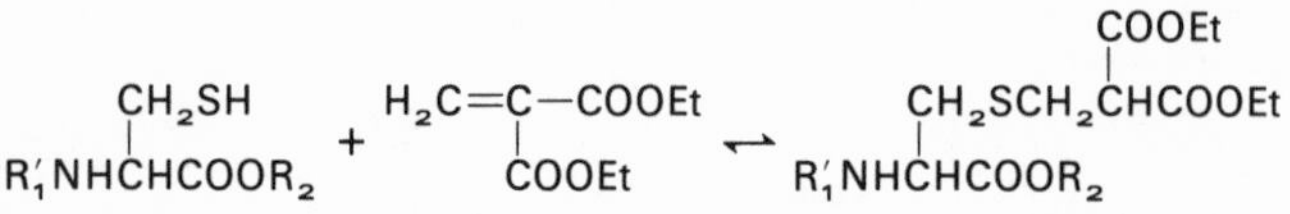

$$
\begin{array}{c}
\text{CH}_2\text{SH} \\
| \\
\text{R}_1'\text{NHCHCOOR}_2
\end{array}
\ +\
\begin{array}{c}
\text{H}_2\text{C}=\text{C}-\text{COOEt} \\
| \\
\text{COOEt}
\end{array}
\ \rightleftarrows\
\begin{array}{c}
\text{COOEt} \\
| \\
\text{CH}_2\text{SCH}_2\text{CHCOOEt} \\
| \\
\text{R}_1'\text{NHCHCOOR}_2
\end{array}
$$

Fig. 17. Introduction and removal of the Dce group.

S-Sulfonate derivatives of cysteine can also be regarded as S-acyl derivatives. They are formed by treatment of disulfides with sodium hydrogen
sulfite followed by tetrathionate, cupric ions, or iodoso benzoic acid (Swan,
1957; Bailey and Cole, 1959; Dixon and Wardlaw, 1960). The regeneration
of the sulfhydryl group can be achieved by treatment with thiols (Fig. 16).
These reactions have been applied in the isolation and reoxidation of the
A and B chains of insulin. However, owing to the great acid sensitivity of the
$S-SO_3H$ group, it has not been used as a protecting group in peptide
synthesis.

Finally another S-protecting group (Fig. 17) may be mentioned which,
although not an acyl group, is alkali sensitive. The S-β,β-diethoxycarbonyl-
methyl group (Dce) (Wieland and Sieber, 1969a) is stable in HBr/AcOH. Its
removal proceeds by a β-elimination reaction using a 1 N solution of KOH
at room temperature for 5–10 min. The same authors (1969b) have used this
group in the synthesis of glutathione.

III. UNSYMMETRICAL CYSTINE PEPTIDES

Many cysteine peptides have been synthesized using S-Tri, S-Dpm,
and S-acyl cysteine intermediates in our laboratory (see the work of Zervas
and Photaki during the years 1960–1970) and also by other research groups
(Sokolovsky et al., 1964; Hiskey et al., 1966; Riniker et al., 1969; Zahn
et al., 1969). The liberation of the sulfhydryl group and its oxidation to the
corresponding symmetrical open-chain cystine peptides present no problem.
There is, however, a problem in the synthesis of unsymmetrical open-chain
cystine peptides, as will be seen from the following discussion.

Attempts by Fischer and Gerngross (1909) to prepare monoglycyl- and
monoleucyl-L-cystine through aminolysis of their respective monohaloacyl-
L-cystine precursors resulted in the formation, in each instance, of a dipeptide
which was not pure, and of appreciable amounts of cystine. This unusual
aminolysis of an α-haloacylamino acid can be explained by the well-known
fact that unsymmetrical open-chain derivatives of cystine are not stable but

rearrange (Fig. 18) to the symmetrical ones (Ryle and Sanger, 1955; Benesch and Benesch, 1958). Such a rearrangement was established, e.g., in the case of monobenzyloxycarbonyl-L-cystine (Zervas *et al.,* 1959), which after treatment with an aqueous solution of diethylamine for 2 hr was converted to a mixture of free cystine and bis-benzyloxycarbonylcystine to the extent of some 40% (Fig. 19). Under less alkaline conditions, e.g., *p*H 7.5, a few days were required for only 25% rearrangement, while no detectable interchange was noted at *p*H 6.5 after 5 days of storage. Decarbobenzoxylation of mono-(benzyloxycarbonylglycyl)-L-cystine with anhydrous HBr/AcOH yielded the desired glycyl-L-cystine in addition to a small amount of free cystine, resulting probably from rearrangement under these strongly acidic conditions. At *p*H 7.5, this unsymmetrical dipeptide rearranges to an extent of approximately 20% on standing at 25°C for a few days.

Various other protected and unprotected unsymmetrical open-chain cystine peptides or derivatives have been synthesized and studied by other workers with regard to their ability to rearrange (Weygand and Zumach, 1962; Hiskey and Tucker, 1962; Zahn and Otten, 1962; Rydon and Serráo, 1964; Veber *et al.*, 1969; Kamber, 1971; Harpp and Back, 1971). In particular, Kamber prepared many free unsymmetrical cystine peptides and studied their stability at various *p*Hs. His results confirm and extend the earlier observations (Zervas *et al.*, 1959) and show that in alkaline or strongly acid solutions rearrangement occurs while in neutral solution rearrangement is minimal. It is, however, not easy during peptide synthesis to always avoid treatment of intermediates with alkaline or strongly acidic reagents.

The existence and the relative stability of unsymmetrical cystine peptides such as oxytocin, vasopressin, and calcitonin may apparently be attributed

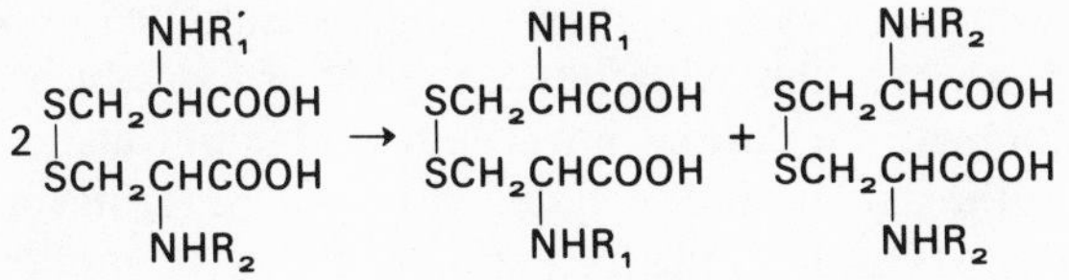

$$2 \begin{array}{c} NHR_1 \\ | \\ SCH_2CHCOOH \\ | \\ SCH_2CHCOOH \\ | \\ NHR_2 \end{array} \rightarrow \begin{array}{c} NHR_1 \\ | \\ SCH_2CHCOOH \\ | \\ SCH_2CHCOOH \\ | \\ NHR_1 \end{array} + \begin{array}{c} NHR_2 \\ | \\ SCH_2CHCOOH \\ | \\ SCH_2CHCOOH \\ | \\ NHR_2 \end{array}$$

Fig. 18. Rearrangement of unsymmetrical open-chain cystine derivatives.

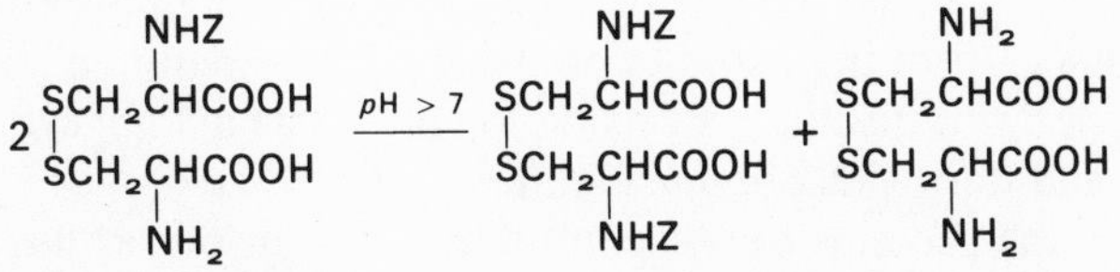

$$2 \begin{array}{c} NHZ \\ | \\ SCH_2CHCOOH \\ | \\ SCH_2CHCOOH \\ | \\ NH_2 \end{array} \xrightarrow{pH > 7} \begin{array}{c} NHZ \\ | \\ SCH_2CHCOOH \\ | \\ SCH_2CHCOOH \\ | \\ NHZ \end{array} + \begin{array}{c} NH_2 \\ | \\ SCH_2CHCOOH \\ | \\ SCH_2CHCOOH \\ | \\ NH_2 \end{array}$$

Fig. 19. Rearrangement of monobenzyloxycarbonyl-L-cystine.

$$\overline{\text{Cys-Tyr-Ile-Gln-Asn-Cys-Pro-Leu-Gly-NH}_2}$$

Fig. 20. Oxytocin.

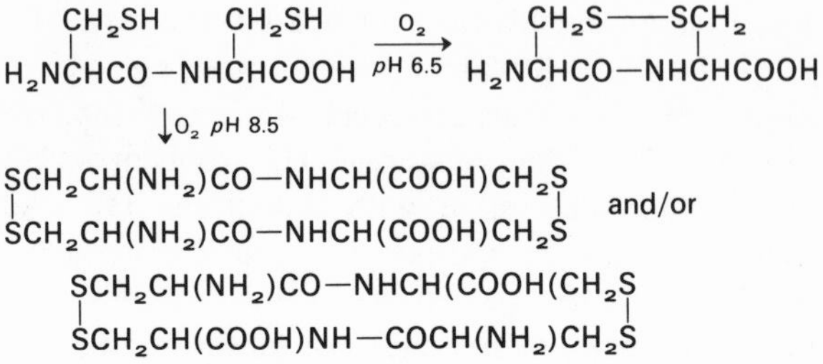

$$
\begin{array}{cc}
\text{CH}_2\text{SH} & \text{CH}_2\text{SH} \\
| & | \\
\text{H}_2\text{NCHCO}-\text{NHCHCOOH}
\end{array}
\xrightarrow[\text{pH 6.5}]{\text{O}_2}
\begin{array}{cc}
\text{CH}_2\text{S}\text{-----}\text{SCH}_2 \\
| \qquad\qquad | \\
\text{H}_2\text{NCHCO}-\text{NHCHCOOH}
\end{array}
$$

$\downarrow$ O$_2$, pH 8.5

$$
\begin{array}{l}
\text{SCH}_2\text{CH(NH}_2\text{)CO}-\text{NHCH(COOH)CH}_2\text{S} \\
| \qquad\qquad\qquad\qquad\qquad\qquad\qquad | \quad \text{and/or} \\
\text{SCH}_2\text{CH(NH}_2\text{)CO}-\text{NHCH(COOH)CH}_2\text{S}
\end{array}
$$

$$
\begin{array}{l}
\text{SCH}_2\text{CH(NH}_2\text{)CO}-\text{NHCH(COOH(CH}_2\text{S} \\
| \qquad\qquad\qquad\qquad\qquad\qquad\qquad\quad | \\
\text{SCH}_2\text{CH(COOH)NH}-\text{COCH(NH}_2\text{)CH}_2\text{S}
\end{array}
$$

Fig. 21. Products of oxidation of L-cysteinyl-L-cysteine.

to the fact that these compounds are of a cyclic nature, the only —S—S— bridge in their structure being implicated in the ring system (Fig. 20).

The preparation by synthesis of such compounds requires the use of one S-protecting group. For the synthesis, e.g., of oxytocin and its analogues, S-benzylcysteine has been mostly used. In addition to the S-benzyl, S-acyl-cysteines, e.g., S-ethylcarbamoyl (Guttmann, 1966), and S-benzoyl or S-benzyloxycarbonyl (Photaki, 1966) have also been used for this purpose. After removal of the S-protecting groups, N-protected oxytoceines have been obtained in crystalline form (Photaki, 1964) as intermediates in this last-mentioned synthesis of oxytocin. The use of S-tritylcysteine for synthesis of such cyclic cystine peptides has the advantage that the oxidative removal of the S-Tri group leads directly to the desired cystine derivative, as proved, e.g., in the synthesis of calcitonin (Riniker et al., 1969).

In the syntheses of cyclic cystine peptides mentioned above, the only problem which arises is the oxidation of the SH group, which can lead not only to formation of the desired intrachain bridge but also to dimers or polymers. Thus the oxidation of L-cysteinyl-L-cysteine (Izumiya and Green-stein, 1954; cf. p. 61) leads to the formation of the cyclic monomer L-cyclo-cystinyl in pure form as well as to the corresponding parallel and antiparallel dimer products (Fig. 21). It is well known that oxidation of the SH-peptide in dilute solution and at neutral pH helps in the formation of the relevant monomer, e.g., cyclocystinyl and oxytocin. In a series of publications, Rydon and coworkers report their studies on the conditions required for formation of monomeric cyclic disulfides by oxidation of L-cysteinyl-(glycyl)$_n$-L-cysteine ($n = 0$–15) and the yields obtained (Hardy et al., 1971).

Most of the proteins can be considered, in principle, as unsymmetrical polypeptides of cystine. These proteins, i.e., insulin, are more or less stable, because in this case more than one cystine —S—S— bridge holds the poly-peptide chains together, forcing them to participate in a multimember ring system. The synthesis of insulin (Katsoyannis et al., 1963; Meienhofer et al., 1963; Kung et al., 1965) and of ribonuclease (Hirshmann et al., 1969; Gutte and Merrifield, 1969) have proved that the incorporation of the same S-protected cysteine in a chain can lead finally to the desired compound, even if the yields cannot be very good. As a matter of fact, natural insulin (Dixon and Wardlaw, 1960) and ribonuclease (White, 1961; Anfinsen and Haber, 1961), which lose activity upon reduction, regain their initial biological activity to a great extent after reoxidation. However, the synthesis of long open-chain cysteine peptides is a tedious job and should not be jeopardized, especially if the final product does not have an easily determined biological activity.

Some years ago (Zervas and Photaki, 1962), we expressed the opinion that an approach to the synthesis, in a controlled way, of unsymmetrical cystine peptides with two or more cystine —S—S— bridges would be facili-tated if the following requirements could be met: (1) Cysteines bearing different S-protecting groups should be available which could be incor-porated into a peptide chain; these S-protecting groups must be selectively removable in such a way that the peptide bond and an already existing bridge in the molecule would not be effected. (2) Two different peptide chains con-taining the S-protected cysteines could be coupled through their amino ends to a polyvalent N-protecting group G (Fig. 22). The same procedure could be applied using another polyvalent group combined with the carboxyl ends of two different peptide chains bearing the S-protected cysteines. The selective removal of two S-protecting groups (one from each of the two chains) and the oxydation of the SH groups thus formed would establish an —S—S— bridge, so that a multimembered ring would be formed and rearrangement of the cystine peptide chains would be prevented. By repetition of the selective splitting off of two or more S-protecting groups, the formation of a second —S—S— bridge and of an additional ring might be achieved. The next step would be the removal of the polyvalent N-protecting group G in such a way that neither the peptide bonds nor the —S—S— bridges would be affected.

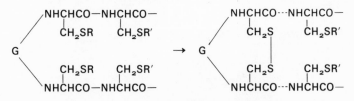

Fig. 22. Use of a polyvalent N-protecting group.

Alternatively, the above polyvalent N-protecting group could be combined with two S-protected derivatives of cysteine. After the formation of the first —S—S— bridge, the cystine derivative could be lengthened at both sides, and at the desired length it could be supplied with new S-protected cysteine derivatives, etc.

The first requirement has been satisfactorily met, since a variety of S-protecting groups selectively removable are now available. This fact has already helped to overcome for the first time (Zervas *et al.*, 1963a, b; Cosmatos *et al.*, 1965; Zervas *et al.*, 1965) the unique difficulties inherent in establishing an —S—S— bridge specifically between two of three cysteine residues of a peptide chain. This has been realized by the synthesis of a fragment of the A chain of sheep insulin bearing the 6–11 intrachain bridge (Fig. 23). The removal of the remaining S-protecting group is carried out as described on pp. 64–68.

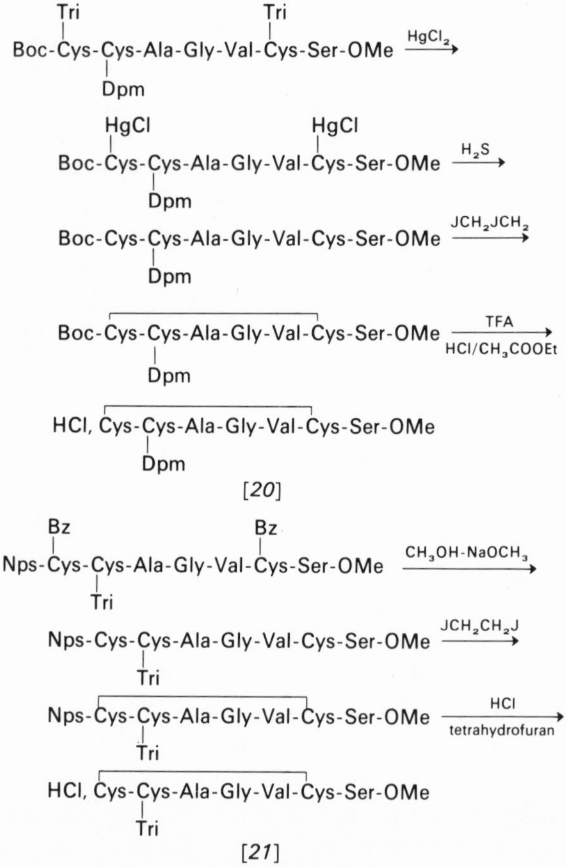

Fig. 23. Synthesis of the 6–11 intrachain bridge of sheep insulin.

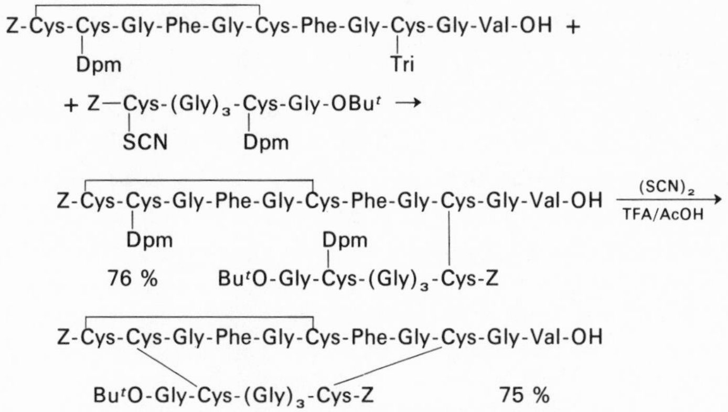

Z-Cys-Cys-Gly-Phe-Gly-Cys-Phe-Gly-Cys-Gly-Val-OH +
  | |
  Dpm   Tri

+ Z—Cys-(Gly)$_3$-Cys-Gly-OBu^t →
     | |
     SCN   Dpm

Z-Cys-Cys-Gly-Phe-Gly-Cys-Phe-Gly-Cys-Gly-Val-OH $\xrightarrow[\text{TFA/AcOH}]{\text{(SCN)}_2}$
  | | |
  Dpm   Dpm |
  76 %   Bu^{t}O-Gly-Cys-(Gly)$_3$-Cys-Z

Z-Cys-Cys-Gly-Phe-Gly-Cys-Phe-Gly-Cys-Gly-Val-OH
      \   /
   Bu^{t}O-Gly-Cys-(Gly)$_3$-Cys-Z   75 %

Fig. 24. Synthesis of a tris-cystine peptide by Hiskey *et al.*

In recent years, Hiskey *et al.* (1969) have reported that, using methods designed only to meet the first requirement mentioned above, they have been able to synthesize peptides bearing more than one —S—S— bridge, and even a peptide containing the ring system of insulin. The satisfactory formation of the 20-membered ring of the intrachain in this synthesis corresponds to the other examples of syntheses of such rings in oxytocin, vasopressin, and the insulin fragment of Fig. 23. Furthermore, surprisingly, the establishment of two —S—S— cross-linkages, one of which is formed by the interaction of two cysteine *S*-protecting groups and the other by cooxidation of two *S*-protected cysteine residues, is reported by Hiskey to give monomeric pure products in very good yield (Fig. 24).

Zervas and his associates still believe that the realization of the second objective mentioned above (p. 75) would provide the safest controlled method of synthesis of a peptide having more than one cross-linkage.

L-Glutamic acid has been used as a polyvalent *N*-protecting group (Yovanidis, 1964; Zervas *et al.*, 1967), the two carboxyl groups being linked to two different peptides bearing *S*-protected cysteines (Fig. 25). In the last stage of the synthesis, the γ-peptide bond could be selectively hydrolyzed so that the polyvalent *N*-protecting group could become a member of the peptide chain.

Attempts to oxidize compound A (Fig. 25) to a monomeric —S—S— derivative have yielded only mixtures from which it has not been possible to isolate workable yields of the required compound, and this scheme has had to be abandoned.

Another polyvalent *N*-protecting group (Fig. 26) which may be considered is monobenzyl phosphoric acid. Substance B of Fig. 26 (Zervas and Sakarellos, unpublished data) has been used as starting material for joining

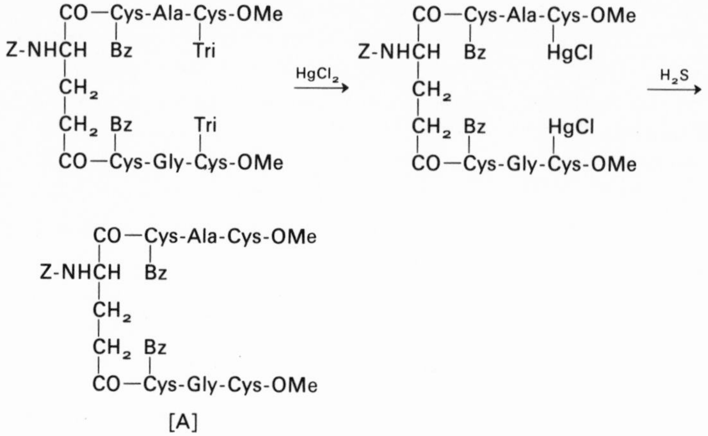

$$
\begin{array}{ccc}
\text{CO}\!-\!\text{Cys-Ala-Cys-OMe} \\
\text{Z-NHCH} \quad \text{Bz} \qquad \text{Tri} \\
\mid \\
\text{CH}_2 \\
\mid \\
\text{CH}_2 \;\; \text{Bz} \qquad \text{Tri} \\
\text{CO}\!-\!\text{Cys-Gly-Cys-OMe}
\end{array}
\xrightarrow{\text{HgCl}_2}
\begin{array}{ccc}
\text{CO}\!-\!\text{Cys-Ala-Cys-OMe} \\
\text{Z-NHCH} \quad \text{Bz} \qquad \text{HgCl} \\
\mid \\
\text{CH}_2 \\
\mid \\
\text{CH}_2 \;\; \text{Bz} \qquad \text{HgCl} \\
\text{CO}\!-\!\text{Cys-Gly-Cys-OMe}
\end{array}
\xrightarrow{\text{H}_2\text{S}}
$$

$$
\begin{array}{cc}
\text{CO}\!-\!\text{Cys-Ala-Cys-OMe} \\
\text{Z-NHCH} \quad \text{Bz} \\
\mid \\
\text{CH}_2 \\
\mid \\
\text{CH}_2 \;\; \text{Bz} \\
\text{CO}\!-\!\text{Cys-Gly-Cys-OMe}
\end{array}
$$

[A]

Fig. 25. L-Glutamic acid as a polyvalent N-protecting group.

$$
\text{C}_6\text{H}_5\text{OP(O)Cl} + 2\ \text{Phe-OBzl} \rightarrow
\begin{array}{c}
\text{C}_6\text{H}_5\text{O} \qquad \text{NHCH(R)COOBzl} \\
\diagdown \ \ \diagup \\
\text{P} \\
\diagup \ \ \diagdown \\
\text{O} \qquad \text{NHCH(R)COOBzl}
\end{array}
\xrightarrow[\text{C}_6\text{H}_5\text{CH}_2\text{OH}]{\text{C}_6\text{H}_5\text{CH}_2\text{ONa}}
$$

$$
\begin{array}{c}
\text{C}_6\text{H}_5\text{CH}_2\text{O} \qquad \text{NHCH(R)COOBzl} \\
\diagdown \ \ \diagup \\
\text{P} \\
\diagup \ \ \diagdown \\
\text{O} \qquad \text{NHCH(R)COOBzl}
\end{array}
\xrightarrow[\text{30 min}]{0.1\ \text{N NaOH}}
\begin{array}{c}
\text{C}_6\text{H}_5\text{CH}_2\text{O} \qquad \text{NHCH(R)COOBzl} \\
\diagdown \ \ \diagup \\
\text{P} \\
\diagup \ \ \diagdown \\
\text{O} \qquad \text{NHCH(R)COOH}
\end{array}
$$

mp 112°C   mp 155°C   (B)

R = $C_6H_5CH_2$

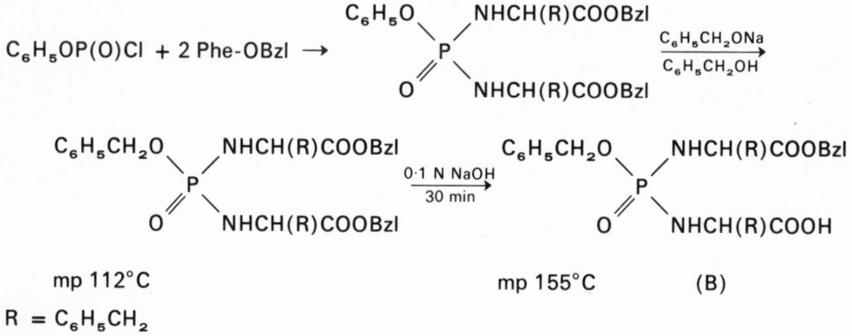

Fig. 26. Monobenzyl phosphoric acid as a polyvalent N-protecting group.

two different cysteine peptides, e.g., the two tripeptides linked to Z-glutamic acid in Fig. 25. This approach has not been pursued further because, on the fortieth anniversary of the carbobenzoxy method, hope has been raised that this method may yet provide a solution to the problem (Zervas–Photaki, Zervas, Stathaki, and Ferderigos, unpublished data).

Phthalyl alcohol* [22] was transformed to its corresponding bis-carbonyl-p-nitrophenyl ester [23] and subsequently to bis-S-trityl-L-cysteine diphenyl methyl ester [24]. Detritylation as usual by Hg^{2+} followed by oxidation or detritylating oxidation by iodine in one operation leads to the

---

*Use of the corresponding bis-*tert*-alcohol would mean the application of the Boc method for this purpose.

monomeric cyclic crystalline cystine derivative [25] in good yield (Fig. 27). As expected, compound [23] is stable to HBr/AcOH, whereas under the same conditions compound [25] loses both the *N*- and the *C*-protecting groups. Selective removal under alkaline or acidic conditions (HCl/CH₃NO₂) of one Dpm group has not been possible. This problem has been met by synthesizing a compound (Fig. 28) that contains two different ester groups such as Dpm

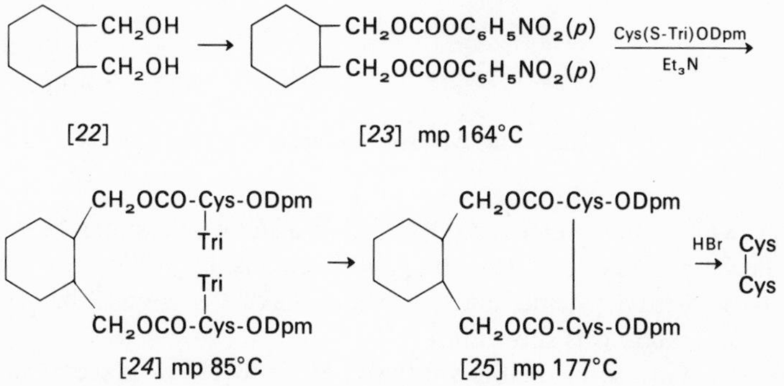

Fig. 27. Phthalyloxycarbonyl as a polyvalent *N*-protecting group.

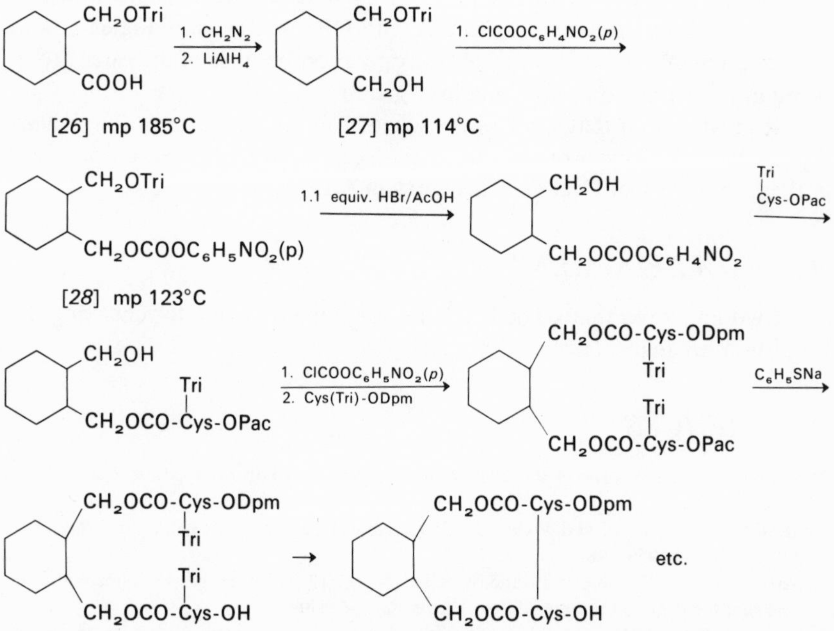

Fig. 28. Scheme for the synthesis of parallel unsymmetrical cystine peptides.

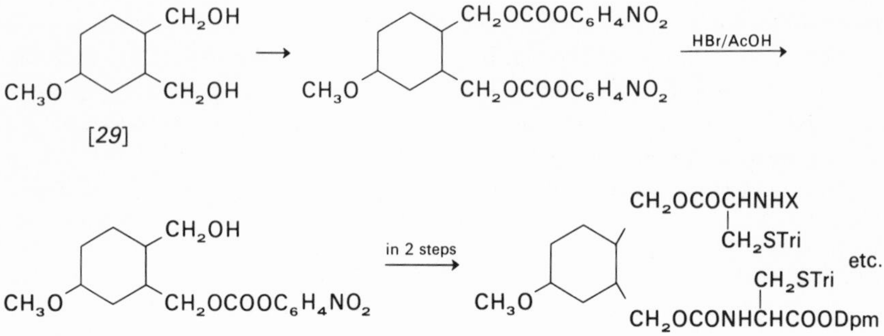

Fig. 29. Scheme for the synthesis of antiparallel cystine peptides.

and Pac, which can be removed selectively in a stepwise manner. In this way, it should be possible to extend the peptide chains, by adding two different peptides of S-tritylcysteine, and to form a second $-S-S-$ bridge. This sequence of reactions is shown in Fig. 28.

Finally, the use of methoxyphthalyl alcohol [29] in place of phthalyl alcohol as starting material should both facilitate the synthesis of parallel unsymmetrical cyclic cystine peptides and also make possible the synthesis of antiparallel cyclic cystine peptides according to the scheme shown in Fig. 29. Treatment of the final product with HBr/AcOH should release both the amino and carboxyl groups in one operation because of the ester group being in the p-position to the methoxy group.

It must be said that some of the schemes for synthesis shown in the last figures are still being worked out as part of the current research program of the Laboratory of Organic Chemistry in Athens.

ACKNOWLEDGMENT

I would like to thank Dr. Joyce Taylor-Papadimitriou for checking the English in my manuscript.

REFERENCES

Albertson, N. F., and McKay, F. C., 1953, Acid-catalyzed decarbobenzoxylation, *J. Am. Chem. Soc.* **75**:5323.

Amiard, G., Heymès, R., and Velluz, L., 1956, Nouvelle synthèse du glutathion, *Bull. Soc. Chim. France* **1956**:698.

Anderson, G. W., Blondinger, J., and Welcher, A. D., 1952, Tetraethyl pyrophosphite as a reagent for peptide synthesis, *J. Am. Chem. Soc.* **74**:5309.

Anfinsen, C. B., and Haber, E., 1961, Studies on the reduction and re-formation of protein disulfide bonds, *J. Biol. Chem.* **236**:1361.

Bailey, J. L., and Cole, R. D., 1959, Studies on the reaction of sulfite with proteins, *J. Biol. Chem.* **234**:1733.

Benesch, R. E., and Benesch, R., 1958, The mechanism of disulfide interchange in acid solution; role of sulfenium ions, *J. Am. Chem. Soc.* **80**:1666.

Ben-Ishai, D., and Berger, A., 1952, Cleavage of *N*-carbobenzoxy groups by dry hydrogen bromide and hydrogen chloride, *J. Org. Chem.* **17**:1564.

Berger, A., Noguchi, J., and Katchalski, E., 1956, Poly-L-cysteine, *J. Am. Chem. Soc.* **78**:4483.

Bergmann, M., and Michalis, G., 1930, Katalytische Hydrierung von 1-Cystin zu 1-Cystein, *Berichte* **63**:987.

Bergmann, M., and Zervas, L., 1932, Über ein allgemeines Verfahren der Peptid-Synthese, *Berichte* **65**:1192.

Berse, C., Boucher, R., and Piché, L., 1957, Preparation of L-cystinyl and L-cysteinyl peptides through catalytic hydrogenation of intermediates, *J. Org. Chem.* **22**:805.

Boissonnas, R. A., and Preitner, G., 1953, Étude comparative de la scission de divers groupes de blocage de la fonction α-amino des acides aminés, *Helv. Chim. Acta* **36**:875.

Brownlee, P. J. E., Cox, M. E., Handford, B. O., Marsden, J. C., and Young, G. T., 1964, Amino-acids and peptides. Part XX. *S*-Benzylthiomethyl-L-cysteine and its use in the synthesis of peptides, *J. Chem. Soc.* **1964**:3832.

Callahan, F. M., Anderson, G. W., Paul, R., and Zimmerman, J. E., 1963, The tertiary butyl group as a blocking agent for hydroxyl, sulfhydryl and amino functions in peptide synthesis, *J. Am. Chem. Soc.* **85**:201.

Camble, R., Purkayastha, R., and Young, G. T., 1968, Amino-acids and peptides. Part XXIX. The use of *S*-benzylthiomethyl-L-cysteine in peptide synthesis: Synthesis of glutathione and homoglutathione, *J. Chem. Soc.* (*C*) **1968**:1219.

Cosmatos, A., Photaki, I., and Zervas, L., 1961, Peptidsynthesen über *N*-Phosphorylaminosäure-phosphorsäure-anhydride, *Chem. Ber.* **94**:2644.

Cosmatos, A., Photaki, I., and Zervas, L., 1965, The synthesis of an oxytocin-type fragment of insulin, *in*: "Peptides: Proceedings of the Sixth European Symposium, Athens, 1963" (L. Zervas, ed.), p. 301, Pergamon Press, Oxford.

Dixon, G. H., and Wardlaw, A. C., 1960, Regeneration of insulin activity from the separated and inactive A and B chains, *Nature* **188**:721.

du Vigneaud, V., and Miller, G. L., 1936, A synthesis of glutathione, *J. Biol. Chem.* **116**:469.

du Vigneaud, V., Ressler, C., Swan, J. M., Roberts, C. W., Katsoyannis, P. G., and Gordon, S., 1953, The synthesis of an octapeptide amide with the hormonal activity of oxytocin, *J. Am. Chem. Soc.* **75**:4879.

du Vigneaud, V., Bartlett, M. F., and Jöhl, A., 1957, The synthesis of lysine vasopressin, *J. Am. Chem. Soc.* **79**:5572.

du Vigneaud, V., Gish, D. T., Katsoyannis, P. G., and Hess, G. P., 1958, Synthesis of the pressor-antidiuretic hormone, arginine-vasopressin, *J. Am. Chem. Soc.* **80**:3355.

Fischer, E., and Gerngross, O., 1909, Synthese von Polypeptiden. XXX. Derivate des L-Cystins, *Berichte* **42**:1485.

Gutte, B., and Merrifield, R. B., 1969, The total synthesis of an enzyme with ribonuclease A activity, *J. Am. Chem. Soc.* **91**:501.

Guttmann, St., 1966, Synthèse du glutathion et de l'oxytocine à l'aide d'un nouveau groupe protecteur de la fonction thiol, *Helv. Chim. Acta* **49**:83.

Hanson, R. W., and Law, H. D., 1965, Substituted diphenylmethyl protecting groups in peptide synthesis, *J. Chem. Soc.* **1965**:7285.

Hardy, P. M., Ridge, B., Rydon, H. N., and dos S. P. Serrão, F. O., 1971, Polypeptides. Part XV. The synthesis and oxidation of some L-cysteinylpolyglycyl-L-cysteines, *J. Chem. Soc.* (*C*) **1971**:1722.

Harington, C. H., and Mead, T. H., 1935, The synthesis of glutathione, *Biochem. J.*, **29**:1602.

Harpp, D. N., and Back, T. G., 1971, The synthesis of some new cystine-containing unsymmetrical disulfides, *J. Org. Chem.* **36**:3828.

Hillmann-Elies, A., Hillmann, G., and Jatzkewitz, H., 1953, *N*-(Triphenylmethyl)amino acids and peptides, *Z. Naturforsch.* **8b**:445.

Hirotsu, Y., Shiba, T., and Kaneko, T., 1967, Synthetic studies of bacitracin. V. Synthesis of thiazoline peptides from cystine peptides by dehydration procedure, *Bull. Chem. Soc. Japan* **40**:2950.

Hirshmann, R., Nutt, R. F., Veber, D. F., Vitali, R. A., Varga, S. L., Jacob, T. A., Holly, F. W., and Denkewalter, R. G., 1969, Studies on the total synthesis of an enzyme. V. The preparation of enzymatically active material, *J. Am. Chem. Soc.* **91**:507.

Hiskey, R. G., and Adams, J. B., Jr., 1965*a*, A convenient preparation of *S*-benzhydryl- and *S*-trityl-L-cysteine, *J. Org. Chem.* **30**:1340.

Hiskey, R. G., and Adams, J. B., Jr., 1955*b*, Sulfur-containing polypeptides. I. Use of the *N*-benzhydryloxycarbonyl group and the benzhydryl ester, *J. Am. Chem. Soc.* **87**:3969.

Hiskey, R. G., and Sparrow, J. T., 1970, Sulfur-containing polypeptides. IX. The use of the *S*-isobutyloxymethyl protective group, *J. Org. Chem.* **35**:215.

Hiskey, R. G., and Tucker, W. P., 1962, Chemistry of aliphatic disulfides. V. Preparation of some open-chain unsymmetrical cystine derivatives from thioethers of cysteine, *J. Am. Chem. Soc.* **84**:4794.

Hiskey, R. G., Mizoguchi, T., and Inui, T., 1966, Sulfur-containing polypeptides. III. *S–N* benzoyl group migration in cysteine derivatives, *J. Org. Chem.* **31**:1192.

Hiskey, R. G., Staples, J. T., and Smith, R. L., 1967, Sulfur-containing polypeptides. VII. Synthesis of *S*-trityl-L-cysteine peptides using acid-labile amino and carboxy protective groups, *J. Org. Chem.* **32**:2772.

Hiskey, R. G., Thomas, A. M., Smith, R. L., and Jones, W. C., Jr., 1969, Sulfur-containing polypeptides. XI. A synthetic route to triscystine peptides, *J. Am. Chem. Soc.* **91**:7525.

Hiskey, R. G., Upham, R. A., Beverly, G. M., and Jones, W. C., Jr., 1970*a*, Sulfur-containing polypeptides. X. A study of *β*-elimination of mercaptides from cysteine peptides. *J. Org. Chem.* **35**:513.

Hiskey, R. G., Davis, G. W., Safdy, M. E., Inui, T., Upham, R. A., and Jones, W. G., Jr., 1970*b*, Sulfur-containing polypeptides. XIII. Bis cystine peptide derivatives, *J. Org. Chem.* **35**:4148.

Holland, G. F., and Cohen, L. A., 1958, Studies on the synthesis of insulin peptides, *J. Am. Chem. Soc.* **80**:3765.

Inukai, N., Nakano, K., and Murakami, M., 1967, The peptide synthesis. I. Use of the *S*-ethylmercapto group for the protection of the thiol function of cysteine, *Bull. Chem. Soc. Japan* **40**:2913.

Izumiya, N., and Greenstein, J. P., 1954, Studies on polycysteine peptides and proteins. I. Isomeric cystinylcystine peptides, *Arch. Biochem. Biophys.* **52**:203.

Kamber, B., 1971, Die Synthese von Insulinfragmenten mit intakter interchenarer Disulfidbrucke A^{20}-B^{19}, *Helv. Chim. Acta* **42**:398.

Kamber, B., and Rittel, W., 1968, Eine neue, einfache Methode zur Synthese von Cystinpeptiden, *Helv. Chim. Acta* **51**:2061.

Katsoyannis, P. G., 1961, Insulin peptides. I. Synthesis of cysteine-containing peptides related to the A-chain of sheep insulin, *J. Am. Chem. Soc.* **83**:4053.

Katsoyannis, P. G., Tometsko, A., and Fukuda, K., 1963, Insulin peptides. IX. The synthesis of the A-chain of insulin and its combination with natural B-chain to generate insulin activity, *J. Am. Chem. Soc.* **85**:2863.

Katsoyannis, P. G., Zalut, C., Tometsko, A., Tilak, M., Johnson, S., and Trakatellis, A. C., 1971, Insulin peptides. XXI. A synthesis of the B-chain of sheep (bovine, porcine) insulin and its isolation as the *S*-sulfonated derivative, *J. Am. Chem. Soc.* **93**:5871.

King, F. E., Clark-Lewis, J. W., and Wade, R., 1957, Syntheses from phthalimido-acids. Part VIII. Synthesis of glutathione by a new route to cysteinyl peptides, *J. Chem. Soc.* **1957**:880.

König, W., Geiger, R., and Siedel, W., 1968, Neue S-Schutzgruppen für Cystein, *Chem. Ber.* **101**:681.

Kung, Y.-T., Du, Y.-C., Cheng, C.-C., Ke, L.-T., Hwang, W.-T., Hu, S.-C., Niu, C.-Y., Hsu, J.-Z., Chang, W.-C., Cheng, L.-I., Li, H.-S., Wang, Y., Chi, A.-H., Loh, T.-P., Shi, P.-T., Li, C.-H., Yieh, Y.-H., Tang, K.-I., and Hsing, C.-Y., 1965, Total synthesis of crystalline bovine insulin, *Sci. Sinica (Peking)* **14**:1710.

Meienhofer, J., Schnabel, E., Bremer, H., Brinkhoff, O., Zabel, R., Sroka, W., Klostermeyer, H., Bradenburg, D., Okuda, T., and Zahn, H., 1963, Synthese der Insulinketten und ihre Kombination zu insulinaktiven Präparaten, *Z. Naturforsch.* **18b**:1120.

Ondetti, M. A., and Bodanszky, M., 1962, On the S-p-Nitrobenzyl Protecting Group, *Chem. Ind. (Lond.)* **1962**:697.

Phocas, I., Yovanidis, C., Photaki, I., and Zervas, L., 1967, New Methods in peptide synthesis. Part IV. N → S transfer of N-o-nitrophenylsulfenyl groups in cysteine peptides, *J. Chem. Soc. (C)* **1967**:1506.

Photaki, I., 1963, Transformation of serine to cysteine. β-Elimination reactions in serine derivatives, *J. Am. Chem. Soc.* **85**:1123.

Photaki, I., 1964, Synthesis of N-protected oxytoceines, *Experientia* **20**:487.

Photaki, I., 1966, A new synthesis of oxytocin using S-acyl cysteines as intermediates, *J. Am. Chem. Soc.* **88**:2292.

Photaki, I., 1967, Some new methods in peptide synthesis, in "Pharmacology of Hormonal Polypeptides and Proteins," p. 1, Plenum Press, New York.

Photaki, I., 1970, On cysteine and cystine peptides. Part VI. S-Acylcysteines in peptide synthesis, *J. Chem. Soc. (C)* **1970**:2687.

Photaki, I., and Bardakos, V., 1965a, Transformation of L-serine to L-cysteine, *Experientia* **21**:371.

Photaki, I., and Bardakos, V., 1965b, Transformation of L-serine peptides to L-cysteine peptides, *J. Am. Chem. Soc.* **87**:3489.

Photaki, I., and Bardakos, V., 1966, Transformation of β-chloro-L-alanine peptides into L-cysteine peptides, *Chem. Commun.* **1966**:818.

Photaki, I., Phocas, I., Taylor-Papadimitriou, J., and Zervas, L., 1968a, On cysteine and cystine peptides, in "Peptides: Proceedings of the Ninth European Symposium, Orsay, 1968" (E. Bricas, ed.), p. 201, North-Holland, Amsterdam.

Photaki, I., Bardakos, V., Lake, A. W., and Lowe, G., 1968b, Synthesis and catalytic properties of the pentapeptide Thr-Ala-Cys-His-Asp, *J. Chem. Soc. (C)* **1968**:1860.

Photaki, I., Taylor-Papadimitriou, J., Sakarellos, C., Mazarakis, P., and Zervas, L., 1970, On cysteine and cystine peptides. Part V. S-Trityl and S-diphenylmethylcysteine and -cysteine peptides, *J. Chem. Soc. (C)* **1970**:2683.

Pimlott, P. J. E., and Young, G. T., 1958, A new method for the protection of thiol groups, *Proc. Chem. Soc.* **1958**:257.

Ressler, Ch., and du Vigneaud, V., 1954, The synthesis of the tetrapeptide amide S-benzyl-L-cysteinyl-L-prolyl-L-leucylglycinamide, *J. Am. Chem. Soc.* **76**:3107.

Riniker, B., Brugger, M., Kamber, B., Sieber, P., and Rittel, W., 1969, Thyrocalcitonin IV. Die Totalsynthese des α-Thyrocalcitonins, *Helv. Chim. Acta* **52**:1058.

Rydon, H. N., and dos S. P. Serrão, F. O., 1964, Polypeptides. Part XI. Studies on the synthesis of unsymmetrical peptides of cystine, *J. Chem. Soc.* **1964**:3638.

Ryle, A. P., and Sanger, F., 1955, Disulfide interchange reactions, *Biochem. J.* **60**:535.

Sakakibara, S., Shimonishi, Y., Kishida, Y., Okada, M., and Sugihara, H., 1967, Use of anhydrous hydrogen fluoride in peptide synthesis. I. Behavior of various protective groups in anhydrous hydrogen fluoride, *Bull. Chem. Soc. Japan* **40**:2164.

Schöberl, A., and Wagner, A., 1956, Untersuchungen zur Frage der Lanthionin-Bildung aus Wolle und Cystin, Z. Physiol. Chem. **304**:97.

Sheehan, J. C., and Yang, D. D. H., 1958, A new synthesis of cysteinyl peptides, J. Am. Chem. Soc. **80**: 1158.

Sifferd, R. H., and du Vigneaud, V., 1935, A new synthesis of carnosine, with some observations on the splitting of the benzyl group from carbobenzoxy derivatives and from benzylthioethers, J. Biol. Chem. **108**: 753.

Sokolovsky, M., Wilchek, M., and Patchornik, A., 1964, On the synthesis of cysteine peptides, J. Am. Chem. Soc. **86**:1202.

Stelakatos, G. C., Paganou, A., and Zervas, L., 1966, New methods in peptide synthesis. Part III. Protection of carboxyl group, J. Chem. Soc. (C) **1966**:1191.

Swan, J. M., 1957: Thiols, disulfides and thiosulfates: Some new reactions and possibilities in peptide and protein chemistry, Nature **180**:643.

Theodoropoulos, D. M., 1953, Doctoral dissertation, Division of Natural Sciences (Chemistry Section), University of Athens, Greece.

Veber, D. F., Milkowski, J. D., Denkewalter, R. G., and Hirshmann, R., 1968, The synthesis of peptides in aqueous medium. IV. A novel protecting group for cysteine, Tetrahedron Letters **26**:3057.

Veber, D. F., Hirshmann, R., and Denkewalter, R. G., 1969, The synthesis of peptides in aqueous medium. VI. The synthesis of an unsymmetrical cystine peptide fragment of insulin, J. Org. Chem. **34**:753.

Velluz, L., Amiard, G., Bartos, J., Goffinet, B., and Heymès, R., 1956, Accès a l'ocytocine de synthèse, a l'aide d'intermediaires S,N-trytilés, Bull. Soc. Chim. France **1956**: 1464.

Weber, V., 1969, Synthese einer Insulin-A-Kette nach der Merrifield-Methode unter Verwendung der S-Äthylmercapto-Schutzgruppe, Z. Physiol. Chem. **350**:1421.

Weygand, F., and Zumach, G., 1962, Cystinepeptide durch Dehydrierung mit vicinalen Dihalogeniden, Z. Naturforsch. **17b**:807.

White, F. H., Jr., 1961, Regeneration of native secondary and tertiary structures by air oxidation of reduced ribonuclease, J. Biol. Chem. **236**:1353.

Wieland, Th., and Sieber, A., 1969a, Der β,β-Diäthoxycarbonyläthyl-Rest (Dce). Eine durch Basen abspaltbare Schutzgruppe für den Cystein-Schwefel, Ann. Chem. **722**:222.

Wieland, Th., and Sieber, A., 1969b, Synthese des Glutathions unter Verwendung des Diäthoxycarbonyläthyl (Dce)-Rests als Schwefel-Schutzgruppe, Ann. Chem. **727**:121.

Wilchek, M., Zioudrou, C., and Patchornik, A., 1966, The conversion of L-β-chloroalanine peptides to L-cysteine peptides, J. Org. Chem. **31**:2865.

Wünsch, E., and Spangenberg, R., 1971, Eine Neue S-Schutzgruppe für Cystein, in "Peptides 1969: Proceedings of the Tenth European Symposium" (E. Scoffone, ed.), p. 30, North-Holland, Amsterdam.

Yovanidis, C., 1964, Doctoral dissertation, Division of Natural Sciences (Chemistry Section), University of Athens, Greece.

Zahn, H., and Hammerström, K., 1969, N-tert.-Butyloxycarbonyl-L-cystein und S-substituierte Derivate, Chem. Ber. **102**:1048.

Zahn, H., and Otten, H. G., 1962, Über unsymmetrische, offenkettige Cystin-peptide, Ann. Chem. **653**:139.

Zahn, H., and Schmidt, G., 1970, Synthese der Insulinsequenz $(B_{17-30})_2$ als symmetrisches Disulfid und der Insulin-B-Kette als polymeres Disulfid, Ann. Chem. **731**:101.

Zahn, H., Danho, W., Klostermeyer, H., Gattner, H. G., and Repin, J., 1969, Eine Synthese der A-Kette des Schafinsulins unter ausschliesslicher Verwendung säurelabiler Schutzgruppen, Z. Naturforsch. **24b**:1127.

Zervas, L., 1955, Phosphorylation et tritylation des acides aminés. Nouvelles methodes de la synthèse peptidique, *in* "Résumés des Communications du XIVe Congrès International de Chimie Pure et Appliquée, Zürich, p. 224.

Zervas, L., 1962, Über Cystein-und Cystinpeptide, *Coll. Czech. Chem. Commun.* **27**:2242.

Zervas, L., 1970, Über die *N*-Trityl- und *N*-*o*-Nitrophenylsulfenyl-methode zur Peptidsynthese, *Z. Naturforsch.* **25b**:322.

Zervas, L., and Ferderigos, N., 1973, Umwandlung von Cysteinylserin in Lanthionin, *Experientia* **29**:262; On Lanthionine and Cyclolanthiovyl, *Isr. J. Chem.* **11**:in press.

Zervas, L., and Hamalidis, Ch., 1965, New methods in peptide synthesis. II. Further examples of the use of the *o*-nitrophenylsulfenyl groups for the protection of amino groups, *J. Am. Chem. Soc.* **87**:99.

Zervas, L., and Photaki, I., 1960, Über Cystein- und Cystinpeptide, *Chimia (Aarau)* **14**:375.

Zervas, L., and Photaki, I., 1962, On cysteine and cystine peptides. I. New *S*-protecting groups for cysteine, *J. Am. Chem. Soc.* **84**:3887.

Zervas, L., and Theodoropoulos, D. M., 1956, *N*-Tritylamino acids and peptides. A new method of peptide synthesis, *J. Am. Chem. Soc.* **78**:1359.

Zervas, L., Benoiton, L., Weiss, E., Winitz, M., and Greenstein, J. P., 1959, Preparation and disulfide interchange reactions of unsymmetrical open-chain derivatives of cystine, *J. Am. Chem. Soc.* **81**:1729.

Zervas, L., Photaki, I., and Ghelis, N., 1963a, On cysteine and cystine peptides. II. *S*-Acylcysteines in peptide synthesis, *J. Am. Chem. Soc.* **85**:1337.

Zervas, L., Photaki, I., Cosmatos, A., and Ghelis, N., 1963b, On cysteine and cystine peptides, *in* "Peptides: Proceedings of the Fifth European Symposium, Oxford, 1962" (G. T. Young, ed.), p. 27, Pergamon Press, Oxford.

Zervas, L., Borovas, D., and Gazis, E., 1963c, New methods in peptide synthesis. I. Tritylsulfenyl and *o*-nitrophenylsulfenyl groups as *N*-protecting groups, *J. Am. Chem. Soc.* **85**:3660.

Zervas, L., Photaki, I., Cosmatos, A., and Borovas, D., 1965, On cysteine and cystine peptides. III. Synthesis of a fragment of insulin containing the intrachain disulfide bridge, *J. Am. Chem. Soc.* **87**:4922.

Zervas, L., Photaki, I., Yovanidis, C., Taylor, J., Phocas, I., and Bardakos, V., 1967, Some problems concerning amino, carboxyl and side-chain protection, *in* "Peptides: Proceedings of the Eighth European Symposium, Noordwijk, The Netherlands, 1966" (H. C. Beyerman *et al.*, ed.), p. 28, North-Holland, Amsterdam.

Zervas, L., Photaki, I., and Phocas, I., 1968, Notiz über *S*-Trityl-L-Cystein, *Chem. Ber.* **101**:3332.

Zioudrou, C., Wilchek, M., and Patchornik, A., 1965, Conversion of the L-serine residue to an L-cysteine residue in peptides, *Biochemistry* **4**:1811.

CHAPTER 5

THE TOSYL AND RELATED
PROTECTING GROUPS IN AMINO ACID
AND PEPTIDE CHEMISTRY

Josef Rudinger

Institut für Molekularbiologie und Biophysik
Eidgenössische Technische Hochschule
Zürich, Switzerland

I. INTRODUCTION

The revolution wrought in peptide chemistry in 1932, when Bergmann and Zervas (1932) introduced their benzyloxycarbonyl protecting group, was so profound that all the earlier procedures for amino-group protection were effectively consigned to the peptide chemist's museum, labeled "of historical interest only." The development of further protecting groups, in particular the range of "urethan" protecting groups descended in direct lineage from benzyloxycarbonyl, or of the sulfenyl protecting groups introduced into peptide chemistry once again by Zervas and his coworkers (Gazis *et al.*, 1963; Zervas *et al.*, 1963; Zervas and Hamalidis, 1965), would seem to have reduced still further our need to use techniques predating the contribution of Bergmann and Zervas (1932).

Yet one group dating back to Emil Fischer (1915) has been repeatedly taken off the shelf, dusted, and used in combination with the most modern techniques: the toluene-*p*-sulfonyl or "tosyl" group. It seems worthwhile to review briefly the history of this group and to examine the reasons why it has been able to occupy at least a modest niche in peptide chemistry to this day against the tough competition of groups such as benzyloxycarbonyl, *t*-butyl-oxycarbonyl, and *o*-nitrophenylsulfenyl.

Among the advantages of the tosyl protecting group are its great stability over a wide range of reaction conditions; the ready availability and negligible cost of the reagent, tosyl chloride; and the generally good crystallinity of simple tosyl derivatives. On the debit side, we must note that the standard conditions for its removal are fairly drastic or insufficiently selective or both, and that the strongly electrophilic character of the tosyl group and in particular the high nucleophilic reactivity it confers on the tosylamino group may favor side-reactions during peptide synthesis (Section VIIA). However, as we shall see, the electrophilic potency of the tosyl group can in turn be put to good use in aminoacid and peptide chemistry (Sections V, VIA).

II. INTRODUCTION OF TOSYL GROUPS

A. Preparation of α-Tosylamino Acids

Little need be said about the methods for introducing tosyl protecting groups into α-amino acids. Tosyl chloride is the universal reagent, and aqueous alkaline conditions are used. Tosylation in the water–ether system has been found to be accelerated by the use of crude tosyl chloride (Rudinger *et al.*, 1959) or by the deliberate addition of toluene-*p*-sulfonic acid (Gut and Poduška, 1971), presumably because the sodium toluene-*p*-sulfonate acts as an emulsifier. Even in water alone the reaction will proceed at a reasonable rate if the reaction mixture is heated above the melting point of tosyl chloride, or if an efficient stirrer (homogenizer) is used (Zaoral and Rudinger, 1959). Special measures may be needed to obtain N^α-tosyl derivatives of amino acids with side-chain functional groups. Amino acid esters can, of course, be tosylated in organic solvents, but under certain conditions side-reactions such as tosylation of hydroxyl groups or dehydration of amides may also take place.

Carboxylate groups can react with tosyl chloride to form mixed anhydrides, and although these would be expected to hydrolyze rapidly in aqueous alkaline solution (in anhydrous media they undergo disproportionation to the symmetrical anhydrides; cf. Brewster and Ciotti, 1955) certain side-reactions observed in the preparation of arenesulfonylamino acids have been traced to their formation. As early as 1922, 1-tosyl-3-benzamido-2-piperidone [2] was identified as a byproduct of the tosylation of N^δ-benzoylornithine [1] (Thomas *et al.*, 1922) (Fig. 1), and later the formation of tosylpyroglutamic acid was found to be responsible for difficulties encountered in the tosylation of glutamic acid—particularly under mild reaction conditions (Rudinger *et al.*, 1959). The reaction has, in fact, been deliberately used to prepare tosyllactams (Rudinger *et al.*, 1959; Gut and Rudinger, 1963). Even where the mixed anhydride formation cannot manifest itself by ring closure, it may obtrude itself by other side-reactions. Thus the formation of dansyl amide as a byproduct during the preparation of dan-

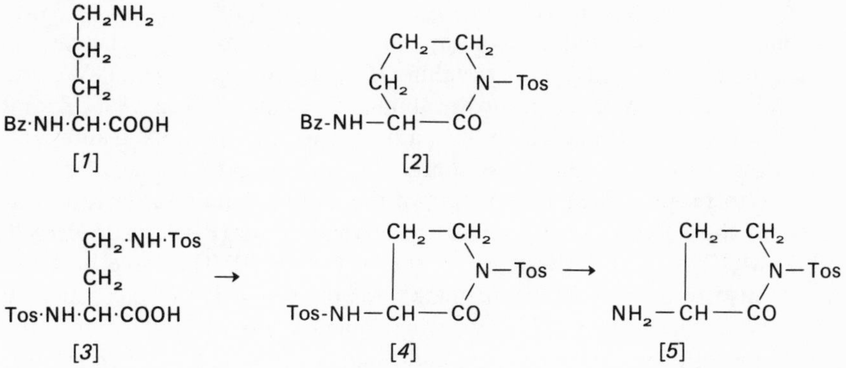

$$
\begin{array}{c}
CH_2NH_2 \\
| \\
CH_2 \\
| \\
CH_2 \\
| \\
Bz\cdot NH\cdot CH\cdot COOH \\
[1]
\end{array}
$$

$$
\begin{array}{c}
CH_2\!-\!CH_2 \\
/ \qquad \backslash \\
CH_2 \qquad N\!-\!Tos \\
\backslash \qquad / \\
Bz\text{-}NH\!-\!CH\!-\!CO \\
[2]
\end{array}
$$

$$
\begin{array}{c}
CH_2\cdot NH\cdot Tos \\
| \\
CH_2 \qquad \rightarrow \\
| \\
Tos\cdot NH\cdot CH\cdot COOH \\
[3]
\end{array}
$$

$$
\begin{array}{c}
CH_2\!-\!CH_2 \\
| \qquad \backslash \\
| \qquad N\!-\!Tos \qquad \rightarrow \\
| \qquad / \\
Tos\text{-}NH\!-\!CH\!-\!CO \\
[4]
\end{array}
$$

$$
\begin{array}{c}
CH_2\!-\!CH_2 \\
| \qquad \backslash \\
| \qquad N\!-\!Tos \\
| \qquad / \\
NH_2\!-\!CH\!-\!CO \\
[5]
\end{array}
$$

Fig. 1

sylamino acids appears to be due to a Beecham-type fragmentation (see Section VIIA) of the mixed sulfonic anhydride (Neadle and Pollitt, 1965).

B. Side-Chain Substitution by Tosyl Groups

The N^{ω}-tosyl derivatives of lysine, ornithine, and α,γ-diaminobutyric acid are most conveniently prepared by tosylation of the amino acid–copper-(II) complexes (Erlanger et al., 1954; Roeske et al., 1956; Christensen and Riggs, 1956; Poduška and Rudinger, 1959), a technique originally developed by Kurtz (1949) for the preparation of N^{ω}-substituted derivatives of the diamino acids, including the benzenesulfonyl derivatives. The same technique can conveniently be used to obtain O-tosyltyrosine (Katsoyannis et al., 1957), which is also accessible through N-acyltyrosine esters (Jackson, 1952) or by selective removal of the N-tosyl group from N,O-ditosyltyrosine with hydrogen iodide (Fischer, 1915) or, presumably, hydrogen bromide.

The access to N^{g}-tosyl derivatives of arginine is through tosylation of N^{α}-benzyloxycarbonylarginine in strongly alkaline solutions ($pH > 12$) (Schwyzer and Li, 1958; Schnabel and Li, 1960; Ramachandran and Li, 1962). Similarly, N^{α}-protected derivatives of histidine are used to introduce the tosyl group into the imidazole ring, but in this case mildly alkaline conditions are required (Sakakibara and Fujii, 1969).

III. GENERAL PROPERTIES OF TOSYLAMINO ACIDS

A. Physical Properties

The tosyl derivatives of the protein-constituent amino acids are all crystalline compounds (see Greenstein and Winitz, 1961a). Some may crystallize as hydrates or, less conventionally, as benzene solvates; thus tosyl-L-proline forms a crystalline monohydrate or hemibenzenate, whereas the unsolvated compound is noncrystalline (Pravda and Rudinger, 1955). The pK_a values of the carboxyl group are about 3.5 in water (Lovén, 1896; Zaoral and Rudinger, 1961) and 5.3 in 80% aqueous methoxyethanol (Zaoral and Rudinger, 1961). The good crystallinity of the acids and their salts led Emil Fischer to propose their use, or that of the naphthalene-β-sulfonyl derivatives, for the isolation and identification of amino acids (Fischer and Bergell, 1902; McChesney and Swan, 1937; see Section VIII). The same property would also seem to make them suitable for the resolution of racemates by way of diastereomeric salts; an early example of such an application is the resolution of tosylalanine (Gibson and Simonsen, 1915), a more recent one the resolution of tosylpyroglutamic acid (Rudinger and Czurbová, 1954).

B. Chemical Reactivity

The outstanding chemical characteristics of arenesulfonamides are their great stability under a variety of conditions and the strong inductive electrophilic effect of the sulfonyl substituent.

Arenesulfonamides are stable to both alkaline and acidic hydrolysis under any but the most vigorous conditions. They easily withstand the conditions commonly used for the saponification of peptide esters with alkali and also resist the action of other nucleophilic reagents such as ammonia and hydrazine. Their stability to acid is such that they not only remain unaffected by the acidic reagents commonly used to remove other protecting groups (hydrogen chloride, hydrogen bromide under moderate conditions, liquid hydrogen fluoride, trifluoroacetic acid) but also survive the total hydrolysis of peptide bonds in proteins or peptides (see Section VIII).

Sulfonamide groups are also stable to catalytic hydrogenation, but they can be cleaved reductively e.g., by sodium in liquid ammonia or by electrolysis. The cleavage with hydrogen bromide, hydrogen iodide, or phosphonium iodide under moderately vigorous conditions also leads to reduction products but probably is not primarily a reduction (see Section IVA).

Because of the electrophilic properties of the sulfonyl groups, primary and secondary sulfonamides are relatively strong acids ($pK_a \sim 11$). The ionized sulfonamide group in turn is an excellent nucleophile and is readily alkylated and acylated. As we shall see, such reactions can be exploited for synthetic purposes (Sections V, VIA), but they can also appear as undesirable side-reactions (Section VIIA). The fragmentation of tosylamino acid chlorides (Section VIIA) is another example of a side-reaction initiated by ionization of the toxylamino group. The difference between a "good," constructive reaction and a "bad" side-reaction is, of course, merely a matter of one's point of view; and several times in the past it has proved possible to exploit originally unexpected and undesired reaction modes of the tosylamino group to good purpose.

IV. REMOVAL OF TOSYL AND RELATED PROTECTING GROUPS

A. Cleavage with Hydrogen Iodide or Hydrogen Bromide

Fischer (1915) demonstrated that arenesulfonamides can be cleaved with hydrogen iodide–phosphonium iodide at 70–100°C with formation of the amines and thiophenols (thionaphthols). Thomas *et al.* (1922) carried out such a cleavage selectively in the presence of an *N*-benzoyl substituent, but it remained for Schönheimer (1926) to demonstrate that these or somewhat milder conditions could be used to deprotect tosylpeptides. Although the procedure did not find favor in its original form, the related cleavage with hydrogen bromide has acquired some limited importance.

Ohle *et al.* (1936) observed that the tosyl group was cleaved from 1,2-isopropylidene-*N*-phenyl-*N*-tosyl-6-amino-6-deoxyglucose by hydrogen bromide under mild conditions, with formation of the amino sugar and of di-*p*-tolyl disulfide. It had also been noted (Snyder and Heckert, 1952) that phenol had a beneficial effect on the cleavage of sulfonamides with hot aqueous hydrobromic acid, evidently by scavenging the bromine formed during reduction of the sulfonic acid moiety. Weisblat *et al.* (1953) developed a procedure for the cleavage of sulfonamides using concentrated hydrogen bromide in acetic acid in the presence of phenol at room temperature during 5 hr to 3 days. We were able to show (Poduška *et al.*, 1955) that these conditions can be used to remove the tosyl group from simple peptides and that shorter reaction times at higher temperatures may also be used. Although several peptides have been prepared by variations of this procedure (Poduška *et al.*, 1955, 1956; Rudinger, 1959*b*; Gut and Rudinger, 1963; Gut *et al.*, 1968), some limitations emerged in the course of this work. Thus it was found that side-chain tosylamino groups, as in N^α,N^δ-ditosylornithylglycine ethyl ester, are considerably more resistant to the action of this reagent than tosyl-amino groups in the α-position (Gut *et al.*, 1968) and that *S*-benzyl groups are attacked under the conditions required to remove *N*-tosyl substituents (Jošt and Rudinger, 1961). The use of hydrogen bromide in acetic acid has therefore largely been confined to the removal of tosyl groups from amino acid derivatives which could not be exposed to more drastic, hydrolytic conditions, as, for instance, in syntheses of glutamine (Rudinger *et al.*, 1959), of α-aminoadipic acid (Rudinger and Farkašová, 1963) and its δ-amide (Gut and Rudinger, 1963), and of numerous derivatives of the

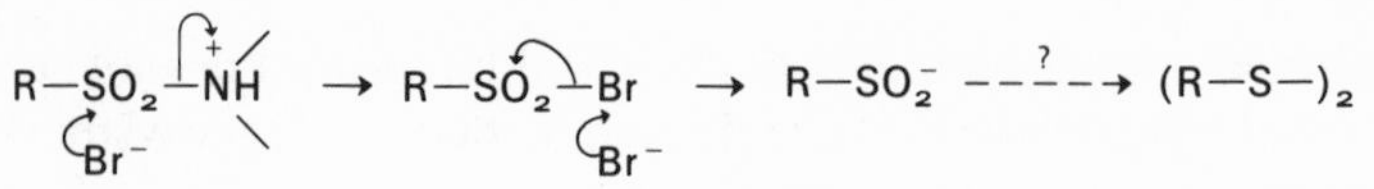

$$R-SO_2-\overset{+}{N}H \quad \rightarrow \quad R-SO_2-Br \quad \rightarrow \quad R-SO_2^- \ ---\overset{?}{-}-\rightarrow \ (R-S-)_2$$
$$\quad Br^- \qquad\qquad Br^-$$

Fig. 2

diamino-monocarboxylic acids (Zaoral and Rudinger, 1957, 1959; Rudinger *et al.*, 1960; Poduška *et al.*, 1965).

Fischer (1915) observed that only the *N*-tosyl group of *N,O*-ditosyltyrosine was attacked by phosphonium iodide and prepared *O*-tosyltyrosine in this way. He also noted that *o*-sulfobenzoic acid imide (saccharin) was— "bemerkenswerterweise"—unattacked by the reagent. We found that other sulfonic-carboxylic acid imides, in particular the tosyllactams derived from glutamic and α,γ-diaminobutyric acids, similarly were stable to hydrogen bromide in acetic acid; moreover, the γ-tosylamino acids actually cyclized to the tosyllactams under the influence of the reagent (Rudinger *et al.*, 1959; Poduška and Rudinger, 1959). These circumstances were exploited to effect the selective removal of the N^α-tosyl group from N^α,N^γ-ditosyl-α,γ-diaminobutyric acid [3], with the formation of the useful synthetic intermediate L-3-amino-1-tosyl-5-pyrrolidone [5] (Fig. 1): After formation of the pyrrolidine ring, the N^γ-tosyl group in [4] is stabilized to the reagent, whereas the N^α-tosyl group is normally removed (Poduška and Rudinger, 1959). A similar reaction sequence was also identified when ditosylornithine was treated with hydrogen bromide in acetic acid, but because of the lower stability of the six-membered ring the reaction mixture was more complex and the procedure was not preparatively useful (Gut *et al.*, 1968).

It seems probable that the cleavage by hydrogen bromide or iodide is initiated by attack of a halide anion at the protonated sulfonamide (Fig. 2) and that the disulfide is formed by reduction of the sulfonyl bromide or iodide initially formed by the excess of hydrogen halide (Poduška and Rudinger, 1959; Rudinger, 1963*b*). Such a mechanism would explain the resistance of sulfonic acid imides and aryl esters (decreased basicity) as well as the ineffectiveness of hydrogen chloride and hydrogen fluoride.

B. Reduction with Metals in Liquid Ammonia

Du Vigneaud and Behrens (1937) found that tosylamine groups can be reductively cleaved with sodium in liquid ammonia, and this has remained the most extensively used method for removing tosyl protecting groups. Initially, separation of the peptide from the sodium ions presented considerable difficulties which had to be solved from case to case by exploiting differential solubilities (du Vigneaud and Behrens, 1937; Bovarnick, 1943; Woolley, 1948; Swan and du Vigneaud, 1954), but with the advent of ion exchange resins more general and convenient procedures became available (Swan and du Vigneaud, 1954; Rudinger, 1954b).

Sodium in liquid ammonia will also cleave benzyloxycarbonyl and substituted benzyloxycarbonyl groups, tosyl esters, benzyl ethers and esters, N^{im}-benzyl groups, and benzyl thioethers, so that the simultaneous deprotection of many functional groups can be accomplished with this reagent. If it is desired to preserve the S-benzyl protection of cysteine while removing tosyl (or benzyloxycarbonyl) groups, the sulfydryl groups formed by complete deprotection can be rebenzylated in situ with benzyl chloride (du Vigneaud et al., 1954; Jošt and Rudinger, 1961). On the other hand, t-butyloxycarbonyl groups are unattacked by sodium in liquid ammonia.

In the course of time, a number of problems were encountered in sodium–ammonia reductions. In particular, removal of tosyl groups was sometimes excessively slow (Meienhofer and Li, 1962) and cleavages of the peptide chain were observed (Hofmann and Yajima, 1961; Guttmann, 1963; Jošt and Rudinger, 1967). The sites of cleavage were identified as X–Pro peptide bonds (Guttmann, 1963; see also Jošt and Rudinger, 1967). Independently, the reductive cleavage of X–Pro bonds by sodium in liquid ammonia had been studied as a possible technique for selective protein cleavage (Wilchek et al., 1965; Benisek and Cole, 1965; Ressler and Kashelikar, 1966). There is some controversy as to whether this reaction is a metal/acid reduction (Wilchek et al., 1965; Ressler and Kashelikar, 1966) or independent of the presence of acids in the ammonia system (Benisek et al., 1967). The studies of Guttmann (1963) show that it is likely that acid is required, and recently Katsoyannis et al. (1971b) have shown that the cleavage can be suppressed by a large excess of sodium amide in the liquid ammonia solution. The dependence of the reaction on the presence and amount of adventitious acids, such as moisture, or acid groupings (carboxyl, hydroxyl, tosylamide) in the peptide itself may explain why some authors have failed to observe such side-reactions even where the structural conditions were met (e.g., Bajusz and Medzihradszky, 1963; Ramachandran et al., 1965; Blake et al., 1970). Unfortunately, the tosylamino group itself can act as an acid for the purposes of this reaction (Guttmann, 1963), so that it constitutes an ever-present danger in the deprotection of tosyl derivatives with sodium in ammonia.

Other side-reactions have occasionally been observed, such as the conversion of arginine to ornithine (Guttmann, 1963) and of (protected) cysteine to alanine (Katsoyannis, 1964, 1966; Zahn et al., 1967), damage to threonine (Brenner, 1959; Zahn et al., 1967), serine (Wünsch, 1959), or tryptophan (Bajusz and Medzihradszky, 1963), fission of the $C_{(5)}$—N bond in a proline derivative (Ramachandran, 1965), $N \rightarrow N'$ acyl migrations (Poduška et al., 1965), and racemization (Hope and Humphries, 1964), but some of these occur only under special structural or experimental conditions.

To reduce the incidence of side-reactions, efforts have been made to control the dosage of sodium more accurately. Du Vigneaud *et al.* (1954) introduced the "sodium stick" technique, in which the metal is gradually dissolved out of the end of a narrow glass tube that can be withdrawn from the reaction mixture when the blue coloration which indicates the end-point has become stable. Nesvadba (1962) proposed the use of an extraction apparatus in which the sodium is carried into the reaction mixture in solution in the refluxing ammonia (Nesvadba and Roth, 1967). Finally, Katsoyannis *et al.* (1971*a*) use what is essentially a burette to add a solution of sodium in liquid ammonia to the reaction mixture in a controlled manner.

The actual course of the reaction remained unclear for some time. It was generally thought that the sulfur-containing product of the reduction was the thiophenol (e.g., Birch and Smith, 1958), and even the stoichiometry appeared uncertain (see Bajusz and Medzihradszky, 1963). In the course of our own work, we identified sulfite and sulfate ions, the latter presumably arising from the former by air oxidation (Rudinger, 1954*b*), and toluene-*p*-sulfinic acid (Jošt and Rudinger, 1961) as products of the sodium–ammonia reduction of tosylamino acids and tosylpeptides, and these findings have been confirmed by Kovacs and Ghatak (1963, 1966).

Using the extraction procedure of Nesvadba (1962), we made a rather more detailed study of the stoichiometry and reaction products in the reduction of tosylamino acids and some model compounds (Zimmermannová *et al.*, 1966; Rudinger and Maassen van den Brink-Zimmermannová, 1973). We were able to show that under these conditions the reduction of tosylglycine requires 2 g-atoms of sodium per mole and that toluenesulfinic acid can be isolated in high yield (see also Hope and Horncastle, 1966). The reduction of the tosyl derivatives of other monoamino-monocarboxylic acids and of glycylglycine takes a similar course. On the other hand, with the amide or methylamide of toluene-*p*-sulfonic acid a clear end-point is reached after only 1 g-atom of sodium has been added, and some starting material is recovered together with a low yield of sulfinic acid. However, reduction of *N*-tosylpiperidine once more consumes 2 g-atoms of sodium per mole and gives sulfinic acid in good yield.

These results can be rationalized by assuming that, in tosylglycine and similar compounds, the two electrons taken up in the reduction are accommodated in the toluenesulfinate and carboxylate anions (Fig. 3A). When toluenesulfonamide is reduced, the second electron "comes to rest" by ionizing the most acidic species present—the sulfonamide itself; and since the sulfonamide anion is evidently resistant to reduction, the reaction comes to a stop when all the remaining sulfonamide is ionized (Fig. 3B). Tosylpiperidine in turn has no ionizable hydrogen and reduction proceeds to completion, presumably with the formation of 1 mole of amide anion (Fig. 3C). In agreement with

(A) $R-SO_2-NH-CH_2-COOH + 2e \rightarrow R-SO_2^- + NH_2-CH_2-COO^-$

(B) $2\ R-SO_2-NH_2 + 2e \rightarrow R-SO_2^- + R-SO_2-NH^- + NH_3$

(C) $R-SO_2-N(CH_2)_5 + NH_3 + 2e \rightarrow R-SO_2^- + NH_2^- + NH(CH_2)_5$

Fig. 3

this explanation, the reduction of toluene-p-sulfonamide can be brought to completion by adding 1 mole of an acid of similar or greater strength in the ammonia system—e.g., ammonium chloride, ethanol, acetamide, or urea—either at the beginning of the reaction or after the first end-point has been reached. There is no doubt that cases of incomplete or slow reduction of tosyl groups can in general be explained by an excess of reducible over acidic groupings in the particular protected derivatives—the peptide of Meienhofer and Li (1962) contained four tosyl and one benzyloxycarbonyl groups as against a single carboxyl—and that these difficulties can be overcome by adding the appropriate amounts of acids in the ammonia system, most conveniently perhaps urea. It should be noted that peptide groups cannot function as acids for this purpose.

Opposite deviations from the two-electron stoichiometry are found when the compounds to be reduced contain additional acidic groups (as in tosylserine or tosylglutamic acid) or when extraneous acid is present. It is under these conditions that thiol and sulfite (sulfate) are found among the reaction products as well as sulfinic acid. The thiol, sulfite, and sulfinic acid are not the products of parallel reaction paths, as has been suggested by Kovacs and Ghatak (1966), but the first two are formed from the last. We have found that whereas sodium toluene-p-sulfinate dissolved in liquid ammonia consumed no sodium, the free acid reacted with sodium to give a mixture which, after the end-point had been reached, contained sulfite and thiol as well as unreacted sulfinate. Now thiol is a plausible reduction product of sulfinic acid, but the oxidation state of sulfite and sulfinate is the same so that the former cannot arise from the latter by reduction. On the other hand, it has long been known (Kraus and White, 1923) that sodium benzenesulfonate is cleaved by sodium in liquid ammonia at the C—S bond, with the formation of sulfite. It is therefore reasonable to assume that the sulfite arises from toluene-p-sulfonate formed, in turn, from the sulfinic acid perhaps by a disproportionation reaction. It must, however, be admitted that so far we have been unable to demonstrate such a disproportionation of toluene-p-sulfinic acid in solutions resembling the reaction mixtures after sodium–ammonia reduction in composition.

Since the more complex reaction course described above is dependent on the presence of acids commensurate in strength with toluenesulfinic acid, it should be possible to prevent it by neutralization. Indeed, we have found that the reduction of monosodium tosylglutamate takes essentially the same course as that of tosyglycine.

These results show that some measure of control over the course of the reduction, as also over the incidence of the (acid- or base-dependent) side-reactions, can be obtained by adjusting the acid–base balance of the reaction mixture, although the methods for doing this are admittedly still imperfect.

The use of lithium, calcium, or barium in place of sodium has been briefly examined (Rudinger, 1954*b*; Zimmermannová *et al.*, 1966; Rudinger and Maassen van den Brink-Zimmermannová, 1973). Lithium gives the same stoichiometry and reaction products as sodium, but with calcium the reduction always proceeds beyond the sulfinate stage, and indeed calcium toluene-*p*-sulfinate itself in ammonia solution consumes sodium in much the same way as the free sulfinic acid. The use of calcium or barium may offer advantages where nearly neutral conditions are to be maintained during working-up, without the introduction of awkward anions. Thus ammonium fluoride may be added to the liquid ammonia solution to give the insoluble calcium fluoride (Zimmermannová *et al.*, 1966; Poduška *et al.*, 1965) or the calcium may be largely removed from aqueous solution with carbon dioxide (Poduška *et al.*, 1969).

In the author's opinion, these findings and their preparative exploitation should help to dispel at least some of the skepticism with which the sodium–ammonia reduction, and consequently the use of the tosyl protecting group, is at present regarded by most peptide chemists.

C. Electrochemical Cleavage

Horner and Neumann (1965) were able to show that arenesulfonamides are cleaved to the amine and sulfinic acid by electrolytic reduction in alcoholic solutions containing tetramethylammonium ions and suggested this as a procedure for removing tosyl protecting groups from peptides. Benzamides and substituted benzamides are reductively cleaved under the same conditions.

This attractive method has so far found little use in preparative peptide chemistry. Its scope may be increased by the observation (E. Kasafírek and J. Rudinger, unpublished; Mairanovskiǐ and Loginova, 1971) that dimethylformamide, alone or in mixture with alcohols, can serve as the solvent for this electrolytic reduction. Pless and Guttmann (1967) have noted that the electrolytic removal of the tosyl group from the arginine side-chain is accompanied by ornithine formation, which they ascribe to the increasing alkalinity of the solution. However, this particular problem has now been solved by the use of liquid hydrogen fluoride, and one would hope for further applications of the electrolytic method to the deprotection of α- and ω-amino groups.

V. TOSYLAMINO GROUPS IN CARBOXYL PROTECTION AND CARBOXYL ACTIVATION

A. N-Tosyllactams

The properties of N-tosylpyroglutamic acid [7] first drew attention to the fact that N-substitution of an amide (lactam) nitrogen by a tosyl group increases the electrophilic reactivity of the carbonyl to such an extent that such derivatives can be regarded as "carboxyl-activated." The preparation and reactions of this versatile intermediate are summarized in Fig. 4 (cf. Rudinger, 1962). Not only is the CO–N bond of the tosyllactam grouping readily hydrolyzed under alkaline conditions (Harington and Moggridge, 1940), but it can also be aminolysed to give γ-amides of tosylglutamic acid [8]. In this way, tosylglutamine, tosylglutamic acid γ-anilide and γ-hydrazide, and a number of similar derivatives have been obtained (Swan and du Vigneaud, 1954; Rudinger, 1954a; Rudinger and Czurbová, 1954; cf. Rudinger, 1962). Reaction with amino acid esters afforded tosyl-γ-glutamyl peptides such as tosyl-γ-glutamylglycine (Rudinger, 1954b; Rudinger and Czurbová, 1954), although rather more vigorous conditions are required for peptide bond formation, particularly when amino esters with bulky side-chains are used (Rudinger, 1954b; Clayton et al., 1956; Rudinger, 1962).

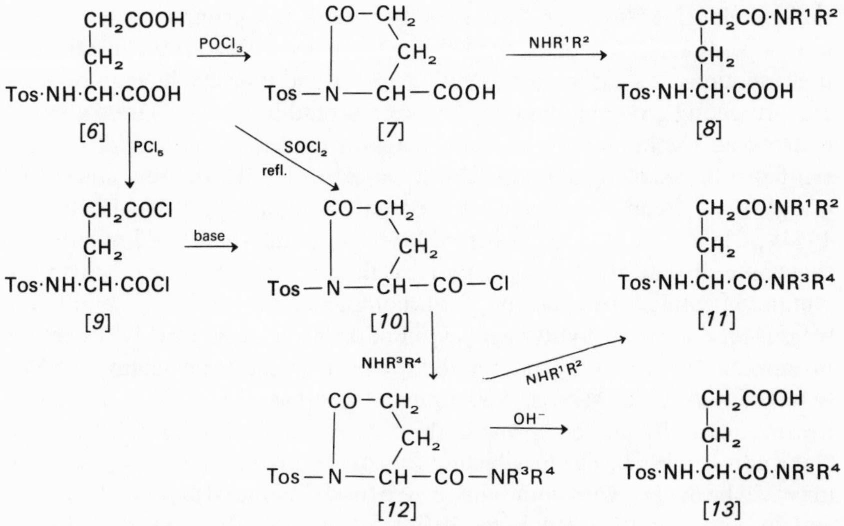

Fig. 4

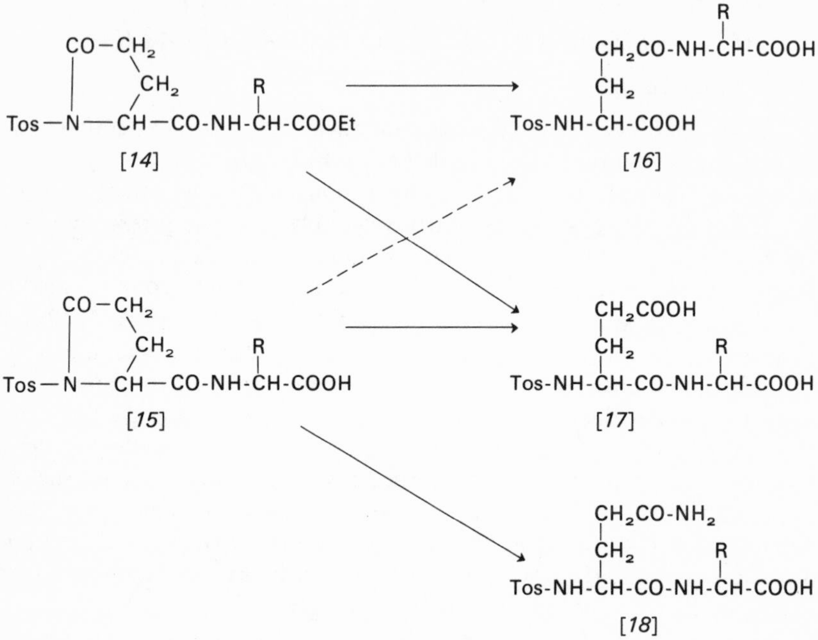

$$
\begin{array}{ll}
\underset{\displaystyle \underset{\text{Tos}-\text{N}-\text{CH}-\text{CO-NH-CH-COOEt}}{}}{\overset{\text{CO}-\text{CH}_2}{}} & \xrightarrow{\hspace{2cm}}
\end{array}
$$

CO−CH₂
     \ CH₂   R
      /
Tos−N——CH−CO-NH-CH-COOEt
          [14]

                    R
CH₂CO-NH-CH-COOH
CH₂
Tos-NH-CH-COOH
          [16]

CO−CH₂
     \ CH₂   R
      /
Tos−N——CH−CO-NH-CH-COOH
          [15]

CH₂COOH
CH₂   R
Tos-NH-CH-CO-NH-CH-COOH
          [17]

CH₂CO-NH₂
CH₂   R
Tos-NH-CH-CO-NH-CH-COOH
          [18]

Fig. 5

On the other hand, when the free (α) carboxyl group of tosylpyro-glutamic acid is converted into more reactive "activated" forms, such as the acid chloride [10], the reactivity of the tosyllactam grouping is sufficiently low to permit selective reaction at this activated α-carboxyl; the tosyllactam thus functions as a carboxyl-protecting group, and it meets the requirements of a protecting group in that the γ-carboxyl is readily set free by mild alkaline hydrolysis. The first example of this reaction sequence was provided by the synthesis of tosylisoglutamine [13, $R^3 = R^4 = H$] (Harington and Mog-gridge, 1940). It has also been used to prepare α-glutamyl peptides (Rudinger, 1954b; Clayton et al., 1956; Morris, 1960), although there is a limitation to this route in that alkaline hydrolysis of the tosyllactam ring in tosylpyro-glutamyl peptide esters [14] may be accompanied by $\alpha \rightarrow \gamma$ transamidation (Fig. 5), giving some γ-glutamyl [16] in addition to α-gutamyl [17] peptide, presumably by attack of the peptide nitrogen at the tosyllactam carbonyl (Clayton et al., 1956; Morris, 1960; Gut and Rudinger, 1963). The correspond-ing free acids [15] are less prone to this side-reaction (Rudinger, 1954b; Gut and Rudinger, 1963). The tosyllactam ring of tosylpyroglutamyl peptides [15] may also be opened with ammonia to give tosylglutaminyl peptides [8] (Swan and du Vigneaud, 1954; Rudinger, 1954b; Swan, 1956; Rudinger et al., 1956a; Kaneko et al., 1957; Chillemi et al., 1957; Rudinger and Pravda, 1958; see section VIIB).

Tosylpyroglutamic acid [7], the key intermediate in these reactions, is formed from tosyglutamic acid [6] under a variety of dehydrating conditions (Harington and Moggridge, 1940; Rudinger, 1954a; Rudinger et al., 1959), most conveniently perhaps by treatment with phosphorus trichloride (Rudinger et al., 1959). Tosylpyroglutamyl chloride [10] is best obtained by refluxing tosylglutamic acid with thionyl chloride (Rudinger, 1954a); phosphorus pentachloride in the cold (Swan and du Vigneaud, 1954) affords the dichloride of tosyglutamic acid [9] (Stedman, 1957), which, however, readily cyclizes to tosylpyroglutamyl chloride [10]. The reaction products obtained from the two chlorides may thus be the same.

Cyclization also occurs readily when the carboxyl group of N^{γ}-tosyl-α,γ-diaminobutyric acid derivatives is activated for peptide synthesis; the resulting derivatives of 3-amino-1-tosyl-2-pyrrolidone [5] can once more be exploited as carboxyl-activated intermediates for the synthesis of peptides of α,γ-diaminobutyric acid (Poduška and Rudinger, 1957, 1959, 1966). The 3-amino derivative [5] itself can be acylated at the amino group, with the tosyllactam grouping in turn playing the role of a carboxyl-protecting group.

Analogous intermediates with a six-membered tosyllactam ring can be formed, although less readily, from tosyl-α-aminoadipic acid (Gut and Rudinger, 1963) or suitable derivatives of ornithine (Gut et al., 1968) and used for the synthesis of α-aminoadipyl and ornithyl peptides.

B. N-Tosylcarboxamides

An obvious extrapolation from the properties of the tosyllactams suggested that N-tosylcarboxamides in general should behave as "activated" carboxyl derivatives (Rudinger, 1959b). Such derivatives are accessible (Fig. 6) by acylation of toluene-p-sulfonamide with protected amino acids (Wieland and Hennig, 1960) or from the free carboxylic acids by reaction with toluene-p-sulfonyl isocyanate (K. Poduška and J. Rudinger, unpublished results; see Rudinger, 1963b). As it turned out, imide derivatives of the type [19], still containing an NH group, undergo aminolysis only with difficulty, presumably because as rather strong acids they are ionized and the carbonyl group is deactivated by the neighboring negative charge (Rudinger, 1959a; Wieland and Hennig, 1960). However, when the imide group is methylated, e.g., with diazomethane (Wieland and Hennig, 1960), the

$$R\text{-COX} + NH_2\text{-Tos} \longrightarrow R\text{-CO-NH-Tos} \quad [19]$$

$$R\text{-COOH} + O{:}C{:}N\text{-Tos} \qquad R\text{-CO-}\underset{\underset{CH_3}{|}}{N}\text{-Tos} \quad [20]$$

Fig. 6

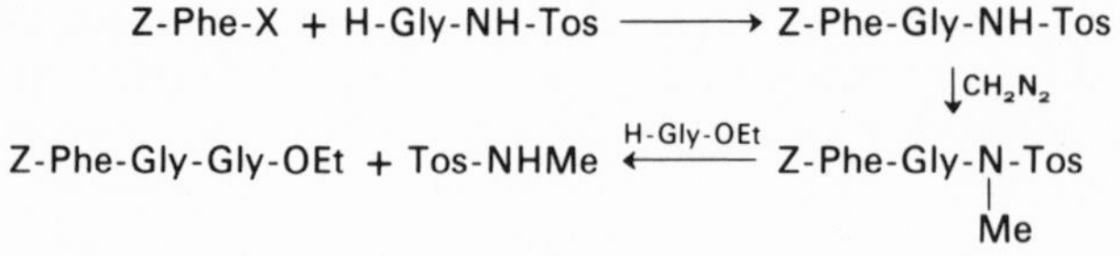

Z-Phe-X + H-Gly-NH-Tos ⟶ Z-Phe-Gly-NH-Tos

$\downarrow$CH₂N₂

Z-Phe-Gly-Gly-OEt + Tos-NHMe $\xleftarrow{\text{H-Gly-OEt}}$ Z-Phe-Gly-N-Tos
                                                                            |
                                                                            Me

Fig. 7

resulting derivatives [20] show about the same reactivity as do tosyllactams. Because of this dual reactivity, the sulfonamide grouping can serve, in principle, both for carboxyl protection, permitting a peptide chain to be built up on a carboxyl-terminal N-(aminoacyl)sulfonamide, and—after methylation—for carboxyl activation, allowing the peptide chain to be extended at the carboxyl end (Fig. 7; K. Poduška and J. Rudinger, unpublished results, see Rudinger, 1963b). Although this "safety-catch" method of carboxyl activation has found no application in classical peptide synthesis, it has recently been exploited in a variant of solid-phase synthesis (Kenner et al., 1971). In this procedure, the carboxyl-terminal amino acid is "anchored" to the sulfonamide carrier resin by acylation of the sulfonamide group, the peptide chain is extended stepwise in the usual way, and, finally, after "activation" of the acyl bond to the resin by methylation of the imide nitrogen, the peptide chain is detached from the resin by mild alkaline hydrolysis, aminolysis, or hydrazinolysis.

The same principle underlies the reactivity of N-acyl derivatives of o-sulfobenzoic acid imide ("mixed anhydrides of saccharin"; Micheel and Lorenz, 1963). Aminolysis of this N,N-diacylsulfonamide occurs preferentially at the exocyclic carbonyl group, presumably for steric reasons.

C. 3-Tosyl-5-oxazolidones

Even when the electron pull of the tosylamino group is transmitted to the carboxyl through an interpolated methylene group, the activating effect is still felt. Micheel and Thomas (1957) showed that tosylamino acids react with formaldehyde and acetic anhydride to give the appropriate 4-substituted 3-tosyl-5-oxazolidones [21] (Fig. 8), which can be regarded as "activated methyl esters" in the same sense as, e.g., cyanomethyl esters. The derivatives

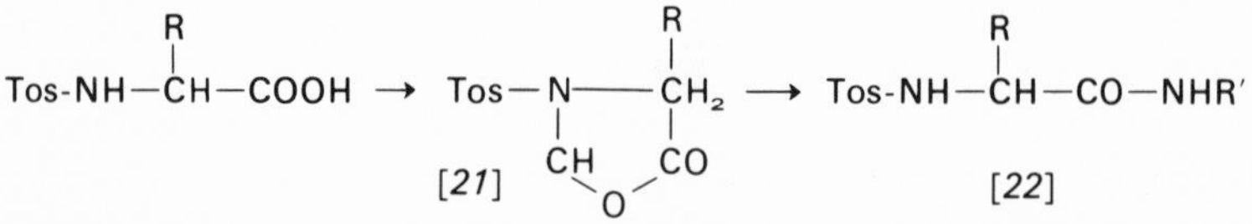

$$
\begin{array}{ccc}
\text{R} & \text{R} & \text{R} \\
| & | & | \\
\text{Tos-NH-CH-COOH} \rightarrow & \text{Tos-N}\underline{\quad}\text{CH}_2 \rightarrow & \text{Tos-NH-CH-CO-NHR}' \\
& | \quad\quad | & \\
[21] & \text{CH} \quad \text{CO} & [22] \\
& \diagdown_\text{O}\diagup &
\end{array}
$$

Fig. 8

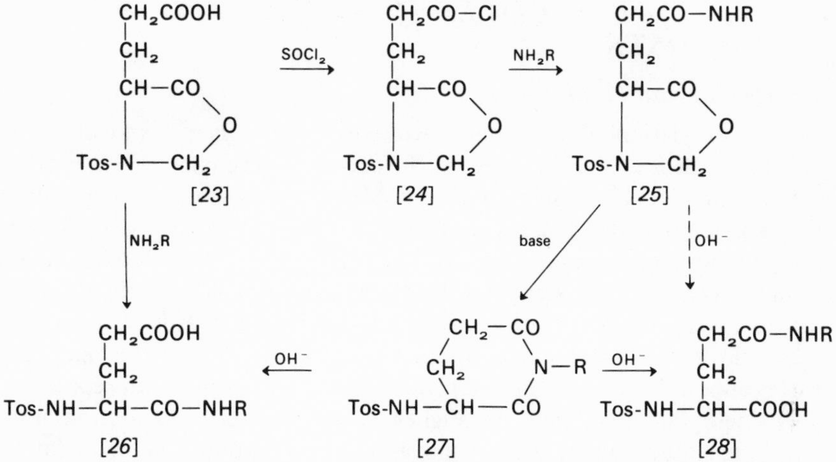

Fig. 9

react with amines, including aminoacid esters, with formation of the amides (peptides) [22], albeit under rather vigorous conditions.

In tosylglutamic acid, steric relations are such that only the α-carboxyl group reacts in this way (Micheel and Haneke, 1959), and the product [23] has been used to prepare α-glutamyl peptides [26] (Fig. 9). Like the tosyllactam, the tosyloxazolidone grouping is only moderately reactive and can also be exploited as a protecting group for the α-carboxyl and tosylamino groups during more vigorous activation and reaction of the γ-carboxyl group (Rudinger and Farkašová, 1963; Gut and Rudinger, 1963). The tosyloxazolidones derived from glutamic and α-aminoadipic acids and their ω-acid chlorides thus form a remarkable counterpart to the tosyllactams and their α-carboxyl-activated derivatives, and can similarly serve for the preparation of both α- and ω-derivatives of the dicarboxylic acids.

This circumstance has been exploited for the chain extension, by a series of Arndt–Eistert syntheses, of L-glutamic to L-α-aminoadipic, L-α-amino-pimelic, and L-α-aminosuberic acids (Rudinger and Farkašová, 1963; Farkašová and Rudinger, 1965; cf. Section VIB, Fig. 14).

The use of this route to γ-peptides of glutamic acid [28] was complicated by a tendency to intramolecular attack of the γ-amide at the oxazolidone grouping of [25], with the formation of cyclic imides [27] (Rudinger and Farkašová, 1963). For steric reasons, this reaction is less obtrusive in the corresponding derivatives of α-aminoadipic acid, and δ-peptides of α-aminoadipic acid have accordingly been synthesized by this route (Gut and Rudinger, 1963; D. Morris, V. Gut, and J. Rudinger, unpublished results; cf. Rudinger, 1963a).

VI. TOSYL DERIVATIVES IN AMINO ACID CHEMISTRY

A. Alkylamino Acids

The alkylation of tosylamino acids, with subsequent removal of the tosyl group, is a classical procedure (Fischer and Bergmann, 1913; Fischer and Lipschitz, 1915; Fischer and von Mechel, 1916) for the preparation of N-alkylamino acids, particularly the N-methyl derivatives of the optically active amino acids (cf. Greenstein and Winitz, 1961b). Methyl iodide or dimethyl sulfate has served as the alkylating agent in aqueous solution, and the tosyl group has been cleaved off either by vigorous hydrolysis with hydrochloric acid or by one of the standard procedures for removing tosyl groups in peptide synthesis (Section IV); in fact, it was in the preparation of N^{α}-methylhistidine that du Vigneaud and Behrens (1937) first used sodium in liquid ammonia for removal of a tosyl group. The methylation procedure has been accused of causing racemization (Zehnder, 1951; Quitt et al., 1963), but the balance of the evidence seems to indicate that, provided unnecessarily vigorous conditions are avoided, the products are optically pure (Greenstein and Winitz, 1961b; see, e.g., Fischer, 1915; Cook et al., 1949; Izumiya, 1952). The intermediate N-tosyl-N-methylamino acids may themselves be used directly in peptide synthesis, as illustrated in the preparation of [1-N-methylhemicystine]oxytocin (Jošt et al., 1961, 1963b). The same procedure has also been used to obtain the N^{ω}-monomethyl derivatives of DL-ornithine (Thomas et al., 1922) and L-lysine (Benoiton, 1964).

From the simpler amino acids, other N-alkyl derivatives have also been prepared through the tosyl or other arenesulfonyl derivatives (e.g., Fischer and von Mechel, 1916; Cocker, 1937). Intramolecular alkylation of tosylamino groups is the basis for the preparation of tosylproline from α-bromo-δ-tosylamino- and δ-bromo-α-tosylaminovaleric acids or their amides (Izumiya, 1953; Pravda and Rudinger, 1955).

B. Side-Chain Modification

Some of the natural amino acids are counted among the most readily available optically active materials and are therefore attractive starting points for the synthesis of other compounds. In particular, operations on suitable side-chain functional groups with preservation of the α-amino carboxylic acid grouping (and its configuration) can serve for the preparation of chiral rare or "unnatural" α-amino acids from readily available ones. For these purposes, the tosyl group has special attraction as the protecting group for the α-amino function because it is robust enough to survive even the fairly vigorous conditions sometimes involved in the side-chain chemistry. In addition, the special properties of the tosylamino group may sometimes be exploited for carboxyl protection or activation (Section V).

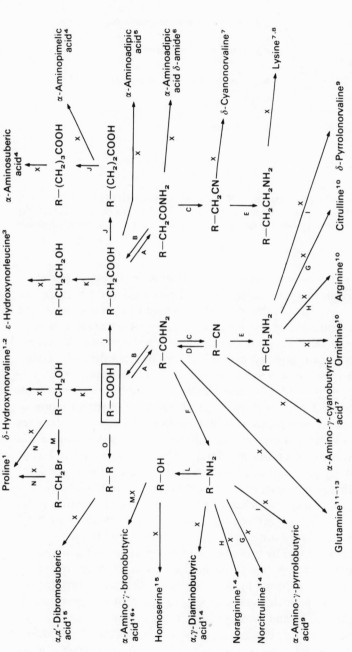

Fig. 10. Syntheses from tosylglutamic acid. $R = Tos \cdot NH \cdot CH(COOH) \cdot CH_2 \cdot CH_2—$ or functional derivatives. [1]Pravda and Rudinger (1955). [2]Goodman and Felix (1964). [3]M. Mühlemann and J. Rudinger (unpublished results). [4]Farkašová and Rudinger (1965). [5]Rudinger and Farkašová (1963). [6]Gut and Rudinger (1963). [7]E. Kasafírek, M. Havránek, and J. Rudinger (unpublished results). [8]K. Eisler and J. Rudinger (unpublished results). [9]Poduška et al. (1969). [10]Zaoral and Rudinger (1957, 1959). [11]Rudinger (1954a,b). [12]Swan and du Vigneaud (1954). [13]Rudinger et al. (1959). [14]Rudinger et al. (1960). [15]O. Keller and J. Rudinger (unpublished results). [16]Jošt and Rudinger (1967).

*This bromo acid has been further used for the synthesis of cystathionine (Jošt and Rudinger, 1967) and of selenomethionine and selenoethionine (Jakubke et al., 1968).

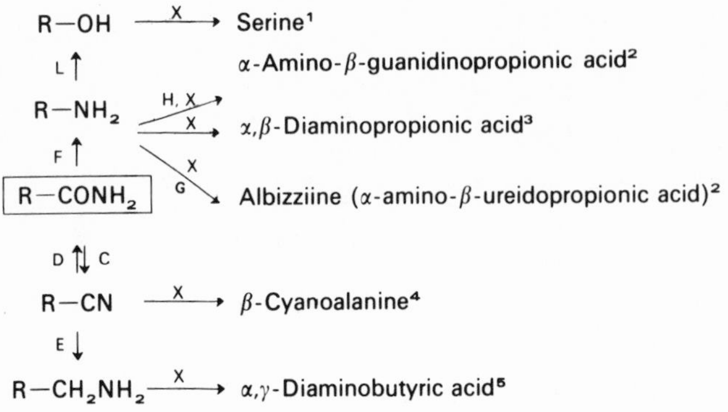

R—OH $\xrightarrow{\quad X \quad}$ Serine[1]

L ↑ $\qquad$ α-Amino-β-guanidinopropionic acid[2]

R—NH$_2$ $\xrightarrow{\quad H, X \quad}$

$\xrightarrow{\quad X \quad}$ α,β-Diaminopropionic acid[3]

F ↑ $\searrow^X$

R—CONH$_2$ $\xrightarrow{\quad G \quad}$ Albizziine (α-amino-β-ureidopropionic acid)[2]

D ↿⇂ C

R—CN $\xrightarrow{\quad X \quad}$ β-Cyanoalanine[4]

E ↓

R—CH$_2$NH$_2$ $\xrightarrow{\quad X \quad}$ α,γ-Diaminobutyric acid[5]

Fig. 11. Syntheses from tosylasparagine. R = Tos·NH·CH(COOH)·CH$_2$— or functional derivatives. [1]Miyoshi *et al.* (1969). [2]Rudinger *et al.* (1960). [3]K. Poduška (unpublished results). [4]E. Kasafírek, M. Havránek, and J. Rudinger (unpublished results). [5]Zaoral and Rudinger (1957, 1959).

Tosyl-L-glutamic acid has proved a most versatile starting material for this approach. Figure 10 summarizes the syntheses which have been carried out, leading to a number of other natural or "unnatural" amino acids. The more restricted field covered by working from tosyl-L-asparagine is summarized in Fig. 11. The reactions used in these transformations include A, conversion of side-chain carboxyl to amide groups by aminolysis of tosyllactams, and B, the reverse selective acid hydrolysis of carboxamide to carboxyl groups; C, dehydration of amides to nitriles, and D, the reconversion of nitriles to amide groups with hydrogen bromide; E, hydrogenation of nitrile to aminomethylene groups; F, Hoffmann degradation of carboxamides to amines; G, carbamylation, and H, amidination of side-chain amino groups; I, conversion of amino to pyrrole substituents; J, Arndt–Eistert syntheses involving preparation of acid chlorides, conversion to diazoketones, and Wolff rearrangement to the homologous acid amides followed by hydrolysis to the acids; K, selective reduction of tosyllactam carbonyl groups in the presence of carboxamide, lithium carboxylate, or *t*-butyl ester groups with lithium borohydride; L, deamination with nitrous acid; M, replacement of hydroxyl by bromine; N, intramolecular alkylation of the tosylamino group; and O, Kolbe electrolytic synthesis. In both schemes, x denotes the removal of the tosyl together with any other protecting groups.

As specific examples, the synthesis of proline from glutamic acid is shown in Fig. 12 (Pravda and Rudinger, 1955); the synthesis of the phenylalanine analogue, β-pyrroloalanine, from asparagine in Fig. 13 (Poduška *et al.*, 1969); the preparation of L-α-aminoadipic from glutamic acid in Fig. 14 (Rudinger and Farkašová, 1963); and the preparation of ornithine

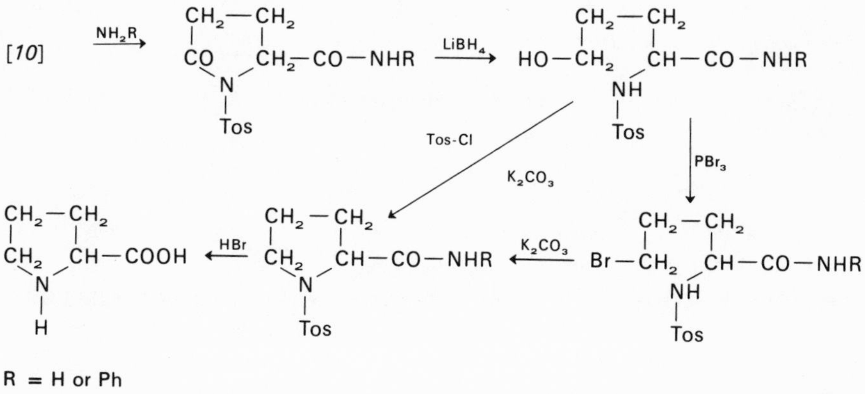

[10]

R = H or Ph

Fig. 12

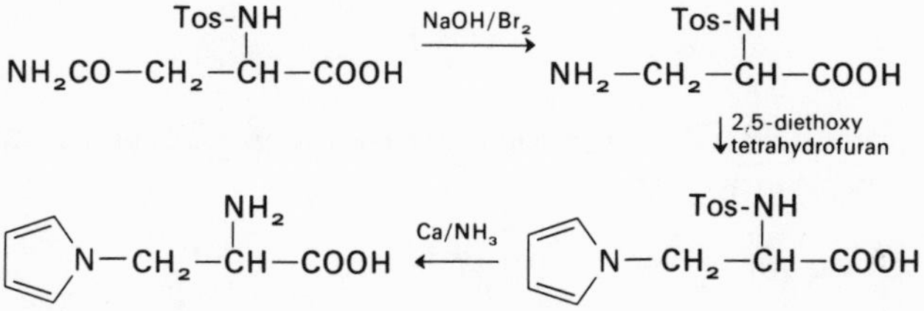

Fig. 13

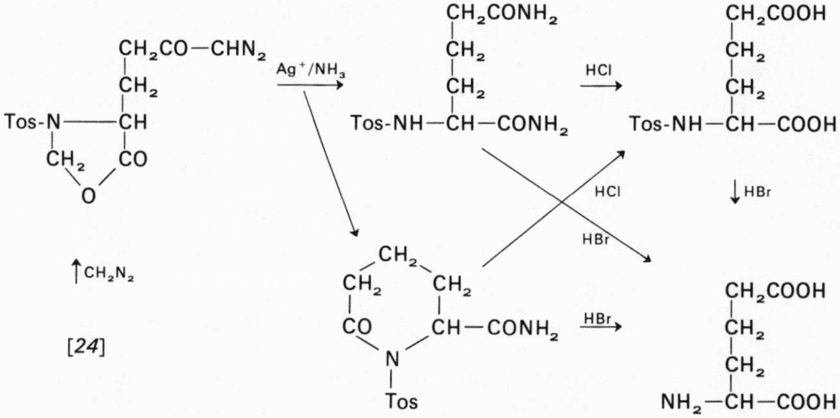

[24]

Fig. 14

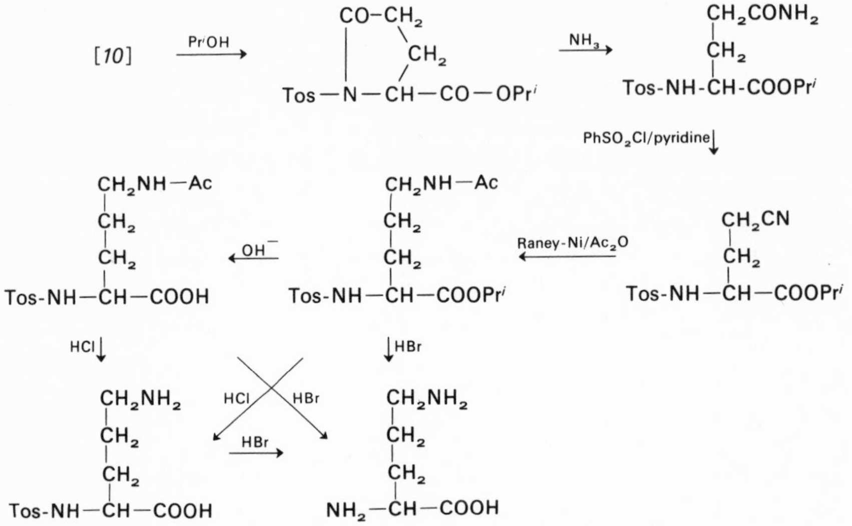

[10] $\xrightarrow{\text{Pr}'\text{OH}}$

$$
\begin{array}{c}
\text{CO-CH}_2 \\
\qquad\quad \diagdown\text{CH}_2 \\
\qquad\quad \diagup \\
\text{Tos}-\text{N}-\text{CH}-\text{CO}-\text{OPr}^i
\end{array}
$$

$\xrightarrow{\text{NH}_3}$

$$
\begin{array}{c}
\text{CH}_2\text{CONH}_2 \\
\text{CH}_2 \\
\text{Tos-NH-CH-COOPr}^i
\end{array}
$$

$\text{PhSO}_2\text{Cl/pyridine}\downarrow$

$$
\begin{array}{c}
\text{CH}_2\text{NH}-\text{Ac} \\
\text{CH}_2 \\
\text{CH}_2 \\
\text{Tos-NH}-\text{CH}-\text{COOH}
\end{array}
$$

$\xleftarrow{\text{OH}^-}$

$$
\begin{array}{c}
\text{CH}_2\text{NH}-\text{Ac} \\
\text{CH}_2 \\
\text{CH}_2 \\
\text{Tos-NH}-\text{CH}-\text{COOPr}^i
\end{array}
$$

$\xleftarrow{\text{Raney-Ni/Ac}_2\text{O}}$

$$
\begin{array}{c}
\text{CH}_2\text{CN} \\
\text{CH}_2 \\
\text{Tos-NH}-\text{CH}-\text{COOPr}^i
\end{array}
$$

$\text{HCl}\downarrow$

$$
\begin{array}{c}
\text{CH}_2\text{NH}_2 \\
\text{CH}_2 \\
\text{CH}_2 \\
\text{Tos-NH}-\text{CH}-\text{COOH}
\end{array}
$$

$\text{HCl} \diagup\!\!\!\!\diagdown \text{HBr}$ $\xrightarrow{\text{HBr}}$

$\downarrow\text{HBr}$

$$
\begin{array}{c}
\text{CH}_2\text{NH}_2 \\
\text{CH}_2 \\
\text{CH}_2 \\
\text{NH}_2-\text{CH}-\text{COOH}
\end{array}
$$

Fig. 15

from glutamic acid (which has been carried out on a kilogram scale; Gut and Poduška, 1971) in Fig. 15.

C. Modification of α-Carboxyl Groups

Tosyl derivatives may also be of advantage when modification of the α-carboxyl group is contemplated. Most such reactions will lead outside the amino acid field which is our present concern. However, we may note the preparation of β-aminoadipic acid starting from tosylglutamic acid (Rudinger and Farkašová, 1963) and of the homologous γ- and δ-amino acids from tosylleucine (Chimiak, 1969) (Fig. 16) as well as the synthesis of tosylamino chloromethyl ketones [30] (Schoellmann and Shaw, 1963; Shaw et al., 1965; Husain and Lowe, 1965), which, being analogues of the amino acid

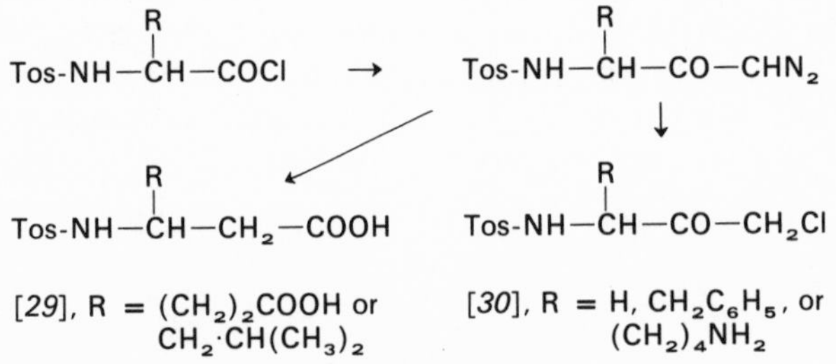

$$
\begin{array}{c}
\text{R} \\
\text{Tos-NH}-\text{CH}-\text{COCl}
\end{array}
\quad \rightarrow \quad
\begin{array}{c}
\text{R} \\
\text{Tos-NH}-\text{CH}-\text{CO}-\text{CHN}_2
\end{array}
$$

$\downarrow$

$$
\begin{array}{c}
\text{R} \\
\text{Tos-NH}-\text{CH}-\text{CH}_2-\text{COOH}
\end{array}
\qquad\qquad
\begin{array}{c}
\text{R} \\
\text{Tos-NH}-\text{CH}-\text{CO}-\text{CH}_2\text{Cl}
\end{array}
$$

[29], R = (CH$_2$)$_2$COOH or　　　　[30], R = H, CH$_2$C$_6$H$_5$, or
　　　　CH$_2$·CH(CH$_3$)$_2$　　　　　　　　　　(CH$_2$)$_4$NH$_2$

Fig. 16

ester substrates, act as irreversible inhibitors of chymotrypsin, trypsin, and papain. Both groups of compounds are obtained from the crystalline diazoketones, which are readily accessible from the tosylamino acid chlorides (Fig. 16). Methyl ketones may be prepared by reduction from the haloketones (Harington and Moggridge, 1940). α-Tosylamino-aldehydes, which can be converted to α-amino-aldehydes, are accessible by reduction of 1-(tosylaminoacyl)-3,5-dimethylpyrazoles with lithium aluminum hydride (Ried and Pfaender, 1961).

VII. TOSYL PROTECTING GROUPS IN PEPTIDE SYNTHESIS

A. Coupling Reactions with Tosylamino Acids

The first peptide coupling reactions with arenesulfonyl-protected amino acids were carried out by Fischer (1903) and Fischer and Bergell (1903) using naphthalene-β-sulfonylamino acid chlorides; the products served as reference compounds for the identification of peptides and were also examined as enzyme substrates. However, it was not until Schönheimer (1926) applied the phosphonium iodide method to the removal of tosyl groups from dipeptides (prepared by the chloride or azide coupling) that arenesulfonyl substituents became established as protecting groups for peptide synthesis. The stable and frequently crystalline tosylamino acid chlorides continued to be favorite derivatives for the preparation of tosyl peptides (e.g., Hillmann and Hillmann, 1951; Katsoyannis and du Vigneaud, 1954; Swan and du Vigneaud, 1954; Rudinger, 1954b; Jošt and Rudinger, 1961; Berse et al., 1961).

However, arenesulfonyl-α-amino acid chlorides and azides have been found to decompose under aqueous alkaline conditions (Wiley et al., 1952; Wiley and Davis, 1954; Beecham, 1955, 1957a,b). The reaction of tosyl derivatives has been studied in detail by Beecham (1957a,b), who proposed the plausible mechanism shown in Fig. 17. The reaction sequence is seen to be initiated by ionization of the sulfonamide group, and in agreement with this mechanism derivatives lacking the acidic proton (e.g., the chlorides of tosylproline, -sarcosine, and -pyroglutamic acid) do not undergo such fragmentation and can safely be used for peptide synthesis in aqueous

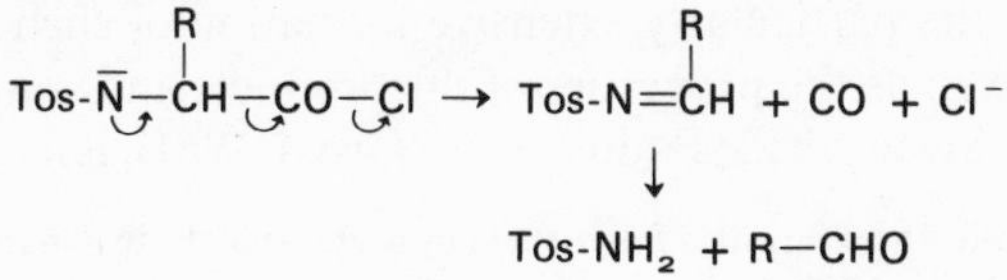

$$\text{Tos-}\overline{\text{N}}\text{—CH—CO—Cl} \rightarrow \text{Tos-N=CH} + \text{CO} + \text{Cl}^-$$

$$\downarrow$$

$$\text{Tos-NH}_2 + \text{R—CHO}$$

Fig. 17

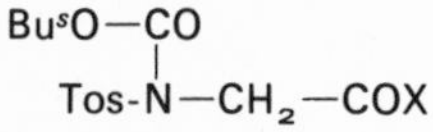

$$\text{Bu}^s\text{O}-\text{CO}$$
$$\text{Tos-N}-\text{CH}_2-\text{COX}$$

[31]

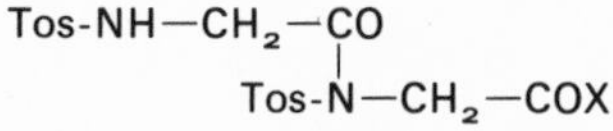

$$\text{Tos-NH}-\text{CH}_2-\text{CO}$$
$$\text{Tos-N}-\text{CH}_2-\text{COX}$$

[32]

Tos
|
N
CO   CH$_2$
| |
CH$_2$ CO
N
|
Tos

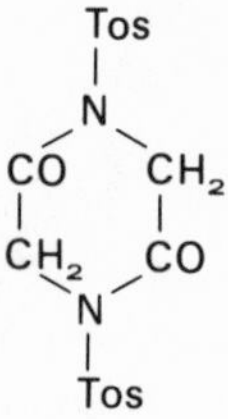

[33]

a, X = OH; b, X = NH−C$_6$H$_5$

Fig. 18

solutions. Tosylglycyl chloride is also relatively stable, and the extent of decomposition increases with the bulk of the side-chain. In the case of the azides, a Curtius rearrangement is assumed to precede fragmentation, since cyanate ion (20%) was identified as one of the reaction products.

It should be understood that this reaction does not take place in organic solvents in the absence of excess base; however, it does in many cases limit the attractive possibilities of coupling tosylamino acid chlorides with free amino acids or peptides in aqueous solution.*

Attempts to use the mixed anhydride procedure with tosyl-α-amino acids under the standard conditions have repeatedly failed (cf. Zaoral and Rudinger, 1961). This failure has been ascribed to steric hindrance (Theodoropoulos and Craig, 1955) or, more plausibly, to the reactivity of the tosylamino group (Hillmann and Hillmann, 1951). Using the synthesis of tosylglycine anilide as a model reaction, we were able to show (Zaoral and Rudinger, 1961) that under the standard conditions (s-butyl chloroformate and N-ethylpiperidine at 0°C) no mixed anhydride was, in fact, formed from tosylglycine. When pyridine was used as the base and the reaction mixture was carefully fractionated, ten crystalline materials were obtained, among them tosylglycine anilide in 16 % yield. Five other products had arisen by acylation of the tosylamide grouping either with s-butyl chloroformate [31a,b] or with tosylglycine [32a,b and 33] (Fig. 18). 1,4-Ditosyl-2,5-piperazinedione [33] had already been isolated as a minor byproduct from acylations with tosylglycine chloride (Hillmann and Hillmann, 1951; see also Ried and Pfaender, 1961), and products of the general type [31] had been obtained in peptide syntheses with other glycine derivatives by a variety of procedures.

To explain the particularly extensive formation of such products from tosylglycine as well as the properties of the products and some other results in the literature (Swan, 1952; Poduška and Gross, 1961), it has been assumed

---

*Arenesulfonyl-α-amino acids are also fragmented by acetic anhydride in pyridine (Wiley et al., 1952; Wiley and Davis, 1954) or by concentrated sulfuric acid (Beecham, 1963), but these reactions have no relevance under the conditions of peptide synthesis.

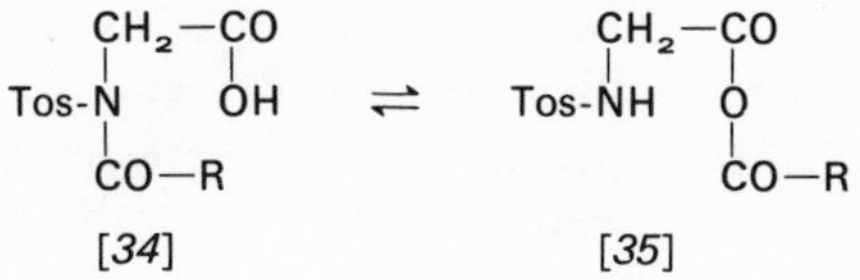

$$\begin{array}{ccc}
\underset{\text{Tos-N}}{\overset{\text{CH}_2-\text{CO}}{|}} \quad \underset{\text{OH}}{|} & \rightleftharpoons & \underset{\text{Tos-NH}}{\overset{\text{CH}_2-\text{CO}}{|}} \quad \underset{\text{O}}{|} \\
\text{CO-R} & & \text{CO-R} \\
[34] & & [35]
\end{array}$$

Fig. 19

that N-acyltosylglycines [34] and the corresponding mixed or symmetrical anhydrides [35] are readily interconvertible under basic conditions (Fig. 19; Zaoral and Rudinger, 1961).

A normal course of the mixed anhydride synthesis with tosylglycine has been achieved by the use of pivaloyl chloride as the reagent (Zaoral, 1962). It has also been noted that 3-tosyl-2,5-oxazoliddiones [36] (Fig. 20), the "N-carboxyanhydrides" derived from tosyl-α-amino acids, combine the structural features of a mixed anhydride and of a doubly substituted (and hence fully protected) amino group, and examples of peptide syntheses with such derivatives have been given (Zaoral and Rudinger, 1961).

Tosylamino acids can be coupled normally using dicyclohexylcarbodiimide (e.g., du Vigneaud et al., 1957) or tetraethyl pyrophosphite (Barrass and Elmore, 1957). Berse et al. (1961) failed in an attempt to prepare the p-nitrophenyl ester from tosyl-S-benzylcysteine by an unspecified procedure.

The increased reactivity of the tosylamino group also shows up in the pronounced tendency to cyclization, with the formation of tosyllactams, of derivatives in which the activated carboxyl is in the γ- or δ-position to the tosylamino group, as in the derivatives of glutamic, α-aminoadipic, and α,γ-diaminobutyric acids and ornithine (see Section VA). Although other N-substituted derivatives of these amino acids also undergo this type of reaction, they do so much less readily. Where these special steric relations are absent, acylation of the tosylamino group does not seem to constitute a serious difficulty in peptide synthesis, as witness the numerous syntheses successfully carried out with N^ε-tosyllysine (Section VIIC).

B. Use of α-Tosylamino Acids in Peptide Synthesis

From time to time, the tosyl group has been used for the protection of α-amino groups (for early examples, see Schönheimer, 1926; Woolley, 1948).

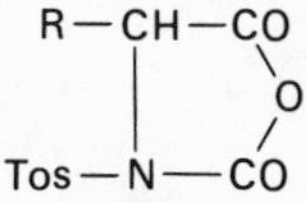

$$\begin{array}{c}
\text{R}-\text{CH}-\text{CO} \\
| \qquad \diagdown \text{O} \\
| \qquad \diagup \\
\text{Tos}-\text{N}-\text{CO} \\
[36]
\end{array}$$

Fig. 20

In spite of recent improvements, procedures for its removal are still rather laborious and attended by other drawbacks (Section IV), so that its use has in general been confined to situations where these disadvantages are minimized or somehow compensated. In particular, with only a few exceptions (e.g., Katsoyannis and du Vigneaud, 1954; Swan and du Vigneaud, 1954; Rudinger et al., 1956a; Rudinger and Pravda, 1958) the tosyl group has been used to protect only the terminal amino acid of the desired sequence so that its removal forms the final stage of the synthesis. Moreover, the introduction of cysteine into a peptide sequence as the S-benzyl derivative in any case made the use of sodium in liquid ammonia mandatory for the final deblocking step until recently, so that in this context the removal of the tosyl group is accomplished without additional complications. It was this consideration which prompted the use of tosyl for the protection of the terminal amino group in syntheses of oxytocin (Honzl and Rudinger, 1955; Rudinger et al., 1956a) and of arginine- (du Vigneaud et al., 1958) and lysine-vasopressin (Meienhofer and du Vigneaud, 1960). The crystallinity of the N^α,N^ω-ditosyl nonapeptide intermediates in the synthesis of the vasopressins has proved a particular asset in their preparation by solid-phase synthesis (Meienhofer and Sano, 1968; Meienhofer et al., 1970). Subsequently, N-terminal tosyl groups have figured in the synthesis of numerous analogues of both oxytocin and the vasopressins (for early examples, see Rudinger et al., 1956b; Jaquenoud and Boissonnas, 1962; Jošt et al., 1961, 1963a,b; Nesvadba et al., 1963).

With the availability of an alternative method for the cleavage of S-benzyl groups, and of alternative S-protecting groups for cysteine, the N-tosyl/S-benzyl combination becomes less attractive. However, there may still be a case for using a terminal N^α-tosyl group where tosyl is also chosen for the protection of side-chain amino groups, and the final step must be a tosyl cleavage in any case.

The tosyl protecting groups may also offer advantages when a vigorous synthetic reaction is to be used, such as the chloride procedure (e.g., Katsoyannis and du Vigneaud, 1954; Swan and du Vigneaud, 1954; Rudinger, 1954; see, however, Section VIIA), or where the protected peptide itself is to be subjected to fairly drastic chemical modification. An example of the latter kind is the application of the dehydration–hydrogenation sequence illustrated in Fig. 15 to peptide derivatives of glutamine and asparagine (Zaoral and Rudinger, 1957).

Another situation in which the tosyl group may be preferred arises when the N^α-tosylated amino acid is available in any case as a synthetic intermediate from the preparation of the amino acid itself or of a particular side-chain derivative. Thus tosyl-L-α-aminoadipic acid, which is readily prepared from tosylglutamic acid by the Arndt–Eistert synthesis (Fig. 14;

Rudinger and Farkašová, 1963), has been used for the synthesis of α- and δ-(α-aminoadipyl) peptides (Gut and Rudinger, 1963; D. Morris, V. Gut and J. Rudinger, unpublished results; see Rudinger, 1963a).

Finally, tosylpeptides may be chosen as intermediates where the tosyl group can render double service by also providing carboxyl activation (Section VA), as in the synthesis of γ-glutamyl peptides (e.g., Rudinger, 1954b; Rudinger and Czurbová, 1954). At one time, when free glutamine was not readily available, this approach as particularly popular for the synthesis of glutaminyl peptides (Swan and du Vigneaud, 1954; Rudinger, 1954b; Swan, 1956; Rudinger et al., 1956a; Kaneko et al., 1957; Chillemi et al., 1957; Rudinger and Pravda, 1958).

C. Use of the ω-Tosyl Protection in Peptide Synthesis

In synthetic practice, the tosyl group has found its widest application as a stable N^ω-protecting group for lysine and ornithine. Its uses for this purpose are, too numerous to be listed here (see, e.g., Schröder and Lübke, 1965, 1966). It may merely be noted that N^ε-tosyllysine has been used in syntheses of lysine-vasopressin by both conventional (Meienhofer and du Vigneaud, 1960) and solid-phase (Meienhofer and Sano, 1968) procedures, in syntheses of α-melanotropin (Blake and Li, 1971) and fragments of corticotropin (e.g., Li et al., 1961), as well as in syntheses of insulin (Meienhofer et al., 1963; Katsoyannis et al., 1964) and in the preparation of numerous analogues of these hormones. Similarly, N^δ-tosylornithine has been used in syntheses of gramicidin S (Schwyzer and Sieber, 1957) and the ornithine analogues of, e.g., melanotropin-active peptides (Li et al., 1960) and vasopressin (Huguenin and Boissonnas, 1963). Our own synthetic work with N^γ-tosyl-α,γ-diaminobutyric acid ended in anticlimax, by the synthesis (Poduška and Rudinger, 1966) of a cyclic decapeptide structure wrongly ascribed to the antibiotic circulin A.

Following the observation that removal of the tosyl group from protected derivatives of the corticotropin–melanotropin family of peptides by sodium in liquid ammonia was attended by side-reactions (Schwyzer et al., 1960; Hofmann and Yajima, 1961), some schools have abandoned the use of tosyl in favor of protecting groups removable by acid cleavage alone (Hofmann and Yajima, 1961; Schwyzer et al., 1963; Schwyzer and Kappeler, 1963), and a similar development has taken place in work on insulin (Zahn et al., 1969) for similar reasons (Zahn et al., 1967). Others, however, have continued to use tosyl-protected intermediates for syntheses both in the corticotropin and in the insulin field (e.g., Ramachandran et al., 1965; Li and Hemassi, 1972; Katsoyannis et al., 1971a,b) with apparent success.

N^ε-Tosylarginine was first applied in the synthesis of corticotropin fragments (Schwyzer and Li, 1958; see also Li et al., 1961) and later in syntheses of arginine-vasopressin (Huguenin and Boissonnas, 1962; Studer, 1963; Meienhofer *et al.*, 1970) and of bradykinin and kallidin (Guttmann *et al.*, 1962; Pless *et al.*, 1962; Mazur and Plume, 1968). The finding that the tosyl group may be removed from the guanidine grouping without recourse to sodium–ammonia treatment by the use of the newly popular reagent hydrogen fluoride (Mazur and Plume, 1968) has favored its continued use (e.g., Blake *et al.*, 1972).

The protection of the histidine imidazole group by tosylation (Sakaki-bara and Fujii, 1969) has been tested in a synthesis of angiotensin II (Fujii and Sakakibara, 1970). N^{im}-Tosylhistidine has found favor in solid-phase synthesis, as witness its use in the preparation of thyreotropin releasing factor and of the angiotensins by this procedure (Stewart *et al.*, 1972).

O-Tosyltyrosine has found application in syntheses of lysine-vasopressin (Katsoyannis *et al.*, 1957), including one of radioactively labeled material (Thomas *et al.*, 1967), and of some protected insulin sequences (Stewart, 1967).

VIII. ANALYTICAL USES OF ARENESULFONYL DERIVATIVES

Emil Fischer developed two analytical applications of arenesulfonyl derivatives: Because of their good crystallinity and defined melting points, the naphthalene-β-sulfonyl amino acids and peptides were considered useful derivatives for the isolation and identification of amino acids and peptides from protein hydrolysates (Fischer and Bergell, 1902; Fischer and Abder-halden, 1907); and because of the stability of the sulfonamide grouping under conditions which hydrolyze peptide bonds, labeling of the free amino group by arenesulfonyl substitution and characterization of the arenesul-fonylamino acid after hydrolysis were used to identify the amino-terminal amino acid of a peptide (Fischer and Abderhalden, 1907). For both these uses, arenesulfonyl derivatives are with us still (or again).

The isolation and characterization of sulfonyl derivatives from mixtures were, of course, extremely laborious and required large amounts of material. The second difficulty was eliminated by the use of radioactively labeled arenesulfonyl groups—p-iodobenzenesulfonyl (pipsyl) chloride labeled with ^{131}I proved a suitable reagent—together with the use of unlabeled carrier derivatives to facilitate isolation (Keston *et al.*, 1946). By using a ^{35}S-labeled derivative as "indicator" rather than a carrier and by separating the labeled amino acids by chromatographic techniques, Keston *et al.* (1947) could also obviate the laborious separation by crystallization (see Keston and Uden-friend, 1949). The further development of this technique was overtaken by the advent of ion exchange chromatography for the separation of amino acids. However, in recent times the very high sensitivity with which the fluorescence of 1-dimethylaminonaphthalene-5-sulfonyl (dansyl) derivatives can be measured has motivated repeated attempts to develop procedures for the quantitative preparation, resolution, and determination of dansyl derivatives from amino acid mixtures (see, e.g., Blackburn, 1968). In principle, such a method should permit the quantitative analysis of amino acid mixtures to be carried out on a scale of 10–100 pmoles, but it must be admitted that to date none of the proposed procedures can seriously rival the automated method of Moore and Stein for general use in protein chemistry.

Following Fischer's early work, naphthalene-β-sulfonyl (Abderhalden and Funk, 1910) and benzenesulfonyl (Abderhalden and Bahn, 1932, 1935) "tags" were used sporadically to identify the amino-terminal residue in small peptides, and Gurin and Clarke (1934) investigated the use of benzenesulfonyl labeling for the determination of free amino groups in proteins. Again, radioactively labeled pipsyl chloride showed considerable promise as a reagent for this purpose (Udenfriend and Velick, 1951) but lost the day to Sanger's fluoro-2,4-dinitrobenzene reagent. And yet again, the advantages of a fluorescent label (Hartley and Massey, 1956) brought another sulfonyl reagent, dansyl chloride, back on the scene for end-group determination (Gray and Hartley, 1963a; Gros and Labousse, 1969), including application, in conjunction with the Edman degradation, to sequence determination (Gray and Hartley, 1963b). For these purposes, the dansyl group is at present in general use.

IX. SULFONYL DERIVATIVES OTHER THAN TOSYL

As must have been apparent from the preceding sections, naphthalene-β-sulfonyl was an early predecessor of the tosyl group in most of its applications, and the benzenesulfonyl group has from time to time been used interchangeably with tosyl. As far as can be judged from the literature, the properties of all three groups are qualitatively similar. The only quantitative comparison which appears to have been made (Horner and Singer, 1969) revealed only slight differences in the electrolytic reduction potentials of variously substituted benzenesulfonamides.

The toluene-ω-sulfonyl derivatives, on the other hand, show some distinct differences. They are reductively cleaved not only by sodium in liquid ammonia and by hydrogen iodide and hydrogen bromide but also by Raney nickel (Milne and Peng, 1957a); the complicated working-up procedure described for this last reaction could no doubt be simplified by the use of ion exchange or chelating resins. The amino acid and peptide derivatives described are usually crystalline (Milne and Peng, 1957a), and the racemates can be resolved through stereoselective enzymic synthesis of the phenylhydrazides (Milne and Peng, 1957b). It would be interesting to determine the course of reduction of toluene-ω-sulfonamides with sodium and ammonia and on electrolysis; it seems likely that the primary site of attack here will be the C—S rather than the S—N bond.

The utility of arenesulfonyl groups with special properties, such as the fluorescent dansyl group and the iodine-labeled pipsyl group, for analytical purposes has already been pointed out (Section VIII). In addition, the pipsyl group may be of interest as a heavy-atom marker for X-ray crystal analysis; the pipsyl analogues of the tosylamino-chloroketone inhibitors [30b,c] of chymotrypsin (J. Rudinger and P. Bützer, unpublished) and of trypsin (Kuranova and Smirnova, 1971) have been prepared in this context.

Quite apart from such special applications, the purely synthetic utility of N-sulfonyl protection should be capable of further improvement. It has been repeatedly pointed out (Sections IIIB, VA, VIIA) that some of the complications attending the use of the tosyl group arise from the presence of the acidic sulfonamide hydrogen. These complications should be eliminated by use of a protecting group which would replace both hydrogens of a primary amino group in much the same way as the phthaloyl group does, but which would at the same time show the high stability of a sulfonyl derivative. With this aim in mind, we have recently prepared several crystalline benzene-o-sulfimide derivatives [37] from amino acids by reaction with benzene-o-disulfonic acid chloride (Fig. 21). The protecting group can be removed with sodium in liquid ammonia or by electrolytic reduction under the conditions given by Horner and Neumann (1965), but it is, not unexpectedly, stable to hydrogen bromide in acetic acid (J. Rudinger and P. Bützer, unpublished results). The potential of these derivatives in amino acid chemistry and peptide synthesis is now being examined.

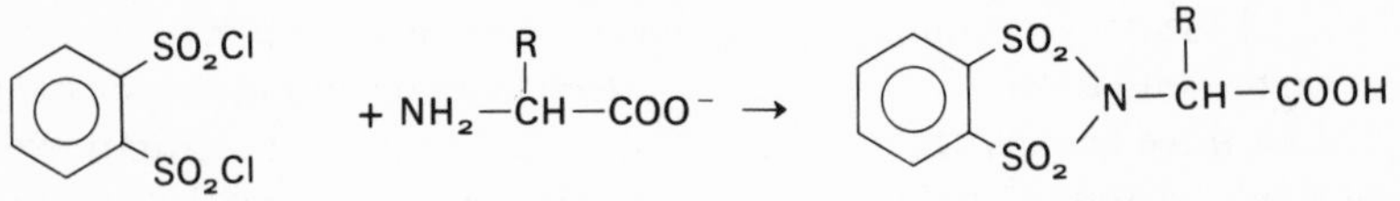

Fig. 21

X. CONCLUSION

The competent peptide chemist will, of course, keep step with progress in his field and, indeed, contribute to this progress. That is not to say that he will abandon all older methods as soon as a new one comes along; it is the mark of a good craftsman that he knows and perfects all the tools of his trade, modern as well as classical. That is why the benzyloxycarbonyl group of Bergmann and Zervas (1932) and, yes, even the tosyl group of Emil Fischer (1915) will not become mere "museum pieces" for many years yet.

ACKNOWLEDGMENT

The recent unpublished work referred to in this review was supported by the Swiss National Foundation for Scientific Research, grant No. 3.424.70.

REFERENCES

Abderhalden, B., and Bahn, A., 1932, Isolierung von Glycyl-seryl-prolyl-tyrosyl-prolin und von Seryl-prolyl-tyrosyl-prolin beim stufenweisen Abbau von Seidenfibroin (*Bombyx mori*), *Z. Physiol. Chem.* **210**:246.
Abderhalden, E., and Bahn, A., 1935, Beitrag zur Synthese von Dipeptiden, in denen Serin die Aminogruppe aufweist, *Z. Physiol. Chem.* **234**:181.
Abderhalden, E., and Funk, C., 1910, Weiterer Beitrag zur Kenntnis der partiellen Hydrolyse von Proteinen, *Z. Physiol. Chem.* **64**:436.
Bajusz, S., and Medzihradszky, K., 1963, *in* "Peptides: Proceedings of the Fifth European Peptide Symposium, Oxford, 1962" (G. T. Young, ed.), pp. 49–52, Pergamon Press, Oxford.
Barrass, B. C., and Elmore, D. T., 1957, The synthesis of peptide derivatives of basic amino-acids, *J. Chem. Soc.* **1957**:3134.
Beecham, A. F., 1955, The alkaline degradation of toluene-*p*-sulphonyl-α-amino-acid chlorides, *Chem. Ind. (Lond.)* **1955**:1120.
Beecham, A. F., 1957a, Tosyl-α-amino acids. I. Degradation of the acid chlorides and azides by aqueous alkali, *J. Am. Chem. Soc.* **79**:3257.
Beecham, A. F., 1957b, Tosyl-α-amino acids. II. The use of acid chlorides for peptide synthesis in the presence of aqueous alkali, *J. Am. Chem. Soc.* **79**:3262.
Beecham, A. F., 1963, Tosyl-α-amino acids. III. The acid-catalysed degradation of α-amino acids and derivatives, *Aust. J. Chem.* **16**:889.
Benisek, W. F., and Cole, R. D., 1965, Sodium/liquid ammonia reduction of proline-containing peptides, *Biochem. Biophys. Res. Commun.* **20**:655.
Benisek, W. F., Raftery, M. A., and Cole, R. D., 1967, Reductive cleavage of acylproline peptide bonds, *Biochemistry* **6**:3780.
Benoiton, L., 1964, Amino acids and peptides. II. Synthesis of ε-*N*-methyl-L-lysine and related compounds, *Can. J. Chem.* **42**:2043.
Bergmann, M., and Zervas, L., 1932, Über ein allgemeines Verfahren der Peptid-Synthese, *Berichte* **65**:1192.
Berse, C., Massiah, T., and Piché, L., 1961, Protected dipeptides containing cysteine, glycine, phenylalanine, and tyrosine, *J. Org. Chem.* **26**:4514.
Birch, A. J., and Smith, H., 1958, Reduction by metal-amine solutions: Applications in synthesis and determination of structure, *Quart. Rev. (Lond.)* **12**:17.
Blackburn, S., 1968, "Amino Acid Determination: Methods and Techniques," pp. 173–179, M. Dekker, New York.
Blake, J., and Li, C. H., 1971, The solid-phase synthesis of alpha-melanotropin, *Internat. J. Protein Res.* **3**:185.

Blake, J. J., Crooks, R. W., and Li. C. H., 1970, Solid-phase synthesis of (5-glutamine)-α-melanotropin, *Biochemistry* 9:2071.

Blake, J., Wang, K.-T., and Li, C. H., 1972, Adrenocorticotropin. Solid-phase synthesis of α^{1-19}-adrenocorticotropic hormone, alanyl-α^{1-19}-adrenocorticotropic hormone, and prolyl-α^{1-19}-adrenocorticotropic hormone and their adrenocorticotropic activity, *Biochemistry* 11:438.

Bovarnick, M. R., 1943, Substitution of heated asparagine–glutamate mixture for nicotinamide as a growth factor for *Bacterium dysenteriae* and other microorganisms, *J. Biol. Chem.* 148:151.

Brenner, M., 1959, Discussion, *in* "Proceedings of the Symposium on Methods of Peptide Synthesis, Prague, 1958," *Coll. Czech. Chem. Commun.* 24:141 (Special Issue).

Brewster, J. H., and Ciotti, C. J., 1955, Dehydrations with aromatic sulfonyl halides in pyridine. A convenient method for the preparation of esters, *J. Am. Chem. Soc.* 7:6214.

Chillemi, D., Scarso, L., and Scoffone, E., 1957, Sintesi di peptidi della catena A dell'insulina, *Gazz. Chim. Ital.* 87:1356.

Chimiak, A., 1969, Synthesis of L-leucine homologues, *Roczniki Chem.* 42:209.

Christensen, H. N., and Riggs, T. R., 1956, Structural evidences for chelation and Schiff's base formation in amino acid transfer into cells, *J. Biol. Chem.* 220:265.

Clayton, D. W., Kenner, G. W., and Sheppard, R. C., 1956, Peptides. Part V. Condensation of the γ-carboxyl group of α-glutamyl peptides with the peptide chain, *J. Chem. Soc.* 1956:371.

Cocker, W., 1937, Preparation of the simpler α-alkylamino acids, *J. Chem. Soc.* 1937:1693.

Cook, A. H., Cox, S. F., and Farmer, T. H., 1949, Production of antibiotics by fungi, Part IV. Lateritiin-I, Lateritiin-II, avenacein, sambucinin, and fructigenin, *J. Chem. Soc.* 1949:1022.

du Vigneaud, V., and Behrens, O. K., 1937, A method for protecting the imidazole ring of histidine during certain reactions and its application to the preparation of 1-amino-*N*-methylhistidine, *J. Biol. Chem.* 117:27.

du Vigneaud, V., Ressler, C., Swan, J. M., Roberts, C. W., and Katsoyannis, P. G., 1954, The synthesis of oxytocin, *J. Am. Chem. Soc.* 76:3115.

du Vigneaud, V., Bartlett, M. F., and Jöhl, A., 1957, The synthesis of lysine-vasopressin, *J. Am. Chem. Soc.* 79:5572.

du Vigneaud, V., Gish, D. T., Katsoyannis, P. G., and Hess, G. P., 1958, Synthesis of the pressor-antidiuretic hormone, arginine-vasopressin, *J. Am. Chem. Soc.* 80:3355.

Erlanger, B. F., Sachs, H., and Brand, E., 1954, The synthesis of peptides related to gramicidin S, *J. Am. Chem. Soc.* 76:1806.

Farkašová, H., and Rudinger, J., 1965, Amino acids and peptides. LVII. A synthesis of L-α-aminopimelic and L-α-aminosuberic acid, *Coll. Czech. Chem. Commun.* 30:3117.

Fischer, E., 1903, Synthese von Derivaten der Polypeptide, *Berichte* 36:2094.

Fischer, E., 1915, Reduktion der Aryl-sulfamide durch Jodwasserstoff, *Berichte* 48:93.

Fischer, E., and Abderhalden, E., 1907, Bildung von Polypeptiden bei der Hydrolyse der Proteine, *Berichte* 40:3544.

Fischer, E., and Bergell, P., 1902, Über die β-Naphthalinsulfoderivate der Aminosäuren, *Berichte* 35:3779.

Fischer, E., and Bergell, P., 1903, Über die Derivate einiger Dipeptide und ihr Verhalten gegen Pankreasfermente, *Berichte* 36:2592.

Fischer, E., and Bergmann, M., 1913, Methylderivate der δ-Aminovaleriansäure und des *dl*-Ornithins, *Annalen* 398:96.

Fischer, E., and Lipschitz, W., 1915, Optisch-aktive *N*-Monomethyl-Derivate von Alanin, Leucin, Phenyl-alanin und Tyrosin, *Berichte* 48:360.

Fischer, E., and von Mechel, L., 1916, Bildung aktiver, sekundärer Aminosäuren aus Halogensäuren und primären Aminen, *Berichte* 49:1355.

Fujii, T., and Sakakibara, S., 1970, The solid-phase synthesis of Ile^5-angiotensin II to demonstrate the use of N^{im}-tosyl-histidine, *Bull. Chem. Soc. Japan* **43**:3954.

Gazis, E., Bezas, B., Stelakatos, G. C., and Zervas, L., 1963, On the protection of α-amino and carboxyl groups for peptide synthesis, *in* "Peptides: Proceedings of the Fifth European Peptide Symposium, Oxford, 1962" (G. T. Young, ed.), pp. 17–21, Pergamon Press, Oxford.

Gibson, C. S., and Simonsen, J. L., 1915, The resolution of externally compensated *p*-toluenesulphonylalanine into its optically active components, *J. Chem. Soc.* **107**:798.

Goodman, M., and Felix, A. M., 1964, Conformational aspects of polypeptide structure. XIII. A nonionic helical polypeptide in aqueous solution, *Biochemistry* **3**:1529.

Gray, W. R., and Hartley, B. S., 1963*a*, A fluorescent end-group reagent for proteins and peptides, *Biochem. J.* **89**:59p.

Gray, W. R., and Hartley, B. S., 1963*b*, The structure of a chymotryptic peptide from *Pseudomonas* cytochrome *c*-551, *Biochem. J.* **89**:379.

Greenstein, J. P., and Winitz, M., 1961*a*, "The Chemistry of the Amino Acids," Vol. 2, pp. 888–889, Wiley, New York.

Greenstein, J. P., and Winitz, M., 1961*b*, "The Chemistry of the Amino Acids," Vol. 3, pp. 2756–2759, Wiley, New York.

Gros, C., and Labouesse, B., 1969, Study of the dansylation reaction of amino acids, peptides and proteins, *Europ. J. Biochem.* **7**:463.

Gurin, S., and Clarke, H. T., 1934, Allocation of the free amino groups in proteins and peptides, *J. Biol. Chem.* **107**:395.

Gut, V., and Poduška, K., 1971, Amino acids and peptides. CII. Large-scale preparation of ornithine from glutamic acid, *Coll. Czech. Chem. Commun.* **36**:3470.

Gut, V., and Rudinger, J., 1963, Amino acids and peptides. XLI. Methods for the synthesis of α-aminoadipic acid peptides, *Coll. Czech. Chem. Commun.* **28**:2953.

Gut, V., Rudinger, J., Walter, R., Herling, P. A., and Schwartz, I. L., 1968, Amino acids and peptides. Part LXXXII. Derivatives of 1-tosyl-3-amino-2-piperidone: Preparation and conversion to derivatives and peptides of ornithine, *Tetrahedron* **24**:6351.

Guttmann, S., 1963, On the use of the tosyl group for the protection of basic amino-acids, *in* "Peptides: Proceedings of the Fifth European Peptide Symposium, Oxford, 1962" (G. T. Young, ed.), pp. 41–47, Pergamon Press, Oxford.

Guttmann, S., Pless, J., and Boissonnas, R. A., 1962, Nouvelle synthèse de la bradykinine, *Helv. Chim. Acta* **45**:170.

Harington, C. R., and Moggridge, R. C. C., 1940, Experiments in the glutamic acid series, *J. Chem. Soc.* **1940**:706.

Hartley, B. S., and Massey, V., 1956, The active site of chymotrypsin. I. Labelling with a fluorescent dye, *Biochim. Biophys. Acta* **21**:58.

Hillmann, A., and Hillmann, G., 1951, Über leicht abspaltbare Acylreste bei der Peptidsynthese, *Z. Naturforsch.* **6b**:340.

Hoffmann, K., and Yajima, H., 1961, Studies on polypeptides. XX. Synthesis and corticotropic activity of a peptide amide corresponding to the *N*-terminal tridecapeptide sequence of the corticotropins, *J. Am. Chem. Soc.* **83**:2289.

Honzl, J., and Rudinger, J., 1955, Amino-acids and peptides. XVIII. Synthetic studies in the oxytocin field. II. The synthesis of some derivatives of L-cysteinyl-L-tyrosylglycine, L-cysteinyl-L-tyrosyl-L-leucine and L-cysteinyl-L-tyrosyl-L-isoleucine, *Coll. Czech. Chem. Commun.* **20**:1190.

Hope, D. B., and Horncastle, K. C., 1966, Synthesis of some dibasic sulphur-containing amino-acids related to L-lysine, *J. Chem. Soc.* (C) **1966**:1098.

Hope, D. B., and Humphries, J. F., 1964, A convenient preparation of L-homocystine from L-methionine, *J. Chem. Soc.* **1964**:869.

Horner, L., and Neumann, H., 1965, Studien zum Vorgang der Wasserstoffübertragung. XIII. Reduktive Spaltung von Säureamiden und Estern mit Tetramethylammonium (TMA). Benzoyl- und Tosylrest als Schutzgruppe bei Peptidsynthesen, *Chem. Ber.* **98**:3462.

Horner, L., and Singer, R.-J., 1969, Studien zum Vorgang der Wasserstoffübertragung, 19. Über die gezielte hydrierende Ablösung einiger Acyl-Gruppen vom Stickstoff an der Quecksilberkathode, *Annalen* **723**:1.

Huguenin, R. L., and Boissonnas, R. A., 1962, Synthèses de la Phé²-arginine-vasopressine et de la Phé²-arginine-vasotocine et nouvelles synthèses de l'arginine-vasopressine et de l'arginine-vasotocine, *Helv. Chim. Acta* **45**:1629.

Huguenin, R. L., and Boissonnas, R. A., 1963, Synthèse de l'Orn⁸-vasopressine et de l'Orn⁸-oxytocine, *Helv. Chim. Acta* **46**:1669.

Husain, S. S., and Lowe, G., 1965, The location of the active centre cysteine residue in the primary sequence of papain, *Chem. Commun.* **1965**:345.

Izumiya, N., 1952, Synthesis of *N*-methylhydroxyamino acids, *Kyushu Mem. Med. Sci.* **3**:1; *Chem. Abst.* **47**:9264i, 1953.

Izumiya, N., 1953, Walden inversion of amino acids. VII. The formation of *N*-*p*-toluenesulfonylproline from *N*δ-*p*-toluenesulfonylornithine, *Bull. Chem. Soc. Japan* **26**:53.

Jackson, E. L., 1952, *O*-*p*-Toluenesulfonyl-L-tyrosine, its *N*-acetyl and *N*-benzoyl derivative, *J. Am. Chem. Soc.* **74**:837.

Jakubke, H.-D., Fischer, J., Jošt, K., and Rudinger, J., 1968, Amino acids and peptides. LXXXVI. Synthesis of L-selenomethionine, L-selenoethionine and their *tert*-butyloxycarbonyl derivatives, *Coll. Czech. Chem. Commun.* **33**:3910.

Jaquenoud, P.-A., and Boissonnas, R. A., 1962, Synthèse de l'Asp(NH₂)⁴-oxytocine, de la Glu(NH₂)⁵-oxytocine et de l'Asp (NH₂)⁴-Glu(NH₂)⁵-oxytocine, *Helv. Chim. Acta* **45**:1601.

Jošt, K., and Rudinger, J., 1961, Amino-acids and peptides. XXXIV. Some peptides of *S*-benzylcysteine, *Coll. Czech. Chem. Commun.* **26**:2345.

Jošt, K., and Rudinger, J., 1967, Amino acids and peptides. LXXIV. Derivatives of L-cystathionine suitable for peptide synthesis, *Coll. Czech. Chem. Commun.* **32**:2485.

Jošt, K., Rudinger, J., and Šorm, F., 1961, Amino-acids and peptides. XXXV. Analogues of oxytocin substituted in positions 1 and 2 of the peptide chain: Protected intermediates, *Coll. Czech. Chem. Commun.* **26**:2496.

Jošt, K., Rudinger, J., and Šorm, F., 1963a, Analogues of oxytocin exerting protracted biological effects, *Coll. Czech. Chem. Commun.* **28**:2021.

Jošt, K., Rudinger, J., and Šorm, F., 1963b, Some structural analogues of oxytocin modified in position 2 of the peptide chain: Preparation and some chemical and biological properties, *Coll. Czech. Chem. Commun.* **28**:1706.

Kaneko, T., Shiba, T., Watarai, S., Imai, S., Shimada, T., and Ueno, K., 1957, Synthesis of eisenine, *Chem. Ind. (Lond.)* **1957**:986.

Katsoyannis, P. G., 1964, The synthesis of the insulin chains and their combination to biologically active material, *Diabetes* **13**:339.

Katsoyannis, P. G., 1966, The chemical synthesis of human and sheep insulin, *Am. J. Med.* **40**:652.

Katsoyannis, P. G., and du Vigneaud, V., 1954, The synthesis of *p*-toluenesulfonyl-L-isoleucyl-L-glutaminyl-L-asparagine and related peptides, *J. Am. Chem. Soc.* **76**:3113.

Katsoyannis, P. G., Gish, D. T., and du Vigneaud, V., 1957, Synthetic studies on arginine-vasopressin: Condensation of *S*-benzyl-*N*-carbobenzoxy-L-cysteinyl-L-tyrosyl-L-glutaminyl-L-asparagine and its *O*-tosyl derivative with *S*-benzyl-L-cysteinyl-L-prolyl-L-arginylglycinamide, *J. Am. Chem. Soc.* **79**:4516.

Katsoyannis, P. G., Fukuda, K., Tometsko, A., Suzuki, K., and Tilak, M., 1964, Insulin peptides. X. The synthesis of the B-chain of insulin and its combination with natural or synthetic A-chain to generate insulin activity, *J. Am. Chem. Soc.* **86**:930.

Katsoyannis, P. G., Ginos, J., Zalut, C., Tilak, M., Johnson, S., and Trakatellis, A. C., 1971a, Insulin peptides. XXII. A synthesis of the human insulin B-chain S-sulfonate, *J. Am. Chem. Soc.* **93**:5877.

Katsoyannis, P. G., Zalut, C., Tometsko, A., Tilak, M., Johnson, S., and Trakatellis, A. C., 1971b, Insulin peptides. XXI. A synthesis of the B chain of sheep (bovine) insulin and its isolation as the S-sulfonated derivative, *J. Am. Chem. Soc.* **93**:5871.

Kenner, G. W., McDermott, J. R., and Sheppard, R. C., 1971, The safety catch principle in solid phase peptide synthesis, *Chem. Commun.* **1971**:636.

Keston, A. S., and Udenfriend, S., 1949, The application of the isotopic-derivative method to the analysis of proteins, *Cold Spring Harbor Symp. Quant. Biol.* **14**:92.

Keston, A. S., Udenfriend, S., and Cannan, K. R., 1946, Microanalysis of mixtures (aminoacids) in the form of isotopic derivatives, *J. Am. Chem. Soc.* **68**:1390.

Keston, A. S., Udenfriend, S., and Levy, M., 1947, Paper chromatography applied to the isotopic derivative method of analysis, *J. Am. Chem. Soc.* **69**:3151.

Kovacs, J., and Ghatak, U. R., 1963, Mode of cleavage of tosyl protecting groups of tosylamino-acids and peptides, *Chem. Ind. (Lond.)* **1963**:913.

Kovacs, J., and Ghatak, U. R., 1966, Investigation of the sodium–liquid ammonia cleavage of a tosyl protecting group of tosylamino acids and peptides, *J. Org. Chem.* **31**:119.

Kraus, C. A., and White, G. F., 1923, Reactions of strongly electropositive metals with organic substances in liquid ammonia solution. I. Preliminary investigations, *J. Am. Chem. Soc.* **45**:768.

Kuranova, I. P., and Smirnova, E. A., 1971, Synthesis of a specific trypsin inhibitor containing a heavy atom, *Izv. Akad. Nauk SSSR, Ser. Khim.* **1971**:2535.

Kurtz, A. C., 1949, Use of copper (II) ion in masking α amino groups of amino acids, *J. Biol. Chem.* **180**:1253.

Li, C. H., and Hemassi, B., 1972, Adrenocorticotropin. 40. The synthesis of a protected nonapeptide and of a biologically active nonadecapeptide related to adrenocorticotropic hormone. [5-Glutamine]adrenocorticotropin-(1-19), *J. Med. Chem.* **15**:217.

Li, C. H., Schnabel, E., and Chung, D., 1960, The synthesis of L-histidyl-L-phenylanalyl-L-ornithyl-L-tryptophyl-glycine and L-histidyl-D-phenylalanyl-L-ornithyl-L-tryptophyl-glycine and their melanocyte-stimulating activity, *J. Am. Chem. Soc.* **82**:2062.

Li, C. H., Meienhofer, J., Schnabel, E., Chung, D., Lo, T.-B., and Ramachandran, J., 1961, Synthesis of a biologically active nonadecapeptide corresponding to the first nineteen amino acid residues of adrenocorticotropins, *J. Am. Chem. Soc.* **83**:4449.

Lovén, J. M., 1896, Affinitätsgrössen einiger organischen Säuren, *Z. Phys. Chem.* **19**:456.

Mairanovskiĭ, V. G., and Loginova, N. F., 1971, Removal of the tosyl protecting group by electrochemical reduction in dimethylformamide, *Zh. Obshch. Khim.* **41**:2581.

Mazur, R. H., and Plume, G., 1968, Improved synthesis of bradykinin, *Experientia* **24**:661.

McChesney, E. W., and Swann, W. K., 1937, The identification of the amino acids: p-Toluenesulfonyl chloride as a reagent, *J. Am. Chem. Soc.* **59**:1116.

Meienhofer, J., and du Vigneaud, V., 1960, Preparation of lysine -vasopressin through a crystalline nonapeptide intermediate and purification of the hormone by chromatography, *J. Am. Chem. Soc.* **82**:2279.

Meienhofer, J., and Li, C. H., 1962, Preparation of L-lysyl-L-lysyl-L-arginyl-L-arginyl-L-proline, *J. Am. Chem. Soc.* **84**:2434.

Meienhofer, J., and Sano, Y., 1968, A solid-phase synthesis of [lysine]-vasopressin through a crystalline protected nonapeptide intermediate, *J. Am. Chem. Soc.* **90**:2996.

Meienhofer, J., Schnabel, E., Bremer, H., Brinkhoff, O., Zabel, R., Sroka, W., Klostermeyer, H., Brandenburg, D., Okuda, T., and Zahn, H., 1963, Synthese der Insulinketten und ihre Kombination zu insulinaktiven Präparaten, *Z. Naturforsch.* **18b**:1120.

Meienhofer, J., Trzeciak, A., Havran, R. T., and Walter, R., 1970, A solid-phase synthesis of [8-arginine]-vasopressin through a crystalline protected nonapeptide intermediate and biological properties of the hormone, *J. Am. Chem. Soc.* **92**:7199.

Micheel, F., and Haneke, H., 1959, Peptidsynthesen nach dem Oxazolidonverfahren, II. Peptide der Glutaminsäure, *Chem. Ber.* **92**:309.

Micheel, F., and Lorenz, M., 1963, Peptidsynthesen mit gemischten Anhydriden aus Aminosäuren und Saccharin, *Tetrahedron Letters* **1963**:2119.

Micheel, F., and Thomas, S., 1957, Eine neue Peptidsynthese, *Chem. Ber.* **90**:2906.

Milne, H. B., and Peng, C.-H., 1957*a*, The use of benzylsulfonyl chloride in peptide syntheses, *J. Am. Chem. Soc.* **79**:639.

Milne, H. B., and Peng, C.-H., 1957*b*, The use of *N*-benzylsulfonyl-α-amino acids in enzymatic syntheses of the L-phenylhydrazides, and for enzymatic resolution, *J. Am. Chem. Soc.* **79**:645.

Miyoshi, M., Fujii, T., Yoneda, N., and Okumura, K., 1969, Reactions of L-α-tosylamino-β-propiolactone. II. A novel synthesis of L-seryl peptides, *Chem. Pharm. Bull.* **17**:1617.

Morris, D., 1960, Synthesis of the α- and γ-isomers of glutamylcystinylvaline, *Biochem. J.* **76**:349.

Neadle, D. J., and Pollitt, R. J., 1965, The formation of 1-dimethylaminonaphthalene-5-sulphonamide during the preparation of 1-dimethylaminonaphthalene-5-sulphonylamino acids, *Biochem. J.* **97**:607.

Nesvadba, H., 1962, Discussion at the Fifth European Peptide Symposium, Oxford, September 1962; see Nesvadba and Roth (1967).

Nesvadba, H., and Roth, H., 1967, Zur reduktiven Abspaltung von Peptidschutzgruppen mittels Natrium in flüssigem Ammoniak, *Monatsh. Chem.* **98**:1432.

Nesvadba, H., Honzl, J., and Rudinger, J., 1963, Amino acids and peptides. XXXVII. Some structural analogues of oxytocin modified in position 3 of the peptide chain: Synthesis and some chemical and biological properties, *Coll. Czech. Chem. Commun.* **28**:1691.

Ohle, H., Friedeberg, H., and Haeseler, G., 1936, Über Derivate des Glucose-(6)-phenylamins (6-Phenyl-amino-chinovose), *Berichte* **69**:2311.

Pless, J., and Guttmann, S., 1967, New results concerning the protection of the guanido group of arginine, *in* "Peptides: Proceedings of the Eighth European Peptide symposium, Noordwijk, 1966" (H. C. Beyermann, A. van de Linde, and W. Maassen van den Brink, eds.), pp. 50–54, North-Holland, Amsterdam.

Pless, J., Stürmer, E., Guttmann, S., and Boissonnas, R. A., 1962, Kallidin, Synthese und Eigenschaften (Vorläufige Mitteilung), *Helv. Chim. Acta* **45**:394.

Poduška, K., and Gross, H., 1961, Über α-Halogenäther. VII. Reaktionen von Dichlormethylalkyl-äthern mit Aminosäurederivaten, *Chem. Ber.* **94**:527.

Poduška, K., and Rudinger, J., 1957, Amino-acids and peptides. XXI. *N*-Substituted pyrrolidones as intermediates in the synthesis of γ-aminobutyric and α,γ-diaminobutyric acid peptides, *Coll. Czech. Chem. Commun.* **22**:1283.

Poduška, K., and Rudinger, J., 1959, Amino-acids and peptides. XXVII. Derivatives of 1-tosyl-L-3-aminopyrrolid-2-one: Preparation and synthetic potentialities, *Coll. Czech. Chem. Commun.* **24**:3449.

Poduška, K., and Rudinger, J., 1966, Amino acids and peptides. LXII. Synthesis of a protected cyclodecapeptide containing α,γ-diaminobutyric acid, *Coll. Czech. Chem. Commun.* **31**:2938.

Poduška, K., Rudinger, J., and Šorm, F., 1955, Amino-acids and peptides. XVI. Peptides of DL-α,β-diaminopropionic acid, *Coll. Czech. Chem. Commun.* **20**:1174.

Poduška, K., Kathrukha, G. S., Silaev, A. B., and Rudinger, J., 1965, Amino acids and peptides. LII. Intramolecular aminolysis of amide bonds in derivatives of α,γ-diaminobutyric acid, α,β-diaminopropionic acid, and ornithine, *Coll. Czech. Chem. Commun.* **30**:2410.

Poduška, K., Rudinger, J., Gloede, J., and Gross, H., 1969, Aminosäuren und Peptide. XC. Synthese des β-Pyrrolo-L-alanins, der L-2-Amino-4-pyrrolobuttersäure und der L-2-Amino-5-pyrrolovaleriansäure, *Coll. Czech. Chem. Commun.* **34**:1002.

Pravda, Z., and Rudinger, J., 1955, Amino-acids and peptides. XIII. A synthesis of L-proline from L-glutamic acid, *Coll. Czech. Chem. Commun.* **20**:1.

Quitt, P., Hellerbach, J., and Vogler, K., 1963, Die Synthese optisch aktiver *N*-Monomethyl-Aminosäuren, *Helv. Chim. Acta* **46**:327.

Ramachandran, J., 1965, Reaction of amino-acid esters with sodium in liquid ammonia: Cleavage of the proline ring, *Nature* **206**: 927.

Ramachandran, J., Chung, D., and Li, C. H., 1965, Adrenocorticotropins. XXXIV. Aspects of structure-activity relationships of the ACTH molecule. Synthesis of a heptadecapeptide amide, an octadecapeptide amide, and a nonadecapeptide amide possessing high biological activities. *J. Am. Chem. Soc.* **87**:2696.

Ramachandran, J., and Li, C. H., 1962, Preparation of crystalline N^g-tosylarginine derivatives, *J. Org. Chem.* **27**:4006.

Ressler, C., and Kashelikar, D. V., 1966, Identification of asparaginyl and glutaminyl residues in endo position in peptides by dehydration-reduction, *J. Am. Chem. Soc.* **88**: 2025.

Ried, W., and Pfaender, P., 1961, α-Aminoaldehyde durch hydrogenolytische Spaltung *N*-tosylierter α-Aminoacyl-dimethylpyrazole, *Ann.* **640**:111.

Roeske, R., Stewart, F. H. C., Stedman, R. J., and du Vigneaud, V., 1956, Synthesis of a protected tetrapeptide amide containing the carboxyl terminal sequence of lysine-vasopressin, *J. Am. Chem. Soc.* **78**:5883.

Rudinger, J., 1954a, Amino-acids and peptides. X. Some derivatives and reactions of 1-*p*-toluenesulphonyl-L-pyrrolid-5-one-2-carboxylic acid, *Coll. Czech. Chem. Commun.* **19**:365.

Rudinger, J., 1954b, Amino-acids and peptides. XI. A new synthesis of peptides of glutamic acid, *Coll. Czech. Chem. Commun.* **19**:375.

Rudinger, J., 1959a, Discussion, *in* "Proceedings of the Symposium on Methods of Peptide Synthesis, Prague, 1958," *Coll. Czech. Chem. Commun.* **24**:29–30 (Special Issue).

Rudinger, J., 1959b, Discussion, *in* "Proceedings of the Symposium on Methods of Peptide Synthesis, Prague, 1958," *Coll. Czech. Chem. Commun.* **24**:81–82 (Special Issue).

Rudinger, J., 1962, Some problems in peptide synthesis with polyfunctional amino-acids, *Rec. Chem. Progr.* **23**:3.

Rudinger, J., 1963a, Synthesis of peptides with some "unnatural" amino acids: General survey, *in* "Peptides: Proceedings of the Fifth European Peptide Symposium, Oxford, 1962" (G. T. Young, ed.), pp. 133–164, Pergamon Press, Oxford.

Rudinger, J., 1963b, Organic chemistry in peptide synthesis, *Pure Appl. Chem.* **7**:335.

Rudinger, J., and Czurbová, H., 1954, Amino-acids and peptides. XII. A new resolution of DL-glutamic acid; synthesis of D-glutamine and γ-D-glutamylglycine, *Coll. Czech. Chem. Commun.* **19**:386.

Rudinger, J., and Farkašová, H., 1963, Amino acids and peptides. XL. Synthesis of L-α-aminoadipic acid and L-β-aminoadipic acid from L-glutamic acid, *Coll. Czech. Chem. Commun.* **28**:2941.

Rudinger, J., and Maassen van den Brink-Zimmermannová, H., 1973, Reduction of tosylamino acids and related compounds with sodium in liquid ammonia: Stoicheiometry and products, *Helv. Chim. Acta*, in press.

Rudinger, J., and Pravda, Z., 1958, Amino-acids and peptides. XXII. Synthesis of some glutamine peptides; the structure of eisenin, *Coll. Czech. Chem. Commun.* **23**:1947.

Rudinger, J., Honzl, J., and Zaoral, M., 1956a, Amino-acids and peptides. XIX. Synthetic studies in the oxytocin field. III. An alternative synthesis of oxytocin, *Coll. Czech. Chem. Commun.* **21**:202.

Rudinger, J., Honzl, J., and Zaoral, M., 1956b, Syntheses in the oxytocin field. IV. Two analogues, *Coll. Czech. Chem. Commun.* **21**:770.

Rudinger, J., Poduška, K., Zaoral, M., and Jošt, K., 1959, Amino-acids and peptides. XXVI. An improved synthesis of L-glutamine, *Coll. Czech. Chem. Commun.* **24**:2013.

Rudinger, J., Poduška, K., and Zaoral, M., 1960. Amino-acids and peptides. XXIX. Synthesis of the lower homologues of L-arginine and L-citrulline, *Coll. Czech. Chem. Commun.* **25**:2022.

Sakakibara, S., and Fujii, T., 1969, Synthesis and use of N^{im}-tosyl-L-histidine, *Bull. Chem. Soc. Japan* **42**:1466.

Schnabel, E., and Li, C. H., 1960, The synthesis of L-histidyl-D-phenylalanyl-L-arginyl-L-tryptophylglycine and its melanocyte-stimulating activity, *J. Am. Chem. Soc.* **82**:4576.

Schoellmann, G., and Shaw, E., 1963, Direct evidence of the presence of histidine in the active center of chymotrypsin, *Biochemistry* **2**:252.

Schönheimer, R., 1926, Ein Beitrag zur Bereitung von Peptiden, *Z. Physiol. Chem.* **154**:203.

Schröder, E., and Lübke, K., 1965, 1966, "The Peptides," Vols. I and II, Academic Press, New York.

Schwyzer, R., and Kappeler, H., 1963, Synthese eines Tetracosapeptides mit hoher corticotroper Wirksamkeit: β^{1-24}-Corticotropin, *Helv. Chim. Acta* **46**:1550.

Schwyzer, R., and Li, C. H., 1958, A new synthesis of the pentapeptide L-histidyl-L-phenylalanyl-L-arginyl-L-tryptophyl-glycine and its melanocyte-stimulating activity, *Nature* **182**:1669.

Schwyzer, R., and Sieber, P., 1957, Die Synthese von Gramicidin S, *Helv. Chim. Acta* **40**:624.

Schwyzer, R., Rittel, W., Kappeler, H., and Iselin, B., 1960, Synthese eines Nonadeka-peptides mit hoher corticotroper Wirksamkeit, *Angew. Chem.* **72**:915.

Schwyzer, R., Iselin, B., Kappeler, H., Riniker, B., Rittel, W., and Zuber, H., 1963, Die Synthese des β-Melanotropins (β-MSH) mit der Aminosäuresquenz des bovinen Hormons, *Helv. Chim. Acta* **46**:1975.

Shaw, E., Mares-Guaia, M., and Cohen, W., 1965, Evidence for an active-center histidine in trypsin through use of a specific reagent, 1-chloro-3-tosylamido-7-amino-2-heptanone, the chloromethyl ketone derived from N^{α}-tosyl-L-lysine, *Biochemistry* **4**:2219.

Snyder, H. R., and Heckert, R. E., 1952, A method for the rapid cleavage of sulphonamides, *J. Am. Chem. Soc.* **74**:2006.

Stedman, R. J., 1957, 1-p-Toluenesulfonyl-L-pyroglutamyl chloride and 1-p-toluenesulfonyl-L-glutamyl dichloride in the preparation of glutamic acid derivatives, *J. Am. Chem. Soc.* **79**:4691.

Stewart, F. H. C., 1967, Insulin peptides. I. Some protected peptides with sequences derived from the A-chain of ovine insulin, *Aust. J. Chem.* **20**:1991.

Stewart, J. M., Knight, M., Paiva, A. C. M., and Paiva, T., 1972, Histidine in solid phase peptide synthesis: Thyrotropin releasing factor and the angiotensins, in "Progress in Peptide Research," Vol. 2: "Proceedings of the Second American Peptide Symposium, Cleveland, 1970," p. 10-1, Gordon and Breach, Ann Arbor.

Studer, R. O., 1963, Vergleich von synthetischem und natürlichem Arginin-vasopressin, *Helv. Chim. Acta* **46**:421.

Swan, J. M., 1952, Thiohydanatoins. III. Anhydride intermediates in the formation of 1-acyl-2-thiohydantoins from acylamino acids, *Aust. J. Sci. Ind. Res.* **A5**:728.

Swan, J. M., 1956, The synthesis of L-glutaminylglycine and glutaminylglycine, in "Proceedings of the International Wool Textile Research Conference, Australia, 1955," Vol. C, pp. 175–180, CSIRO, Melbourne.

Swan, J. M., and du Vigneaud, V., 1954, The synthesis of L-glutaminyl-L-asparagine, L-glutamine and L-isoglutamine from p-toluenesulfonyl-L-glutamic acid, *J. Am. Chem. Soc.* **76**:3110.

Theodoropoulos, D., and Craig, L. C., 1955, The synthesis of several isoleucyl peptides and certain of their properties, *J. Org. Chem.* **20**:1169.

Thomas, K., Kapfhammer, J., and Flaschenträger, B., 1922, Über δ-Methylornithin und δ-Methylarginin. Zur Frage der Herkunft des Kreatins. IV. Mitteilung, *Z. Physiol. Chem.* **124**:75.

Thomas, P. J., Havránek, M., and Rudinger, J., 1967, Amino acids and peptides. LXXII. Synthesis of 3-phenylanine-[U-^{14}C]-8-lysine-vasopressin, *Coll. Czech. Chem. Commun.* **32**:1767.

Udenfriend, S., and Velick, S. F., 1951, The isotope derivative method of protein amino end-group analysis, *J. Biol. Chem.* **190**:733.

Weisblat, D. I., Magerlein, B. J., and Myers, D. R., 1953, The cleavage of sulphonamides, *J. Am. Chem. Soc.* **75**:3630.

Wieland, T., and Hennig, H. J., 1960, Aminosäure-sulfimide, *Chem. Ber.* **93**:1236.

Wilchek, M., Sarid, S., and Patchornik, A., 1965, Use of sodium in liquid ammonia for cleavage of *N*-proline peptides, *Biochim. Biophys. Acta* **104**:616.

Wiley, R. H., and Davis, R. F., 1954, Base-catalyzed decomposition of substituted α-(benzene-sulfonamido)-carboxylic acids and their chlorides, *J. Am. Chem. Soc.* **76**:3496.

Wiley, R. H., Davis, H. L., Gensheimer, D. E., and Smith, N. R., 1952, Base-catalyzed decomposition of α-(benzene- and p-toluenesulfonamido)-phenylacetyl chloride, *J. Am. Chem. Soc.* **74**:936.

Woolley, D. W., 1948, Strepogenin activity of derivatives of glutamic acid, *J. Biol. Chem.* **172**:71.

Wünsch, E., 1959, Discussion, *in* "Proceedings of the Symposium on Methods of Peptide Synthesis, Prague, 1958," *Coll. Czech. Chem. Commun.* **24**:141 (Special Issue).

Zahn, H., Okuda, T., and Shimonishi, Y., 1967, Merrifield synthesis of human insulin chains and their alteration during sodium treatment, *in* "Peptides: Proceedings of the Eighth European Peptide Symposium, Noordwijk, 1966" (H. C. Beyerman, A. van de Linde, and W. Maassen van den Brink, eds.), pp. 108–112, North-Holland, Amsterdam.

Zahn, H., Danho, W., Klostermeyer, H., Gattner, H. G., and Repin, J., 1969, Eine Synthese der A-Kette des Schafinsulins unter ausschliesslicher Verwendung säurelabiler Schutzgruppen, *Z. Naturforsch.* **24b**:1127.

Zaoral, M., 1962, Amino-acids and peptides. XXXVI. Pivaloyl chloride as a reagent in the mixed anhydride synthesis of peptides, *Coll. Czech. Chem. Commun.* **27**:1273.

Zaoral, M., and Rudinger, J., 1957, A new synthesis of derivatives and peptides of ornithine and α,γ-diaminobutyric acid, *Proc. Chem. Soc. (Lond.)* **1957**:176.

Zaoral, M., and Rudinger, J., 1959, Amino-acids and peptides. XXV. Dehydration of derivatives of asparagine and glutamine; a new route to α,γ-diaminobutyric acid, ornithine, and arginine, *Coll. Czech. Chem. Commun.* **24**:1993.

Zaoral, M., and Rudinger, J., 1961, Amino-acids and peptides. XXXI. Products formed from tosylglycine under conditions of a mixed carbonic anhydride synthesis, *Coll. Czech. Chem. Commun.* **26**:2316.

Zehnder, K., 1951, Über die Konstitution der *N*-Methyl-aminosäuren aus Enniatin A und B, Doctoral thesis, Eidgenössische Technische Hochschule, Zürich.

Zervas, L., and Hamalidis, J., 1965, New methods in peptide synthesis. II. Further examples of the use of the *o*-nitrophenylsulfenyl group for the protection of amino groups, *J. Am. Chem. Soc.* **87**:99.

Zervas, Borovas, D., and Gazis, E., 1963, New methods in peptide synthesis. I. Tritylsulfenyl and *o*-nitrophenylsulfenyl groups as *N*-protecting groups, *J. Am. Chem. Soc.* **85**:3660.

Zimmermannová, H., Katrukha, G. S., Poduška, K., and Rudinger, J., 1966, Fission of tosyl-amide groups with metals in liquid ammonia, *in* "Peptides: Proceedings of the 6th European Peptide Symposium, Athens 1963" (L. Zervas, ed.), pp. 21–26, Pergamon Press, Oxford.

CHAPTER 6

TACTICS FOR MINIMAL PROTECTION IN PEPTIDE SYNTHESIS

Ralph Hirschmann and Daniel F. Veber

Merck Sharp & Dohme Research Laboratories
Division of Merck & Co., Inc.
West Point, Pennsylvania

Viel allgemeinerer Anwendung fähig ist, wie wir im folgenden zeigen, der Rest $C_6H_5 \cdot CH_2 \cdot O \cdot CO$ der Benzylester-kohlensäure, kurz Carbobenzoxy-Rest (Cbzo) gennant; denn er lässt sich mit Hilfe des leicht zugänglichen Chlorids $C_6H_5 \cdot CH_2O \cdot CO \cdot Cl$ unschwer in Amino-säuren der verschiedensten Art einführen ... und—was das Wesentliche ist—durch einfache katalytische Hydrierung im offenen Gefäss in Form von Toluol und Kohlendioxyd wieder abspalten. (Max Bergmann and Leonidas Zervas, *Chem. Ber.* **65**:1192, 1932)

I. THE CONCEPT OF THE MINIMAL USE OF PROTECTING GROUPS

This chapter will review primarily those aspects of our work in peptide synthesis which relate to protecting groups. Some of the concepts to be presented are novel, others are not novel *per se*. However, we believe that *in toto* they represent a unique approach to peptide synthesis. This appears to us an appropriate subject matter for an article written in honor of Professor Leonidas Zervas, who—with the late Max Bergmann—brought about a renaissance of peptide chemistry through the introduction of the benzyloxycarbonyl protecting group in 1932.

We have sought to restrict the use of protecting groups to a minimum. In our first report on the use of Leuchs' anhydrides (*N*-carboxy-α-amino acid anhydrides; NCAs) in peptide synthesis (Denkewalter *et al.*, 1966), in which we described coupling reactions involving all of the 20 coded amino acids, protection of the third functional group of polyfunctional amino acids was restricted to the ε-amino nitrogen of lysine and to the sulfhydryl of cysteine in the nucleophiles; in the NCAs, protection of the third functionality was restricted to the hydroxyl groups of serine and threonine, the ε-amino group of lysine, the heterocyclic ring of histidine, and the sulfhydryl group of cysteine. In subsequent reports (Dewey *et al.*, 1968, 1971), an analogue of the NCA of histidine was described which did not require protection of the imidazole ring. This analogue was a 2,5-thiazolidinedione (an *N*-thio-carboxyanhydride; NTA).

The coupling reactions with NCAs and NTAs were carried out in water, a solvent which appears to be well suited for minimal protection in peptide synthesis because we have not observed reactions of the acylating agent with the hydroxyl groups of serine and threonine or with the carboxyl groups of aspartic and glutamic acids in this hydroxylic solvent. We were even able to employ the Leuchs' anhydride of glutamic acid and the novel crystalline NCA of aspartic acid without protection of the ω-carboxyl group (Denkewalter *et al.*, 1966; Hirschmann *et al.*, 1971). Because arginine is protonated on its guanidino group in aqueous medium in the *p*H range generally employed in peptide synthesis, protection other than protonation was not necessary for the guanidino group. The use of protonated arginine in peptide synthesis had already been described by Anderson (1953) and by Gish and Carpenter (1953).

We have not protected the third functionality of asparagine and glutamine. Even though the primary amide group of these amino acids is known to be more labile than the typical secondary amide bond which forms the backbone of polypeptides, we have preferred to avoid the experimental conditions which would necessitate protection of any amide bond. For example, saponification has not been part of any of our synthetic schemes, because saponification has been known to cause side-reactions with unprotected asparagine or glutamine (Hirschmann *et al.*, 1967; Dewey *et al.*, 1969).

Having employed the principle of minimal protection when we began to study peptide synthesis with NCAs in 1963, we retained this concept when we subsequently employed also more conventional methods for the formation of peptide bonds, as in the synthesis of angiotensin, calcitonin, ribonuclease S-protein, ribonuclease S-peptide, and LH-RH. All of these compounds, except the S-protein, have been obtained pure and biologically fully active. We have chosen minimal protection for several reasons:

1. We believe that in many instances protecting groups for the third functionality are as likely to cause side-reactions as to prevent them; two obvious exceptions are lysine and cysteine, where protection

of the nucleophilic side-chain heteroatoms is indispensable (see below).

2. In the synthesis of polypeptides of high molecular weight, minimal use of protecting groups is advantageous because in conventional peptide synthesis solubility problems represent major obstacles. Many of the commonly employed protecting groups have hydrocarbon residues which decrease the solubility of such protected polypeptides both in aqueous medium and in other polar solvents such as DMF and HMP.

3. Minimal use of protecting groups often simplifies purification. For example, those chromatographic purification techniques which require a mixture to be soluble in aqueous medium are more likely to be applicable when few nonpolar blocking groups are in the molecule. Furthermore, minimal use of protecting groups lowers molecular weights and thus permits more effective use of gel filtration for purification of large peptides. Minimal use of protecting groups is also advantageous for purification not only from solubility and molecular weight considerations, but also because the side-chains lend unique character to each peptide. This individuality tends to be overwhelmed by bulky hydrocarbon protecting groups. For example, the presence of unprotected aspartic acid, glutamic acid, histidine, and arginine residues permits one to take advantage of the ionic character of these amino acids in ion exchange chromatography. More subtle differences, such as those between tyrosine and phenylalanine which often permit separation by adsorption chromatography, may be lost by protection. Similarly, the presence of unprotected acidic and basic groups makes possible the use of electrophoresis both as an analytical tool and for preparative purification procedures. Finally, the fact that arginine is not protected permits use of trypsin to effect cleavage at arginine with trypsin. This can be helpful in the evaluation of the purity of high molecular weight synthetic peptides.

Minimal use of protecting groups will, of course, affect the tactics of the coupling reactions. Thus the presence of unprotected β- and γ-carboxyl groups of aspartic and glutamic acids, respectively, prevents the use of reagents such as carbodiimides for the synthesis of fragments or the subsequent coupling of fragments (Denkewalter, 1969). However, as far as we are aware, lack of protection for the third functionality of arginine, asparagine, glutamine, methionine, serine, threonine, tyrosine, tryptophan, and histidine has not created serious problems in our hands. We have found the use of unprotected tyrosine to create problems in coupling reactions with NCAs in aqueous medium, but the phenolic hydroxyl has not been a problem to us under anhydrous conditions, partly because hydrazinolysis has been found to be a useful technique for breaking any ester bonds which might be formed between the acylating agent and a hydroxyamino acid (Jenkins *et al.*, 1969).

The above arguments in favor of minimal protection apply only in part to solid-phase synthesis. The need for protecting groups is presumably greater when solid-phase peptide synthesis is employed (Merrifield, 1969). Furthermore, because the solid-phase method provides a unique solution to the solubility problem encountered in conventional synthesis, the more extensive use of protecting groups does not generate a solubility problem. Also, to the extent that carbodiimides are employed as coupling reagents in solid-phase synthesis, protection of the additional carboxyl function of aspartic and glutamic acids in solid-phase synthesis is a necessity. In our somewhat limited experience with solid-phase synthesis, we have generally combined it with azide fragment coupling techniques. In this approach, the side-chain protection of the dibasic acids need not be retained after the introduction of that residue if proper tactics are utilized. We consider this to be desirable because the presence of ω-esters in polypeptides can lead to side-reactions in solid-support synthesis (Merrifield, 1969; Nitecki, 1971) as well as in conventional peptide synthesis (Hirschmann *et al.*, 1971).

II. INDISPENSABLE PROTECTION

While we have had considerable success in reducing the use of protecting groups to a minimum, protection is in fact indispensable in peptide synthesis for the following:

1. The α-amino nitrogen of a carboxy-activated amino acid or peptide to prevent random polymerization.
2. The ε-amino group of lysine to prevent the formation of peptide bonds involving the ε- rather than the α-amino nitrogen.
3. The sulfhydryl group of cysteine to prevent the formation of thio-esters.

The remainder of this chapter will deal with "indispensable protection."

A. Nitrogen-Protecting Groups

1. Temporary Protecting Groups. In the synthesis of a polypeptide, the *N*-protecting group of the carboxy-activated amino acid or peptide is generally to be removed following the coupling reaction. For this reason, this protecting group is often called the "temporary" protecting group.

The following sections will describe our experiences with certain temporary blocking groups.

   a. *Carbamate Anion:* As early as 1905, Siegfried (1905a,b, 1906) reported that amino acids form stable carbamates when treated with carbon dioxide in aqueous medium at alkaline pH. It was shown by Leuchs (1906) that stable carbamates can be obtained from a Leuchs' anhydride at 0°C on treatment with barium hydroxide. The equilibrium between amino acid carbamates and the amino acid and carbon dioxide was studied extensively by Stadie and O'Brien (1936). More recently, Lemieux and Barton (1971) have used NMR spectroscopy to study the carbamate reaction of amino acids and peptides. When Leuchs' anhydrides are used in peptide synthesis, the carbamate becomes a nitrogen-protecting group, but it differs from other protecting groups in that it is generated *in situ* when the anhydride reacts with a nucleophile. The anhydride serves simultaneously to activate the carboxyl group and to protect the α-amino nitrogen. Since carbamates have limited stability at alkaline pH even at low temperatures, the carbamate anion may be regarded as the most acid-labile protecting group which has been employed in peptide synthesis. The NCAs, and therefore, the carbamate protecting group, had been employed in controlled peptide synthesis by Hunt and du Vigneaud (1938) and by Bailey under anhydrous conditions (1949, 1950) and in aqueous medium by Wesseley (1925) and by Bartlett and his associates (Bartlett and Jones, 1957; Bartlett and Dittmer, 1957) prior to our own work with Leuchs' anhydrides. The shortcomings of the use of NCAs in aqueous medium have been studied in Bartlett's laboratory (1957) and in our own (Hirschmann et al., 1967). We showed that the carbamates have reasonable stability in the pH range of 10.2–10.5 at 0°C but that the protecting group is generally not sufficiently stable to permit one to execute, without purification of intermediates, more than five sequential steps. When a reaction product separates after acidification, the method affords the desired product more rapidly than is feasible by any other method. For example, the synthesis of essentially pure renin inhibitor Leu-Leu-Val-Phe is easily accomplished by the NCA method in half a day (Strachan, unpublished). Recently reported studies (Iwakura et al., 1970; Katakai et al., 1971) suggest that the usefulness of this method may be broadened.

   b. *Thiocarbamate Anion:* We have previously pointed out (Hirschmann et al., 1967) that the use of the α-amino acid N-carboxyanhydride in the synthesis of peptides in aqueous solution is complicated by the fact that below pH 11 the instability of peptide carbamates leads to overreactions via decarboxylation, whereas at pH 11 overreaction via the NCA anion, the formation of hydantoic acids, and hydrolysis become troublesome side-reactions. A more stable carbamate analogue would permit peptide condensation to be carried out at lower pH. It was thought that analogues of the NCAs in which the singly bonded oxygen is replaced by sulfur should overcome side-reactions due to loss of the protecting group because we expected thiocarbamates to show greater stability at a given pH than carbamates. This proved indeed to be the case, as shown, for example, by the fact that salts of amino acid thiocarbamates were stable to electrophoresis at pH 11 (Dewey et al., 1971). Moreover, it was in some instances possible to effect *complete* disappearance of the starting amino acid on reaction with an excess of the 2,5-thiazolidinedione, indicative of complete suppression of thiocarbamate exchange. This had not been achieved with the Leuchs' anhydrides. The NTAs have proven to be useful in peptide synthesis only in limited cases, primarily because of racemization (Dewey et al., 1968, 1971). In addition, in special situations the thiocarbamate anion may prove to be of use as a temporary protecting group in alkaline aqueous medium.

　　c. Butyloxycarbonyl and 1-Methylcyclobutyloxycarbonyl Protecting Groups: The butyloxycarbonyl protecting group of McKay and Albertson (1957) is probably the amine-blocking group most extensively used in the synthesis of polypeptides at the present time. The only limitation which we have observed in its use as a temporary nitrogen-protecting group is the fact that it is not as stable in 50% aqueous acetic acid as we would like it to be. As we had reported earlier (Hirschmann, 1971), dansylation studies indicated partial loss of the butyloxycarbonyl group, especially from the amino-terminal serine in 50% acetic acid, a solvent system which we used for gel filtrations in connection with the synthesis of the S-protein of ribonuclease S. The stability of butyloxycarbonyl derivatives of amino acids in aqueous acetic acid has been studied by Sieber and Iselin (1968b). These workers observed that the stability decreases with increasing water content and reported a half-life of about 10 days for Boc-glycine in 60% acetic acid at room temperature. The fact that our protected peptides were reasonably stable in 50% acetic acid is doubtless due to the fact that butyloxycarbonyl derivatives of peptides are considerably more acid stable than those of amino acids because of the greater basicity of the protected amino nitrogen in the latter. We have sought to find a substitute for the butyloxycarbonyl group which would be slightly more acid stable, because 50% aqueous acetic acid is a very useful solvent for gel filtration. Moreover, we felt that the protecting group should not through its size or bulk generate excessive solubility problems, and we felt that the group should not introduce an additional asymmetrical center.

　　It seemed reasonable to us (Veber et al., 1972b) that the development of a slightly more acid-stable urethan protecting group could be designed using the rates of solvolysis of the alcohol tosylates as a guide. In fact, Bláha and Rudinger (1965) had demonstrated the applicability of such an approach and pointed out that "the existence of a linear correlation between the rates of fission of these [urethan-type protecting] groups and the rates of tosylate solvolysis makes it possible to use the extensive rate data available for the latter reaction [and other reactions similarly correlated with it] to predict the stability to acid-catalyzed fission of carbamates derived from a range of further alcohols, and facilitates the rational choice of protecting groups of any desired stability within the limits of validity of this correlation." We quickly discovered, however, that the correlation was not sufficiently sensitive to permit the solution of a protecting group with an acid stability only slightly greater than that of butyloxycarbonyl. With the restrictions outlined above, the choice of alcohols for a new protecting group becomes quite restricted. Of the alcohols in this category, the tosylate solvolyses have not been studied under a single set of conditions. Thus some of the studies had been carried out under neutral conditions in aqueous dioxane or ethanol, while others were done in acetic acid. Variations were found also in the leaving groups studied, e.g., nitrobenzoate, chloride, and tosylate. These variations required us to make an educated guess based on the available data. Therefore, we determined the half-lives in acid media of the N-protected derivatives of phenylalanine and of the methyl ester of phenylalanylalanine (see Table I). The results led us to conclude that the 1-methylcyclobutyloxycarbonyl group should have sufficient stability in 50% acetic acid for use in gel filtration and yet be sufficiently acid labile for removal under fairly mild conditions. We found that the protecting group was completely removed in less than 30 min in trifluoroacetic acid. On the other hand, there was no detectable loss of protection after 48 hr in 50% aqueous acetic acid. Under the same conditions, the butyloxycarbonyl group was removed to the extent of about 10–15%.

We do not anticipate using the 1-methylcyclobutyloxycarbonyl protecting group routinely in place of the butyloxycarbonyl protecting group. We believe, however, that in the synthesis of high molecular weight polypeptides the more stable 1-methylcyclobutyloxycarbonyl protecting group should serve a useful function in the N-terminal protection of those fragments which will require purification by gel chromatography and which are likely to lack adequate solubility in water.

2. The Concept of "Stable Protecting Groups." As indicated above, a temporary protecting group is generally desired for the nitrogen of the carboxy-activated amino acid. In contrast, "a more permanent" protecting group is desired both for the ε-amino group of lysine and the sulfhydryl of cysteine.

Ideally, one would like to have the protecting groups for the α-amino nitrogen and for the ε-amino function of lysine designed in such a way that either one can be removed without affecting the other. The concept of such

Table I. Acid Cleavage of Various N-Protecting Groups of Phenylalanine and of Phenylalanylalanine Methyl Ester

Protecting group	$t_{1/2}$(X-phe) (min) 25°		$t_{1/2}$(X-phe-ala-OMe) (min)	
	TFA	Formic	TFA	Formic
(benzyl-CH₂-O-CO-) $C_6H_5-CH_2-O-CO-$	300	—	—	—
(cyclopropyl-CH₂-O-CO-)	40	—	—	—
(1-methylcyclobutyl-O-CO-) CH₃	2	—	3	180
(tert-butyl, $(CH_3)_3C-O-CO-$)	Complete 1 min.	4	Complete 1–2 min	10
(1-methylcyclohexyl-O-CO-) CH₃	Complete 1 min	2	—	—
(1-cyclopropylethyl-O-CO-) CH₃	—	1.5	—	—

two-directional selectivity has been nicely summarized by Bodanszky and Ondetti (1966). We should like to comment on it further. Total lack of reactivity of a protecting group to conditions employed for the removal of another protecting group for the same functionality (in this case amino nitrogen) is relatively rare. Often one is dealing not with absolutes (complete stability) but with relative differences in rate (see below). Clearly, a point can be reached where the *difference in the rates* of removal of two protecting groups becomes so great that for all practical purposes the difference in reactivity becomes indistinguishable from the "absolute" difference mentioned above.

Bodanszky and Ondetti (1966) referred to the benzyloxycarbonyl and butyloxycarbonyl protecting groups as meeting the requirement of complete two-directional selectivity. Indeed, the combination of these two protecting groups has been successfully employed in peptide synthesis. An outstanding example is the synthesis of ACTH by Schwyzer and his associates (Schwyzer and Rittel, 1961; Schwyzer and Sieber, 1966), in which the butyloxycarbonyl group was employed as the permanent and truly "stable" protecting group and the benzyloxycarbonyl group provided temporary protection for the α-amino nitrogen. The latter could be removed selectively by catalytic hydrogenation, conditions to which the butyloxycarbonyl group is completely inert.

When the roles of the above protecting groups are reversed, i.e., when the butyloxycarbonyl group becomes the temporary protecting group and the benzyloxycarbonyl functions as the permanent protecting group, the pair becomes an example of *relative* difference in stability and is now less satisfactory. These two blocking groups differ in their acid-catalyzed cleavage rates only by about three orders of magnitude (Losse *et al.*, 1968). In the synthesis of relatively small peptides, this difference in acid stability of the benzyloxycarbonyl and the butyloxycarbonyl groups—although not absolute—is sufficiently great to permit use of the former as a lysine-protecting group because purification of the final product and/or the intermediates is generally readily accomplished. In the synthesis of larger peptides or of proteins, the fact that the rates of acid-catalyzed hydrolysis of these two urethans differ only by a factor of about 2000 becomes objectionable because the separation of impurities lacking the ε-benzyloxycarbonyl blocking group is likely to be difficult. If a coupling reaction is subsequently effected at the ε-amino group rather than at the desired α-nitrogen, the resulting impurities may be very difficult to separate. This pitfall in the use of the benzyloxycarbonyl group as a permanent blocking group has been pointed out, e.g., by Halpern and Nitecki (1967). We have previously reported the results of quantitative studies in connection with the synthesis of ribonuclease S-protein which indicated that about 2–3 % of an ε-amino benzyloxycarbonyl protecting group might be lost during conventional procedures used for removal of the butyloxycarbonyl blocking group (Hirschmann, 1971). This was true even when trifluoroacetic acid in methylene chloride was used. That this is also a problem in solid-phase synthesis has been reported by Wang and Merrifield (1969).

It has not always been recognized that, because of the inherent chemical differences between the mechanisms of removal of the benzyloxycarbonyl and the butyloxycarbonyl blocking groups, a change in reaction conditions can alter the relative acid stabilities of these two blocking groups. Thus we had reported the observation by Dr. Walton that the addition of dimethylsulfide as a scavenger (to trap t-butylium ions which might otherwise alkylate methionine or aromatic amino acids) during acid-catalyzed removal of the butyloxycarbonyl blocking groups will lead to increased loss of the benzyloxycarbonyl group (Hirschmann, 1971). This is due to the fact that dimethylsulfide acting as a nucleophile will *enhance* the rate of decomposition of the benzyloxycarbonyl derivative but will—because of reduced solvent polarity—actually *decrease* the rate of decomposition of the butyloxy-carbonyl group. This can be explained by the fact that the former reaction shows considerable S_N2 character, whereas the latter is largely S_N1. These observations remind us that seemingly trivial changes in reaction conditions can markedly affect the relative stabilities of two protecting groups. Conversely, the absence of a good nucleophile can be used to rationalize the observation (Schnabel *et al.*, 1971; Hiskey *et al.*, 1971) that selective removal of the butyloxycarbonyl protecting group in the presence of the benzyloxy-carbonyl group can be effected by boron trifluoride ethereate in acetic acid. It is not known whether this method will be suitable for higher molecular weight polypeptides of limited solubility. Similarly Loffet and Dremier (1971) have reported the use of mercaptoethanesulfonic acid, also in glacial acetic acid, for the selective removal of the butyloxycarbonyl protecting group. It will also be of interest to know whether the presence of scavengers such as dimethyl sulfide would affect the selectivity of these reactions.

   a. Alternatives to the Use of the Benzyloxycarbonyl Protecting Group: Several alternatives could be considered to the use of the butyloxycarbonyl group as the temporary protecting group and the benzyloxycarbonyl group as the permanent protecting group. Reversing the role of the groups, which had served so well in the synthesis of ACTH, is not useful in the synthesis of proteins having sulfur-containing amino acids (ACTH has only one methionine). It is possible to utilize a more acid-labile protecting group in the place of the butyloxycarbonyl function. Indeed, Sieber and Iselin (1968*a,b*) have introduced the 2-(*p*-diphenyl-isopropyloxycarbonyl) group into peptide chemistry. The rate of acid-catalyzed cleavage proceeds up to 10,000 times faster than that of the butyloxycarbonyl group. This increased lability to acid permits the use of this protecting group in connection with benzyloxy-carbonyl because the difference in the relative rates of acid-catalyzed hydrolysis is great enough to make it an "absolute" difference. This new blocking group has therefore been widely and successfully used in peptide synthesis. We have, however, looked for an alternative solution because—as indicated above—we have found that in the purification of high molecular weight polypeptides on Sephadex it is often highly desirable to be able to use aqueous acetic acid as a solvent. The protecting group of Sieber does not have adequate stability to permit chromatography in this acidic solvent system.

It is also possible to increase the stability to acid of the benzyloxy-carbonyl protecting group. For example, negatively substituted analogues of the benzyloxycarbonyl group have been described (Noda et al., 1970). We suspect that these blocking groups are likely to aggravate further the solubility problems encountered with larger peptides. They may prove useful, however, in the synthesis of low molecular weight peptides in solution or in solid-phase synthesis where, as mentioned above, solubility is not a problem.

The use of the trifluoroacetyl protecting group in peptide chemistry was proposed by Weygand (Weygand and Csendes, 1952). Use of this protecting group is not practical when the synthetic scheme requires treatment of the protected peptide with hydrazine. The observation by Yajima et al. (1968) that the N-formyl blocking group is cleavable with hydrazine acetate places the same restriction on the use of this lysine-blocking group. The same argument applies also to the use of acetoacetylation (Marzotto et al., 1968) for protection of the ε-amino group.

b. Desiderata for a New Lysine-Protecting Group: We therefore searched for a new permanent protecting group for lysine which should satisfy, if possible, all of the following conditions:

1. It should be stable under all conditions commonly employed in peptide synthesis.
2. It should be *completely* stable to acid under conditions adequate for the acid-catalyzed removal of a temporary protecting group. The latter should be stable in 50 % aqueous acidic acid, a solvent system which we have employed for gel filtration.
3. For maximum flexibility, it should also be completely stable under the conditions required for the removal of the acetamidomethyl blocking group for cysteine (Veber et al., 1968, 1972a) [Hg (II) at pH 4] (see below) to permit one to remove the latter protecting group selectively.
4. Conversely, it should also be removable under conditions to which the acetamidomethyl blocking group is stable. This would permit removal of the lysine-blocking group without removing the acetamidomethyl blocking group.
5. The conditions required for the removal of the new protecting group should be mild enough not to cause unwanted side-reactions.
6. The new protecting group should increase rather than decrease the solubility of large peptides in polar solvents.
7. Peptides containing the new blocking group on the ε-amino nitrogen of lysine should, after hydrolysis at 110°C in 6 N HCl, give lysine or a lysine derivative detectable in amino acid analysis.

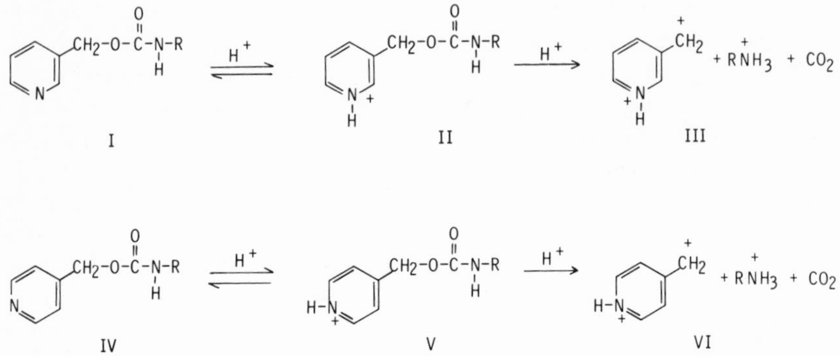

Fig. 1. Rationale for relative acid stabilities of the nicotinyloxycarbonyl and isonicotinyloxycarbonyl protecting groups.

8. The new protecting group should be useful also for the derivatization of proteins. This would permit comparison of derivatized natural proteins with synthetic intermediates.

c. The Isonicotinyloxycarbonyl Protecting Group: We tried to design a urethan protecting group which would be "completely" stable under the acidic conditions required for the removal of an α-amino-protecting group such as butyloxycarbonyl. A plausible solution to this problem was an analogue of the benzyloxycarbonyl blocking group in which the aromatic ring contains a basic nitrogen, on the assumption that nitrogen protonation would greatly enhance the acid stability of such a protecting group. Thus we studied the nicotinyloxycarbonyl (NOC) [I] and the isonicotinyloxycarbonyl (INOC) [IV] groups (see Fig. 1) (Veber *et al.*, 1972b). We anticipated that the presence of a full positive charge on the heteroatom of the aromatic ring would provide the "absolute" stability in acid which we desired. We expected that this should be particularly true of the isonicotinyloxycarbonyl protecting group in which the electron deficiency at the site of the developing carbonium ion should be greater [III vs. VI] (see Fig. 1).

We hoped, moreover, that the isonicotinyloxycarbonyl group might be removable with zinc.* The nicotinyloxycarbonyl group was expected to be less readily reducible under these conditions because of the formation of a charge-separated intermediate [VIII] in the reduction of this derivative. The corresponding intermediate [VII] in the removal of the isonicotinyloxy-carbonyl group would be neutral (see Fig. 2). Finally, we anticipated that both protecting groups should be removable by catalytic hydrogenation,†

*Camble *et al.* (1969) have reported the chemical reduction of 4-picolyl esters.
†The removal of both protecting groups from α-amino nitrogen by catalytic hydrogenation has recently been reported in a Belgian patent (No. 754,412).

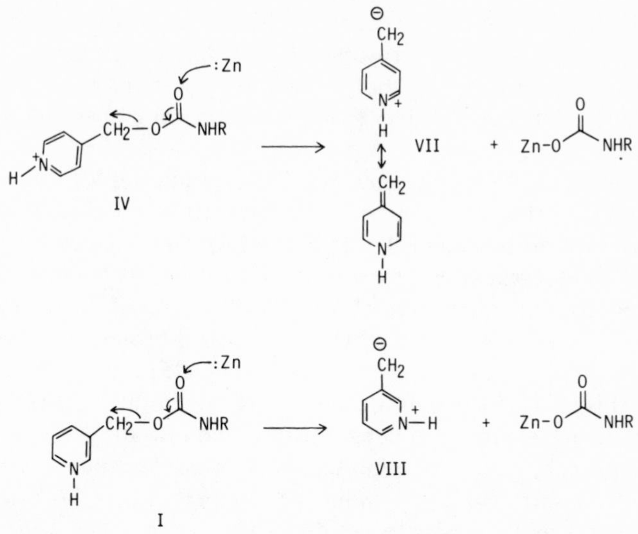

Fig. 2. Intermediates in the zinc reduction of [IV] and [I].

provided that the peptide contains no or only few sulfur-containing amino acids and that the molecular weight is not significantly higher than that of ACTH. Because of these restrictions, the proposed protecting groups would not have been of sufficiently broad usefulness if catalytic hydrogenation were the only cleavage method available.

Our expectations that these protecting groups would be highly acid stable were confirmed. Both protecting groups were found to be stable in anhydrous acids, including trifluoroacetic acid and hydrogen fluoride (Veber et al., 1972b). They are only slowly cleaved even when the counter ion of the proton is a good nucleophile, as with HBr in either acetic acid or ethyl acetate. As expected, the nicotinyloxycarbonyl group was cleaved more rapidly. The high degree of acid stability of [I] and especially of [IV] has thus given us the desired "absolute" stability in acid.

We have also found that the isonicotinyloxycarbonyl group is removed smoothly by zinc in 50% aqueous acetic acid. Both groups are removed by catalytic hydrogenation. However, the nicotinyloxycarbonyl group is only slowly removed by zinc under acidic conditions. Thus the isonicotinyloxy-carbonyl group appeared to be the protecting group of choice for the ε-amino group of lysine.

ε-Isonicotinyloxycarbonyl-lysine has been used in these laboratories (Holly et al., unpublished) for the preparation of lysine-containing peptides by both solution and solid-phase techniques. The syntheses proceeded well, and the removal of the protecting group went smoothly using either zinc or catalytic hydrogenation. In the course of these studies, it was found that the solubility properties of the protected lysine derivatives are highly satisfactory. The isonicotinyloxycarbonyl derivatives were more soluble in both organic and aqueous media than corresponding benzyloxycarbonyl derivatives.

We wished to check the generality of the conditions for removal of the isonicotinyloxycarbonyl group with zinc. Accordingly reduced acetamidomethylated natural S-protein (Veber *et al.*, 1969) was treated with isonicotinyl-succinimidocarbonate; this introduced a total of six out of possible nine tritium-labeled isonicotinyloxycarbonyl blocking groups on the free amino groups of S-octa-acetamidomethyl-octahydro S-protein (Paleveda, unpublished). No attempt was made to obtain complete protection of the amino groups in this study. Enzymatic degradation of the resulting protein showed that the radioactive label had been incorporated into the ε-amino group of lysine and probably into the amino-terminal serine. As expected, amino acid analysis of the enzymic digest indicated the presence of both lysine and ε-isonicotinyloxycarbonyl-lysine residues. The latter eluted with histidine on the standard amino acid analyzer columns. Total removal (more than 99%) of the radioactive label could be accomplished by the use of a large excess of zinc dust in 50% aqueous acetic acid using high-speed stirring (Palevada, unpublished). Although the removal of the protecting group from a protein is a slower process and requires a larger excess of zinc than is the case with smaller peptides, it can be carried to completion, and thus the use of the isonicotinyloxy group appears to be useful in protein chemistry.

Figure 3 outlines the chemistry involved in the preparation of ε-INOC-lysine [IX] and α-Boc-ε-INOC-lysine [X]. The experimental details of these preparations will be reported elsewhere.

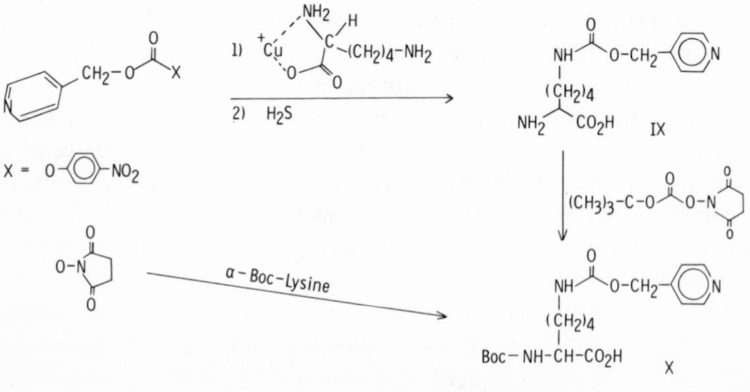

Fig. 3. Preparation of α-Boc-ε-INOC-Lysine.

B. Protection of the Thiol of Cysteine

The sulfhydryl of cysteine resembles the ε-amino group of lysine in that its protecting group should be stable throughout a synthesis. In a preliminary communication (Veber et al., 1968), we described the acetamidomethyl protecting group for cysteine which we have employed in the synthesis of ribonuclease S-protein (Denkewalter et al., 1969). A detailed description of our experiences with this sulfhydryl-protecting group has appeared elsewhere (Veber et al., 1972a). In the latter paper, we concluded that the acetamidomethyl blocking group satisfies all of the following requirements:

1. It can be used for peptide synthesis either in solution or in the solid-phase method.
2. It may be introduced in high yield under mild conditions into cysteine-containing proteins such as reduced S-protein, and it can be removed from such derivatized proteins.
3. It is stable to both the mild acidic conditions required to remove, e.g., the butyloxycarbonyl blocking group, and to treatment with liquid HF at 0°C.
4. It is stable also to conditions required to convert esters to hydrazides and hydrazides to azides, and to the weakly basic conditions employed in typical peptide coupling reactions.
5. The blocking group may be removed smoothly with Hg (II) under mild conditions.

More recently, we have also been able to show that this protecting group is stable under the conditions required for the removal of the isonicotinyloxycarbonyl protecting group via either chemical or catalytic reduction. In the synthesis of a lysine- and cysteine-containing peptide or protein, it is therefore possible to remove selectively one of the following groups: the benzyloxycarbonyl, the isonicotinyloxycarbonyl, or the acetamidomethyl. Thus it should be possible, for example, to generate a disulfide bridge either prior to or after the removal of the lysine-protecting groups.

III. CONCLUSION

We feel that the tactics resulting from the use of the protecting groups described herein are suitable for protein synthesis. In this scheme, the α-amino group would be protected using such acid-labile groups as carbamate, thiocarbamate, butyloxycarbonyl, or 1-methylcyclobutyloxycarbonyl. The latter two blocking groups are stable under the conditions suitable for removal of either isonicotinyloxy or acetamidomethyl blocking groups. The ε-amino group of lysine would be protected by isonicotinyloxycarbonyl, which is stable to acidic conditions and to Hg (II) ion but can be removed reductively. Similarly, the acetamidomethyl group is stable under the various conditions suitable for the removal of the other protecting groups mentioned above. We believe that tactics based on the use of protecting groups removable by *chemically distinct* methods should serve to reduce side-reactions in the synthesis of proteins.

ACKNOWLEDGMENT

This review is based on the work of all of the members of the Merck peptide group. We wish to acknowledge the unpublished data cited herein by Dr. F. W. Holly, Mr. W. J. Paleveda, Mrs. R. F. Nutt, Dr. G. Gal, and Mr. R. G. Strachan. It is a pleasure to acknowledge the studies of Dr. S. F. Brady on the novel acid labile protecting groups discussed herein prior to full publication.

REFERENCES

Anderson, G. W., 1953, The synthesis of an arginyl peptide, *J. Amer. Chem. Soc.* **75**:6081.

Bailey, J. L., 1949, A new peptide synthesis, *Nature* **164**:889.

Bailey, J. L., 1950, The synthesis of simple peptides from anhydro-*N*-carboxy-amino-acids, *J. Chem. Soc.* **1950**:3461.

Bartlett, P. D., and Dittmer, D. C., 1957, A kinetic study of the Leuchs anhydrides in aqueous solution. II., *J. Am. Chem. Soc.* **79**:2159.

Bartlett, P. D., and Jones, R. H., 1956, A kinetic study of the Leuchs anhydrides in aqueous solution. I., *J. Am. Chem. Soc.* **79**:2153.

Bláha, K., and Rudinger, J., 1965, Amino acids and peptides. XLVIII. Rates of fission of some cycloalkyloxycarbonylglycines with hydrogen bromide in acetic acid, *Coll. Czech. Chem. Commun.* **30**:599.

Bodanszky, M., and Ondetti, M. A., 1966, "Peptide Synthesis," Interscience, New York.

Camble, R., Garner, R., and Young, G. T., 1969, Amino-acids and peptides. Part XXX. Facilitation of peptide synthesis by the use of 4-picolyl esters for carboxy-group protection, *J. Chem. Soc. (C)* **1969**:1911.

Denkewalter, R. G., Schwam, H., Strachan, R. G., Beesley, T. E., Veber, D. F., Schoenewaldt, E. F., Barkemeyer, H., Paleveda, W. J., Jr., Jacob, T. A., and Hirschmann, R., 1966, The controlled synthesis of peptides in aqueous medium. I. The use of α-amino acid *N*-carboxyanhydrides, *J. Am. Chem. Soc.* **88**:3163.

Denkewalter, R. G., Veber, D. F., Holly, F. W., and Hirschmann, R., 1969, Studies on the total synthesis of an enzyme. I. Objective and strategy, *J. Am. Chem. Soc.* **91(2)**:502.

Dewey, R. S., Schoenewaldt, E. F., Joshua, H., Paleveda, W. J., Jr., Schwam, H., Barkemeyer, H., Arison, B. H., Veber, D. F., Denkewalter, R. G., and Hirschmann, R., 1968, Synthesis of peptides in aqueous medium. V. Preparation and use of 2,5-thiazolidinediones (NTA's). Use of the ^{13}C-H nuclear magnetic resonance signal as internal standard for quantitative studies, *J. Am. Chem. Soc.* **90**:3254.

Dewey, R. S., Barkemeyer, H., and Hirschmann, R., 1969, Use of the *N*-hydroxysuccinimide ester of α-butyloxycarbonylglutamine in peptide synthesis, *Chem. Ind.* **1969**:1632.

Dewey, R. S., Schoenewaldt, E. F., Joshua, H., Paleveda, W. J., Jr., Schwam, H., Barkemeyer, H., Arison, B. H., Veber, D. F., Strachan, R. G., Milkowski, J., Denkewalter, R. G., and Hirschmann, R., 1971, The synthesis of peptides in aqueous medium. VII. The preparation and use of 2,5-thiazolidinediones in peptide synthesis, *J. Org. Chem.* **36(1)**:49.

Gish, D. T., and Carpenter, F. H., 1953, Preparation of arginyl peptides, *J. Am. Chem. Soc.* **75**:5872.

Halpern, B., and Nitecki, D. E., 1967, The deblocking of *t*-Butyloxycarbonyl-peptides with formic acid, *Tetrahedron Letters* **1967**(31):3031.

Hirschmann, R., 1971, Synthesis of an enzyme. Accomplishments and remaining problems, *Intra-Sci. Chem. Rep.* **5**(3):203.

Hirschmann, R., Strachan, R. G., Schwam, H., Schoenewaldt, E. F., Joshua, H., Barkemeyer, B., Veber, D. F., Paleveda, W. J., Jr., Jacob, T. A., Beesley, T. E., and Denkewalter, R. G., 1967, The controlled synthesis of peptides in aqueous medium. III. Use of Leuchs' anhydrides in the synthesis of dipeptides. Mechanism and control of side reactions, *J. Org. Chem.* **32**:3415.

Hirschmann, R., Schwam, H., Strachan, R. G., Schoenewaldt, E. F., Barkemeyer, H., Miller, S. M., Conn, J. B., Garsky, V., Veber, D. F., and Denkewalter, R. G., 1971. The controlled synthesis of peptides in aqueous medium. VIII. The preparation and use of novel α-amino acid N-carboxyanhydrides, *J. Am. Chem. Soc.* **93**(11):2746.

Hiskey, R. G., Beacham, L. M., Matl, V. G., Smith, J. N., Williams, E. B., Jr., Thomas, A. M., and Walters, E. T., 1971, Sulfer-containing polypeptides. XIV. Removal of the *tert*-butyloxycarbonyl group with boron trifluoride etherate, *J. Org. Chem.* **36**:488.

Holly, F. W., Paleveda, W. J., Nutt, R. F., and Gal, G., Unpublished observation from these laboratories.

Hunt, M., and du Vigneaud, V., 1938: The preparation of *d*-alanyl-*l*-histidine and *l*-alanyl-*l*-histidine and an investigation of their effect on the blood pressure in comparison with *l*-carnosine, *J. Biol. Chem.* **124**:699.

Iwakura, Y., Uno, K., Oya, M., and Katakai, R., 1970, Stepwise synthesis of oligopeptides with N-carboxy α-amino acid anhydrides, *Biopolymers* **9**:1419.

Jenkins, S. R., Nutt, R. F., Dewey, R. S., Veber, D. F., Holly, F. W., Paleveda, W. J., Jr., Lanza, T., Jr., Strachan, R. G., Schoenewaldt, E. F., Barkemeyer, H., Dickinson, M. J., Sondey, J., Hirschmann, R., and Walton, E., 1969, Studies on the total synthesis of an enzyme. III. Synthesis of a protected hexacontapeptide corresponding to the 65–24 sequence of ribonuclease A, *J. Am. Chem. Soc.* **91**(2):505.

Katakai, R., Oya, M., Uno, K., and Iwakura, Y., 1971, Stepwise synthesis of oligopeptides with N-carboxy α-amino acid anhydrides. II. Oligopeptides with some polar side chains, *Biopolymers* **10**:2199.

Lemieux, R. U., and Barton, M. A., 1971, Peptide conformations. I. Nuclear magnetic resonance study of the carbamate reaction of amino acids and peptides, *Can. J. Chem.* **49**(5):767.

Loffet, A., and Dremier, C., 1971, A new reagent for the cleavage of the tertiary butyloxycarbonyl protecting group, *Experientia* **27**:1003.

Leuchs, H., 1906, Über die Glycin-carbonsäure, *Berichte* **39**:857.

Losse, G., Zeidler, D., and Grieshaber, T., 1968, Kinetik der säuren Abspaltung von N- und C-Schutzgruppen bei Peptiden, *Annalen* **715**:196.

Marzotto, A., Pajetta, P., Galzigna, L., and Scoffone, E., 1968, Reversible acetoacetylation of amino groups in proteins, *Biochim. Biophys. Acta* **154**:450.

McKay, F. C., and Albertson, N. F., 1957, New amino-masking groups for peptide synthesis, *J. Am. Chem. Soc.* **79**:4686.

Merrifield, R. B., 1969, Solid-phase peptide synthesis, *in* "Advances in Enzymology," Vol. 32 (F. F. Nord, ed.), pp. 221–296, Interscience, New York.

Nitecki, D., 1971, Immunologically active peptides of glucagon, *Intra-Sci. Chem. Rep.* **5**(4):295.

Noda, K., Terada, Sh., and Izumiya, N., 1970, Modified benzyloxycarbonyl groups for protection of ε-amino group of lysine, *Bull. Chem. Soc. Japan* **43**:1883.

Paleveda, W. J., Jr., Unpublished observation from these laboratories.

Schnabel, E., Klostermeyer, H., and Berndt, H., 1971, Zur selektiven acidolytischen Abspaltbarkeit der tert-Butyloxycarbonyl-Gruppe, *Ann. Chem.* **749**:90.

Schwyzer, R., and Rittel, W., 1961, Synthese von Peptid-Zwitschenprodukten für den Aufbau eines corticotrop wirksamen Nonadecapeptids. I. N^ε-t-Butyloxycarbonyl-L-lysin, N^α-(N^ε-t-Butyloxycarbonyl-L-lysyl)-N^ε-t-butyloxycarbonyl-L-lysin, N^ε-t-Butyloxycarbonyl-L-lysyl-L-prolyl-L-valyl-glycin und Derivate, *Helv. Chim. Acta* **44**:159.

Schwyzer, R., and Sieber, P., 1966, Die Totalsynthese des β-Corticotropins (adrenocorticotropes Hormon; ACTH), *Helv. Chim. Acta* **49**:134.

Sieber, P., and Iselin, B., 1968a, Use of new N-aralkyloxycarbonyl protecting groups in peptide synthesis, in "Peptides 1968" (E. Bricas, ed.), p. 85, North-Holland, Amsterdam.

Sieber, P., and Iselin, B., 1968b, 77. Selektive acidolytische Spaltung von Aralkyloxycarbonyl-Aminoschutzgruppen, *Helv. Chim. Acta* **51**:614.

Siegfried, M., 1905a, Über die Bindung von Kohlensäure durch amphotere Amidokorper. II. Mitteilung, *Z. Physiol. Chem.* **44**:85.

Siegfried, M., 1905b, Über die Bindung von Kohlensäure durch amphotere Amidokorper, *Z. Physiol. Chem.* **46**:401.

Siegfried, M., 1906, Über die Abscheidung von Amidosäuren, *Chem. Ber.* **39**:358.

Stadie, W. C., and O'Brien, H., 1936, The carbamate equilibrium. I. The equilibrium of amino acids, carbon dioxide, and carbamates in aqueous solution; with a note on the Ferguson–Roughton carbamate method, *J. Biol. Chem.* **112**:723.

Strachan, R. G., Unpublished observation from these laboratories.

Veber, D. F., Milkowski, J. D., Denkewalter, R. G., and Hirschmann, R., 1968, The synthesis of peptides in aqueous medium. IV. A novel protecting group for cysteine, *Tetrahedron Letters* **26**:2057.

Veber, D. F., Varga, S. L., Milkowski, J. D., Joshua, H., Conn, J. B., Hirschmann, R., and Denkewalter, R. G., 1969, Studies on the total synthesis of an enzyme. IV. Some factors affecting the conversion of protected S-protein to ribonuclease S', *J. Am. Chem. Soc.* **91(2)**:506.

Veber, D. F., Milkowski, J., Varga, S. L., Denkewalter, R. G., and Hirschmann, R., 1972a, Acetamidomethyl. A novel thiol protecting group for cysteine, *J. Am. Chem. Soc.* **94**:5456.

Veber, D. F., Brady, S. F., and Hirschmann, R., 1972b, Some novel amine protecting groups, in "Chemistry and Biology of Peptides, Proceedings of the Third American Peptide Symposium" (J. Meienhofer, ed.), Ann Arbor Science Publishers, Ann Arbor, Mich.

Wang, S. S., and Merrifield, R. B., 1969, Preparation of some new biphenylisopropyloxycarbonyl amino acids and their application to the solid phase synthesis of a tryptophan-containing heptapeptide of bovine parathyroid hormone, *Internat. J. Protein Res. I* **1969**:235.

Wesseley, F., 1925, Untersuchungen über α-Amino-N-Carbonsäureanhydride. I., *Z. Physiol. Chem.* **146**:72.

Weygand, F., and Csendes, E., 1952, N-Trifluoroacetyl-aminosäure, *Angew. Chem.* **64(5)**:136.

Yajima, H., Kawasaki, K., Okada, Y., Minami, H., Kubo, K., and Yamashite, I., 1968, Studies on peptides. XVI. Regeneration of lysine from N^ε-formyllysine by aqueous hydrazine or hydroxylamine and their application to the synthesis of α-melanocyte-stimulating hormone, *Chem. Pharm. Bull.* **16(5)**:919.

CHAPTER 7

PEPTIDE SYNTHESIS AND
THE SPECIFICITY OF PROTEINASES

Joseph S. Fruton

Kline Biology Tower
Yale University
New Haven, Connecticut

It was my good fortune to share a laboratory with Leonidas Zervas at the Rockefeller Institute for Medical Research after he came there in 1934 to rejoin Max Bergmann. The year before, the Nazis had obliged Bergmann to resign as Director of the Kaiser Wilhelm Institute for Leather Research in Dresden; after finding a haven in New York, he secured the help of the Rockefeller Foundation in bringing Zervas to the United States. I cannot express adequately my gratitude to Professor Zervas for the instruction he gave me in the art of peptide synthesis during his 2-year stay in New York. This essay is not only an act of homage to a great chemist, but also an acknowledgment of the debt owed him by one of his students.

Among the first fruits of the carbobenzoxy method announced by Bergmann and Zervas in 1932 was its application to the study of the specificity of peptidases. The initial papers (published in 1933) on this subject dealt with dipeptidase and carboxypeptidase, and the views of the Bergmann laboratory about these enzymes were strongly influenced by the work of Grassmann and Waldschmidt-Leitz, two former students of Willstätter. It should be recalled that during the early 1930s, Waldschmidt-Leitz was fighting a rear-guard action in defense of Willstätter's view that enzymes are small molecules of unknown constitution adsorbed on catalytically inactive protein carriers. Sumner's claim (in 1926) to have obtained urease in the form of a crystalline protein was only then becoming accepted, as a consequence of Northrop's crystallization of pepsin in 1930 and his clear demonstration that the catalytic activity of this enzyme is a property of a protein molecule.

My first assignment in Bergmann's American laboratory was to continue the work begun in Dresden on the dipeptidase of intestinal mucosa. In the paper (Bergmann et al., 1935) describing these results, the "polyaffinity" theory of enzyme action was presented, and it was elaborated in a succeeding paper: "If an enzyme catalyzes only one of two antipodes, then it must contain at least three different atoms or atomic groups which are fixed in space with respect to one another, these groups entering during the catalysis into relation with a similar number of different atoms or atomic groups of the substrate" (Bergmann et al., 1936). An extension of this hypothesis appeared 12 years later, when Ogston (1948) explained the asymmetrical conversion of a symmetrical molecule during an enzyme-catalyzed reaction in terms of a three-point attachment of the substrate. During the course of our work on dipeptidase, we showed that peptides such as L-leucyl-D-alanine are cleaved, and we explained this result by assuming that, in its interaction with the enzyme, this peptide is in the conformation shown. Many years later, my work on the specificity of dipeptidyl transferase (cathepsin C), which cleaves dipeptidyl units from the amino terminus of polypeptide chains, led me to revive this hypothesis and to assume that a synthetic substrate such as D-alanyl-L-phenylalaninamide has a similar conformation in the ezyme–substrate complex (Izumiya and Fruton, 1956) (Fig. 1).

After our work on peptidases, we turned to the specificity of papain, which Grassmann had partially purified, and the dried papaya latex needed for the purification was readily available to Bergmann through his close friend Leo Wallerstein. By 1936, Kunitz and Northrop had described the crystallization of chymotrypsin and trypsin, and synthetic peptide substrates were soon found for these proteinases as well (Bergmann and Fruton, 1937; Bergmann et al., 1939). The success achieved in these efforts encouraged us to undertake the search for synthetic substrates for pepsin.

In the mid-1930s, pepsin was still considered by many biochemists to effect a physical deaggregation of peptides thought to be associated by noncovalent bonds to form macromolecular proteins. This denial that pepsin catalyzes the hydrolysis of peptide bonds was based on the argument that no well-defined peptide substrates had yet been found for this enzyme. More-

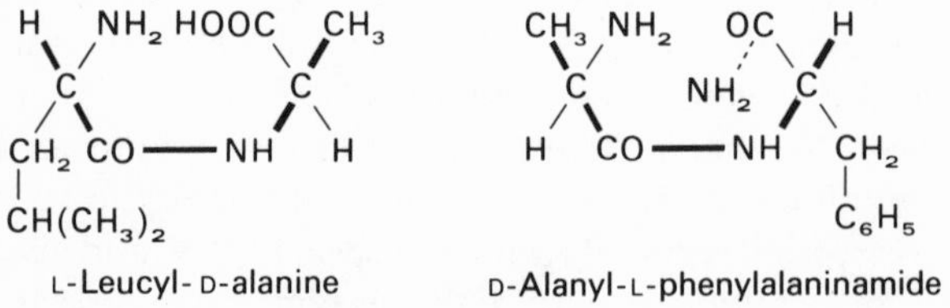

L-Leucyl-D-alanine   D-Alanyl-L-phenylalaninamide

Fig. 1

over, it was also believed that because of its pH optimum near 2, pepsin interacts with cationic groups in the protein substrate. With the finding (Fruton and Bergmann, 1938, 1939) that Z-Glu-Tyr and Z-Glu-Tyr-NH$_2$ are hydrolyzed by pepsin, these views had to be abandoned. These synthetic substrates were cleaved very slowly, however, and were insoluble at the acid end of the pH range of peptic activity. A decisive advance was made by Lillian Baker, who worked in Bergmann's laboratory during World War II, in showing that acetyl dipeptides such as Ac-Phe-Tyr are hydrolyzed by pepsin much more rapidly than Z-Glu-Tyr. Her work provided the first indication that the action of pepsin on small synthetic substrates is favored by the presence of aromatic amino acids on both sides of the sensitive peptide bond. Acetyl dipeptides of the kind introduced by Baker have been widely used in recent studies on the kinetics of pepsin action (for a review, see Fruton, 1971). After 1950, the extensive studies on the amino acid sequence of proteins led to the view that pepsin is an enzyme possessing a broad side-chain specificity (Hill, 1965). This conclusion should be seen, however, in relation to the fact that in most sequence studies large amounts of enzyme and prolonged incubation periods are employed, thus obscuring differences in the rates of the enzymic cleavage of various kinds of peptide bonds.

The uncertain state of the problem of the specificity of pepsin prompted me, in 1965, to resume research on the action of this enzyme on synthetic substrates. This work was greatly facilitated by the notable improvements made in the art of peptide synthesis during the preceding two decades. Moreover, the dramatic advances in the study of the three-dimensional structure of enzyme proteins by means of X-ray crystallography had provided a clearer view of enzyme–substrate interaction. Although a detailed model of the structure of pepsin has not yet been proposed, our "substrate-oriented" approach has defined some aspects of the interaction of pepsin with its substrates and may help in the future interpretation of the model when it becomes available. Also, our recent results have suggested more general considerations relating to the action of enzymes that act on oligomeric substrates. In the remainder of this chapter, I shall attempt to summarize the present state of our work on the specificity of pepsin action.

Our first objective was to prepare peptide substrates whose structure could be varied systematically and whose rate of cleavage was considerably greater than that of the acetyl dipeptides of the kind introduced by Baker, the most sensitive of which was Ac-Phe-TyrI$_2$ [k_{cat} = 0.2 sec^{-1}, K_M = 0.08 mM; Jackson et al., 1965). We wished initially to avoid the presence of a carboxyl group in our substrates, because the pK_a of this group falls in the range of pepsin activity (pH 1–5). To study pepsin action in aqueous solution, however, it was necessary to introduce a charged hydrophilic group; such groups included the imidazolium group of a histidyl residue (Inouye and

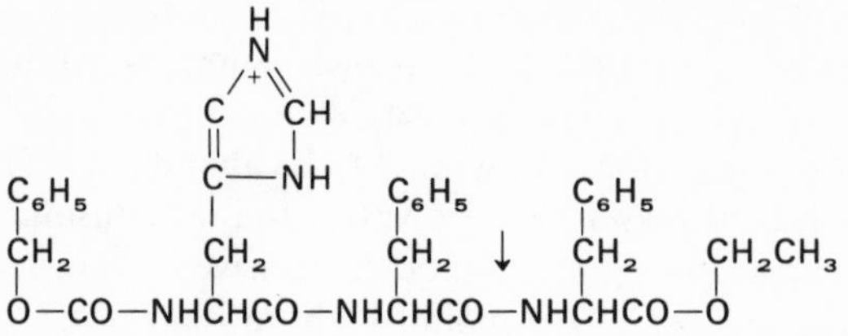

Benzyloxycarbonyl-L-histidyl-L-phenylalanyl-L-phenylalanine
ethyl ester

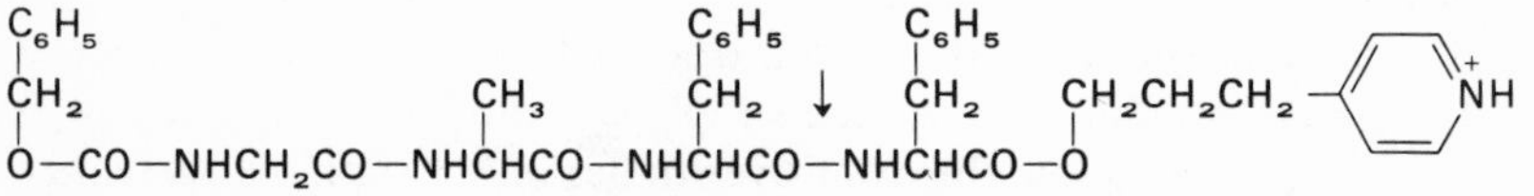

Benzyloxycarbonyl-glycyl-L-alanyl-L-phenylalanyl-L-phenylalanine
4(3-hydroxypropyl)pyridyl ester

Fig. 2

Fruton, 1967) or the pyridinium group at the carboxyl or amino terminus of the peptide (Sachdev and Fruton, 1969). Two typical substrates are shown in Fig. 2; at 37°C, Z-His-Phe-Phe-OEt was found to be hydrolyzed optimally by pepsin near pH 4, where $k_{cat} = 9.3\,\text{sec}^{-1}$ and $K_M = 0.2\,\text{mM}$ (Hollands et al., 1969), and under the same conditions Z-Gly-Ala-Phe-Phe-OP4P is cleaved with $k_{cat} = 400\,\text{sec}^{-1}$ and $K_M = 0.1\,\text{mM}$ (Sachdev and Fruton, 1970). In the formulas, the vertical arrow denotes the site of enzymic cleavage; the other peptide bonds of the substrates were not attacked to a measurable extent under the conditions of our studies.

To define more precisely the side-chain specificity of pepsin, a series of peptide derivatives of the general structure Z-His-X-Y-OMe was synthesized, where X and Y are L-amino acid residues (or glycyl), and the kinetic parameters of the enzymic hydrolysis of the X—Y bond were determined (Trout and Fruton, 1969). In all cases, either X or Y was an L-phenylalanyl residue. Among the substrates of the type Z-His-X-Phe-OMe, the two that were cleaved most rapidly were those in which X is Phe or p-nitro-L-phenylalanyl [Phe(NO$_2$)]. The favorable effect of the aromatic and planar substituent at the β-carbon of the X-residue was emphasized by the finding that when X is β-cyclohexyl-L-alanyl the value of k_{cat} is approximately one-tenth that when X is Phe, and similar to that for substrates in which the X-position is occupied by an aliphatic amino acid residue larger than alanyl (norvalyl, norleucyl, leucyl, methionyl). Apparently, the side-chains of these amino acids can interact with a portion of the enzymic region that binds planar aromatic groups. Of special interest was the finding that under conditions where Z-His-Gly-Phe-OMe is hydrolyzed at a measurable rate (about 1 % that when X is Phe), the comparable compound with X being Val or Ile is completely resistant to peptic action. This suggests that steric hindrance is operative and invites the hypothesis that, when the X-position is occupied by a residue that is branched at the β-carbon, one of the catalytic groups of pepsin may be prevented from attacking the carbonyl group of the sensitive bond. As for the substrates of the type Z-His-Phe-Y-OMe, all the substrates with an aliphatic side-chain in the Y-residue (including those with branching at the β-carbon) are cleaved slowly at rates that are quite similar. These are hydrolyzed at about 1 % the rate for the most sensitive among this group of substrates, Z-His-Phe-Trp-OMe; those with Y being Phe or Tyr are cleaved somewhat less rapidly. That this preference for Trp, Phe, and Tyr in the Y-position may be related to their aromatic character is suggested by the relative resistance of the compound in which Y is β-cyclohexyl-L-alanyl and speaks for a preference for a planar aromatic side-chain over the similarly hydrophobic but nonplanar alicyclic side-chain. It may be concluded, therefore, that with small synthetic substrates of the type AX-YB, where the X—Y bond is broken, pepsin shows a preference for Phe in the X-position and for Trp, Phe, or Tyr in the Y-position. Only limited data are available on the effect of changing the A and B groups on the side-chain specificity of pepsin, but it would appear that this conclusion holds for substrates in which the A group is changed from Z-His to Ac or Gly-Gly. As will be seen later, the effect of changes in the A and B groups of a synthetic substrate A-Phe-Phe-B may be very large, thus offering an explanation for the apparently broad specificity of pepsin when it is allowed to act on proteins for a long time.

In its action on the Phe–Phe bond of substrates of the type A-Phe-Phe-B, pepsin exhibits an absolute requirement for the L-enantiomer in both the X- and Y-positions of the substrate. Diastereoisomeric compounds, such as Z-His-D-Phe-Phe-OEt or Z-His-Phe-D-Phe-OEt, are competitive inhibitors of the cleavage of the L-L-L substrate, and the K_I values found (0.2–0.3 mM) are the same as the kinetically determined value of K_M for Z-His-Phe-Phe-OEt under comparable conditions (Inouye and Fruton, 1967). It should be added that the Phe-Phe-OEt also inhibits pepsin action competitively with a K_I of about 0.2 mM, thus indicating that this unit contributes the principal binding energy to the interaction of Z-His-Phe-Phe-OEt with pepsin. The resistance of the D-Phe-Phe or Phe-D-Phe unit of a diastereoisomer of this substrate does not arise from weaker binding but rather from faulty positioning of the Phe–Phe bond with respect to the catalytic groups of the enzyme. A contribution to the positioning of this bond for productive attack may be provided by the carboxyl-terminal CO group of the Phe–Phe unit, since Z-His-Phe-L-phenylalaninol and its acetate were found to be resistant to pepsin action (Inouye and Fruton, 1967, 1968). Further work is needed to explain this finding; in particular, it will be of interest to examine the susceptibility of a substrate analogue in which the $CO-OC_2H_5$ (or $CO-OCH_3$) group has been replaced by $CO-C_2H_5$ (or $CO-CH_3$).

As noted above, for a series of pepsin substrates of the type A-Phe-Phe-B, changes in the A and B groups can have large effects on the rate of the enzymic cleavage of the Phe–Phe bond. For example, at pH 4 and 37°C, Z-His-Phe(NO$_2$)-Phe-OMe is hydrolyzed with $k_{cat} = 0.3\ sec^{-1}$ and $K_M = 0.4\ mM$; replacement of the OMe group by Ala-Ala-OMe gives a substrate for which $k_{cat} = 28\ sec^{-1}$ and $K_M = 0.1\ mM$ (Medzihradszky et al., 1970). Even more striking differences in the catalytic efficiency (as measured by k_{cat}) have been observed with substrates of the type A-Phe-Phe-OP4P (Sachdev and Fruton, 1969, 1970). For the series where A is Z, Z-Gly, or Z-Gly-Gly, the k_{cat} values (in sec^{-1}) are 0.7, 3.1, and 72, respectively, with corresponding K_M values (in mM), of 0.2, 0.4, and 0.4. Moreover, when the Gly-Gly unit of Z-Gly-Gly-Phe-Phe-OP4P is replaced by other dipeptidyl units, very large differences in k_{cat} are found. To cite only three of the many peptide derivatives tested, when A is Z-Gly-Ala, $k_{cat} = 400\ sec^{-1}$ and $K_M = 0.1\ mM$, when A is Z-Gly-D-Ala, $k_{cat} = 0.2\ sec^{-1}$ and $K_M = 0.3\ mM$, and when A is Z-Gly-Pro, $k_{cat} = 0.06\ sec^{-1}$ and $K_M = 0.1\ mM$. It will be seen that the K_M values for these various substrates are very similar, and the available data are consistent with the view that $K_M \cong K_S$ (the dissociation constant of the rate-limiting enzyme–substrate complex) in the cleavage of peptide substrates by pepsin. It would appear, therefore, that the large differences in catalytic efficiency are not a consequence of differences in the total binding energy in the interaction of the enzyme and its substrate but that other factors play a significant role. Since the variations in k_{cat} are a consequence of structural alterations in the substrate at loci other than the sensitive Phe–Phe unit, it may be concluded that the A and B groups of A-Phe-Phe-B participate in "secondary interactions" that can modify the catalytic efficiency to an important extent.

The data in hand indicate that for pepsin substrates of the type A-Phe-Phe-OP4P, where A is a Z-dipeptidyl group, this group interacts with the enzyme as a unit and that the benzyl portion of the benzyloxycarbonyl group participates significantly in this interaction. To study more directly the secondary interaction of such an amino-terminal substituent with pepsin, we have recently turned to the fluorescent-probe technique (Edelman and McClure, 1968; Stryer, 1968), and have replaced the hydrophobic benzyloxycarbonyl group by the 1-dimethylaminonaphthalenesulfonyl (dansyl, Dns) group. When the dansyl group is transferred from an aqueous environment to a nonpolar environment, or is bound to proteins, its fluorescence is enhanced and the emission maximum is shifted to a shorter wavelength.

A series of Dns-peptide esters of the type A-Phe-Phe-B was synthesized and found to be hydrolyzed by pepsin at the Phe–Phe bond with k_{cat} values roughly one-tenth of those found for the corresponding Z-peptide esters; structural changes in the A group (e.g., replacement of Dns-Gly-Gly by Dns-Gly-Ala or Dns-Gly-Pro) produced the same changes in k_{cat} as those observed in the Z-peptide series (Sachdev *et al.*, 1972).

When a relatively poor substrate such as Dns-Gly-Pro-Phe-Phe-OP4P (15.6 μM) is mixed with pepsin (12.6 μM) at pH 3.1, there is a shift in the uncorrected emission maximum of about 18 nm (from 536 to 518 nm) and a nearly threefold increase in the fluorescence intensity at the maximum. From a series of experiments in which the pepsin concentration was varied, the value of the dissociation constant of the peptide–pepsin complex was estimated and found to be the same (within the precision of the measurements) as the kinetically determined K_M value for the enzymic hydrolysis of this substrate. That the binding of the dansyl group of Dns-Gly-Pro-Phe-Phe-OP4P depends in large part on the interaction of the Phe-Phe-OP4P portion of the substrate with the catalytic site of the enzyme is indicated by the finding that the enhancement of fluorescence was greatly decreased when active pepsin was replaced by pepsin that had been stoichiometrically inactivated with the active-site-directed inhibitor tosyl-L-phenylalanyl-diazomethane (Delpierre and Fruton, 1966). Moreover, under the conditions mentioned above, compounds such as dansylamide, Dns-Phe, and Dns-Gly-Gly-Phe are bound to pepsin weakly or not at all. Consequently, when a good substrate such as Dns-Gly-Gly-Phe-Phe-OEt is used, there is at first an enhancement of fluorescence and a shift of the emission maximum to a shorter wavelength, but within a few minutes (as the hydrolysis proceeds) the fluorescence intensity and maximum change to those characteristic of the split product Dns-Gly-Gly-Phe in water.

It is evident from these results that the striking affinity of pepsin for the dansyl group when it is part of a peptide such as Dns-Gly-Gly-Phe-Phe-OEt is largely a consequence of the specific interaction of the Phe-Phe-OEt portion of the substrate with the active site of the enzyme and that the dansyl group is "dragged" into a hydrophobic region of the enzyme whose intrinsic affinity for the dansyl group may be low. In this connection, it should be added that fluorescence measurements with the resistant diastereoisomer Dns-Gly-Gly-D-Phe-Phe-OEt gave a value for the dissociation constant of the peptide–pepsin complex similar to the K_M value for the sensitive substrate Dns-Gly-Gly-Phe-Phe-OEt. As was noted above, the D-Phe–Phe unit is bound at the catalytic site of pepsin in a conformation that renders the peptide bond resistant to enzymic attack, but the extent of the interaction of the amino-terminal dansyl group with a hydrophobic region of the protein appears to be unaffected.

The conclusions drawn thus far from the studies on the role of secondary enzyme-substrate interactions in the action of pepsin on oligomeric substrates raise several questions of wider significance. The synthetic peptide substrates found for endopeptidases during the 1930s indicated the side-chain preferences of individual proteinases with respect to the amino acid residues contributing the CO or NH group to the sensitive bond (e.g., the preference of trypsin for an arginyl—Y or lysyl—Y bond). More recently, indications of other specific interactions have come from work with streptococcal proteinase (Gerwin et al., 1966) and papain (Schechter and Berger, 1967), whose action is·favored by the presence of a hydrophobic group (Z or Phe) on the amino side of a sensitive X-Y unit. Indeed, with these two enzymes of apparently broad side-chain specificity with respect to the X residue, the catalytic action on a substrate AX-YB depends so importantly on the nature of the A group as to outweigh, in some cases, the favorable effect of a preferred X-unit. Despite the extensive kinetic studies on chymotrypsin, the question of the importance of secondary interactions in its cleavage of moderately long oligopeptides has not been explored systematically; for example, it will be of interest to examine the interaction of the dansyl group of Dns-peptide substrates of chymotrypsin with this enzyme in a manner similar to that described above for pepsin. Although much more work is needed, it would seem that the cleavage of proteins by proteinases should not only be viewed in terms of the amino acid residues that flank the sensitive peptide bond, but that attention should also be given to interactions at loci somewhat removed from the site of catalytic action. These secondary interactions, because of their cooperative nature, may be the predominant factors in the positioning of sensitive bonds when a particular proteinase (especially one of broad side-chain specificity at the catalytic site) acts on a given protein. This view of specificity thus becomes one that embraces all the structural elements of the peptide substrate and of the enzyme protein that interact with each other, and the specificity relation between enzyme and substrate may be considered to represent a mutual conformational adjustment so as to achieve optimum fit. Obviously, the structural constraints imposed on a region of the folded peptide chain of an enzyme protein are much greater than those on the extended peptide chain of the substrate. Nonetheless, it appears reasonable to consider significant conformational flexibility in the enzyme at the sites of such secondary interaction. It may be added that secondary interactions are clearly operative in the action of other types of enzymes that act on oligomeric substrates. This has been strikingly demonstrated for lysozyme, where secondary interactions direct the cleavage of oligosaccharides via transglycosylation rather than direct hydrolysis (Chipman and Sharon, 1969), thus reemphasizing the importance of the transferase action of "hydrolases" in their action on oligomeric substrates (Fruton, 1957).

In addition to these general problems of enzyme specificity, the recent data obtained with pepsin raise the question of the relation of binding energy (as measured by K_M when K_M is believed to approximate K_S) to catalytic efficiency. It was noted above that large differences in k_{cat} among closely related pepsin substrates are not accompanied by significant changes in K_M. A similar result has also been encountered with liver esterase (Hofstee, 1967; Goldberg and Fruton, 1969), where changes in the nature of the R group of RCO—OR' substrates are reflected in large differences in k_{cat} and relatively small differences in K_M (which may or may not approximate K_S). Examples of this kind have been cited (see Koshland and Neet, 1968) in support of hypotheses designed to explain the specific catalytic efficiency of enzymes in terms of the utilization of potential binding energy in the enzyme–substrate interaction to lower the free energy of activation in the catalytic process.

One of the major aspirations of present-day enzyme chemistry is to describe the productive interaction of an enzyme with its substrate in terms of a molecular model based on the determination of the detailed three-dimensional structure of the enzyme by means of X-ray crystallography. This approach has already yielded significant information and is likely to be even more fruitful in the future, although the question still seems open whether the structure of the enzyme is as rigid in solution as it is in the crystal. This question may have considerable importance for the interpretation of the effect of secondary interactions on the catalytic efficiency of enzymes that act on oligomeric substrates, and it may be more difficult than previously thought to define the total area of enzyme–substrate interaction in terms of fixed "subsites" one of which contains the catalytic groups of a given enzyme. At present, therefore, model building has not yet replaced the older experimental approaches to the study of enzymic catalysis. One of these has been the specific chemical modification of enzymes; when coupled with modern analytical techniques of protein chemistry, this approach has given decisive knowledge about the nature of the catalytically important groups of numerous enzymes and the location of these groups in the linear amino acid sequence of the individual proteins. Another has been the "substrate-oriented" approach, which has depended heavily on the development of new methods of synthetic organic chemistry so as to permit the systematic and unambiguous variation in the structure of substrates and substrate analogues. When added to the other modes of attack, and supplemented by the use of modern spectroscopic techniques, this approach is also likely to continue to yield useful knowledge about the many unsolved problems of enzymic catalysis.

In concluding this essay, it should be recalled that the first incisive study of enzyme specificity was described in 1894 by Bergmann's teacher, Emil Fischer. It was Fischer's pioneer synthetic work in the sugar field that yielded the diastereoisomeric α- and β-methylglucosides and permitted his demonstration that they were acted on differently by a yeast-enzyme preparation and by emulsin. The enzyme studies on peptidases and proteinases that followed the invention of the carbobenzoxy method by Bergmann and Zervas were therefore in the tradition arising from the work of Fischer. Those of us who have continued in this tradition, and have sought to learn something about the enzymes that act on peptides, owe much to those who have further elaborated the armamentarium of peptide synthesis. Among these colleagues, Professor Zervas is preeminent, because of the scope, precision, and elegance of his work.

ACKNOWLEDGMENT

The recent research of my laboratory, described above, was aided by grants from the National Institutes of Health (GM-6452 and GM-18172) and from the National Science Foundation (GB-5212X and GB-18628).

REFERENCES

Bergmann, M., and Fruton, J. S., 1937, On proteolytic enzymes. XIII. Synthetic substrates for chymotrypsin, *J. Biol. Chem.* **118**:405.

Bergmann, M., Zervas, L., Fruton, J. S., Schneider, F., and Schleich, H., 1935, On proteolytic enzymes. V. On the specificity of dipeptidase, *J. Biol. Chem.* **109**:325.

Bergmann, M., Zervas, L., and Fruton, J. S., 1936, On proteolytic enzymes. XI. The specificity of the enzyme papain peptidase I, *J. Biol. Chem.* **115**:593.

Bergmann, M., Fruton, J. S., and Pollok, H., 1939, The specificity of trypsin, *J. Biol. Chem.* **127**:643.

Chipman, D. M., and Sharon, N., 1969, Mechanism of lysozyme action, *Science* **165**:454.

Delpierre, G. R., and Fruton, J. S., 1966, Specific inactivation of pepsin by a diazoketone, *Proc. Natl. Acad. Sci.* **56**:1817.

Edelman, G. M., and McClure, W. O., 1968, Fluorescent probes and the conformation of proteins, *Accounts Chem. Res.* **1**:65.

Fruton, J. S., 1957, Enzymic hydrolysis and synthesis of peptide bonds, *Harvey Lectures* **51**:64.

Fruton, J. S., 1971, Pepsin, *in* "The Enzymes," 3rd ed., Vol. 3 (P. D. Boyer, ed.), p. 119, Academic Press, New York.

Fruton, J. S., and Bergmann, M., 1938, The specificity of pepsin action, *Science* **87**:557.

Fruton, J. S., and Bergmann, M., 1939, The specificity of pepsin, *J. Biol. Chem.* **127**:627.

Gerwin, B. I., Stein, W. H., and Moore, S., 1966, On the specificity of streptococcal proteinase, *J. Biol. Chem.* **241**:3331.

Goldberg, M. I., and Fruton, J. S., 1969, Beef liver esterase as a catalyst of acyl transfer to amino acid esters, *Biochemistry* **8**:86.

Hill, R. L., 1965, Hydrolysis of proteins, *Advan. Protein Chem.* **20**:37.

Hofstee, B. H. J., 1967, Substrate hydrophobic groups and the maximal rate of enzyme reactions, *Nature* **213**:42.

Hollands, T. R., Voynick, I. M., and Fruton, J. S., 1969, Action of pepsin on cationic synthetic substrates, *Biochemistry* **8**:575.

Inouye, K., and Fruton, J. S., 1967, Studies on the specificity of pepsin, *Biochemistry* **6**:1765.

Inouye, K., and Fruton, J. S., 1968, The inhibition of pepsin action, *Biochemistry* **7**:1611.

Izumiya, N., and Fruton, J. S., 1956, The specificity of cathepsin C, *J. Biol. Chem.* **218**:59.

Jackson, W. T., Schlamowitz, M., and Shaw, A., 1965, Kinetics of the pepsin-catalyzed hydrolysis of *N*-acetyl-L-phenylalanyl-L-diiodotyrosine, *Biochemistry* **4**:1537.

Koshland, D. E., and Neet, K. E., 1968, The catalytic and regulatory properties of enzymes, *Ann. Rev. Biochem.* **37**:359.

Medzihradszky, K., Voynick, I. M., Medzihradszky-Schweiger, H., and Fruton, J. S., 1970, Effect of secondary enzyme–substrate interactions on the cleavage of synthetic substrates by pepsin, *Biochemistry* **9**:1154.

Ogston, A. G., 1948, Interpretation of experiments on metabolic processes, using isotopic tracer elements, *Nature* **162**:963.

Sachdev, G. P., and Fruton, J. S., 1969, Pyridyl esters as synthetic substrates for pepsin, *Biochemistry* **8**:4231.

Sachdev, G. P., and Fruton, J. S., 1970, Secondary enzyme–substrate interactions and the specificity of pepsin, *Biochemistry* **9**:4465.

Sachdev, G. P., Johnston, M. A., and Fruton, J. S., 1972, Fluorescence studies on the interaction of pepsin with its substrates, *Biochemistry* **11**:1080.

Schechter, I., and Berger, A., 1967, On the size of the active site in proteases. I. Papain, *Biochem. Biophys. Res. Commun.* **27**:157.

Stryer, L., 1968, Fluorescence spectroscopy of proteins, *Science* **162**:526.

Trout, G. E., and Fruton, J. S., 1969, The side-chain specificity of pepsin, *Biochemistry* **8**:4183.

CHAPTER 8

STRUCTURAL STUDIES OF NATURALLY OCCURRING CYCLIC POLYPEPTIDES AT ROCKEFELLER UNIVERSITY

L. C. Craig

Rockefeller University
New York, N.Y.

I consider it a great privilege to write an essay for a book on polypeptides that is to honor Professor L. Zervas. Every peptide chemist is familiar with the important contributions he made many years ago in opening the peptide field to synthesis and the many important contributions since then from his laboratory.

I first became acquainted with Dr. Zervas in 1934 when he came to Rockefeller Institute as a member of the scientific staff in the laboratory of Dr. Max Bergmann. At that time, I was a young assistant of Dr. W. A. Jacobs and had been assigned the problem of determining the structure of the alkaloids of ergot. We had just succeeded in isolating a large amphoteric fragment by hydrolytic procedures from one of the alkaloids, ergotinine, to which we gave the name "lysergic acid" (Jacobs and Craig, 1934). Three other fragments had resulted from the hydrolysis, L-phenylalanine, D-proline, and isobutyryl formic acid (Jacobs and Craig, 1935b). The last arose from the unstable α-hydroxyvaline which was stabilized in the intact molecule by ester and amide bonds.

A dipeptide containing phenylalanine and proline was also obtained. Reductive splitting (Jacobs and Craig, 1935a,b) gave a mixture of products which were all separated and shown to be consistent with the above-named fragments only. Structural studies with lysergic acid (Craig *et al.*, 1938) permitted us to propose formula [1] (Fig. 1) for ergotinine, although at the time we had no evidence for the order of phenylalanine and proline. This was

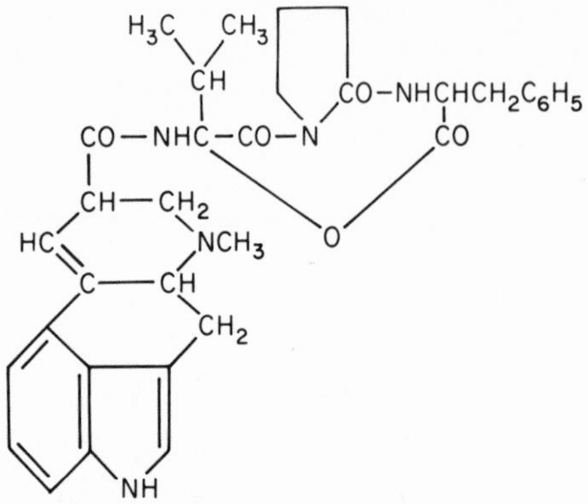

Fig. 1. Structural formula [1].

supplied much later by Stoll *et al.* (1951) and found to be that shown in formula [1] and in formula [2] (Fig. 2). They also postulated the transannular linkage of the two residues.

Stoll and Hoffmann (1943) found that their preparation of ergotoxine was a mixture of three alkaloids, with the main component having the amino acid composition we had earlier found (Jacobs and Craig, 1935c) but with the other two having amino acid replacements, ergokryptine with L-leucine in place of phenylalanine and ergocornine with valine in place of phenylalanine. Since our careful work on the isolation of the fragments of the

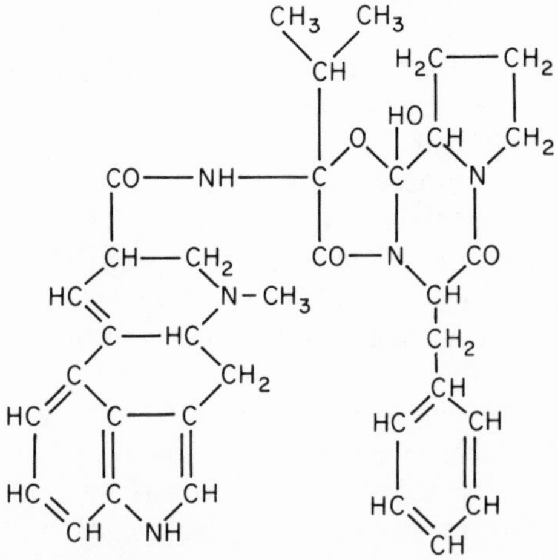

Fig. 2. Structural formula [2].

hydrolysis and butanol reduction had failed to show even a suggestion of leucine or valine in our preparation of ergotinine, that name should have been retained for the alkaloid Stoll and Hoffmann arbitrarily changed to "ergocristine." However, their name appears to be the one now used.

Irrespective of the name, it was shown in 1935 that our preparation of ergotinine was in fact a unique tetrapeptide. Since at that time very few naturally occurring peptides of this size or larger were known, our work was of considerable interest to Drs. Zervas and Bergmann. They were then synthesizing linear peptides with their famous carbobenzoxy method. It was never my good fortune to have been a collaborator of Dr. Zervas nor even to have done extensive peptide synthesis. This account, therefore, will deal only with the contributions to peptide chemistry made by Dr. W. A. Jacobs, myself, and our collaborators. The major thrust of our laboratory through the years has involved the determination of structures and the development of experimental methods suitable for the separation and characterization of polypeptides of all sizes.

The need for separation methods applicable to milligram amounts and also for greater resolution was immediately obvious in the earlier ergot work. Our success in a large part was due to our ability to scale down conventional procedures of distillation (Craig, 1936, 1937) and fractional crystallization (Craig and Post, 1944). It is now amusing to remember that on one occasion in 1938 after I had given a review of our work describing the determination of the structure of lysergic acid at a symposium, a leading alkaloid chemist made the following criticism during the discussion: "This work sounds reasonable but of course it must be repeated on a gram level. One cannot do reliable work on a smell." He was unaware that our work was supported by one of the most meticulous microanalysts of his day, Mr. Demetrious Rigakos, a close friend of Dr. Zervas. Were this critic living today, he would find that studies of comparable complexity are now done on microgram amounts of material.

Not long after the major features (Jacobs and Craig, 1936; Craig et al., 1938) of the ergot alkaloids exclusive of the stereochemistry had been established, Dubos (1939) showed that a selection of soil bacteria could be made which would produce a polypeptide mixture, tyrothrycin, which was highly toxic to many pathogenic bacteria. Hotchkiss and Dubos (1941) separated the mixture into two types of polypeptides, the gramicidins and the tyrocidines. Thus was initiated a period of great research activity directed toward the discovery of other antibiotic polypeptides. These included the penicillins, the polymyxins, the bacitracins, the actinomycins, and many others too numerous to mention. They represent a whole class of cyclic polypeptide natural products from microorganisms with unusual structures resembling those of the ergot alkaloids. They were found to contain amino acids of the dextro series, as was first found in ergot (Jacobs and Craig, 1935*b*) when D-proline was isolated. The ergot alkaloids thus can be considered the first representatives of a numerous group of peptides produced by lower organisms which are atypical to those produced by higher organisms.

In the 1940s, the gramicidins and tyrocidines became popular models for testing many different possibilities regarding separation methods, peptide analysis, characterization, degradation procedures, theories of structure, etc., which it was hoped might later be applied to proteins. They have thus played a role in the development of the present experimental methodology of peptide and protein chemistry and even recently have been used as models for conformational studies including the testing of rules for interpretation of NMR and other spectroscopic data.

Since these substances were used for testing possible experimental methods long before their structures in complete detail were known, it follows that the establishment of their structures evolved along with the development of the methodology. It is not possible in this short essay to cover the many advances made since the gramicidins and tyrocidines became of interest, but a short review of selected observations may be appropriate.

The evolution of the methodology at first progressed along the lines set forth in the dogma of the older classical organic chemistry, which required the isolation of a "pure" individual compound and proper documentation of purity before proceeding with the degradation step. With the polypeptides, this posed a serious difficulty because before 1940 fractionation of mixtures of substances too large to be distilled depended largely on fractional crystallization combined with melting point determination, optical rotation, and ultimate analysis to indicate purity. A preparation was considered pure when repeated recrystallization no longer changed the melting point or other physical property or elemental composition. This reasoning remains the correct one today with the difference that we can now apply separation methods which are vastly more selective and spectroscopic methods such as nuclear magnetic resonance and infrared and ultraviolet spectroscopy which are much more diagnostic for purity determination and characterization. This problem in organic chemistry would be almost completely solved were it not for the fact that detailed structural investigation has moved to much larger and more intricate molecules.

As a class of substances, polypeptides are difficult to crystallize. The gramicidins and tyrocidines, however, crystallized easily, one reason why their availability in 1940 made them so eagerly accepted as "pure" models with which to develop methods. Confidence in their purity was not to last long. Although fractional crystallization, chromatography, and other techniques failed to show them to be mixtures, countercurrent distribution (CCD) (Craig *et al.*, 1950) resolved both gramicidin and tyrocidine into several individual peptides, and the new methods of Moore and Stein (1948) for amino acid analysis revealed that the individual peptide fractions differed only by substitution of specific amino acid residues.

Martin and Synge (1941) used the tyrocidines and gramicidins as models on which to develop their now famous procedures and theories of "partition chromatography." These procedures held great promise, especially for the problem of the determination of the amino acid sequence in polypeptides. Gramicidin S, isolated by Gause and Brazhnikova (1944), was shown to contain five different amino acids (Synge, 1945). Consden *et al.* (1947), using paper chromatographic methods, were able to determine the amino acid sequence in gramicidin S (actually one of the tyrocidine group but unfortunately misnamed).

Gramicidin S had been shown (Synge, 1945) to lack a carboxyl- or amino-terminal group and thus was a cyclic peptide, but whether the ring contained only Val·Orn·Leu·Phe·Pro or this sequence repeated once or twice to give a higher molecular weight structure remained in doubt. The question of the molecular weight was unequivocally solved by a method based on CCD and called the "partial substitution method" (Battersby and Craig, 1951).

In this method, following the usual distribution by CCD the peptide is treated with an insufficient amount of a functional reagent such as the DNP end-group reagent developed by Sanger (1945) and then redistributed. As was shown by Battersby and Craig (1951) with gramicidin S the number of substitutable amino groups present can then be derived. Two were shown to be present, which, with the sequence data of Consden *et al.* (1947), established the complete sequence of the peptide.

The exact shape of the peptide then became of interest, and since it is crystalline, X-ray diffraction methods could be applied. Hodgkin and Oughton (1957) carried out such a study at the 6-Å level and proposed two possible conformations for the molecule, one of which was an antiparallel pleated sheet. Meanwhile, Schwyzer and Sieber (1957) had synthesized the decapeptide and independently proposed the antiparallel pleated sheet model as the most probable one.

These results came at a time when methods for estimating molecular weights and determining sequence and synthetic capability had been greatly improved. It was realized, however, that much more was needed than the determination of primary and secondary structure. The problem of the tertiary structure, particularly the conformation in solution, was equally important. Definitive methods for deriving information in this area were lacking.

In the early 1950s, zone electrophoresis (Kunkel, 1954) came into use and along with paper chromatography became a standard tool of the peptide chemist. The electrophoretic method was a particularly attractive tool for identification purposes because the parameter responsible for migration was clearly the fixed charge on the solute. By comparing the migration behavior of an unknown and certain of its derivatives with a known, the numbers and signs of charges could be estimated. Many different variations of zone electrophoresis have been developed with advantages for special purposes. The more recently developed variation known as "electrofocusing" (Awdeh *et al.*, 1968) offers one of the most critical homogeneity tests known today. In zone electrophoresis, the distance a solute travels in a given electric field is a function of the strength of the field and the resistance offered by the supporting medium. If this medium is a gel with a critical degree of cross-linking, selectivity due to size is superimposed on that due to relative fixed charge and it is possible to derive information concerning the shape of the molecule. The relation of shape to diffusibility has long been known. Gel electrophoresis (Crambach and Rodbard, 1971) has proven to be a critical test of homogeneity and is widely used.

In the mid-1950s, several groups of investigators whose main objective was the improvement of separation and identification methods independently turned their attention towards methods based on relative rates of diffusion. These included myself and coworkers (Craig and King, 1956; Craig, 1967), Porath, Flodin, and coworkers (Porath and Flodin, 1959), and others. There were several reasons for this choice. One was the need for methods which would be as gentle as possible so that fragile solutes would not be altered during the fractionation. Another was related to the need for better methods for studying molecular interactions and the effect of the solvent environment on these interactions. Still another was the desirability of having separation methods whose basis was clearly related to a single parameter, e.g., the partition ratio in CCD, the charge on the solute as in zone electrophoresis, or the sedimentation rate in ultracentrifugation.

Porath and coworkers followed an earlier suggestion of Lathe and Ruthven (1956), who found that a chromatographic column filled with starch as the absorbent appeared to separate mixtures of solutes according to their relative size. The Swedish investigators, however, chose a different carbohydrate support, a dextran modified by a cross-linking reagent, epichlorhydrin, so that a matrix of varying porosity would be presented to the solutes as they passed through the column. Their technique was called "gel filtration." Theories concerning the mechanism and the basis for the separations have been proposed (Determan, 1968), none of which entirely satisfy all experimental experience. Nonetheless, an array of different kinds of porous gels made from dextran, polyacrylamide, and other materials designed for gel filtration (also called "permeation chromatography") have become commercially available and afford one of the most useful and widely used separation techniques known today.

Gel filtration does not always afford a true measure of relative diffusional size, although with careful calibration (Ackers, 1970) of a column with known model solutes it has been widely used to estimate molecular size. When the solutes are of such type that a true reflection of size is obtained, the selectivity is not high as compared to other chromatographic procedures. On the other hand, when adsorption plays a role, perhaps weakly as is often the case, it can be highly selective.

I and my coworkers began a study of simple dialysis with the original objective of making this old familiar separation technique more useful for biochemical studies. We soon found that with Visking dialysis casing it could be made surprisingly selective by adjustment of the porosity of the membrane to a critical pore size for a given range of molecular size, and we enlarged the original objective of the study to include conformation. The technique was called "thin film dialysis." As in the case of gel filtration, size estimations were based entirely on comparison with solutes of known size and shape. In the case of gel filtration comparisons were made from the position of the emerging band on elution, while with thin film dialysis they were made by comparing half escape times, the latter being inversely proportional to the rate of diffusion through the membrane. It was possible to modify membranes and incorporate them into a standard design of cell which on the basis of well-documented models would discriminate between molecular sizes to a limit approximating 2–3% of Stokes radius (Craig and Pulley, 1962). Unfortunately, the method does not lend itself easily to a multistage or countercurrent mode of operation and thus is not well suited for the separation of multicomponent mixtures as is gel filtration. On the other hand, it has certain advantages over gel filtration for characterization of given preparations with respect to size and for study of the influence of the solvent environment. The two techniques complement one another well and are capable of giving mutually supporting information.

While these diffusive techniques were being improved, much effort in other laboratories was being put forth to develop spectroscopic approaches to the problem of conformation. One of these was the use of optical rotatory dispersion (ORD) and circular dichroism (CD). Certain proteins and carbohydrates had been shown by X-ray diffraction studies to have helical conformations, and extensive theories were put forward relating certain types of rotatory dispersion to helicity. One of these was particularly appealing because advanced and very sensitive spectropolarimeters capable of measuring ORD and CD to short wavelengths became commercially available, and with one of these Simmons and Blout (1961) found that helical proteins gave similar spectra with a strong cotton effect at 220 mμ. It was proposed that the strength of the minimum could be used to calculate the degree of helicity of a protein. This idea was so attractive that it became the basis of countless numbers of research papers and dominated the field of research on conformation for several years. Conformation studies became equated with helicity based on interpretation of ORD and CD studies without questioning whether or not the type of spectra given by the helix could also be given by some other conformation.

Our early experience with the cyclic antibiotic polypeptides led us to have doubts about the reliability of the popular interpretation because we had found a number of them to give anomalous rotatory dispersions. These doubts were fully confirmed by the first measurements we were able to make on gramicidin SA in a Cary spectropolarimeter. They gave exactly the type of rotatory dispersion behavior (Ruttenberg *et al.*, 1965) which was supposed to be exclusively given by a helical conformation. We felt strongly that the antiparallel pleated sheet structure earlier proposed by Schwyzer and by Hodgkin and Oughton was correct.

The result stimulated considerable speculation about the conformation of gramicidin SA, and a number of other conformations were proposed (Liquori *et al.*, 1966; Scott *et al.*, 1967), most of which attempted to retain elements of the helix. These possibilities emphasized the importance of independent evidence for or against the antiparallel pleated sheet structure. Fortunately, a considerable advance in the resolution and sensitivity of nuclear magnetic resonance came at that time through the introduction of the 220 MHz spectrometers with supercooled magnets. Strong direct evidence for the antiparallel pleated sheet conformation (Stern *et al.*, 1968) was obtained with this equipment. In addition, the presence of four hydrogen bonds was indicated both by NMR and by tritium exchange (Laiken *et al.*, 1969). Since other cyclic antibiotics, the tyrocidines and bacitracins, also gave similar ORD and CD patterns, it became obvious that ORD and CD measurements alone could not be taken as a reliable indication of a helical structure (Craig, 1968). It can be concluded further that even when shown to be present in the crystal by X-ray diffraction, quantitative correlation of helicity in solution would not necessarily be reliable because of the possibility of other local conformations with strong rotations.

Great interest is now being shown in the use of NMR for conformation studies, but much work will be required with model solutes before reliable rules for interpretation of all the data can be established. The cyclic antibiotic polypeptides are ideal for this purpose because the possibilities are severely restricted by the covalent ring structure. The results will require support by all other techniques giving conformational information: IR, ORD, CD, tritium exchange, thin film dialysis, model building, energy minimization calculations, fluorometric techniques, etc.

Two other types of cyclic antibiotic polypeptides have been intensively studied in this laboratory: the bacitracins (Craig *et al.*, 1958, 1968) and the

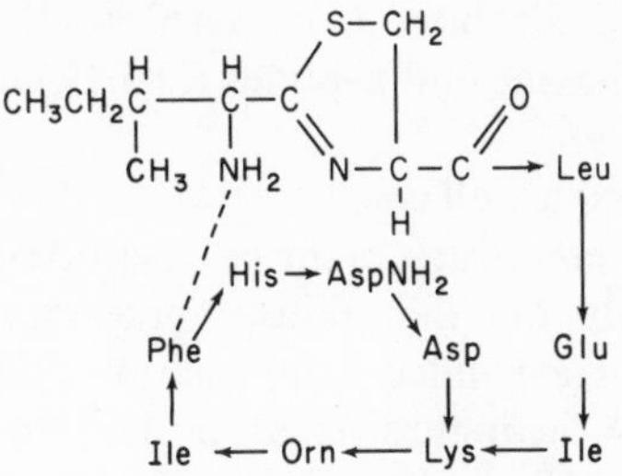

Fig. 3. Structural formula [3].

polymyxins (Hausman, 1956). Both contain polypeptide rings with a polypeptide tail. However, the thin film dialysis technique has clearly shown that the tail does not extend from the ring but is folded compactly over the ring and apparently held by secondary forces. With polymyxin B_1, the sequence was restricted to four possibilities (Hausman, 1956), and the correct one was chosen by synthesis in other laboratories (Suzuki et al., 1964; Vogler et al., 1964). Both bacitracin and polymyxin contain D-amino acids, which required that only partial hydrolysis methods could be used for establishing their sequence.

The structure of bacitracin (formula [3], Fig. 3), is particularly interesting because approximately half its amino acid residues are those known to be important in active sites of enzymes. Actually, formula [3] does not truly represent its structure since there appear to be tautomeric and/or resonance forms, depending on the pH, temperature, and environment. It binds metals and has barely sufficient stability to permit isolation by CCD.

It has been customary to speak of polypeptides without covalently bonded ring structures as "random coils." This term indicates complete lack of preferred conformations in a given solution. However, we know from bond angles, bond distances, the planar nature of the peptide bond, steric hindrance, and model building that the conformation can be random only within certain "allowed" dihedral angles for each covalent bond. Even beyond this limitation, the thin film dialysis technique has indicated that many so-called linear or random-coil polypeptides assume either a definite conformation or a very narrow statistical distribution of mobile conformations (Craig et al., 1971) in favorable solvents. This conclusion is based on the fact that they give straight-line escape plots in certain solvent environments but not in others. This conformation, however, is easily perturbed by a change in solvent environment, as indicated by a change in half escape time or deviation from straight-line behavior. The thin film dialysis behavior is particularly sensitive to a change in temperature, pH, or ionic strength and addition of hydrophobic bond breaking solutes such as urea, dimethylformamide, guanidinium salts, and alcohols. Contrasted to that of rigid covalently linked peptides, the behavior of dipeptides is particularly striking (Burachik *et al.*, 1970).

The results obtained are all consistent with the theory that hydrophobic interaction between different parts of the polypeptide chain usually promotes a compact conformation, one that diffuses more rapidly through the membrane than elliptical or expanded conformations. This interaction can be enhanced by increasing the ionic strength of the solvent environment or by increasing the temperature to about 40°C. Thereafter, further temperature increase seems to weaken the interaction. If the charges along the chain are alternatively positive and negative a more compact conformation will be promoted by the electrostatic effect, but if they are of the same sign the opposite effect due to the shielding of the charges. Where the charges are of promote a less compact conformation, while in the latter it will have the opposite effect due to the shielding of the charges. Where the charges are of the same sign, a straight-line escape pattern may be obtained only in the presence of salt. Thus in the case of alternating charges the effect of increasing the ionic strength may be opposed by the effect on hydrophobic interaction. These effects are strikingly apparent even with dipeptides.

If the experience with the thin film dialysis has been interpreted correctly, it is obvious that linear peptides can have unique preferred conformations depending on the solvent environment and that the behavior of each peptide under a change of environment is highly individualistic and delicately balanced. Increase of the dielectric constant by salt addition increases hydrophobic interaction so that a more favorable conformation for van der Waals forces may be reached. On the other hand, an increase in hydrophobic environment strengthens electrostatic interaction.

Since the shape of a linear polypeptide in solution is determined by the relative affinities each part of the chain holds for the solvent molecules as compared to the affinities each residue or part has for other residues or parts of the chain (within the limits allowed by covalent bond length, steric hindrance, peptide bond planarity, etc.), it follows that in favorable solvents intermolecular interaction must be considered as well as intramolecular interaction. The tyrocidines associate strongly and are therefore interesting models for the study of quaternary structure. Extensive studies with these substances have been made in the ultracentrifuge (Laiken *et al.*, 1971), by thin film dialysis (Burachik *et al.*, 1970), and by NMR (Stern *et al.*, 1969). All these studies are consistent with the view that the monomers aggregate mostly through hydrophobic interaction but that a specific rigid conformation is required for the hydrophobic forces to exert sufficient influence to cause the strong reversible association. The association behavior of the tyrocidines seems to be particularly interesting because it is one of the smallest polypeptide structures which has shown the observed degree of specific interaction with another molecule of its own size. Tyrocidine A interacts to form heteropolymers with tyrocidine B or C (Williams and Craig, 1967). A, B, and C differ by simple replacements of tryptophan for phenylalanine.

The studies of methods with polypeptides have been carried out with the ultimate objective of their improvement to the point of meaningful application to proteins. In fact, a number of attempts have already been made. Insulin was successfully fractionated by CCD (Harfenist and Craig, 1952a), and the first correct amino acid analysis (Harfenist, 1953) reported on material purified this way. The method of partial substitution was then applied and clear evidence given that the monomer was in the 6000 mol. range (Harfenist and Craig, 1952b). Although this result was strongly opposed at the time by investigators using sedimentation and diffusion methods, it was confirmed by the sequence studies of Sangar and collaborators (Ryle et al., 1955). The thin film dialysis method and CCD further indicated that more than one interchangeable conformational form could be isolated (Craig et al., 1960).

When applied to beef pancreatic ribonuclease, the thin film dialysis technique and CCD provided evidence that two interconvertible forms of ribonuclease could be isolated (Craig et al., 1963). From the data, it appears most likely that they are conformational forms with an energy barrier barely sufficient to permit their separation in favorable solvent environments.

The largest protein studied in the Craig laboratory has been hemoglobin. It yielded to study by CCD (Hill et al., 1961), and this afforded the first clear separation of the two polypeptide chains in quantity adequate for sequence study. This set the stage for accomplishing the determination of the complete amino acid sequence of both chains (Konigsberg and Hill, 1962; Konigsberg et al., 1963). The sequence was independently determined in other laboratories (Braunitzer et al., 1961) as well. The sequence information came at a time when it could support the now famous structural studies of Perutz and collaborators on crystalline hemoglobin by X-ray diffraction (Muirhead et al., 1967). There remained, however, the question of how closely the exact structure determined by X-ray measurements in the crystal represented that in solution. Although numerous opinions on this subject have been expressed (von Hippel and Schleich; Rupley, 1969; Perutz et al., 1964), the dialysis technique gave a strong early indication in 1963 (Guidotti and Craig, 1963) that in dilute solution it was a highly mobile dissociable complex whose precise shape was highly influenced by concentration, ionic strength, and specific ions. This behavior would, however, be different in the concentration of hemoglobin in the cell. Perutz et al. (1968) have shown that in the crystal there is considerable difference between oxy- and deoxyhemoglobin, and Simon in this laboratory has shown differences in diffusional size by thin film dialysis.

REFERENCES

Ackers, G. M., 1970, Analytical gel chromatography, *in* "Advances in Protein Chemistry," Vol. 24 (C. B. Anfinsen, Jr., M. L. Anson, J. T. Edsall, and F. M. Richards, eds.), p. 343, Academic Press, New York.

Awdeh, Z. L., Williamson, A. R., and Askonas, B. A., 1968, Isoelectric focusing in polyacrylamide gel and its application to immunoglobulins, *Nature* **219**:66.

Battersby, A. R., and Craig, L. C., 1951, The molecular weight determination of polypeptides, *J. Am. Chem. Soc.* **73**:1887.

Braunitzer, G., Gehring-Muller, R., Hilschmann, N., Hilse, K., Hobom, G., Rudloff, V., and Wittman-Liebold, B., 1961, Die Konstitution des normalen adulten Humanhamoglobins, *Z. Physiol. Chem.* **325**:283.

Burachik, M., Craig, L. C., and Chang, J., 1970, Studies of self-association and conformation of peptides by thin film dialysis, *Biochemistry* **9**:3293.

Consden, R., Gordon, A. H., Martin, A. J. P., and Synge, R. L. M., 1947, Gramicidin S: The sequence of the amino-acid residues, *Biochem. J.* **41**:596.

Craig, L. C., 1936, A microdistillation apparatus, *Ind. Eng. Chem. Anal. Ed.* **8**:219.

Craig, L. C., 1937, A fractional-distillation micro-apparatus, *Ind. Eng. Chem. Anal. Ed.* **9**:441.

Craig, L. C., 1967, Techniques for the study of peptides and proteins by dialysis and diffusion, *in* "Methods in Enzymology," Vol. XI (C. H. W. Hirs, ed.), p. 870, Academic Press, New York.

Craig, L. C., 1958, Conformation studies with polypeptides by rotatory dispersion and thin-film dialysis, *Proc. Nat. Acad. Sci. U.S.* **61**:152.

Craig, L. C., and King, T. P., 1956, Fractional dialysis with cellophane membranes, *J. Am. Chem. Soc.* **78**:4171.

Craig, L. C., and Post, O. W., 1944, Improved apparatus for solubility determination or for small-scale recrystallization, *Ind. Eng. Chem. Anal. Ed.* **16**:413.

Craig, L. C., and Pulley, A. O., 1962, Dialysis studies. IV. Preliminary experiments with sugars, *Biochemistry* **1**:89.

Craig, L. C., Shedlovsky, T., Gould, R. G., Jr., and Jacobs, W. A., 1938, The ergot alkaloids. XIV. The positions of the double bond and the carboxyl group in lysergic acid and its isomer. The structure of the alkaloids, *J. Biol. Chem.* **125**:289.

Craig, L. C., Konigsberg, W., and Hill, R. J., 1958, Bacitracin, *in* "Amino Acids and Peptides with Antimetabolic Activity" (G. E. W. Wolstenholme and C. M. O'Connor, eds.), p. 226, and A. Churchill, London.

Craig, L. C., King, T. P., and Konigsberg, W., 1960, Homogeneity studies with insulin and related substances, *Ann. N.Y. Acad. Sci.* **88**:533.

Craig, L. C., King, T. P., and Crestfield, A. M., 1963, Dialysis studies. V. The behavior of different preparations of bovine ribonuclease, *Biopolymers* **1**:231.

Craig, L. C., Phillips, W. F., and Burachik, M., 1968, Bacitracin A. Isolation by counter double current distribution and characterization, *Biochemistry* **7**:2348.

Craig, L. C., Kac, H., Chen, H. C., and Printz, M. P., 1971, Studies with synthetic ACTH analogs and other linear peptides, *in* "Structure Activity Relationships of Protein and Polypeptide Hormones," Part 1 (M. Margoulies and F. C. Greenwood, eds.), p. 176, Excerpta Medica, Amsterdam.

Craig, L. C., Gregory, J. D., and Barry, G. T., 1950, Studies on polypeptides and amino acids by countercurrent distribution, *Cold Spring Harbor Symposia on Quantitative Biology*, Vol. XIV.

Crambach, A., and Rodbard, D., 1971, Polyacrylamide gel electrophoresis, *Science* **172**:440.

Determan, H., 1968, "Gel Chromatography," p. 63. Springer-Verlag, New York.

Dubos, R. J., 1939, Bactericidal effect of an extract of a soil bacillus on gram positive cocci, *Exptl. Biol. Med.* **40**:311.

Gause, G. F., and Brazhnikova, M. G., 1944, Gramicidin, S., *Lancet* **247**:715.

Guidotti, G., and Craig, L. C., 1963, Dialysis studies, VIII. The behavior of solutes which associate, *Proc. Natl. Acad. Sci.* **50**:46.

Harfenist, E. J., 1953, The amino acid compositions of insulins isolated from beef, pork and sheep glands, *J. Am. Chem. Soc.* **75**:5528.

Harfenist, E. J., and Craig, L. C., 1952a, Countercurrent distribution studies with insulin, *J. Am. Chem. Soc.* **74**:3083.

Harfenist, E. J., and Craig, L. C., 1952b, The molecular weight of insulin, *J. Am. Chem. Soc.* **74**:3087.

Hausman, W., 1956, The amino acid sequence of polymyxin B, *J. Am. Chem. Soc.* **78**:3663.

Hill, J. H., Konigsberg, W., Guidotti, G., and Craig, L. C., 1961, The structure of human hemoglobin. The separation of the α and β chains and their amino acid composition, *J. Biol. Chem.* **237**:1549.

Hodgkin, D. C., and Oughton, B. M., 1957, Possible molecular models for gramicidin S and their relationship to present ideas of protein, *Biochem. J.* **65**:752.

Hotchkiss, R. D., and Dubos, R. J., 1941, The isolation of bactericidal substances from cultures of bacillus brevis, *J. Biol. Chem.* **141**:155.

Jacobs, W. A., and Craig, L. C., 1934, The ergot alkaloids. II. The degradation of ergotinine with alkali: Lysergic acid, *J. Biol. Chem.* **104**:547.

Jacobs, W. A., and Craig, L. C., 1935a, The ergot alkaloids. IV. The cleavage of ergotinine with sodium and butyl alcohol, *J. Biol. Chem.* **108**:595.

Jacobs, W. A., and Craig, L. C., 1935b, The ergot alkaloids. V. The hydrolysis of ergotinine, *J. Biol. Chem.* **110**:521.

Jacobs, W. A., and Craig, L. C., 1935c, The structure of the ergot alkaloids, *J. Am. Chem. Soc.* **57**:383.

Jacobs, W. A., and Craig, L. C., 1936, The ergot alkaloids. X. On ergotamine and ergoclavine, *J. Org. Chem.* **1**:245.

Konigsberg, W., and Hill, R. J., 1962, The structure of human hemoglobin. V. The digestion of the α chain of human hemoglobin with pepsin, *J. Biol. Chem.* **237**:3157.

Konigsberg, W., Goldstein, J., and Hill, R. J., 1963, The structure of human hemoglobin. VII. The digestion of the β chain of human hemoglobin with pepsin, *J. Biol. Chem.* **238**:2028.

Kunkel, H. G., 1954, Zone electrophoresis, *Meth. Biochem. Anal.* **1**:141.

Laiken, S. L., Printz, M. P., and Craig, L. C., 1969, Tritium–hydrogen exchange studies of protein models. I. Gramicidin S-A, *Biochemistry* **8**:519.

Laiken, S. L., Printz, M. P., and Craig, L. C., 1971, Studies on the mode of self-association of tyrocidine B, *Biochem. Biophys. Chem. Commun.* **43**:595.

Lathe, G. H., and Ruthven, C. R. J., 1956, The separation of substances and estimation of their relative molecular sizes by the use of columns of starch in water, *Biochem. J.* **62**:665.

Liquori, A. M., de Santis, P., Kovacs, A. L., and Mazzarella, L., 1966, Stereochemical code of amino-acid residues: The molecular conformation of gramicidine S, *Nature* **211**:1039.

Martin, A. J. P., and Synge, R. L. M., 1941, A new form of chromatogram employing two liquid phases. 1. A theory of chromatography. 2. Application to the microdetermination of the higher amino acids in proteins, *Biochem. J.* **25**:1358.

Moore, S., and Stein, W. H., 1948, Partition chromatography of amino acids on starch. *Ann. N.Y. Acad. Sci.* **49**:265.

Muirhead, H., Cox, J. M., Mazzarella, L., and Perutz, M. F., 1967, Structure and function of hemoglobin. III. A three dimensional Fourier synthesis of human deoxyhemoglobin at 5.5 Å resolution, *J. Mol. Biol.* **28**:117.

Perutz, M. F., Bolton, W., Diamond, R., Muirhead, H., and Watson, H. C., 1964, Structure of hemoglobin. An x-ray examination of reduced horse hemoglobin, *Nature* **203**:687.

Perutz, M. F., Muirhead, H., Cox, J. M., and Goaman, L. C. G., 1968, Three-dimensional fourier synthesis of horse oxyhaemoglobin at 2.8 Å resolution: The atomic model, *Nature* **219**:131.

Porath, J., and Flodin, P., 1959, Gel filtration: A method for desalting and group separation, *Nature* **183**:1657.

Rupley, J. A., 1969, The comparison of protein structure in the crystal and in solution, *in* "Structure and Stability of Biological Molecules" (S. M. Timashef and G. D. Fasman, eds.), pp. 291–352, Marcel Dekker, New York.

Ruttenberg, M. A., King, T. P., and Craig, L. C., 1965, The use of the tyrocidines for the study of conformation and aggregation behavior, *J. Am. Chem. Soc.* **87**:4196.

Ryle, A. P., Sanger, F., Smith, L. F., and Kitai, R., 1955, The disulphide bonds of insulin, *Biochem. J.* **60**:541.

Sanger, F., 1945, The free amino groups of insulin, *Biochem. J.* **39**:507.

Schwyzer, R., and Sieber, P., 1957, Die Synthese von Gramicidin S, *Helv. Chim. Acta* **40**:624.

Scott, R. A., Vanderkooi, G., Tuttle, R. W., Shames, P. M., and Scheraga, H. A., 1967, Minimization of polypeptide energy. III. Application of a rapid energy minimization technique to the calculation of preliminary structures of gramicidin-S, *Proc. Natl. Acad. Sci.* **58**:2204.

Simmons, N. S., and Blout, E. R., 1961, The structure of tobacco mosaic virus and its components: Ultraviolet optical rotatory dispersion, *Biophys. J.* **1**:55.

Stern, A., Gibbons, W. A., and Craig, L. C., 1968, A conformational analysis of gramicidin S-A by nuclear magnetic resonance, *Proc. Natl. Acad. Sci.* **61**:734.

Stern, A., Gibbons, W. A., and Craig, L. C., 1969, The effect of association on nuclear magnetic resonance spectra of tyrocidine B, *J. Am. Chem. Soc.* **91**:2794.

Stoll, A., and Hofmann, A., 1943, Die Alkaloide der Ergotoxingruppe: Ergocristin, Ergokryptin und Ergocornin, *Helv. Chim. Acta* **26**:1570.

Stoll, A., Hofmann, A., and Petrzilka, T., 1951, Die Konstitution der Mutterkornalkaloide. Struktur des Peptidteils. III. *Helv. Chim. Acta* **34**:1544.

Suzuki, T., Hayashi, K., Fujikawa, K., and Tsukamoto, K., 1964, Contribution to the elucidation of the chemical structure of polymyxin B_1, *J. Biochem. Japan* **56**:335.

Synge, R. L. M., 1945, Gramicidin S: Overall chemical characteristics and amino-acid composition, *Biochem. J.* **39**:363.

von Hippel, P. H., and Schleich, T., 1969, The effects of neutral salts on the structure and conformational stability of macromolecules in solution, *in* "Structure and Stability of Biological Molecules" (S. M. Timashef and G. D. Fasman, eds.), pp. 417–574, Marcel Dekker, New York.

Vogler, K., Studer, R. O., Lanz, P., Lergier, W., and Böhni, E., 1964, Total synthesis of the antibiotic polymyxin B, *Experientia* **20**:365.

Williams, R. C., and Craig, L. C., 1967, A method of calculating countercurrent distribution curves of nonideal solutes, *Separation Sci.* **2**:487.

CHAPTER 9

THE CONFORMATIONS OF CYCLOPEPTIDES IN SOLUTION

Yu. A. Ovchinnikov

Shemyakin Institute for Chemistry of Natural Products
USSR Academy of Sciences
Moscow, USSR

I. INTRODUCTION

The importance and diversity of the biological functions of cyclopeptides and the new possibilities for investigating this extensive class of peculiar and often intricately built natural compounds have made them an alluring object of ever-increasing study. Interest is at present centering on their stereochemical features, which is not surprising in view of the importance of spatial structure to the biological properties of peptides and proteins: the conformation–activity relations are the key to their mode of action. The limited conformational flexibility of the cyclopeptides, usually resulting in the existence of a relatively small number of well-defined preferential conformations, facilitates their theoretical and physicochemical conformational analysis in solution.

As with proteins, X-ray analysis, pioneered in 1957 by Hodgkin and coworkers in their study of gramicidin S (Schmidt *et al.*, 1957; Hodgkin and Oughton, 1957), marked the first attempt to obtain an insight into the spatial structure of cyclopeptides. A considerable step forward in these early studies was made by Karle and Karle (1963) when they introduced the direct method (not requiring heavy atom derivatives) of analyzing the crystalline state conformations of cyclohexaglycyl. Subsequent years have been witness to other such interesting works (Zalkin *et al.*, 1966; Konnert and Karle, 1969; Karle *et al.*, 1970; Sobell *et al.*, 1971), the X-ray method as yet being unsurpassed in precision for determination of atomic coordinates.

It is therefore the more amazing that at present this method for stereo-chemical study of cyclopeptides has found a strong competitor in a fundamentally different approach whereby conformational states in solution are studied by a combination of physicochemical methods. While less precise than X-ray analysis, such an approach has been found in many cases to yield more complete information on the spatial structures of a cyclopeptide. For instance, only by this means can one explore the conformational dynamics of the cyclopeptides as a function of environmental conditions, often of decisive significance in understanding the nature and mechanism of their biological action.

The rapid development of physicochemical techniques has given the researcher a wealth of methods from which to draw for the solution of a given conformational problem, and it has placed before him the task of how best to make use of the available techniques. Much can be achieved with proper usage, as is obvious from the considerable progress made, especially in the last 2 or 3 years, in structural studies of cyclopeptides.

We have been attempting to devise a general approach to conformational studies of cyclopeptides/depsipeptides in solution since 1967, based on the combined use of the methods of NMR and IR spectroscopy, optical rotatory dispersion, circular dichroism, dipole moments, and conformational energy calculations. The rationale behind such a "composite" approach is that whereas no one of these methods, even the highly powerful nuclear magnetic resonance, is able by itself to give unequivocal information on the complete spatial structure of peptides, when data obtained from the rational use of each of them are analyzed together one may, as a rule, obtain a sufficiently complete and reliable structural picture with minimum expenditure of time and energy.

Quite naturally, the question of how to utilize each of the above methods most rationally in composite conformational study frequently poses problems of a methodological and theoretical character. Such, for instance, was establishment of a relationship between the specific integral intensities (intensity per NH group) and frequencies of the NH absorption bands which greatly enhanced the effectiveness of IR spectroscopy in structural studies of the cyclopeptides (Ivanov et al., 1971e), as also was the necessity of determining the stereochemical dependence of the $^3J_{NH-CH}$ coupling constants of the peptide $NH-C^\alpha H$ fragments (Bystrov et al., 1969; Bystrov, 1972) before NMR could be utilized to all its capacity for conformational studies of peptides. Another prerequisite for success is systematic study by means of the constituent methods of model compounds such as amides and the simpler peptides.

In this chapter, the power of the composite approach for studying the solution structures of cyclic peptides will be demonstrated. It is the firm belief of the author that this approach will also prove highly fruitful in stereochemical work on highly complex linear and cyclic peptides and the simpler proteins.

First, brief consideration will be given to the general principles of the composite approach, and then the application of these principles to conformational studies of cyclohexapeptides and of valinomycin, enniatins, gramicidin S, and antamanide will be discussed.

II. GENERAL PRINCIPLES OF COMPOSITE PHYSICOCHEMICAL STUDY OF THE SOLUTION CONFORMATION OF CYCLOPEPTIDES

The basic methods for study of the spatial structure of peptides in solution have been developed and refined using cyclopeptide systems as an example. In this way, the scope and limitations of the methods have been revealed, and the experience gained can be used to advantage for extending this approach to linear systems. Less fruitful in this respect have been studies of macromolecular polypeptides and proteins, usually resulting in rather formal descriptions in terms of α- and β-structures and random coils.

In the process of study of the cyclopeptides, a specific "ideology" has taken shape according to which it is considered pointless to strive to obtain and treat all the data that can be gleaned from a given method; rather, each method should be used for solving a concrete, even though at times quite minor, task, with the proviso of maximum reliability of the results. Such information obtained piecemeal from the various methods can then be put together like a sort of jigsaw puzzle to reveal much of, if not the entire, structural picture.

A. Optical Rotatory Dispersion (ORD) and Circular Dichroism (CD)

The parameters of the ORD and CD curves of cyclopeptides (cyclodepsipeptides) are determined mainly by the mutual orientation of the amid/ester chromophores, so the spectropolarimetric method has turned out to be particularly sensitive to changes in the conformational states of the molecules, and from this standpoint it has no peers. Bearing in mind the frequent complexity of obtaining the required stereochemical information, such as orientation of the individual fragments within the peptide chain, recourse to the CD and ORD curves is therefore made when one needs to know whether conformational changes occur with change in medium, temperature, and other conditions or to see whether any shift in the conformational equilibrium of several coexisting forms or other such dynamic events has taken place.

B. Infrared (IR) Spectroscopy

IR spectroscopy proved to be most reliable in studies of intramolecular hydrogen bonding. Of essential importance is the fact that with its help one can determine the ratio of hydrogen-bonded and free CONH groups in the cyclopeptide molecule (Ivanov et al., 1971e). This particularly pertains to solutions in nonpolar solvents that themselves are very weak (if at all) proton donors or acceptors and manifest little tendency to engage in hydrogen bonding (CCl_4, $CHCl_3$, etc.). Usually, the presence of NH bands in the 3430–3480 cm^{-1} (amide A) region is evidence of the presence of free NH groups, whereas the 3300–3380 cm^{-1} region corresponds to hydrogenbonded NH groups. Care must be taken in interpreting bands in the 3380–3420 cm^{-1} region, for they can indicate either free or weakly hydrogenbonded NH groups. It should be stressed that from the integral intensity of the amide A bands one can estimate the number of amide groups in the different types.

C. Nuclear Magnetic Resonance (NMR)

NMR spectroscopy is one of the principal sources of information on cyclopeptide structural characteristics. From the spectral pattern itself, one often obtains an immediate answer as to the number of conformers in equilibrium, their symmetry, etc. Analysis of the amide NH signals permits determination of the number and, of particular importance, the position of the intramolecular hydrogen bonds in the cyclopeptide molecule. For this, two procedures are usually employed:

1. Measurement of the temperature dependence of the NH chemical shifts (δ) in hydrogen-bonding solvents (dimethylsulfoxide, methanol; Ohnishi and Urry, 1969; Urry, 1970). Here large $\Delta\delta/\Delta T$ values $(6-12) \times 10^{-3}$ ppm/degree) correspond to solvated NH groups, whereas small values $(10-2) \times 10^3$ ppm/degree) refer to intramolecular hydrogen bonded NH (Llinas et al., 1970; Urry et al., 1970; Brewster and Bovey, 1971; Cary et al., 1971; van Dreele et al., 1971; Feeney et al., 1971; Ivanov et al., 1971b, 1973a; Portnova et al., 1971a, Weinkam and Jorgensen, 1971c).

2. Differences in deuterium-exchange rates of the NH groups as determined from the rate of intensity decrease of the corresponding signals (Stern et al., 1968; Ovchinnikov et al., 1970; Cary et al., 1971; Kopple, 1971; Ivanov, 1971c, 1973a; Portnova et al., 1971a,b). A sharp

difference in the hydrogen-exchange rates (half-life $\tau_{1/2}$ differences of an order of magnitude or more) makes possible the unequivocal assignment of free and H-bonded NH groups.

Of fundamental importance in conformational studies of peptide systems is analysis of the spin–spin coupling of the NH and $C^\alpha H$ protons, since the vicinal $^3J_{NH-CH}$ constant depends on the dihedral angle θ between the $H-N-C^\alpha$ and $N-C^\alpha-H$ planes. The stereochemical dependence determined in our laboratory makes it possible from the $^3J_{NH-CH}$ values to obtain the range of possible θ values and hence to assess the conformational parameter ϕ (Bystrov et al., 1969; Bystrov, 1972). The most unequivocal results are obtained with systems of limited flexibility, namely, cyclopeptides. With several coexisting equilibrium forms, immediate results are obtained only for values $3\,\text{Hz} > {}^3J_{NH-CH} > 9\,\text{Hz}$, whereas for values of $3\,\text{Hz} < {}^3J_{NH-CH} < 9\,\text{Hz}$ one must know the relative content of the individual conformers before the data can be interpreted.

It stands to reason that the $C^\alpha H - C^\beta H$ coupling constants can shed light on the conformations of the individual amino and hydroxy acid side-chains in the cyclopeptide/cyclodepsipeptide.

The above-described methods (ORD, CD, IR, NMR) are often sufficient to determine the preferential conformation of a cyclopeptide. In other cases, they can strictly limit the number of possible conformations, thus facilitating the use of theoretical conformational analysis for choice of the energetically most advantageous structure and determination of its ϕ, ψ, ω, χ parameters.

D. Theoretical Conformational Analysis

The first stage of theoretical conformational analysis is often calculation of the conformational energy maps of individual peptide/depsipeptide fragments in the cyclopeptide system, the energy of a given conformation being considered as the sum of the energies of the individual fragments (see, for instance, Ivanov et al., 1971b). The results are then refined by minimization of the total energy of the system with respect to several variables (usually with respect to all the angles ϕ, ψ, ω, and χ and the valence angles at C^α) (Popov et al., 1970b). Naturally, direct calculation of the energy of the entire cyclopeptide molecule gives more reliable results, but often this is impractical because it requires too much computer time and also because it may present certain methodological difficulties.

E. Dipole Moments

After determination of the preferred cyclopeptide conformation, one may calculate the dipole moment of the peptide backbone by vector summation of the dipoles of the individual amide (ester) and other polar groups.

Agreement of the results with those found experimentally is an important criterion for the correctness of the suggested conformation (Ivanov et al., 1971b, 1973a; Popov et al., 1970b).

In most cases, the above methods are sufficient for complete description of the conformational states of cyclopeptides in solution. Independent valuable information can also be obtained from electron spin resonance spectra of spin-labeled peptides (Weinkam and Jorgensen, 1971a,b; Ivanov et al., 1973b), ultrasonic absorption (Grell et al., 1971a,b), etc. It is noteworthy that when the spatial structure of a cyclopeptide was first determined by these methods it was later confirmed by X-ray analysis.

III. CYCLOHEXAPEPTIDES

Eighteen-membered cyclic peptide systems have attracted particular attention because it has been theoretically shown that, in contrast to cyclo-tetra- (Ramakrishnan and Sarathy, 1968; Popov et al., 1970a), cyclopenta-, and cycloheptapeptides, they can occur in unstrained conformations with all six amide bonds transplanar (Dale, 1963; Ramakrishnan and Sarathy, 1969; Sarathy and Ramakrishnan, 1972), thus bringing them closer to linear peptides. We therefore have selected cyclohexapeptides not only for studying the nature of the conformational states of these systems in polar and non-polar media but also as models with the ultimate aim of seeing how such conformational states are affected by the nature and configuration of individual amino acid residues, their ring sequence, etc. For this purpose, a whole series of appropriate compounds has been synthesized (Ivanov et al., 1970, 1971f).

The structures of the compounds [1] to [21] are presented in Fig. 1. Of these, the first series, [1] to [14], consists of cyclopeptides comprising all possible combinations of glycine and L-alanine residues, and the second series, [14] to [21], contains all possible diastereomeric cyclohexaalanyls (without the antipodes).

Information aiding elucidation of the conformational state of the cyclo-peptides [1] to [21] came first from UV spectroscopy (Ivanov et al., 1971a). A comparison of the UV spectra of these compounds in aqueous solutions with those of the model amides Ac-Gly-NHMe [22] and Ac-L-Ala-NHMe [23] and of the "random" polypeptides revealed some hypochromism in the region of $\pi \rightarrow \pi^*$ transitions of the amide that in general increases with the number of alanine residues in the ring. The hypochromic effect is particularly noticeable in cyclohexa-L-alanyl [14], where the decrease in intensity as compared with compound [2] amounts to about 45% (Fig. 2). On the basis of these data, one may conclude that in the cyclohexapeptides exciton interaction occurs to split the absorption band and decrease its intensity.

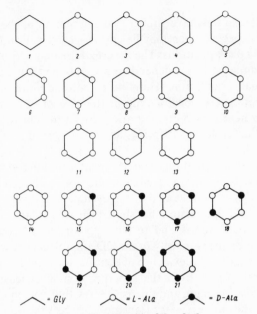

Fig. 1. Cyclopeptides [*1*] to [*21*].

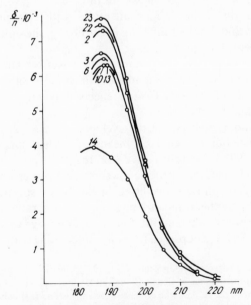

Fig. 2. UV spectra of cyclopeptides [*2*], [*3*], [*6*], [*10*], [*13*], and [*14*] and of diamides [*22*] and [*23*]. In the figure, n equals the number of amide bonds in the molecule; $n = 6$ for (*1*) to (*21*), $n = 2$ for (*22*) and (*23*).

Theoretical calculations have shown that interaction of the chromophoric groups can be expected when they lie close together and differ very little in their absorption peak positions. The nature of the interaction between these groups is, of course, strongly dependent not only on their distance but also on their mutual orientation. Heretofore, chromophore interaction was revealed in peptide UV spectra only in the case of polyamino acids and proteins with α-helical or β-configuration. From the above, the conclusion can be drawn that cyclic hexapeptides are a new type of ordered interacting amide chromophore systems.

More detailed information about this conformational system came from optical rotatory dispersion and circular dichroism studied (Ivanov *et al.*, 1971a,g). It was by these means that the cyclopeptides were shown to exist in a conformational equilibrium that shifted with the polarity of the medium; in other words, they were found to have "polar" and "nonpolar" spatial forms. For instance, the similarity of the CD and ORD curves of compounds [14]–[18] in water (Fig. 3) is evidence of their being in similar conformational states. The curves result from the superposition of at least three Cotton effects, whose position, sign, and intensity show that one, in the 210–215 nm region, refers to the amide $n \rightarrow \pi^*$ transition, whereas the oppositely directed intensive effects at 198–200 and 185 nm refer to the components of the split $\pi \rightarrow \pi^*$ transition. The polar media are thus seen to give rise to a specific rigid conformation of the cyclohexapeptides, with very probably its own characteristic ϕ and ψ values differing from those of the other known spatial forms of polypeptides and proteins.

A decrease in polarity of the medium causes certain conformational rearrangements of the cyclohexapeptides which are manifested in intensity redistribution of the various Cotton effects, but with retention of their position, sign, and number.

Conformational analysis of the cyclopeptides [1]–[21] in polar media has been carried out in detail by means of nuclear magnetic resonance (Ivanov *et al.*, 1971d; Portnova *et al.*, 1971a,b).* In polar solvents such as dimethylsulfoxide, trifluoroacetic acid, and water,† two groups of NH signals (usually at 7.3–8.0 ppm) are displayed in the NMR spectra at room temperature—the higher field signals being due to two intramolecular hydrogen bonded NH protons, whereas the four remaining signals (at 7.9–8.6 ppm) are due to "free" NH groups, i.e., to groups interacting with the solvent. This

---

*The NMR spectral pattern of cyclopeptides [1] to [21], primarily its "homogeneity," i.e., the correspondence of the number of signals in the spectra to the structural groups of the compounds, together with the IR data (the presence of an intense amide II band), shows that these compounds have no *cis* amide bonds (cf. Karle and Karle, 1963; Karle *et al.*, 1970).

†Compounds [1] to [21] proved to be too insoluble in the other common solvents for ordinary NMR spectral work.

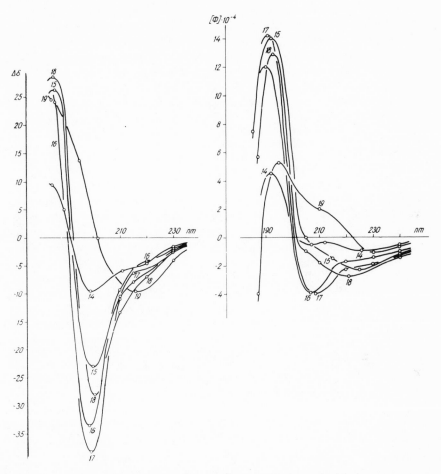

Fig. 3. CD and ORD curves of compounds [*14*] to [*19*] in water.

is well brought out by study of the temperature dependence of these signals (Portnova *et al.*, 1971*a*, *b*).

The findings are in good accord with the "pleated sheet" conformation proposed for these compounds by Schwyzer on the basis of chemical data (Schwyzer *et al.*, 1958, 1964; Schwyzer, 1959). As one can see from Fig. 4, in such a conformation two "transannular" hydrogen bonds of the type 4 → 1 close two ten-membered rings.

In some cases, the existence of such structures (Fig. 4) can be confirmed by study of the deuterium-exchange rates of the NH groups. However, most of the cyclohexapeptides investigated display practically the same NH-exchange rate, apparently due to rapid conformational interchange of the

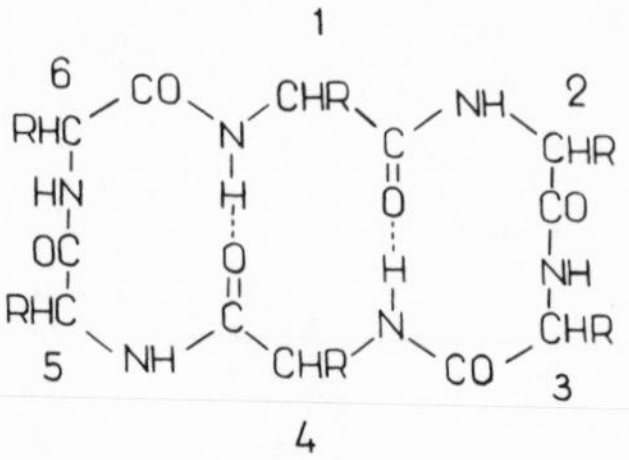

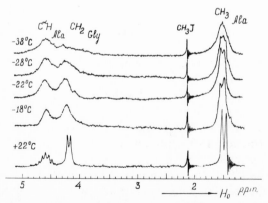

Fig. 4. Intramolecular hydrogen bonding in the "pleated sheet" conformation of cyclic hexapeptides.

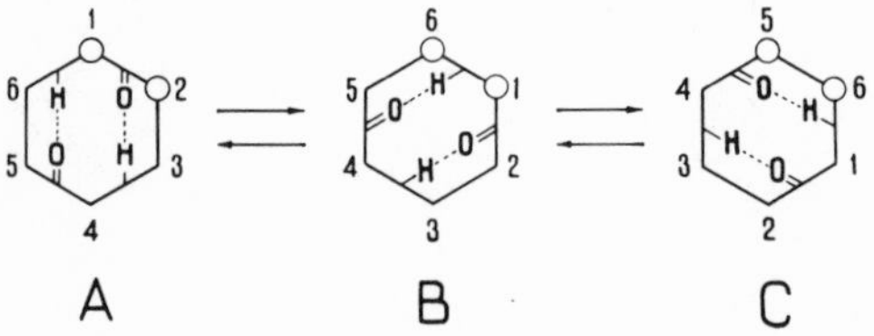

Fig. 5. Equilibrium of the "pleated sheet" structures of a cyclohexapeptide (the example is cyclo-Gly-Gly-Gly-Gly-L-Ala-L-Ala [3]), showing the migration of transannular hydrogen bonds.

Fig. 6. Temperature dependence of the NMR spectrum of cyclo-(L-Ala-Gly)$_3$ [9] in CF_3COOH plus H_2O (7:1).

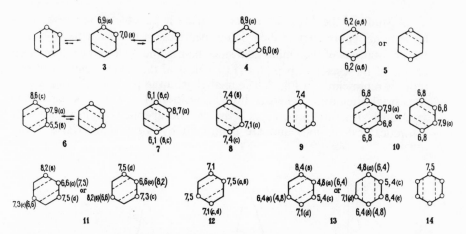

Fig. 7. Position of the transannular H-bonds in "pleated sheet" structures of cyclopeptides [3] to [14] and $^3J_{\mathrm{NH-CH}}$ values in the alanine residues.

type $A \rightleftharpoons B \rightleftharpoons C$ (Fig. 5). Indeed, cooling of a solution of compound [9] in trifluoroacetic acid–water mixture (7:1) causes broadening of the CH_2 signals, evidence of the existence of interconverting conformers (Fig. 6).

Analysis of the NH chemical shifts and their temperature gradients made it possible to define the preferred position of the intramolecular hydrogen bonds in most of the cyclohexapeptides; the predominant structures are given in Fig. 7. It can be readily seen that there seem to be no regularities regarding the position of the hydrogen bonds, apparently due to the small energy differences of the forms A, B, and C.

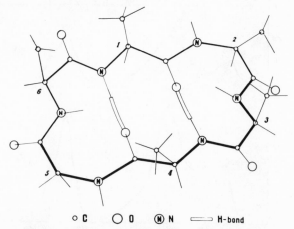

○ C   ○ O   Ⓝ N   ⸺ H-bond

Fig. 8. Predominant conformation of the "pleated sheet" structure of cyclopeptide [13].

A study of the $^3J_{NH-CH}$ constants made it possible to refine the conformational parameters of the cylohexapeptides, in particular to determine the orientations of the amino acid side-chains. The preferred conformation of one of the cyclohexapeptides [13] typical of the other members of this series is represented in Fig. 8. The methyl group in position 2 is of pseudo-equatorial orientation, whereas those in positions 1, 3, 4, and 6 are pseudo-axial. Such structural preferences in the cyclohexapeptides are in accord with theoretical calculations and are confirmed by X-ray data (Karle *et al.*, 1970).

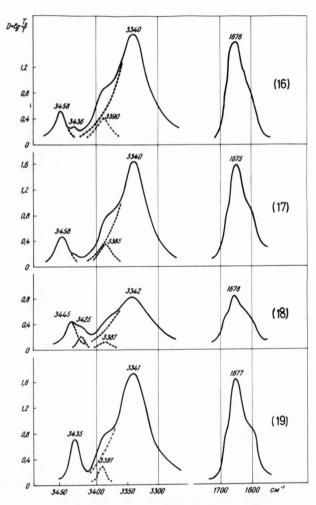

Fig. 9. IR spectra of cyclopeptides [16] to [19].

In summary, it should be emphasized that in polar solvents cyclohexapeptides do not possess fixed spatial structures but exist in a complex conformational equilibrium, with the pleated sheet as the most preferable form (Fig. 8).

In nonpolar media, the conformational states of the cyclohexapeptides have been investigated mainly by IR spectroscopy (Ivanov et al., 1971e,f). As one can see from Fig. 9, cyclohexapeptides [16] to [19] have very similar IR spectra in chloroform solution. There is a strong band in the amide A region at 3340 cm⁻¹ and several bands in the 3410–3460 cm⁻¹ range. An asymmetrical band in the 1670–1675 cm⁻¹ (amide I) region cannot be separated into its components. Detailed study of the IR spectra of a model system (Efremov et al., 1973) showed that the 3340 cm⁻¹ band can be assigned to H-bonded and the 3410–3460 cm⁻¹ bands to non-H-bonded NH groups. Quantitative data from the integral intensities of the corresponding NH bands based on the correlations established in this laboratory have shown that in nonpolar media the preferable conformation of the cyclopeptides [16]–[19] contain at least three or four intramolecular hydrogen bonds. From the above, it follows that in nonpolar solvents the cyclic hexapeptides retain the general form of the "polar" conformation and the 4 → 1 hydrogen bonds but that the equilibrium shifts in favor of conformers somewhat different with ϕ and ψ values from the "polar forms" and with additional intramolecular hydrogen bonds.

Theoretical analysis assuming *trans* amide bond configurations has led to the conclusion that in nonpolar media the cyclohexapeptides are preferentially in conformations of either type A or B (Figs. 10 and 11).

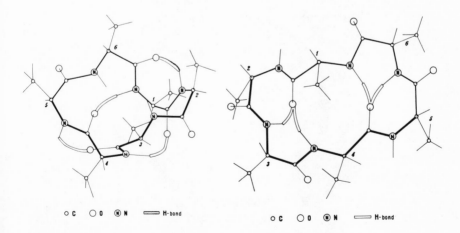

Fig. 10. Form A of cyclo-hexa-L-alanyl.   Fig. 11. Form B of cyclo-hexa-L-alanyl.

The choice falls on type A when the experimental dipole moments for the cyclohexapeptides [*16*] to [*19*] are compared with values calculated for differing conformers. Hence in nonpolar solvents cyclic hexapeptides built up of L- and D-alanine residues are preferentially in type-A or (less probably) type-B conformations, both containing four intramolecular hydrogen bonds. Possibly there are also minor amounts of forms with two hydrogen bonds (4 → 1) and also of the rigid conformation, with six hydrogen bonds of the type 4 → 1 and 3 → 1 also taking part in the equilibrium.

IV. VALINOMYCIN

The cyclodepsipeptide valinomycin (Fig. 12) is unique in its biological importance, being the classic representative of the ionophores selectively increasing the potassium ion permeability of artificial and biological membranes. Valinomycin is the first macrocyclic compound of peptide nature whose spatial structure has been precisely and unequivocally defined (Ivanov *et al.*, 1969, 1971*b*). Such an achievement has been due to the above-described composite physicochemical approach.

Valinomycin is in the form of a 36-membered ring, consisting of alternating amino and hydroxy acid residues and endowed with a wealth of conformational possibilities. Spectral methods showed the conformational states of this antibiotic to be highly dependent on the solvent species. In particular, this follows from the sharp changes in its ORD curves on passing from heptane to ethanol and then to aqueous solutions (Fig. 13).

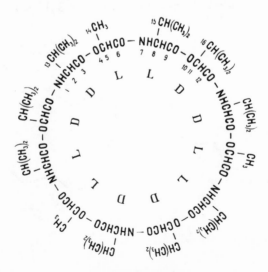

Fig. 12. Valinomycin.

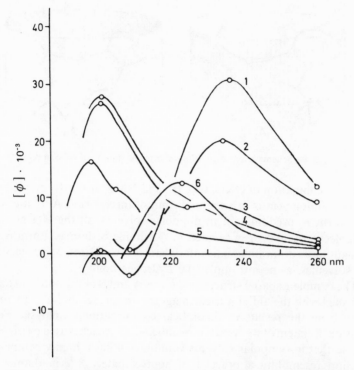

Fig. 13. ORD curves of valinomycin and its K^+-complex, 1, Heptane; 2, heptane–dioxane, 10:1; 3, ethanol; 4, acetonitril; 5, trifluoroethanol–water, 1:2; 6, ethanol plus KBr.

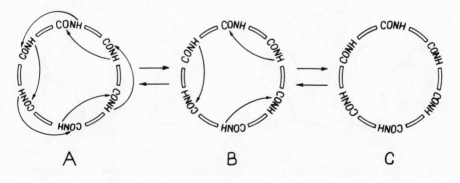

Fig. 14. Schematic representation of $A \rightleftarrows B \rightleftarrows C$ equilibrium of valinomycin.

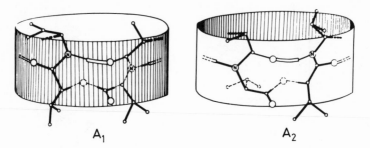

Fig. 15. Schematic representation of forms A_1 and A_2 of valinomycin.

The combination of ORD, CD, IR, and NMR methods revealed that in solution valinomycin is in an equilibrium of three forms, A, B, and C (Fig. 14). In form A, predominant in nonpolar solvents, all the NH groups are intramolecularly hydrogen bonded to the amide carbonyls. Form B is predominant in medium-polar solvents, whereas form C (important in strongly polar solvents) has no intramolecular hydrogen bonds.

The complete spatial structure of forms A and B of valinomycin has been established with the aid of nuclear magnetic resonance and also by drawing heavily from the results of theoretical conformational analysis. Form A constitutes a system of six fused ten-membered rings each closed by a hydrogen bond, so that in nonpolar solvents valinomycin has a highly compact conformation resembling a bracelet of approximately 8 Å in diameter and approximately 4 Å high. In principle, form A can be assigned to two arrangements of the depsipeptide chain, A_1 and A_2, differing in ring chirality and side-chain orientation (Fig. 15). A detailed theoretical analysis of the A_1 and

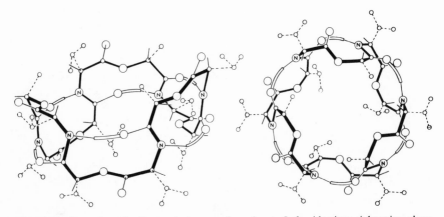

Fig. 16. Conformation of valinomycin in nonpolar solvents. Left: side view; right: view along the symmetry axis.

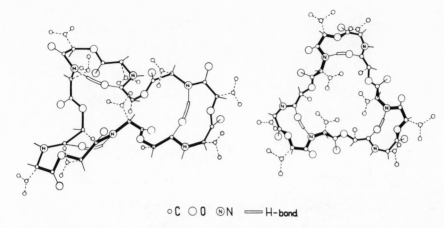

○C ○O Ⓝ N ══ H-bond

Fig. 17. Conformation of valinomycin in solvents of medium polarity. Left: side view; right: view along the symmetry axis.

A_2 variants differing in orientation of the free ester carbonyls (inside or outside the ring) and in the magnitude of the $^3J_{NH-CH}$ constant has led to the conclusion that in nonpolar media valinomycin is preferentially in the A_1 conformation shown in Fig. 16.

In this conformation, all the ester carbonyls are of "outside" orientation and the ϕ and ψ values are approximately as follows*:

	D-Val	L-Lac	L-Val	D-HyIv
ϕ	−40	−100	25	100
ψ	−70	40	70	−30

In form B, valinomycin has a "propeller" type of conformation (Fig. 17) wherein the hydrophobic kernel of the D-valyl and L-lactyl aliphatic sidechains is surrounded by the ten-membered hydrogen-bonded rings.

It is the existence in solution of a conformational equilibrium of these different forms of valinomycin, shifting with change in the environment, which is the decisive factor in the specificity of its biological action, in particular its ability to transport metal ions through membranes.

The ability of valinomycin to form specific complexes with alkali metals and its unusually high K/Na selectivity in this reaction have been an incentive for our studying the spatial structure of the K^+-complex in solution, with the objective of elucidating the details of the complexing reaction that could shed light on the high ion selectivity. The approach to this problem was much the same as that used for free valinomycin.

---

*Conformational nomenclature proposed by the IUPAC-IUB Commission (Kendrew *et al.*, 1970) is used in this chapter.

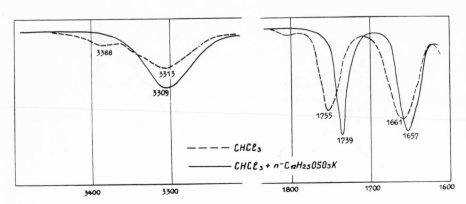

Fig. 18. IR spectra of valinomycin and its K^+-complex in $CHCl_3$.

Thus comparison of the IR spectra of valinomycin and its K^+-complex in $CHCl_3$ (Fig. 18) showed that the complex retains the "bracelet" system of hydrogen bonds (a single band in the amide A region at $3309\ cm^{-1}$), all ester carbonyls being engaged in ion-dipole interaction with the cation (shift of the CO stretch frequency from $1755\ cm^{-1}$ for the free valinomycin to $1739\ cm^{-1}$ for the complexed valinomycin). The comparatively low $^3J_{NH-CH}$ values of the valinomycin·K^+-complex (approximately 5 Hz) show the NH—CH protons to be in *gauche* orientation. It follows from this that in solution the K^+-complex of valinomycin is preferably in the A_2 conformation shown in Fig. 19.

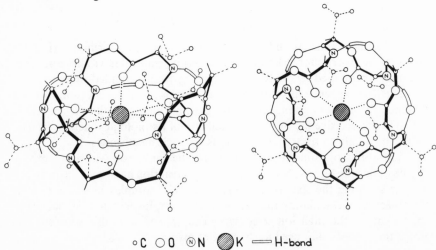

∘C ◯O ⓃN ⬤K ═H-bond

Fig. 19. Conformation of the valinomycin·K^+-complex. Left: side view; right: view along the symmetry axis.

One must take note of the fact that highly valuable information on the structure of the K^+-complex of valinomycin and similar compounds, particularly regarding the mode of interaction of the carbonyls with the metal ion, can be obtained from an analysis of the paramagnetic shift of the ester and amide $^{13}C=O$ signals in the ^{13}C NMR spectra of these compounds (Bystrov et al., 1972; Ohnishi et al., 1972).

The data obtained in this laboratory have been completely confirmed by X-ray analysis of the valinomycin·$KAuCl_4$ complex (Pinkerton et al., 1969). Characteristic of the conformation of this complex is the effective screening of the metal cation by the ester carbonyls, the hydrogen bonding system, and the pendant isopropyl groups. The lipophilic character of the exterior of this complex explains its ready solubility in neutral organic solvents and in the lipophilic regions of the membrane.

A comparison of the conformations of the free and complexed valinomycin (Figs. 16 and 19) readily shows that the complexing reaction is accompanied by major conformational changes in the depsipeptide chain. In particular, very striking is the reorientation of the ester carbonyls from "all outside" (A_1) in the free compound to "all inside" (A_2) in the complex. Inasmuch as interconversion of these two forms is impossible without the rupture of at least three hydrogen bonds, complexation in nonpolar media should in all probability proceed via B as intermediate.

The data presented here on the solution conformations of valinomycin and its K^+-complex have served as the starting point for structure–function studies in the course of which the directed synthesis of a number of membrane-active valinomycin analogues with unique properties has been achieved.

V. ENNIATINS

Enniatins A, B, and C and beauvericin (Fig. 20), 18-membered cyclo-hexadepsipeptides biologically very closely allied to valinomycin, are highly optically active. A definitely manifested dependence of the CD and ORD curves on the solvent polarity is evidence of considerable conformational flexibility in these compounds (see, e.g., Fig. 21). The presence of isosbestic points on the CD and ORD curves of enniatin B indicates participation of two basic forms in the conformational equilibrium—a "polar" form (P) and a "nonpolar" form (N).

The spatial structure of form P (the same conformation as seen from curves in Fig. 21 is also possessed by the enniatin B · K^+-complex) has been determined in both the crystalline state (Dobler et al., 1969) and solutions

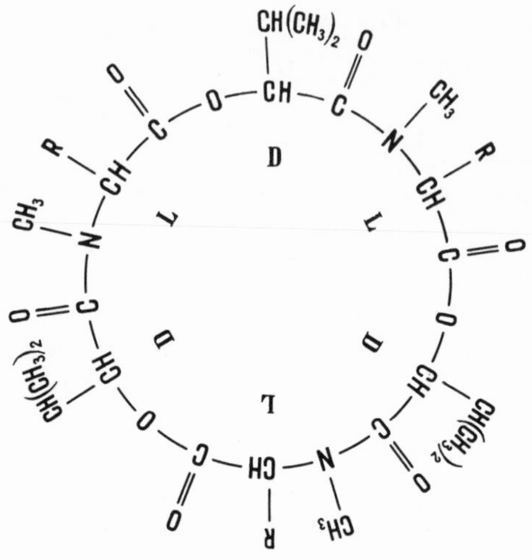

Fig. 20. Antibiotics of the enniatin group.
Enniatin A, $R = CH(CH_3)C_2H_5$; enniatin B,
$R = CH(CH_3)_2$; enniatin C, $R = CH_2CH\cdot$
$(CH_3)_2$; beauvericin, $R = CH_2C_6H_5$.

(Ovchinnikov et al., 1969; Shemyakin et al., 1969), with completely accordant results. The K^+-complex (Fig. 22), resembling a charged discus with lipophilic rim, has $K^+ \ldots O$ distances of 2.6–2.8 Å and the following approximate conformational parameters:

	L-MeVal	D-HyIv
ϕ	−60	60
ψ	120	−120

NMR study of the interaction of enniatin B (or its analogues) with differing alkali ions in solution not only permitted following of the complexing dynamics and the accompanying conformational "induced fit" rearrangements but also brought to light fine differences in the spatial structure of the enniatin complexes, depending on the cation species. Thus the complexing of (tri-N-desmethyl)enniatin B with Li^+, Na^+, K^+, and Cs^+ causes a corresponding increase in the $^3J_{NH-CH}$ constant from 4.9 to 8.5 Hz. This shows that on passing from smaller to larger cations there is a steady increase in size of the internal cavity due to orientation changes in the cation-binding carbonyls, much as in the opening of a flowerbud. In form P, these changes are accompanied by simultaneous turning of the amide and ester planes and, naturally, by changes in the NH—CH dihedral angle (Fig. 23).

In contrast to form P, form N of enniatin B has no symmetry elements (Ovchinnikov *et al.*, 1969; Shemyakin *et al.*, 1969). Such a conclusion follows from temperature study of the enniatin B NMR spectra (Fig. 24). As the temperature is lowered, first the NMR signals broaden and then at

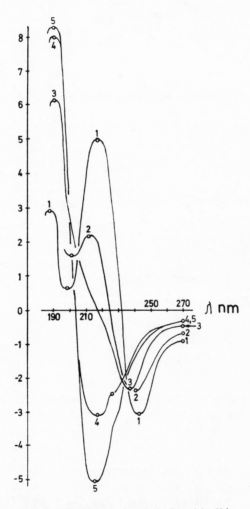

Fig. 21. ORD curves of enniatin B and its K⁺-complex. 1, Heptane; 2, 96% ethanol; 3, aceto-nitril; 4, water–trifluoroethanol, 2:1; 5, 10^{-2} mole/liter KCl in 96% ethanol (tenfold excess of salt).

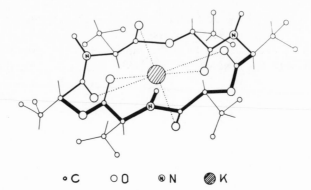

Fig. 22. Conformation of the K⁺-complex of enniatin B.

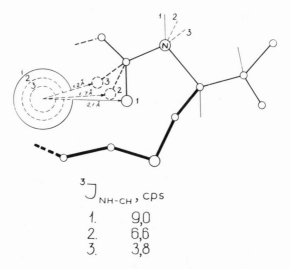

$^3J_{\text{NH-CH}}$, cps

1.   9,0
2.   6,6
3.   3,8

Fig. 23. Effect of the size of the interval cavity on $^3J_{\text{NH-CH}}$ coupling constants of (tri-N-desmethyl)-enniatin B complexes.

−120°C the spectrum reveals the existence in the molecule of three non-equivalent L-MeVal-D-HyIv fragments; this can be particularly seen in the N-methyl region (2.3–3.4 ppm) and in the region of the C$^\alpha$H protons (4.2–5.6 ppm). In form N, each fragment assumes the three different conformations with equal probability, and their rapid equilibration at room temperature, causing averaging of the chemical shifts, is the reason for the simplicity of the spectra.

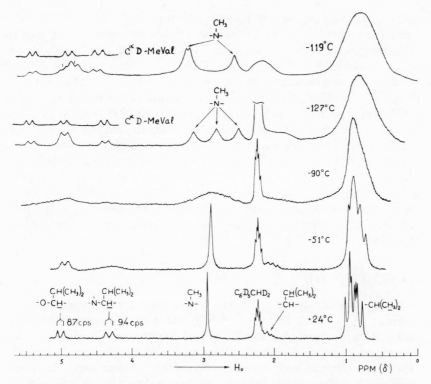

Fig. 24. NMR spectra of enniatin B in CS_2 at $-119°C$ (upper spectrum) and in CS_2–$CD_3C_6D_5$ (2:1) at different temperatures (lower spectra).

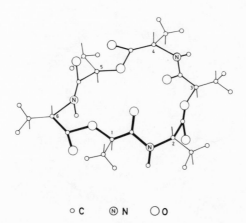

○ C   Ⓝ N   ◯ O

Fig. 25. Conformation of enniatin B in nonpolar
solvents.

A comparison of the NMR data at differing temperatures for enniatin B and its selectively deuterated analogue, synthesized expressly for spectral assignments (Shemyakin *et al.*, 1972), with the results of theoretical conformational analysis (Popov *et al.*, 1970*b*) led to the conclusion that the preferred enniatin B conformation in nonpolar media is that shown in Fig. 25 Ivanov *et al.*, 1973). The dipole moment calculated for this conformation (3.8 D) is in good accord with the experimental value (3.35 D in CCl_4).

The conformational analysis of the enniatins showed that form P, more advantageous than form N from the standpoint of nonbonding interactions, is destabilized by electrostatic interaction of the spatially neared carbonyls. The driving force of the $N \rightarrow P$ transformation is solvation of the polar groups. In the light of this, one can understand the strong similarity between the polar form P and the complex conformation, the solvent and the cation fulfilling essentially the same function *viz.* elimination of dipole–dipole repulsion.

VI. GRAMICIDIN S

The cyclodecapeptide gramicidin S (Fig. 26) is a well-known antibiotic of clinical importance. In recent years, it has become an ever increasingly used tool for biochemical research, although knowledge of its mode of action is confined to vague statements about its ability to "damage" biological membranes (Hunter and Schwarz, 1967). On the other hand, its stereochemistry has been widely investigated, and in this sense it has been the favorite cyclopeptide and probably peptide in general for study.

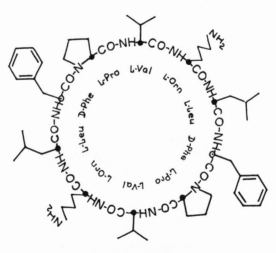

Fig. 26. Gramicidin S.

Beginning in 1953, several spatial structures have been proposed for gramicidin S, often of fundamentally differing types. Hodgkin and Oughton (1957) were the first to propose a pleated sheet model as best conforming to the X-ray data (Schmidt *et al.*, 1957). The same conclusion was made by Schwyzer on the basis of an elegant chemical investigation (Schwyzer *et al.*, 1958; Schwyzer, 1959) and later from NMR studies (Schwyzer and Ludescher, 1968, 1969; Ludescher and Schwyzer, 1971). Subsequently, several other conformations were proposed for gramicidin S, belonging to the β-type but differing considerably from each other (Scheraga *et al.*, 1965; Vanderkooi *et al.*, 1966; Scott *et al.*, 1967; Stern *et al.*, 1968; Momany *et al.*, 1969). Attempts have also been made to experimentally substantiate other structural types (Abbott and Ambrose, 1953; Warner, 1961, 1967), including some with α-helical regions (Liquori *et al.*, 1966; Liquori and Conti, 1968). However, in no case has the spatial structure of gramicidin S been rigorously proved, and the reported data, including those from physical methods and from theoretical calculations, have always left room for an alternative model.

In an attempt to obtain unequivocal structural information and also with the objective of extending the boundaries of the composite approach, we carried out the simultaneous physicochemical study of gramicidin S and its N,N'-diacetyl derivative, lacking ionogenic groups and somewhat more soluble than the antibiotic in nonpolar solvents (Ovchinnikov *et al.*, 1970). A favorable circumstance was the fact that both these compounds possess very rigid and quite similar conformations, as could be surmised from the similarity of the corresponding ORD curves and the practical independence of their shape from the solvents employed (Fig. 27).

The IR spectrum of N,N'-diacetylgramicidin S in chloroform displays an intensive band at $3314 \, cm^{-1}$, belonging to NH groups participating in intramolecular hydrogen bonding, and a weak band at $3426 \, cm^{-1}$, corresponding to free NH groups. An estimate of the integral intensities showed that six NH groups are taking part in the system of intramolecular hydrogen bonds. Comparison of the deuterium-exchange rates of the NH groups (by means of the NMR method) showed that four hydrogen bonds formed by the NH groups of the valine and leucine residues are very strong, whereas the remaining two, due to the ornithyls NH groups, are much less stable.

Further, analysis of the $^3J_{NH-CH}$ spin–spin coupling constants of the valine, leucine, and ornithine residues (nonoverlapping doublets with $^3J_{NH-CH}$ values of 9.4, 9.4, and 9.0 Hz, respectively) showed that the corresponding NH—CH fragments are all *trans*. In the same way, the NH—CH fragment of the D-phenylalanine residue ($^3J_{NH-CH} = 3.5$ Hz) was found to be *gauche*.

The sum total of the evidence discussed here, together with the assumption (following from IR data) of *trans* configuration of the peptide bonds and

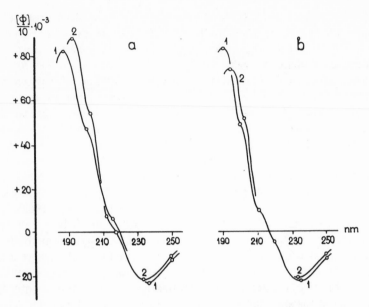

Fig. 27. ORD curves of gramicidin S (a) and N,N'-diacetylgramicidin S (b) in 12:1 heptane–ethanol (1) and 2:3 ethanol–0.1 N HCl (2) mixtures.

a theoretical analysis of the gramicidin S molecule, in which particular account has been made of all possible configurations of the tertiary amide groupings, has very definitely shown that the only conformation satisfying all requirements is the one with approximately the following coordinates:

	L-Val	L-Orn	L-Leu	D-Phe	L-Pro
ϕ	−120	−110	−120	55	−60
ψ	120	110	110	−110	−40

This conformation, which is represented in Fig. 28, is of the β-pleated type, its distinguishing features being rigidity of the framework and the presence of four strong "transannular" hydrogen bonds, the hydrophobic side-chains of the valine and leucine residues being on one side of the plane of the ring and the side-chains of the ornithine residues with free NH_2 groups on the other (in nonpolar solvents, the NH groups of the acetylornithine side-chains participate in hydrogen bonding with the carbonyls of the neighboring peptide groupings of the backbone in the manner shown in Fig. 28). Such topography of the gramicidin S side-chains, recently confirmed by independent methods, is in all probability an essential factor in the manifestation of biological activity by this antibiotic. In this connection, it is highly noteworthy that the antipode of gramicidin S (*enantio*-gramicidin S)

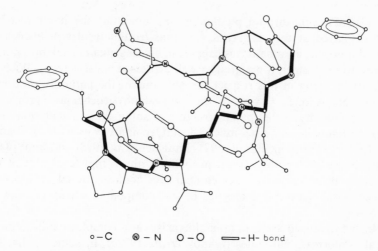

o — C   Ⓝ — N   O — O   ⊏⊐ — H- bond

Fig. 28. Conformation of *N,N'*-diacetylgramicidin S in nonpolar solvents.

recently synthesized in our laboratory has exactly the same antimicrobial activity as the natural antibiotic.

VII. ANTAMANIDE

The cyclodecapeptide antamanide (Fig. 29), isolated in 1968 by Wieland (Wieland *et al.*, 1968) from extracts of the green mushroom *Amanita phalloides*, is in many respects a unique compound. First of all, by a still unelucidated mechanism it completely inhibits the action of the main toxic

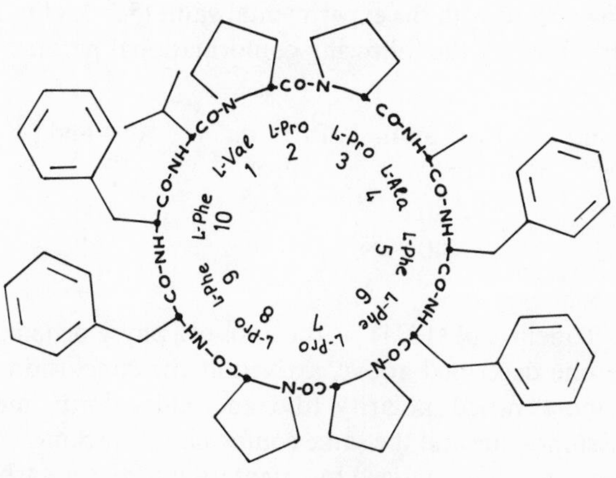

Fig. 29. Antamanide.

principles of this mushroom, phalloidine and amanitins, the inhibiting effect being displayed at very low concentrations (equimolecular or even less). Like valinomycin, antamanide is capable of binding alkali metal ions, but, in contrast to the former, it manifests a well-expressed Na selectivity (Wieland *et al.*, 1970), being in this respect the only one of the naturally occurring peptide complexones. Owing to the unusual physicochemical properties and the peculiarity of the spatial structure of antamanide, elucidation of its spatial structure has required much effort (Ovchinnikov *et al.*, 1971; Ivanov *et al.*, 1971*c*, 1973*a*; Tonelli *et al.*, 1971; Faulstich *et al.*, 1972), and some of the details are as yet unclear. In all probability, the complete picture of the structure will be obtained after comparative conformational studies of a number of the synthetic analogues of this compound, which are now in progress.

In the dissolved state, antamanide exists in a complex equilibrium of several conformational states (Ivanov *et al.*, 1972*b*). In some media, say $CHCl_3$ or heptane–dioxane 5:2, there is a predominant conformation (form A) in which according to IR data all six NH groups take part in the formation of intramolecular hydrogen bonds. In more polar solvents, other forms in which the intramolecular hydrogen bonds are partially (form B) or completely (form C) disrupted begin to predominate. Regarding form A, theoretical conformational analysis with account of all possible combinations of the six intramolecular hydrogen bonds and also the most probable configurations of the Pro fragments and orientation of the NH—CH protons (from NMR data) has made possible the quite reliable assumption initially proposed by Ovchinnikov *et al.* in 1971 that its conformation is that shown in Fig. 30.

The dipole moment calculated for such a conformation (approximately 4.5 D) is in good accord with the experimental value (5.2–5.8 D). The structure possesses approximately the following conformational parameters:

	Val[1] and Phe[6]	Pro[2] and Pro[7]	Pro[3] and Pro[8]	Ala[4] and Phe[9]	Phe[5] and Phe[10]
ϕ	-80	-60	-55	-100	60
ψ	165	-40	-40	10	-70

Recently, Tonelli *et al.* (1971), on the basis of a physicochemical approach similar to the one described above, arrived at the conclusion that in solutions of the most varied polarity (dioxane, chloroform, methanol, etc.) antamanide assumes one and the same conformation, lacking intramolecular hydrogen bonds and characterized by orientation of all the carbonyls on one side of the average plane of the ring. The approximate coordinates of this

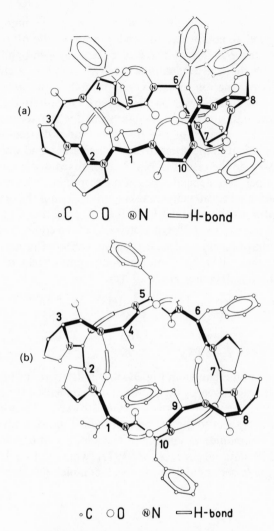

Fig. 30. Conformation of antamanide in nonpolar
solvents. (a) side view; (b) view along the pseudo-
symmetry axis.

conformation are as follows:

	Val[1] and Phe[6]	Pro[2] and Pro[7]	Pro[3] and Pro[8]	Ala[4] and Phe[9]	Phe[5] and Phe[10]
ϕ	−90	−80	−55	−90	−90
ψ	90	130	−55	−60	150

However, this is in disharmony with the large dipole moment (16.6 D) calculated for such a conformation, and moreover the IR spectra unequivocally show the existence of intramolecular hydrogen bonds.*

In contrast to the free antamanide, its Na^+-complex is characterized by a more rigid conformation, as can be seen from the weak solvent dependence of the CD and ORD curves on the solvent. Quantitative examination of the IR spectra indicates *trans* configuration for the secondary amide groups and also the presence of two free and four hydrogen-bonded NH groups. The assignment of the latter to Val^1, Phe^5, Phe^6, and Phe^{10} residues was achieved based on the results of NMR-determined deuterium-exchange rates and the temperature dependencies of the respective NH proton chemical shifts. Further theoretical analysis, taking into account the orientation of the NH—CH protons (determined from the $^3J_{NH-CH}$ coupling constants), the presence of a pseudo-twofold symmetry axis in the complexed molecule, and the formation of an internal cavity by the carbonyl groups and other factors, yielded the following conformational parameters of the Na^+-complex of antamanide (Ivanov *et al.*, 1971*c*):

	Val^1 and Ph^6	Pro^2 and Pro^7	Pro^3 and Pro^8	Ala^4 and Phe^9	Phe^5 and Ph^{10}
φ	-100	-60	-60	150	70
ψ	-160	-50	90	-30	-70

The conformation is shown in Fig. 31. Its similarity to the bracelet form of valinomycin characterized by the presence of a condensed system of intramolecular hydrogen bonded rings is striking. However, the orientation of the carbonyls with respect to the cation and their distance from each other differ greatly in antamanide as compared to valinomycin (a fact supported also by ^{13}C NMR data; Bystrov *et al.*, 1972), which is to a large degree responsible for the lesser stability of the Na^+-complex of antamanide than of valinomycin.†

---

*A similar structure but with nonplanar Val^1-Pro^2 and Phe^6-Pro^7 bonds was discussed by Faulstich *et al.* (1972). However, recently recorded NMR-^{13}C spectra of antamanide (D. Patel, *Biochemistry* 12:667) have shown that two of the four X-Pro bonds of the free molecule probably have the cis configuration.

†An investigation of new antamanide analogs has shown that their Na^+ and Ca^{++} complexes have two, rather than four, hydrogen bonds (V. T. Ivanov, A. A. Kozmin, L. B. Senyavina, N. N. Uvarova, A. I. Miroshnikov, V. F. Bystrov, and Yu. A. Ovchinnikov, *Khim. Prirod. Soed.*, in press). NMR-^{13}C spectra have revealed two cis X-Pro bonds (D. Patel, *Biochemistry* 12:677). The same conclusions were drawn from an X-ray study of the crystalline Li^+ complex of antamanide and the Na^+-complex of (Phe^4, Val^6) antamanide (I. L. Karle, J. Karle, T. Wieland, W. Bürgermeister, H. Faulstich, B. Witkop, *Proc. Natl. Acad. Sci. USA*, 70:1836).

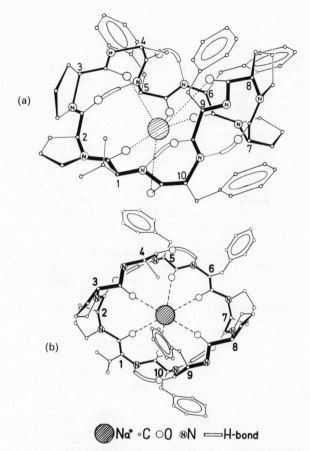

Fig. 31. Conformation of the Na$^+$-complex of antamanide. (a) Side view; (b) view along the pseudo-
symmetry axis.

VIII. CONCLUSION

In conclusion, it may be mentioned that the fruitfulness of the composite approach to study of the conformational states of peptide systems in
solution that has been developed in the study of cyclopeptides has been
demonstrated by other authors working with oxytocin (Walter *et al.*, 1968,
1971; Urry *et al.*, 1968, 1970; Johnson *et al.*, 1969; Urry, 1970; Feeney *et al.*,
1971; Urry and Walter, 1971), angiotensin (Weinkam and Jorgansen, 1971c),
and other biologically important peptides. Further progress in the composite
approach and in the techniques of the individual methods has already made
real the possibility of determining the solution conformations of highly
complex peptide systems, even up to the simpler proteins.

REFERENCES

Abbott, N. B., and Ambrose, E. J., 1953, The conformation of the polypeptide chain in small peptides such as gramicidin S, *Proc. Roy. Soc.* **219A**:17.

Brewster, A. I., and Bovey, F. A., 1971, Conformation of cyclolinopeptide A observed by nuclear resonance spectroscopy, *Proc. Natl. Acad. Sci.* **68**:1199.

Bystrov, V. F., 1972, Spin–Spin interaction between geminal and vicinal protons, *Uspekhi Khimii (Chem. Rev., Russian)* **41**:512.

Bystrov, V. F., Portnova, S. L., Tsetlin, V. I., Ivanov, V. T., and Ovchinnikov, Yu. A., 1969, Conformational studies of peptide systems. The rotational states of the NH—CH fragment of alanine dipeptides by nuclear magnetic resonance, *Tetrahedron* **25**:493.

Bystrov, V. F., Ivanov, V. T., Koz'min, S. A. Mikhaleva, I. I., Khalilulina, K. Kh., Ovchinnikov, Yu. A., Fedin, E. I., and Petrovskii, P. V., 1972, Biologically active alkali metal complexones. A ^{13}C-NMR study of ion–dipole interaction, *FEBS Letters* **21**:34.

Cary, L. W., Takita, T., and Ohnishi, M., 1971, A study of the secondary structure of ilamycin B₁ by 300 MHz proton magnetic resonance, *FEBS Letters* **17**:145.

Dale, J., 1963, Macrocyclic compounds. III. Conformations of cycloalkanes and other flexible macrocycles, *J. Chem. Soc.* **1963**:93.

Dobler, M., Dunitz, J. D., and Krajewski, F., 1969, Structure of the K^+ complex with eniatin B, a macrocyclic antibiotic with K^+-transport properties, *J. Mol. Biol.* **42**:603.

Efremov, E. S., Senyavina, L. B., Zheltova, V. N., Ivanova, A. G., Kostetsky, P. V., Ivanov, V. T., Popov, E. M., and Ovchinnikov, Yu. A., 1973, Conformational states of N-acetyl α-amino acids methylamides and their N-methylated derivatives. I. Infra-red spectra, *Khim. Prir. Soed. (Chem. Nat. Prod., Russian)* **1973**:322.

Faulstich, H., Burgermeister, W., and Wieland, Th., 1972, Antamanide. Evidence for a conformational change and nonplanar amide groups, *Biochem. Biophys. Res. Commun.* **47**:975.

Feeney, J., Roberts, G. C. K., Rockey, J. H., and Burgen, A. S. V., 1971, Conformational studies of oxytocin and lysine vasopressin in aqueous solution using high resolution NMR spectroscopy, *Nature New Biol.* **232**:108.

Grell, E., Eggers, F., and Funck, T., 1971a, Dynamic properties and membrane activity of ion specific antibiotics, in "Abstracts, Symposium on Molecular Mechanisms of Antibiotic Action on Protein Biosynthesis and Membranes, University of Granada, Spain (June 1–4, 1971)," p. 69.

Grell, E., Eggers, F., and Funck, T., 1971b, Konformationsanalyse und Kinetik der Konformationsänderungen von Cyclodepsipeptid-Antibiotika, *Angew. Chem.* **83**:903.

Hodgkin, D. C., and Oughton, B. M., 1957, Possible molecular models for gramicidin S and their relationship to present ideas of protein structure, *Biochim. J.* **65**:752.

Hunter, F., and Schwarz, L., 1967, Tyrocidines and gramicidin S (J₁, J₂), in "Antibiotics," Vol. 1: "Mechanism of Action" (D. Gottlieb and P. D. Shaw, eds.), p. 599, Printing House "Mir," Moscow, 1969.

Ivanov, V. T., Laine, I. A., Abdullaev, N. D., Senyavina, L. B., Popov, E. M., Ovchinnikov, Yu. A., and Shemyakin, M. M., 1969, The physicochemical basis of the functioning of biological membranes. The conformation of valinomycin and its K^+ complex in solution, *Biochem. Biophys. Res. Commun.* **34**:803.

Ivanov, V. T., Shilin, V. V., and Ovchinnikov, Yu. A., 1970, Synthesis of cyclic hexapeptides containing L(D)-alanine and glycine residues, *Zhurn. Obshch. Khim. (J. Gen. Chem., Russian)* **40**:924.

Ivanov, V. T., Kogan, G. A., Meshcheryakova, E. A., Shilin, V. V., and Ovchinnikov, Yu. A. 1971a, Conformational states of cyclopeptide systems. III. Cyclohexapeptides as a system of interacting amide chromophores, *Khim. Prir. Soed. (Chem. Nat. Prod., Russian)* **1972**:309.

Ivanov, V. T., Laine, I. A., Abdullaev, N. D., Pletnev, V. Z., Lipkind, G. M., Arkhipova, S. F., Senyavina, L. B., Meshcheryakova, E. N., Popov, E. M., Bystrov, V. F., and Ovchinnikov, Yu. A., 1971*b*, Conformational states of valinomycin and its complexes with alkaline cations in solution, *Khim. Prir. Soed.* (*Chem. Nat. Prod.*, Russian) **1971**:221.

Ivanov, V. T., Miroshnikov, A. I., Abdullaev, N. D., Senyavina, L. B., Arkhipova, S. F., Uvarova, N. N., Khalilulina, K. Kh., Bystrov, V. F., and Ovchinnikov, Yu. A., 1971*c*, Conformation of the Na⁺ complex of antamanide in solution, *Biochem. Biophys. Res. Commun.* **42**:654.

Ivanov, V. T., Portnova, S. L., Balashova, T. A., Bystrov, V. F., Shilin, V. V., Biernat, J., and Ovchinnikov, Yu. A., 1971*d*, Conformational studies of cyclopeptide systems. V. NMR spectra of cyclohexapeptides built up of alanine and glycine residues. Spin–spin coupling constants of the NH—CH protons and the "pleated sheet" structure, *Khim. Prir. Soed.* (*Chem. Nat. Prod.*, Russian) **1971**:339.

Ivanov, V. T., Senyavina, L. B., Efremov, E. S., Shilin, V. V., and Ovchinnikov, Yu. A., 1971*e*, Conformational studies of cyclopeptide systems. VI. Infrared spectra and dipole moments of diastereomeric cyclohexaalanyls, *Khim. Prir. Soed.* (*Chem. Nat. Prod.*, Russian) **1971**:347.

Ivanov, V. T., Shilin, V. V., Biernat, J., and Ovchinnikov, Yu. A., 1971*f*, Conformational studies of cyclopeptide systems. II. Synthesis of cyclic hexapeptides containing L(D)-alanine and glycine residues, *Zh. Obshch. Khim.* (*J. Gen. Chem.*, Russian) **41**:2318.

Ivanov, V. T., Shilin, V. V., Kogan, G. A., Meshcheryakova, E. N., Senyavina, L. B., Efremov, E. S., and Ovchinnikov, Yu. A., 1971*g*, Circular dichroism, infrared spectra and dipole moments of diastereomeric cyclohexaalanyls, *Tetrahedron Letters* **1971**:2841.

Ivanov, V. T., Evstratov, A. V., Mikhaleva, I. I., Abdullaev, N. D., Bystrov, V. F., and Ovchinnikov, Yu. A., 1973, Conformation of enniatin B in non-polar solvents, *Khim. Prir. Soed.* (*Chem. Nat. Prod.*, Russian) (in press).

Ivanov, V. T., Microshnikov, A. I., Kozmin, S. A., Uvarova, N. N., Khalilulina, K. Kh., Bystrov, V. F., and Ovchinnikov, Yu. A., 1973*a*, Conformational states of antamanide and its analogs in solution, *Khim. Prir. Soed.* (*Chem. Nat. Prod.*, Russian) **1973**:378.

Ivanov, V. T., Miroshnikov, A. I., Snezhkova, E. G., Ovchinnikov, Yu. A., Kulikov, A., and Lichtenstein, H. I., 1973*b*, Electron paramagnetic resonance spectroscopy in the conformational studies of peptides. Gramicidin S, *Khim. Prir. Soed.* (*Chem. Nat. Prod.*, Russian) **1973**:91.

Johnson, L. F., Schwartz, I. L., and Walter, R., 1969, Oxytocin and neurohypophyseal peptides: Spectral assignment and conformational analysis by 220 Mc nuclear resonance, *Proc. Natl. Acad. Sci.* **64**:1269.

Karle, I. L., and Karle, J., 1963, An application of a new phase determination procedure to the structure of cyclo(hexaglycyl) hemihydrate, *Acta Crystallogr.* **16**:969.

Karle, I. L., Gibson, J. W., and Karle, J., 1970, The conformation and crystal structure of the cyclic polypeptide ⌐Gly-Gly-D-Ala-Gly-Gly⌐3H₂O, *J. Am. Chem. Soc.* **92**:3755.

Kendrew, J. C., Klyne, W., Lifson, S., Miyazawa, T., Némethy, G., Phillips, D. C., Ramachandran, G. N., and Scheraga, H. A., 1970, Abbreviations and symbols for the description of the conformation of polypeptide chains, *J. Mol. Biol.* **52**:1; *Biochemistry* **9**:3471.

Konnert, J., and Karle, I. L., 1969, The conformation and crystal structure of the cyclotetradepsipeptide ⌐D-HyIv-L-MeIleu-D-HyIv-L-MeVal⌐, *J. Am. Chem. Soc.* **91**:4888.

Kopple, K. D., 1971, Conformations of cyclic peptides. V. A proton magnetic resonance study of evolidine, cyclo-Ser-Phe-Leu-Pro-Val-Asn-Leu, *Biopolymers* **10**:1139.

Liquori, A. M., and Conti, F., 1968, NMR studies of gramicidin S in solution, *Nature* **217**: 635.

Liquori, A. M., de Santis, P., Kovacs, A. L., and Mazzarella, L., 1966, Stereochemical code of amino acid residues: The molecular conformation of gramicidine, S, *Nature* **211**:1039.

Llinas, M., Klein, M. P., and Neilands, J. B., 1970, Solution conformation of ferrichrome, a microbial iron transport cyclohexapeptide, as deduced by high resolution proton magnetic resonance, *J. Mol. Biol.* **52**:399.

Ludescher, U., and Schwyzer, R., 1971, On the chirality of the cystine disulfide group: Assignment of helical sense in a model compound with a dihedral angle greater than ninety degrees using NMR and CD, *Helv. Chim. Acta* **54**:1637.

Miroshnikov, A. I., Khalilulina, K. Kh. Uvarova, N. N., Ivanov, V. T., and Ovchinnikov, Yu. A., 1973, Synthesis and properties of the symmetric analogs of antamanide, *Khim. Prir. Soed.* (*Chem. Nat. Prod.*, Russian) **1973**:214.

Momany, F. A., Vanderkooi, G., Tuttle, R. W., and Scheraga, H. A., 1969, Minimization of polypeptide energy. IV. Further studies of gramicidin S, *Biochemistry* **8**:744.

Ohnishi, M., and Urry, D. W., 1969, Temperature dependence of amide proton chemical shifts: The secondary structures of gramicidin S and valinomycin, *Biochem. Biophys. Res. Commun.* **36**:194.

Ohnishi, M., Fedarco, M. C., Baldeschwieler, J. D., and Johnson, L. F., 1972, Fourier transform C-13 NMR analysis of some free and potassium-ion complexed antibiotics, *Biochem. Biophys. Res. Commun.* **46**:312.

Ovchinnikov, Yu. A., Ivanov, V. T., Evstratov, A. V., Bystrov, V. F., Abdullaev, N. D., Popov, E. M., Lipkind, G. M., Arkhipova, S. F., Efremov, E. S., and Shemyakin, M. M., 1969, The physicochemical basis of the functioning of biological membranes: Dynamic conformational properties of enniatin B and its K$^+$ complex in solution, *Biochem. Biophys. Res. Commun.* **37**:668.

Ovchinnikov, Yu. A., Ivanov, V. T., Bystrov, V. F., Miroshnikov, A. I., Shepel, E. N., Abdullaev, N. D., Efremov, E. S., and Senyavina, L. B., 1970, The conformation of gramicidin S and its *N,N'*-diacetylderivative in solutions, *Biochem. Biophys. Res. Commun.* **39**:217.

Ovchinnikov, Yu. A., Ivanov, V. T., and Shkrob, A. M., 1971, The chemistry and membrane activity of peptide ionophores, *in* "Proceedings of the International Symposium on Molecular Mechanisms of Antibiotic Action on Protein Biosynthesis and Membranes, June 1–4, 1971, Granada" (D. Vazquez, ed.), Elsevier, Amsterdam, 1972.

Pinkerton, M., Steinrauf, L. L., and Dawkins, P. L., 1969, The molecular structure and some transport properties of valinomycin, *Biochem. Biophys. Res. Commun.* **35**:512.

Popov, E. M., Lipkind, G. M., and Arkhipova, S. F., 1970a, Theoretical conformational analysis of cyclotetraglycyl, *in* "Conformational Calculations of Complex Molecules" (A. I. Kitaigorodski and L. T. Perelman, eds.), pp. 120–127, Institute of Heat and Mass Exchange Press, Minsk.

Popov, E. M., Pletnev, V. Z., Evstratov, A. V., Ivanov, V. T., and Ovchinnikov, Yu. A., 1970b, Theoretical conformational analysis of cyclic hexadepsipeptides. Enniatins, *Khim. Prir. Soed.* (*Chem. Nat. Prod.*, Russian) **1970**:616.

Portnova, S. L., Balashova, T. A., Bystrov, V. F., Shilin, V. V., Biernat, J., Ivanov, V. T., and Ovchinnikov, Yu. A., 1971a, Conformational studies of cyclopeptide systems. IV. NMR spectra of cyclohexapeptides built up of alanine and glycine residues: Chemical shifts and intramolecular hydrogen bonding, *Khim. Prir. Soed.* (*Chem. Nat. Prod.*, Russian) **1971**:323.

Portnova, S. L., Shilin, V. V., Balashova, T. A., Biernat, J., Bystrov, V. F., Ivanov, V. T., and Ovchinnikov, Yu. A., 1971b, Conformational studies of cyclic peptides in solution. NMR spectra of cyclohexapeptides consisting of L(D)-alanine and glycine residues, *Tetrahedron Letters* **1971**:3085.

Ramakrishnan, C., and Sarathy, K. P., 1968, Stereochemical studies on cyclic peptides. III. Conformational analysis of cyclotetrapeptides, *Biochim. Biophys. Acta* **168**:402.

Ramakrishnan, C., and Sarathy, K. P., 1969, Stereochemical studies on cyclic peptides. V. Conformational analysis of cyclohexapeptides, *Internat. J. Protein Res.* **1**:103.

Sarathy, K. P., and Ramakrishnan, C., 1972, Stereochemical studies on cyclic peptides. VII. Effect of different types of energies on the hydrogen-bonded conformations of cyclic hexapeptides, *Internat. J. Protein Res.* **4**:1.

Scheraga, H. A., Leach, S. J., Scott, R. A., and Nemethy, G., 1965, Intramolecular forces and protein conformation, *Disc. Faraday Soc.* **40**:268.

Schmidt, G. M., Hodgkin, D. C., and Oughton, B. M., 1957, A crystallographic study of some derivatives of gramicidin S, *Biochem. J.* **65**:744.

Schwyzer, R., 1959, Synthetic routes to natural polypeptides, *Rec. Chem. Progr.* **20**:146.

Schwyzer, R., and Ludescher, U., 1968, Conformational study of gramicidin S using the phthalimide group as nuclear magnetic resonance marker, *Biochemistry* **7**:2519.

Schwyzer, R., and Ludescher, U., 1969, Untersuchungen über die Konformation des cyclischen Hexapeptides cyclo-Glycyl-L-proplyl-glycyl-glycyl-L-prolyl-glycyl mittels protonenmagnetischer Resonanz und Parallelen zum Cyclodecapeptid Gramicidin S, *Helv. Chim. Acta* **52**:2033.

Schwyzer, R., Sieber, P., and Gorup, B., 1958, Synthese for Polypeptide-Wirkstoffen, *Chimia* **12**:90.

Schwyzer, R., Carrion, J. P., Gorup, B., Nolting, H., and Tun-Kyi, A., 1964, Verdoppelungserscheinungen beim Ringschluss von Peptiden. V. Relative Bedeutung der sterischen Hinderung und der Assoziation über Wasserstoff-Brücken bei Tripeptiden. Spectroscopische Versuche zur Konformationbestimmung, *Helv. Chim. Acta* **47**:441.

Scott, R. A., Vanderkooi, G., Tuttle, R. W., Shames, P. M., and Scheraga, H. A., 1967, Minimization of polypeptide energy. III. Application of a rapid minimization technique to the calculation of preliminary structures of gramicidin S, *Proc. Natl. Acad. Sci.* **58**:2204.

Shemyakin, M. M., Ovchinnikov, Yu. A., Ivanov, V. T., Antonov, V. K., Vinogradova, E. I., Shkrob, A. M., Malenkov, G. M., Evstratov, A. V., Laine, I. A., Melnik, E. I., and Ryabova, I. D., 1969, Cyclodepsipeptides as chemical tools for studying ionic transport through membranes, *J. Membrane Biol.* **1**:402.

Shemyakin, M. M., Ovchinnikov, Yu. A., Ivanov, V. T., Evstratov, A. V., Mikhaleva, I. I., and Ryabova, I. D., 1973, Synthesis and antimicrobial activity of the analogs of enniatin antibiotics, *Zhurn. Obshch. Khim. (J. Gen. Chem., Russian)* (in press).

Sobell, H. M., Jain, S. C., Sakore, T. D., and Nordman, C. E., 1971, Stereochemistry of actinomycin–DNA binding, *Nature New Biol.* **231**:200.

Stern, A., Gibbons, W. A., and Craig, L. C., 1968, Conformational analysis of gramicidin S-A by nuclear magnetic resonance, *Proc. Natl. Acad. Sci.* **61**:734.

Tonelli, A., Patel, D. J., Goodman, M., Naider, F., Faulstich, H., and Wieland, Th., 1971, Experimental and calculated conformational characteristics of the cyclic decapeptide antamanide, *Biochemistry* **10**:3211.

Urry, D. W., 1970, "Spectroscopic Approaches to Biomolecular Conformation," p. 263, American Medical Association, Chicago.

Urry, D. W., and Walter, W., 1971, Proposed conformation of oxytocin in solution, *Proc. Natl. Acad. Sci.* **68**:956.

Urry, D. W., Quadrifoglio, F., Walter, R., and Schwarz, I. L., 1968, Conformational studies on the neurohypophyseal hormones. The disulfide bridge of oxytocin, *Proc. Natl. Acad. Sci.* **60**:967.

Urry, D. W., Ohnishi, M., and Walter, R., 1970, Secondary structure of the peptide hormone oxytocine and its deamino analog, *Proc. Natl. Acad. Sci* **66**:111.

van Dreele, P. H., Brewster, A. I., Scheraga, H. A., Ferger, M. F., and du Vigneaud, V., 1971, Nuclear magnetic resonance spectrum of Lys-vasopressin and its structural implications, *Proc. Natl. Acad. Sci.* **68**:1028.

Vanderkooi, G., Leach, S. J., Némethy, G., Scott, R. A., and Scheraga, H. A., 1966, Initial attempts at a theoretical calculation of the conformation of gramicidin S, *Biochemistry* **5**:2991.

Walter, R., Gordon, W., Schwarz, I. L., Quadrifoglio, F., and Urry, D. W., 1968, Conformation studies on the neurohypophyseal oxytocin and its analogs, *in* "Peptides—1968: Proceedings of the Ninth European Peptide Symposium" (E. Bricas, ed.), p. 50, North-Holland, Amsterdam.

Walter, R., Havran, R. T., Schwartz, I. L., and Johnson, L. F., 1971, Interaction of D_2O, Co^{2+}, Ni^{2+} and Cu^{2+} with oxytocin, *in* "Peptides, 1969. Proceedings of the Tenth European Peptide Symposium, Abano Terme, Italy, 1969" (E. Scoffone, ed.), p. 255, North-Holland, Amsterdam.

Warner, D. T., 1961, Proposed molecular models of gramicidin S and other polypeptides, *Nature* **190**:120.

Warner, D. T., 1967, The use of molecular models in evaluating protein and peptide conformations, *J. Am. Oil Chem. Soc.* **44**:593.

Weinkam, R. J., and Jorgensen, E. C., 1971a, Free radical analogs of histidine, *J. Am. Chem. Soc.* **93**:7028.

Weinkam, R. J., and Jorgensen, E. C., 1971b, Angiotensin II analogs. VIII. The use of free radical containing peptides to indicate the conformation of the carboxyl terminal region of angiotensin II, *J. Am. Chem. Soc.* **93**:7033.

Weinkam, R. J., and Jorgensen, E. C., 1971c, Angiotensin II analogs. IX. Conformational studies of angiotensin II by proton magnetic resonance, *J. Am. Chem. Soc.* **93**:7038.

Wieland, Th., Lüben, G., Ottenheim, H., Faesel, J., de Vries, J. X., Konz, W., Prox, A., and Schmid, J., 1968, Antamanid. Seine Entdeckung, Isolierung, Strukturaufclärung und Synthese, *Angew. Chem.* **80**:209.

Wieland, Th., Faulstich, H., Burgermeister, W., Otting, W., Möhle, W., Shemyakin, M. M., Ovchinnikov, Yu. A., Ivanov, V. T., and Malenkov, G. G., 1970, Affinity of antamanide for sodium ions, *FEBS Letters* **9**:89.

Zalkin, A., Forrester, J. D., and Templeton, D. H., 1966, Ferrichrome-A tetrahydrate. Determination of crystal and molecular structure, *J. Am. Chem. Soc.* **88**:1810.

CHAPTER 10

SURVEY OF SYNTHETIC WORK IN THE FIELD OF THE BACTERIAL CELL WALL PEPTIDES

Evangelos Bricas

Equipe de Recherche No. 15, C.N.R.S. and Laboratoire des Peptides
Institut de Biochimie
Université Paris-Sud
Orsay, France

> Liberty, let others despair of you—I never despair of you.
> WALT WHITMAN (Europe, 1848), *Leaves of Grass*

I. INTRODUCTION

Through work over the last two decades, many details concerning the chemical structure of the cell wall of bacteria have been elucidated. The rapid expansion of knowledge in this field is reflected in the exhaustive surveys published by Salton (1964), Weidel and Pelzer (1964), Martin (1966), Ghuysen *et al.* (1968b), Rogers and Perkins (1968), and Schleifer and Kandler (1972).

In all the bacterial species so far examined, the main structural component of the cell wall is a huge macromolecule called "peptidoglycan," "mucopeptide," "glycopeptide," or "murein"; the name "peptidoglycan" best describes the chemical nature of this bipolymer and will be used in this chapter.

The glycan moiety of this heteropolymer consists of polysaccharide chains containing alternating units of two acetamido sugars, N-acetyl-D-glucosamine and N-acetylmuramic acid (characterized as the 3-O-ether of D-glucosamine and D-lactic acid), joined together by β-1-4 glycoside linkages, just as are the N-acetylglucosamine residues in chitin.

The peptide moiety consists of rather short peptide chains containing a very limited number of amino acids, some of which are present in the less usual D configuration. These peptide subunits are joined to the glycan strains by an amide linkage between the α-NH$_2$ group of the amino-terminal peptide end (generally a L-alanine residue) and the D-lactyl carboxyl group of the muramic acid residue.

The peptide subunits of the various peptidoglycans are composed generally of D-glutamic acid, D-alanine, L-alanine, and either α,α'-meso-diaminopimelic acid (meso-DAP)* or lysine. Glycine, L-serine, L-homoserine, L-ornithine, and D-aspartic and D-glutamic acid α-amides are also found in the peptidoglycans of some species of bacteria (Fig. 1).

An important number of these peptide subunits (30–90% according to species) are cross-linked to each other. In some cases, the cross-linkage is a direct peptide bond between the carboxyl of the C-terminal D-alanine residue of one peptide subunit and the free ω-amino group of the lysine or diaminopimelic acid residue of another peptide subunit. In other cases, the cross-linkage between two peptide subunits is formed indirectly through a peptide bridge containing variable number of amino acid residues.

The resulting overall structure formed by glycan strands, peptide subunits and cross-linkage bridges can be conceived in a general way as a three-dimensional macromolecular network which completely surrounds the bacterial cell (Weidel and Pelzer, 1964). This rigid matrix of the peptidoglycan polymer located among the extracytoplasmic layers of the bacterial envelope is responsible for the defined shape and mechanical resistance of the bacterial cell. Mechanical or sonic disruption of Gram-positive bacterial cells yields preparations of walls which appear in the electron microscope as empty bags retaining the shape and the size of the original cells.

It is noteworthy that rigidity and insolubility of the peptidoglycan matrix are specific properties only of the intact network. Loss of integrity by breakdown of either the glycan or the peptide moiety brings about the solubilization of this insoluble polymer.

The study of the composition and structure of the components isolated after chemical degradation of the purified bacterial peptidoglycans was improved and extended by the development of enzymatic techniques for controlled degradation of these biopolymers (Ghuysen, 1968).

*The usual abbreviations of protein and peptide chemistry are used: Z, benzyloxycarbonyl; BOC, tert-butyloxycarbonyl; OBzl, benzyl ester; OBut, tertiobutyl ester; OSu, N-hydroxy-succinimide ester; HOSu, N-hydroxysuccinimide; DCC, dicyclohexylcarbodiimide; Mix-Anh, mixed carbonic-carboxylic anhydride; DNP, dinitrophenyl; FDNB, fluorodinitrobenzene; WRK, Woodward reagent K; isoglutamine residue, isoGln or Glu $\rightarrow$ NH$_2$; GNAc, N-acetyl-glucosamine; MurNAc, N-acetylmuramic acid; Cpase, carboxypeptidase A; LAP, leucine aminopeptidase. The α,α'-meso-diaminopimelic acid is abbreviated as meso-DAP and not as Dapim, Dpm, or A$_2$pm, recently proposed, in order to avoid confusion, because this abbreviation has been exclusively used in all the papers concerning mesodiaminopimelic acid reviewed here.

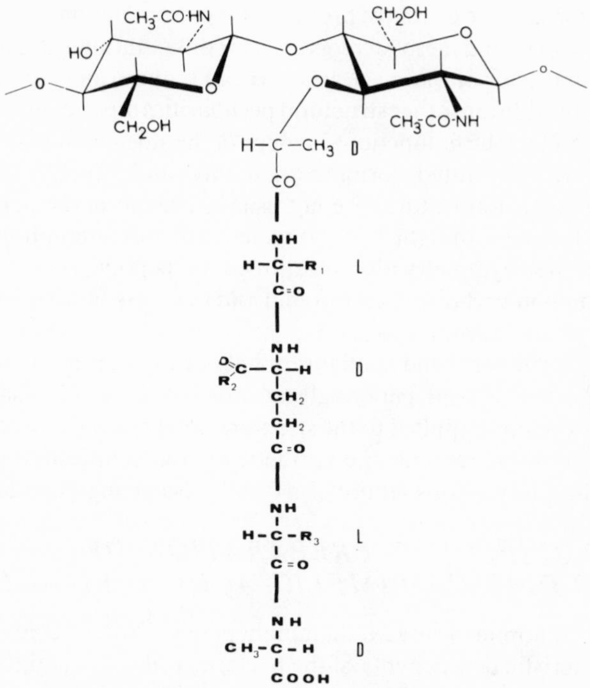

Fig. 1. Repetitive structural pattern common to all cell wall peptidoglycans. Note the alternating D-L-D-L-D stereochemical configuration of the components and the presence of a γ-peptide linkage. R$_1$ = —CH$_3$ (L-Ala), —CH$_2$·OH (L-Ser), —H (Gly). R$_2$ = —OH (D-Glu), —NH$_2$ (D-isoGln, —NHCH$_2$COOH α (D)-Glu-Gly. R$_3$ = —(CH$_2$)$_3$·CH(NH$_2$)·COOH (meso-DAP or LL-DAP), —(CH$_2$)$_4$·NH$_2$ (L-Lys), (CH$_2$)$_{3 \text{ or } 2}$ (L-Orn or αγ-L-Dab).

The concept of a common basic structure characteristic of the cell wall peptidoglycan of various bacterial species was developed as the converging point of different lines of research: thus the identification by Park (1952) of nucleotide *N*-acetylmuramyl peptides which accumulate in penicillin-treated *Staphylococcus aureus* cells as precursors of the biosynthesis of the cell wall peptidoglycan, the detailed studies of the biosynthesis of this biopolymer by Strominger and coworkers (for review, see Strominger, 1970), and many other structural studies established the outlines of the chemical structure of this gigantic bag-shaped macromolecule.

However, the presence of a number of structural peculiarities in the peptide chains of this peptidoglycan made it quite difficult to elucidate the detailed structural features. In the peptide subunits of this polymer, in contrast to the peptide chains of proteins, are present not only L-amino acids but also those of D- and meso-stereochemical configurations. Peptide linkages between amino acids of L and L, L and D, D and L, and D and D configurations are present. In addition, other structural peculiarities are encountered in these bacterial peptides: the ω-functional groups of the side-chain of glutamic acid and lysine are substituted, forming γ-glutamyl and ε-lysyl branched peptides. These structural features are not usually present in the peptide chains of proteins. Furthermore, the L and D moieties of meso-diaminopimelic acid residues are not symmetrically substituted in peptidoglycans containing this α,α'-diaminodicarboxylic amino acid and the cross-linkage thus presents an unusual configurational structure.

All these structural and configurational peculiarities in the peptide part of the bacterial cell wall peptidoglycan are not easily elucidated by the analytical procedures applied to the structural analysis of the protein peptide chains. For all these reasons, the synthetic approach appeared particularly useful for the study of this family of naturally occurring peptides.

II. STEREOSPECIFIC PREPARATION OF MESO-DIAMINOPIMELIC ACID DERIVATIVES

α,α'-Diaminopimelic (or 2,6-diaminoheptane-1,7-dioic) acid is one of the most characteristic components of the bacterial cell wall peptidoglycan in a great number of species. It has never been found as a protein constituent in these microorganisms nor elsewhere. This symmetrical molecule containing two amino groups in α- and α'-positions to the two carboxylic groups is usually present in the peptidoglycan in the meso form.

The main problem concerning the stereospecific synthesis of meso-diaminopimelic acid peptides was to determine how to introduce selectively a substituent on a functional (NH_2 or $COOH$) group adjacent either to its L-asymmetrical α-carbon or to its D-asymmetrical α'-carbon.

A first partial answer to this question was given by Bricas et al. (1962) and Nicot and Bricas (1963a), who applied a procedure associating both chemical and enzymatic methods.

According to this procedure, in a first step "symmetrical" bisubstituted meso-diaminopimelic peptides (Nicot et al., 1965) and derivatives (Bricas et al., 1965) were prepared by the current methods of peptide synthesis. In a second step, these symmetrical peptides were submitted to stereoselective hydrolysis by various peptidases (Fig. 2).

A number of "symmetrical" meso-diaminopimelic tri- and pentapeptides (X = L-Ala, Y = L-Ala or L-Glu-D-Ala) were synthesized by this procedure:

bis-Z-meso-DAP-bis-(L-Ala)   [1]

meso-DAP-bis-(L-Ala)   [2]

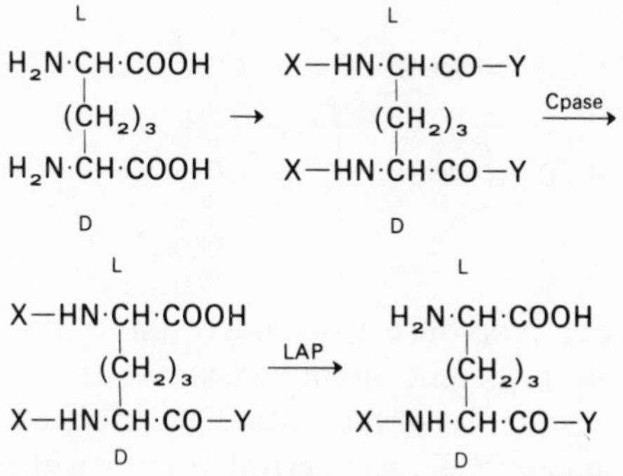

$$
\begin{array}{ccc}
\overset{\text{L}}{\underset{}{\text{H}_2\text{N·CH·COOH}}} & \overset{\text{L}}{\underset{}{\text{X—HN·CH·CO—Y}}} & \\
\underset{}{(\text{CH}_2)_3} \quad \rightarrow & \underset{}{(\text{CH}_2)_3} & \xrightarrow{\text{Cpase}} \\
\underset{\text{D}}{\text{H}_2\text{N·CH·COOH}} & \underset{\text{D}}{\text{X—HN·CH·CO—Y}} &
\end{array}
$$

$$
\begin{array}{ccc}
\overset{\text{L}}{\underset{}{\text{X—HN·CH·COOH}}} & & \overset{\text{L}}{\underset{}{\text{H}_2\text{N·CH·COOH}}} \\
\underset{}{(\text{CH}_2)_3} & \xrightarrow{\text{LAP}} & \underset{}{(\text{CH}_2)_3} \\
\underset{\text{D}}{\text{X—HN·CH·CO—Y}} & & \underset{\text{D}}{\text{X—NH·CH·CO—Y}}
\end{array}
$$

Fig. 2. Preparation of monosubstituted meso-DAP compounds by stereoselective enzymatic hydrolysis of bis-substituted meso-DAP derivatives.

bis-(L-Ala)-meso-DAP [3]

bis-(L-Ala)-meso-DAP-bis-(L-Ala) [4]

meso-DAP-bis-(L-Glu-D-Ala) [5]

By the action of bovine carboxypeptidase A (Cpase) on the symmetrical derivative [1], only the peptide linkage between the carboxyl of the L moiety of the meso-diaminopimelic acid residue and the amino group of the C-terminal L-alanine was hydrolyzed. After removal of the benzyloxycarbonyl groups from the resulting bis-Z-meso-DAP·(D)-L-Ala* derivative, the "unsymmetrical" dipeptide

meso-DAP·(D)-L-Ala [6]

was obtained (Bricas et al., 1962).

The same unsymmetrical dipeptide [6] was also obtained by stereo-specific hydrolysis of the free symmetrical tripeptide [2] by hog kidney leucine aminopeptidase (LAP) (Nicot and Bricas, 1963a). By the same procedure, the following unsymmetrical tripeptides (Bricas and Nicot, 1965) were prepared stereospecifically:

L-Ala-(D)·meso-DAP·(D)-L-Ala [7]

meso-DAP·(D)-(L-Glu-D-Ala) [8]

*For the designation of the substitution on the amino acid carboxyl groups of meso-diaminopimelic acid, we use the following abbreviations (Bricas et al., 1962): In order to specify on which asymmetrical carbon of meso-DAP a substituted *amino* group is located, we advocate the use of the notation (L) or (D) following the designation of the substituents and *before* the abbreviation meso-DAP; D-Ala·(D)-meso-DAP indicates a D-alanyl-meso-diaminopimelic acid dipeptide, the alanyl residue being bound to the amino function of the D moiety of meso-DAP. Similarly, we propose to write (L) or (D) before the substituents and *after* meso-DAP in order to distinguish between the carboxyl-substituted groups; meso-DAP·(L)-D-Ala indicates a dipeptide in which the D-alanine residue is bound to the carboxyl of the L moiety of meso-DAP.

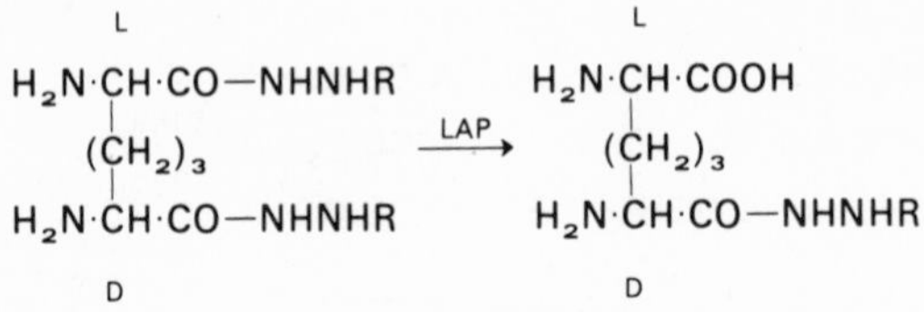

$$\underset{\text{D}}{\overset{\text{L}}{\left.\begin{array}{l} H_2N{\cdot}CH{\cdot}CO-NHNHR \\ (CH_2)_3 \\ H_2N{\cdot}CH{\cdot}CO-NHNHR \end{array}\right.}} \xrightarrow{\text{LAP}} \underset{\text{D}}{\overset{\text{L}}{\left.\begin{array}{l} H_2N{\cdot}CH{\cdot}COOH \\ (CH_2)_3 \\ H_2N{\cdot}CH{\cdot}CO-NHNHR \end{array}\right.}}$$

Fig. 3

The scope of this procedure is obviously limited to the preparation of meso-diaminopimelic peptides containing L-amino acids directly attached to the COOH or NH_2 functions of the meso-diaminopimelic residue. In order to obtain meso-diaminopimelic acid derivatives which offer the possibility of synthesizing any kind of unsymmetrical diaminopimelic acid peptides, this method had to be further improved.

The unsymmetrical compound for these syntheses was prepared by the stereoselective enzymatic splitting of a bis-hydrazide derivative of meso-diaminopimelic acid (Bricas and Nicot, 1965) (Fig. 3). The preparation of these unsymmetrical derivatives of meso-diaminopimelic acid was developed on the basis of the observation of Nicot and Bricas (1963b) concerning the rapid stereospecific hydrolysis of free L-amino acid carboxylic hydrazide and N-α-L-aminoacyl-N'-acylhydrazine derivatives by leucine amino-peptidase (Fig. 4).

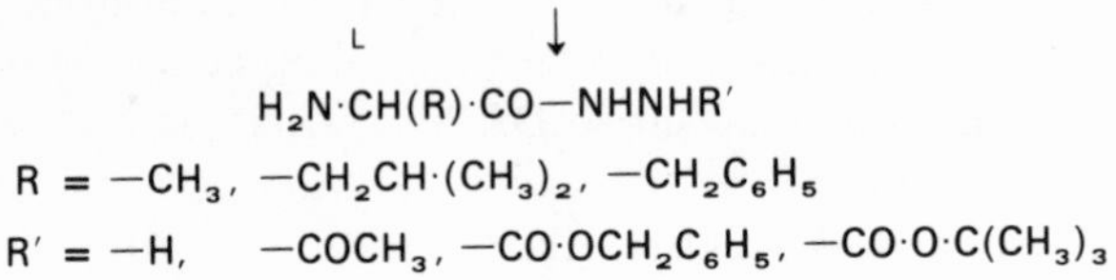

$$\overset{\text{L}\qquad\downarrow}{H_2N{\cdot}CH(R){\cdot}CO-NHNHR'}$$

$R = -CH_3, \quad -CH_2CH{\cdot}(CH_3)_2, \quad -CH_2C_6H_5$

$R' = -H, \quad -COCH_3, \quad -CO{\cdot}OCH_2C_6H_5, \quad -CO{\cdot}O{\cdot}C(CH_3)_3$

Fig. 4

$$\underset{\text{D}}{\overset{\text{L}}{\left.\begin{array}{l} Z{\cdot}HN{\cdot}CH{\cdot}CO-NHNH{\cdot}BOC \\ (CH_2)_3 \\ Z{\cdot}NH{\cdot}CH{\cdot}CO-NHNH{\cdot}BOC \end{array}\right.}} \xrightarrow{H_2/Pd} \underset{\text{D}}{\overset{\text{L}}{\left.\begin{array}{l} H_2N{\cdot}CH{\cdot}CO-NHNHBOC \\ (CH_2)_3 \\ H_2N{\cdot}CH{\cdot}CO-NHNHBOC \end{array}\right.}} \xrightarrow{\text{LAP}}$$

$$\underset{\text{D}}{\overset{\text{L}}{\left.\begin{array}{l} H_2N{\cdot}CH{\cdot}COOH \\ (CH_2)_3 \\ H_2N{\cdot}CH{\cdot}CO-NHNH{\cdot}BOC \end{array}\right.}} \xrightarrow{Z{\cdot}Cl} \underset{\text{D}}{\overset{\text{L}}{\left.\begin{array}{l} Z{\cdot}HN-CH{\cdot}COOH \\ (CH_2)_3 \\ Z{\cdot}NH{\cdot}CH{\cdot}CO-NHNH{\cdot}BOC \end{array}\right.}}$$

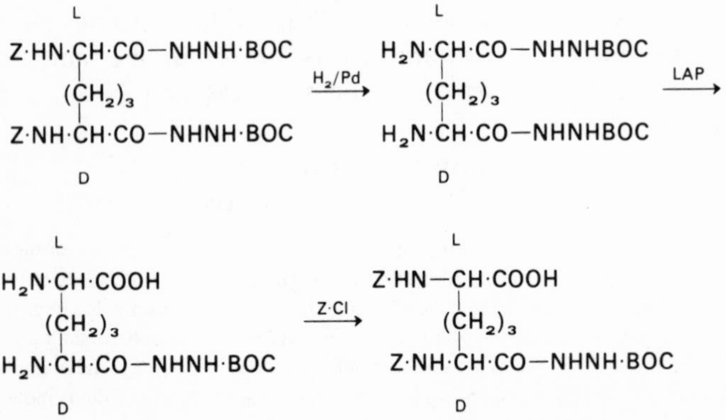

[9]   [10]

Fig. 5

The hydrolysis of L-leucine hydrazide and its derivatives by calf lens leucine aminopeptidase was also reported by Wergin (1965).

The steps in the preparation of the key compound bis-Z-meso-DAP-(D)-NHNH-BOC as described by Dezélée and Bricas (1967) are shown in Fig. 5.

The unsymmetrical dipeptide meso-DAP·(L)-D-Ala [11] was prepared as described by Dezélée and Bricas (1967) by coupling the unsymmetrical derivative [10] with a D-alanine benzylester and by removing the protecting groups Z and OBzl by catalytic hydrogenolysis. The substituted hydrazide NHNH·BOC was removed by oxidation with activated MnO_2 according to Kelly (1963).

Compound [11]

$$meso\text{-}DAP\cdot(L)\text{-}D\text{-}Ala \qquad\qquad [11]$$

is the enantiomorph of meso-DAP·(D)-L-Ala [6] previously synthesized by Bricas et al. (1962).

Van Heijenoort et al. (1969a) prepared stereospecifically a mono-ester and a mono-hydrazide of meso-DAP using another variant of the stereoselective hydrolysis of symmetrical substituted meso-DAP. Starting off with the diamide of meso-DAP and after the action of leucine aminopeptidase or amidase as described by Work et al. (1955), the mono-L-ester and mono-L-hydrazide were obtained as summarized in Fig. 6. The removal of the amide groups was carried out by the chemical deamidation method introduced by Nefkens and Nivard (1965) without important elimination of the methyl ester groups.

The mono-ester derivative [12] synthesized by this method can be used as an intermediate for the stereospecific syntheses of meso-DAP peptides.

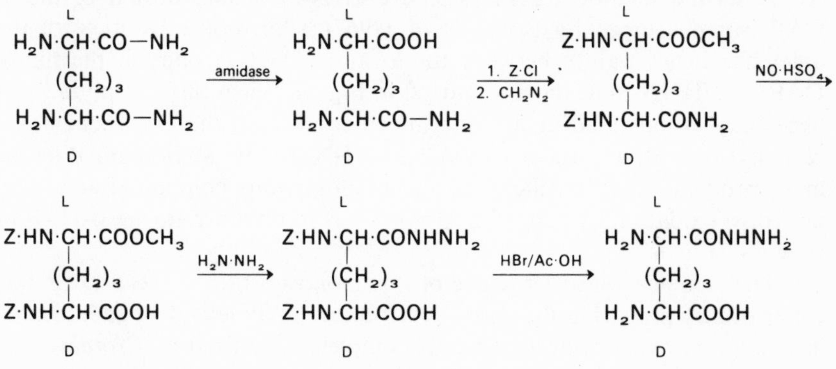

L

$H_2N\cdot CH\cdot CO-NH_2$
|
$(CH_2)_3$ $\xrightarrow{\text{amidase}}$
|
$H_2N\cdot CH\cdot CO-NH_2$

D

L

$H_2N\cdot CH\cdot COOH$
|
$(CH_2)_3$ $\xrightarrow[2.\ CH_2N_2]{1.\ Z\cdot Cl}$
|
$H_2N\cdot CH\cdot CO-NH_2$

D

L

$Z\cdot HN\cdot CH\cdot COOCH_3$
|
$(CH_2)_3$ $\xrightarrow{NO\cdot HSO_4}$
|
$Z\cdot HN\cdot CH\cdot CONH_2$

D

L

$Z\cdot HN\cdot CH\cdot COOCH_3$
|
$(CH_2)_3$ $\xrightarrow{H_2N\cdot NH_2}$
|
$Z\cdot NH\cdot CH\cdot COOH$

D

[12]

L

$Z\cdot HN\cdot CH\cdot CONHNH_2$
|
$(CH_2)_3$ $\xrightarrow{HBr/Ac\cdot OH}$
|
$Z\cdot HN\cdot CH\cdot COOH$

D

[13]

L

$H_2N\cdot CH\cdot CONHNH_2$
|
$(CH_2)_3$
|
$H_2N\cdot CH\cdot COOH$

D

[14]

Fig. 6

The same derivative, after hydrazinolysis and removal of the protecting groups, yielded meso-DAP·(L)-monohydrazide [14] the stereochemical configuration of which was proved by its hydrolysis by leucine aminopeptidase. In contrast, meso-DAP·(D)-monohydrazide [9] (Fig. 5) previously described, was resistant toward this enzyme. Both these compounds were useful for structural study of natural DAP peptides.

The stereoselective substitution of one of the two NH_2 groups of meso-DAP was carried out in three different ways. A first method was described by Nicot and Bricas (1965), who obtained a monoacylated derivative by chelation and benzyloxycarbonylation of the unsymmetrical dipeptide meso-DAP·(D)-L-Ala [6] under determined conditions: the NH_2 group of the meso-DAP adjacent to the free carboxyl group was selectively substituted (Fig. 7).

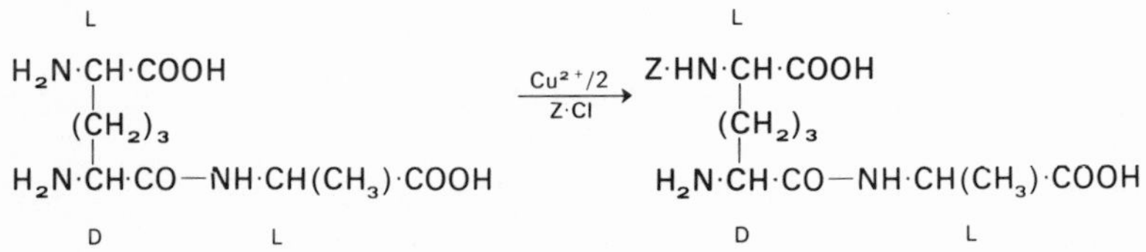

```
        L                                              L
  H₂N·CH·COOH           Cu²⁺/2        Z·HN·CH·COOH
       |              ─────────▶           |
     (CH₂)₃              Z·Cl            (CH₂)₃
       |                                   |
  H₂N·CH·CO—NH·CH(CH₃)·COOH       H₂N·CH·CO—NH·CH(CH₃)·COOH
      D            L                      D            L

     [6]                                     [15]
```

Fig. 7

By coupling the monobenzyloxycarbonyl dibenzyl ester derivative of compound [15] with benzyloxycarbonyl L-alanine, Nicot and Bricas (1965) synthesized the tripeptide L-Ala-(D)·meso-DAP·(D)-L-Ala. The stereochemical structure of this tripeptide was identical to that of the tripeptide [7] previously prepared by a different way by Bricas and Nicot (1965) (Fig. 2). The optical properties of these two tripeptides were identical.

A second method for the selective NH_2-monosubstitution of meso-DAP was developed by Bricas et al. (1967), who took advantage of the difference in pK values between the α- and α'-NH_2 groups in the meso-DAP·(D)-NHNH·BOC (compound [9] in Fig. 5). When this compound was benzyloxycarbonylated under determined conditions, the monobenzyloxycarbonyl derivative Z-(L)·meso-DAP·(D)-NHNH·BOC was obtained as the main product. After purification, the homogeneous compound was submitted to oxidation by activated MnO_2 so as to obtain Z-(L)·meso-DAP as shown in Fig. 8.

The stereochemical structure of this monosubstituted meso-DAP was confirmed by preparing the mono-DNP-meso-DAP derivative as indicated in Fig. 8 and by showing that it was completely oxidized by Crotalus adamanteus L-amino acid oxidase (Dezélée and Bricas, 1968). This indicated that the α-NH_2 function adjacent to the L asymmetrical carbon was free in

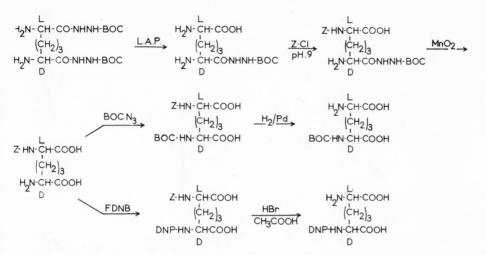

Fig. 8. Preparation of the unsymmetrical derivatives mono-Z-(L)-meso-DAP, mono-BOC-(D)-meso-DAP, and mono-DNP-(D)-meso-DAP.

the mono-DNP-(D)-meso-DAP and thus the starting compound was really mono-Z-(L)-meso-DAP.

The value of the molar rotation of the mono-DNP-(D)-meso-DAP derivative thus synthesized was found to be $[M]_D + 238 \pm 5°$. Furthermore, the di-DNP-D,D-DAP derivative has a molar rotation $[M]_D + 444°$, and the $[M]_D$ value of the mono-DNP-(D)-meso-DAP corresponds practically to half of the *positive* $[M]_D$ value of the D,D-DAP derivative (Diringer and Jusic, 1966).

The optical properties (molar rotation and optical rotatory dispersion) of the mono-DNP-(D)-meso-DAP thus obtained enabled us to determine the stereochemical structure of the meso-DAP moiety bearing a free amino group in the bacterial cell wall peptidoglycan. It was found that the mono-DNP-meso-DAP isolated after dinitrophenylation and acid hydrolysis of a fragment of the peptidoglycan of *Bacillus megaterium* according to Bricas *et al.* (1967a) gave a molar rotation value of $[M]_D + 248 \pm 6°$ and that of *Escherichia coli* according to Diringer and Jusic (1966) a value of $[M]_D + 250 \pm 10°$. From these results, it can be concluded that in these peptidoglycans the other amino group of meso-DAP which is substituted by the γ-D-glutamyl residue is adjacent to the L asymmetrical carbon atom of the meso-DAP residue.

Another useful monosubstituted derivative of meso-DAP bearing a *tert*-butyloxycarbonyl group on its amino group adjacent to the D asymmetrical carbon was prepared as shown in Fig. 8. Using this compound as

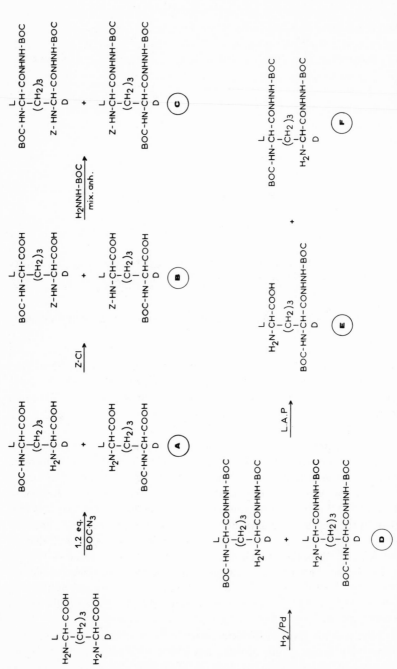

Fig. 9. Stereospecific preparation of *N*-monoacylated derivatives of meso-DAP. Synthesis of the BOC-(D)-meso-DAP-(D)-NHNH-BOC intermediate: compound E.

intermediate, Dezélée and Bricas (1970, 1971) synthesized various unsymmetrical meso-DAP peptides, as described in the next section.

In order to carry out stringent stereospecific mono-NH_2-substitution of meso-DAP, Dezélée and Bricas (1970, 1971) have more recently developed a third method. In this procedure, summarized in Fig. 9, the racemic mixture of mono-BOC-meso-DAP derivative (obtained by partial chemical substitution without stereoselectivity) was benzyloxycarbonylated and the N,N'-bis-substituted meso-DAP (racemic mixture) was coupled with 2 equivalents of tert-butylcarbazate. After removal of the benzyloxycarbonyl group by catalytic hydrogenation, the racemic mixture was submitted to stereoselective enzymatic hydrolysis. Only the enantiomorph with a free amino function adjacent to the L-asymmetrical carbon atom was digested by leucine aminopeptidase.

The resulting two products were now easily separated by fractional crystallization, and the key compound, mono-BOC-(D)·meso-DAP·(D)-NHNH-BOC, was obtained in high yield. Its stereochemical structure and homogeneity were confirmed by quantitative determination of the oxygen absorbed during its oxidation by C. adamanteus L-amino acid oxidase (Dezélée and Bricas, 1970, 1971). This key compound was used as an intermediate for the stereospecific syntheses of the peptide subunits of the peptidoglycans of E. coli and B. megaterium, as described in the next section.

III. SYNTHESIS OF MESO-DIAMINOPIMELIC ACID-CONTAINING PEPTIDE SUBUNITS OF THE ESCHERICHIA COLI CELL WALL PEPTIDOGLYCAN

The main structural features of the cell wall peptidoglycan of E. coli had been established by Weidel and coworkers (for a general review, see Weidel and Pelzer, 1964). According to the results of these authors, the repeating unit in this peptidoglycan, schematically summarized in Fig. 10, is represented by the disaccharide tetrapeptide GNAc-MurNAc-L-Ala-D-Glu-meso-DAP-D-Ala (subunit A). Disaccharide tripeptide subunits differing from the tetrapeptide subunits by the lack of a C-terminal D-alanine residue are also present (subunit B). These tripeptide subunits are produced by "endogenous" enzymatic degradation of the peptidoglycan. A dimer of the disaccharide tetrapeptide (A) is formed by a peptide linkage between the COOH of the C-terminal D-alanine residue of one peptide subunit and the free amino group of the meso-DAP residue of another (subunit C).

A very similar overall structure was observed in the peptidoglycan of B. megaterium (Bricas et al., 1967a; van Heijenoort et al., 1969b). In the structure of the tetrapeptide subunit proposed by these authors, three points

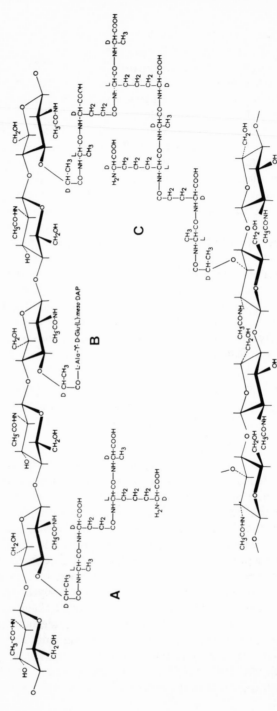

Fig. 10. Structure and stereochemical configuration of a fragment of the *E. coli* peptidoglycan representing (A) a disaccharide tetrapeptide which is the repeating unit monomer, (B) a disaccharide tripeptide, and (C) the cross-linked bis-disaccharide octapeptide.

remained unknown:

1. Which amino group of meso-DAP is attached to the D-glutamic acid residue.
2. Which carboxyl group of D-glutamic acid is linked to the meso-DAP residue.
3. Which carboxyl group of meso-DAP is linked to the D-alanine residue.

Concerning point 1, the results of Diringer and Jusic (1966) and those of Bricas *et al.* (1967*a*) previously described (Section II) had given an unequivocal answer: the α-NH$_2$ group adjacent to the L asymmetrical carbon is bound to the D-glutamyl residue in the peptide subunits of *E. coli* and *B. megaterium* peptidoglycans. In the meantime, Diringer (1968) and van Heijenoort *et al.* (1969*b*), by using the hydrazinolysis technique introduced by Akabori, established that the γ-COOH of the glutamic acid is linked to meso-DAP and that its α-COOH is free in these peptidoglycans. To provide further proof concerning point 2, Dezélée and Bricas (1968) synthesized the α- and γ-isomeric tripeptide L-Ala-D-Glu-(L)·meso-DAP corresponding to the amino acids sequence of the tripeptide subunit (B) of the *E. coli* and *B. megaterium* peptidoglycans (Fig. 10).

The unsymmetrical intermediate for these syntheses was the mono-BOC-(D)·meso·DAP compound prepared as shown in Fig. 8 (see Section II). These tripeptides were synthesized by condensation of the corresponding BOC-L-Ala-D-Glu α- or γ-benzyl esters with the unsymmetrical mono·BOC-(D)-meso-DAP compound. The electrophoretic migrations of the two isomeric tripeptides at pH 4.0 are very different, and the mobility of the natural L-Ala-D-Glu-meso-DAP tripeptide isolated from the corresponding fragments of the *B. megaterium* and *E. coli* peptidoglycans is the same as that of the synthetic γ-isomer L-Ala-γ-D-Glu-(L)·meso-DAP.

The answer to point 3 by a synthetic approach became easier after the elucidation of the structural features concerning points 1 and 2.

Van Heijenoort *et al.* (1969*b*), on the basis of analytical determinations, postulated a structure for the peptide moiety of the repeating unit (A) (Fig. 10), characteristic of the peptidoglycans of these two bacteria. In order to confirm the advocated structure, Dezélée and Bricas (1970, 1971) synthesized the tetrapeptide L-Ala-γ-D-Glu-(L)·meso-DAP·(L)-D-Ala (Fig. 11a), in which the *C*-terminal D-alanine substituted the carboxyl adjacent to the L asymmetrical carbon of the meso-DAP residue.

It was found that the tetrapeptide subunits isolated from *E. coli* and *B. megaterium* peptidoglycans had the same chromatographic properties and the same electrophoretic mobility at pH 4 as the synthetic tetrapeptide L-Ala-γ-D-Glu-(L)·meso-DAP·(L)-D·Ala (a). However, the further comparison of the natural tetrapeptide and the synthetic peptide (a) with an isomeric

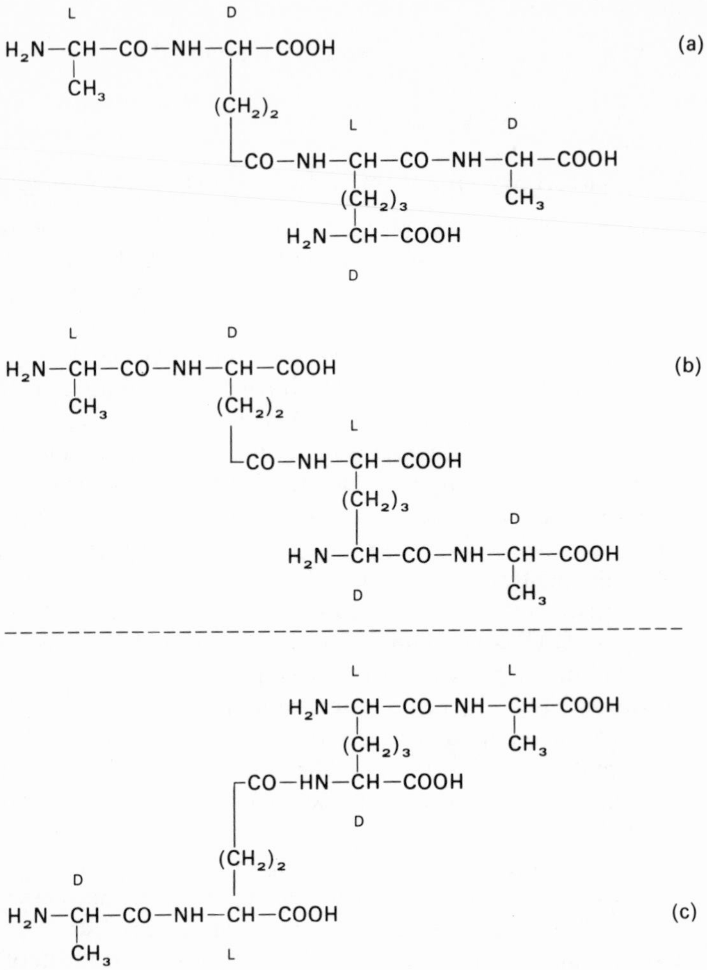

```
       L                D
H₂N—CH—CO—NH—CH—COOH                                                    (a)
       |                |
       CH₃            (CH₂)₂              L                D
                         └CO—NH—CH—CO—NH—CH—COOH
                                   |              |
                                 (CH₂)₃          CH₃
                         H₂N—CH—COOH
                                   D


       L                D
H₂N—CH—CO—NH—CH—COOH                                                    (b)
       |                |
       CH₃            (CH₂)₂              L
                         └CO—NH—CH—COOH
                                   |
                                 (CH₂)₃              D
                         H₂N—CH—CO—NH—CH—COOH
                                   D              CH₃
```

- -

```
                                   L                L
                         H₂N—CH—CO—NH—CH—COOH
                                   |                |
                                 (CH₂)₃            CH₃
                    ┌CO—HN—CH—COOH
                    |           D
                    |
                  (CH₂)₂
       D            |
H₂N—CH—CO—NH—CH—COOH                                                    (c)
       |            L
       CH₃
```

Fig. 11. Structure of the tetrapeptide subunit (a) and those of two enantiomorphic tetrapeptides (b and c), the b differing from the a only in the position of the C-terminal D-alanine.

tetrapeptide in which the C-terminal alanine is bound to the other carboxyl group of meso-DAP was necessary in order to assess unambiguously their identity.

In Fig. 11 are represented schematically the structures of three isomeric tetrapeptides corresponding to this requirement. Peptides b and c are enantiomorphic compounds, and their chromatographic and electrophoretic behavior should be identical. Since peptide c (which is enantiomorphic to peptide b) was easier to synthesize than b, we preferred to use it for comparison with the natural tetrapeptide subunit and the synthetic tetrapeptide a.

The difference in the position of the C-terminal alanine in peptides a and b is the same as in peptides a and c from a structural point of view. If peptides a and b can be separated by paper electrophoresis at an adequate pH value, the same difference must be found also between a and c peptides.

The natural tetrapeptides isolated from the two bacterial peptidoglycans are clearly separated after paper electrophoresis at pH 4.0 from the isomeric synthetic tetrapeptide c and have the same mobility as the synthetic tetrapeptide L-Ala-γ-D-Glu-(L)·meso-DAP·(L)-D-Ala (a).

The structure proposed by van Heijenoort et al. (1969b) for the tetrapeptide subunit of these peptidoglycans was therefore confirmed. These results clearly demonstrate that the C-terminal D-alanine is linked to the carboxyl group of the meso-DAP which is adjacent to the L-asymmetrical carbon of this amino acid. Thus point 3 was clarified definitively.

More recently, Whittle and Anderson (1969) provided an analytical indication concerning the structure around the meso-DAP residue of the peptide subunit of B. megaterium peptidoglycan, confirming the results obtained by the synthetic approach.

For the synthesis of peptide A, Dezélée and Bricas (1970) used as an intermediate compound the unsymmetrical BOC-(D)·meso-DAP·(D)-NHNH·BOC previously prepared stereospecifically as described in Section II (Fig. 9E). The synthesis of this tetrapeptide is summarized in Fig. 12 and that of the second tetrapeptide (c) in Fig. 13.

In the synthesis of tetrapeptide c, the unsymmetrical compound Z-(L)·meso-DAP-(D)-NHNH·BOC (for the preparation of this monosubstituded derivative, see Section II, Fig. 8) was used as an intermediate. The other steps of the synthesis of these two tetrapeptides are the same except that the stereochemical configuration of the amino acids alanine and glutamic acid used in each case differs.

After the determination of the detailed configurational structure of the peptide subunit in the disaccharide peptide monomer A (Fig. 10), the stereochemical configuration of the linkage -D-Ala-(D)·meso-DAP- present in the cross-linked dimer C (Fig. 10) was clarified by van Heijenoort et al. (1969b). It is noteworthy that in E. coli this unusual peptide linkage between two amino acid residues of D and D stereochemical configuration is split, as demonstrated by Bogdanovsky et al. (1969), by a specific endogenous peptidase, the so-called D-D-carboxypeptidase, which is highly sensitive to penicillin.

Thus the synthetic work summarized here has enabled us to confirm some of the important structural and configurational features of the peptide moiety of this type of peptidoglycan as represented in Fig. 10.

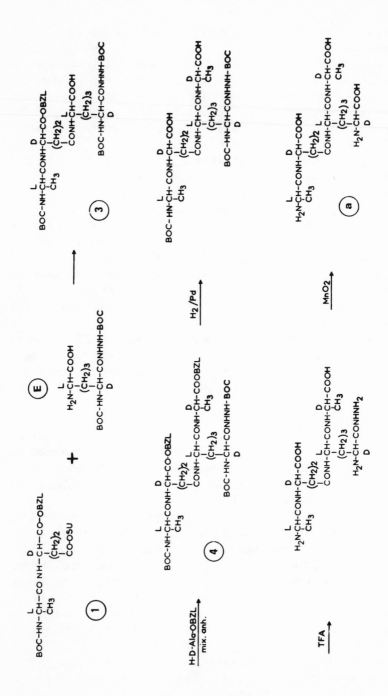

Fig. 12. Steps of the synthesis of tetrapeptide a : L-Ala-γ-D-Glu-(L)-meso-DAP-(L)-D-Ala.

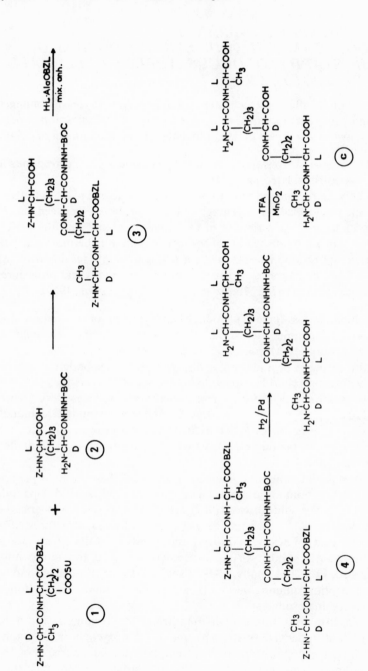

Fig. 13. Steps of the synthesis of tetrapeptide c: D-Ala-γ-L-Glu-(D):meso-DAP·(L)-L-Ala.

IV. SYNTHESIS OF ω-BRANCHED ISOGLUTAMINYL AND LYSYL PEPTIDES OF THE STAPHYLOCOCCUS AUREUS CELL WALL PEPTIDOGLYCAN

The results of different structural investigations by Ghuysen, Strominger, Tipper, and coworkers indicated that the structure of the cell wall peptogly-can of *Staphylococcus aureus* differs from that of *E. coli* in a number of points:

1. The meso-DAP residue is replaced by L-lysine, which corresponds to a ω-decarboxylated meso-DAP.
2. The *St. aureus* peptidoglycan presents a high degree of cross-linkage, which is formed by a pentaglycine peptide bridge between the C-terminal D-alanine residue of one peptide subunit and the ε-NH$_2$ group of a L-lysine residue of another subunit. Furthermore, the γ-D-glutamyl residue is replaced by a D-isoglutaminyl residue (isoGln). This last point was clarified by synthetic and analytical procedures (Bricas *et al.*, 1967b; Munoz *et al.*, 1966; Tipper *et al.*, 1967).

The highly cross-linked peptidoglycan network of *St. aureus* and *M. roseus* was dismantled stepwise by the action of different specific glyco-sidases and peptidases (Munoz *et al.*, 1966), as indicated in Fig. 14.

By this stepwise degradation, a disaccharide tetrapeptide (*St. aureus*) together with a disaccharide tripeptide in the case of *M. roseus* was formed. After new enzymatic digestions, these disaccharide peptides were further degraded to a tripeptide Glu(amide)-Lys-Ala and to a dipeptide Glu(amide)-Lys. By a similar procedure applied to *St. aureus* cells grown in the presence of penicillin, a tetrapeptide was isolated corresponding to Glu-(amide)-Lys-Ala-Ala.

In order to determine the detailed structure of these residual peptide fragments which contain D-amino acid residues and α- and γ-peptide linkages, Lefrancier and Bricas (1967) synthesized various isomeric di-, tri-, and tetrapeptides corresponding to the indicated structures. The dipeptides α- and γ-Glu-Lys, Gln-Lys, and isoGln-Lys, the tripeptides α- and γ-Glu-Lys-Ala, Gln-Lys-Ala, and isoGln-Lys-Ala, the tetrapeptide isoGln-Lys-Ala-Ala, and finally L-Ala-D·isoGln-L-Lys-D-Ala corresponding to the tetrapeptide subunit, were obtained as chromatographically homo-geneous crystalline products.

All the structural isomeric di- and tripeptides synthesized can be separate as dinitrophenyl derivatives by silica gel thin layer chromatography after repeated development in two solvent systems. Comparison of the chromatographic properties of each of the di-, tri-, and tetrapeptide frag-ments isolated from the enzymatically degraded peptidoglycans with those

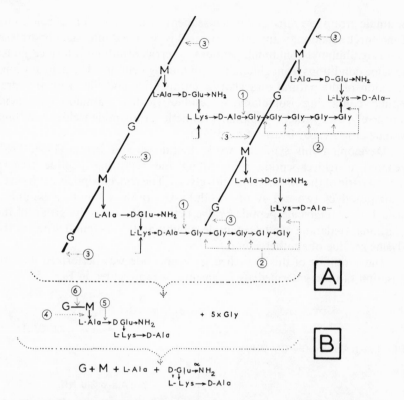

Fig. 14. Stepwise enzymatic degradation of the *St. aureus* cell wall peptidoglycan. A: Preparation of a disaccharide peptide subunit. (1) Site of action of the *Streptomyces* "SA" endopeptidase, (2) degradation of the open pentaglycine bridges with *Streptomyces* aminopeptidase, (3) site of action of *Streptomyces* endo-*N*-acetylmuramidase. B: Further degradation of the isolated disaccharide peptide. (4) Site of action of *Streptomyces* *N*-acetylmuramyl-L-alanine amidase, (5) site of action of *Streptomyces* aminopeptidase, (6) site or action of exo-B-*N*-acetylglucosaminidase (from pig epidymis). G = *N*-Acetylglucosamine; M = *N*-acetylmuramic acid.

of the isomeric synthetic compounds leads to the conclusion that the tetrapeptide subunit of the *St. aureus* peptidoglycan has the structure α-(L-alanyl-D-isoglutaminyl)-L-lysyl-D-alanine (Bricas *et al.*, 1967b). The peptide subunits of the peptidoglycans of several other bacteria have the same structure (Ghuysen *et al.*, 1967). Thus the presence of an amide group on the α-carboxyl group of the D-glutamyl residue in these peptidoglycans was definitely proved.

A further proof of the presence of a glutamyl-α-amide group in the peptidoglycans of these bacteria was obtained by the dehydration–reduction procedure introduced by Ressler and Kashelikar (1966) for identification of glutaminyl or isoglutaminyl residues in the *endo* position. By dehydration of the amide group to a nitril and by subsequent reduction under the conditions of the Birch reaction, γ-aminobutyric acid is obtained after acid hydrolysis from isoglutaminyl-containing peptides, whereas ornithine is formed under the same conditions from glutaminyl-containing peptides. The dehydration–reduction of the natural disaccharide tetrapeptide and the synthetic tetrapeptide containing isoglutaminyl residue yielded γ-aminobutyric acid. Conversely, the synthetic glutaminyl isomeric tetrapeptide yielded ornithine (Munoz *et al.*, 1966; Bricas *et al.*, 1967*b*).

Developing this synthetic work, Lefrancier and Bricas (1968, 1969) prepared a tridecapeptide representing the repeating peptide pattern characteristic of the *St. aureus* peptidyglycan. The tridecapeptide synthesized is composed of two α-(L-Ala-D-isoGlu)-L-Lys-D·Ala subunits cross-linked through a pentaglycine peptide bridge between the carboxyl group of the *C*-terminal D-alanine of a tetrapeptide subunit and the ε-amino group of the L-lysine residue of another subunit.

The synthesis of this branched tridecapeptide was performed by condensation of three intermediate fragments, as represented in Fig. 15.

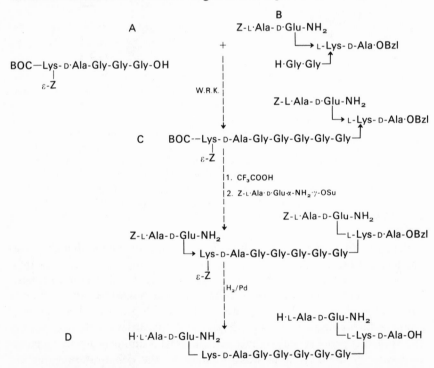

Fig. 15. Fragment condensation by synthesis of the *St. aureus* tridecapeptide representing two tetrapeptide subunits bound by a pentaglycine peptide bridge.

Fragment A was prepared by the usual methods. In order to obtain fragment B, the following steps were necessary starting from the dipeptide derivative α-NPS,ε-BOC-L-Lys-D-Ala-OBzl: (1) removal of the α-NPS protecting group by treatment with thioglycolic acid, (2) coupling with Z-Ala-D-Glu-NH₂, (3) selective elimination of the ε-BOC protecting group from the tetrapeptide derivative α-[Z-L-Ala-D-Glu-NH₂]-Lys-(ε-BOC)-D-Ala-OBzl, (4) coupling of the tetrapeptide derivative thus obtained with NPS-Gly-Gly-OSu, and (5) removal of the NPS protecting group by HCl/CH₃COOH treatment. The coupling of fragments A and B by the Woodward reagent yielded the undecapeptide derivative C. After removal of the BOC group and coupling with Z-L-Ala-D-Glu-α(NH₂),γ-(OSu), the final protected tridecapeptide was obtained. The free tridecapeptide (D) was prepared by hydrogenolysis of the protecting groups.

The synthetic tridecapeptide submitted to the action of the specific endopeptidase isolated from *Streptomyces albus* by Petit *et al.* (1966) was hydrolyzed and yielded two fragments by the selective splitting of the -D-Ala-Gly- peptide bond. One fragment is identical to the tetrapeptide subunit L-Ala--D-isoGln-L-Lys-D-Ala and the other to the ε-branched nonapeptide α-(L-Ala--D-isoGln)-L-Lys-ε-(Gly-Gly-Gly-Gly-Gly)-D-Ala. The disaccharide peptide bearing the open pentaglycine chain attached to the ε-amino function of the lysine residue was isolated from the peptidoglycan of penicillin-treated *St. aureus* cells (Munoz *et al.*, 1966). After further enzymatic degradation of this peptidoglycan fragment, a branched decapeptide bearing the open pentaglycine chain attached to the ε-amino group of the lysine residue was obtained, differing only by one supplementary D-alanine residue.

In earlier synthetic work concerning peptides related to the peptide moiety of the nucleotide-*N*-acetylmuramyl peptide precursor, Garg *et al.* (1962) and Tesser and Nivard (1964) prepared the fully protected pentapeptide derivative corresponding to the sequence of L-Ala-D-(and L)-α-Glu-L-Lys-D-Ala-D-Ala. This peptide derivative is the α-glutamyl isomer of the peptide moiety of the precursor.

At about the same time, both α and γ-unprotected pentapeptide of the same amino acid sequence and also the *N*-acetylmuramyl pentapeptide were synthesized by Lanzilotti *et al.* (1964). The γ-glutamyl synthetic isomer was shown to be identical to the natural product derived from the uridine disphosphate-*N*-acetylmuramyl pentapeptide. Khosla *et al.* (1965) also synthesized the two isomeric α- and γ-L-Ala-D-Glu-L-Lys-D-Ala-D-Ala pentapeptides. All these compounds were related to the nucleotide peptide precursor of the biosynthesis of the *St. aureus* cell wall peptidoglycan.

They differ from the peptide subunits of the peptidoglycan in lacking the α-amide group of the glutamic acid residue and by a supplementary

D-alanine residue present in C-terminal position. Only Jarvis and Strominger (1967) synthesized a tetra- and a pentapeptide, L-Ala-D-isoGln-L-Lys-D-Ala and L-Ala-D·isoGln-L-Lys-(ε-Gly)-D-Ala, which reproduce fragments of the St. aureus peptidoglycan.

V. SYNTHESIS OF LINEAR AND ω-BRANCHED GLUTAMYL AND LYSYL PEPTIDES OF THE MICROCOCCUS LYSODEIKTICUS CELL WALL PEPTIDOGLYCAN

The peptide subunits of the M. lysodeikticus cell wall peptidoglycan contain in addition to the L-alanyl, γ-D-glutamyl, L-lysyl, and D-alanine residues a glycine residue which is bound to the α-COOH of the glutamyl residue. However, the most important difference between this and the previously described peptidoglycan structures is in the cross-linkage between the peptide subunits.

In M. lysodeikticus peptidoglycan, many of the muramyl carboxylic groups are free and the number of those which are substituted by peptide subunits is much lower with respect to the total number of muramic acid residues. In spite of this fact, an equal total number of muramic acid residues and peptide subunits (the last represented by the number of glutamic acid residues) are found in this peptidoglycan. The presence of only a direct cross-linking between two adjacent peptide subunits by a D-alanyl-ε-lysyl linkage, as established by Petit et al. (1966), does not explain this "anomalous" ratio. The possibility still exists that other types of cross-linkages might also occur. As a matter of fact, Ghuysen et al. (1968a) by stepwise enzymatic degradation of this peptidoglycan and Schleifer and Kandler (1967) by the partial acid hydrolysis of the same peptidoglycan clearly demonstrated the presence of another indirect cross-linkage through a peptide chain, as indicated in Fig. 16.

Thus the structure of the peptidoglycan of M. lysodeikticus is unique in that it contains both a directly ε-D-Ala-Lys cross-linkage and an indirect cross-linkage in which pentapeptide subunits are linked in a "head-to-tail" sequence.

In order to give confirmatory proof of the proposed structure, Mulliez et al. (1973) synthesized the pentapeptide subunit which is the monomer and two decapeptides, one bearing the ε-(D-Ala-L-Lys) cross-linkage and the other the "head-to-tail" linkage.

The presence of γ-glutamyl linkages in these branched peptides made it possible to synthesize them by coupling intermediate peptide fragments without racemization using the activation of the γ-COOH of the glutamyl peptides. Thus a small number of di-, tri-, and tetrapeptide derivatives were first prepared by stepwise synthesis. In order to obtain the pentapeptide and

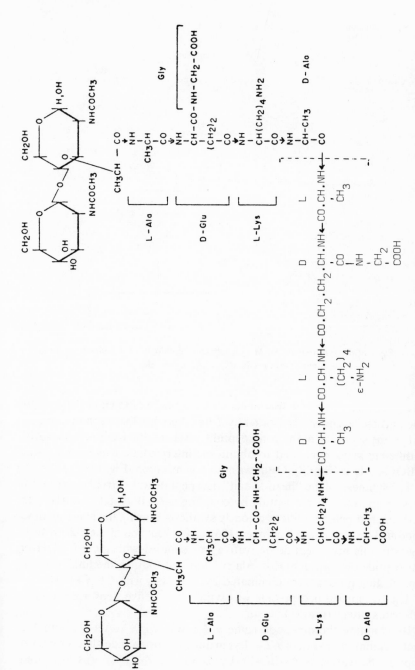

Fig. 16. Structure of a fragment of *M. lysodeikticus* peptidoglycan representing a "head-to-tail" cross-linkage of two disaccharide pentapeptide subunits through a peptide bridge containing a pentapeptide subunit.

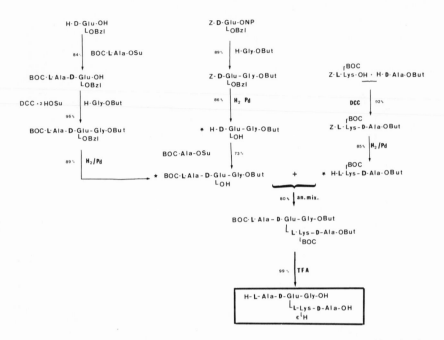

Fig. 17. Steps of the synthesis of the *M. lysodeikticus* peptidoglycan pentapeptide subunit monomer: L-Ala-D-Glu-α(Gly)-γ-(L-Lys-D-Ala).

the two decapeptides, these fragments were coupled as shown in Figs. 17–19. Some of these fragments, indicated by a star, were used more than once.

For the synthesis of the pentapeptide L-Ala-D-Glu-(α-Gly)-L-Lys-D-Ala, two different ways were used to obtain the intermediate tripeptide derivative BOC-L-Ala-D-Glu-Gly-OBut without racemization (Fig. 17).

The synthesis of the "head-to-tail" decapeptide was carried out (Fig. 18) by first preparing a heptapeptide by coupling α-Z,ε-BOC-L-Lys-D-Ala with the ε-branched pentapeptide previously synthesized using the azide method as modified by Honzl and Rudinger (1961). After removal of the α-Z protecting group, this heptapeptide derivative was coupled with the γ-COOH of the tripeptide derivative BOC-L-Ala-D-Glu-Gly-OBut. In the final step, all the protecting groups were eliminated by acidolysis with CF_3COOH.

The ε-branched decapeptide was synthesized in five steps according to a different strategy (Fig. 19).

By coupling the dipeptide azide α-Z-ε-BOC-L-Lys-D-Ala-N_3 with the tetraethyl ammonium salt of α-Z-L-Lys in dimethylformamide, the ε-branched tripeptide derivative α-Z-ε-BOC-L-Lys-D-Ala-ε-(α-Z-L-Lys) was prepared with a very satisfactory yield.

This derivative was coupled with D-Ala-OBut. The ε-branched tetra-peptide derivative thus obtained after removal of the benzyloxycarbonyl groups from the α-NH$_2$ functions of the two lysine residues was condensed with 2 equivalents of the tripeptide derivative BOC-L-Ala-D-Glu-Gly-OBut containing a free γ-COOH. This rather unusual coupling between two γ-glutamyl and two α-lysyl residues was accomplished with a very satisfactory

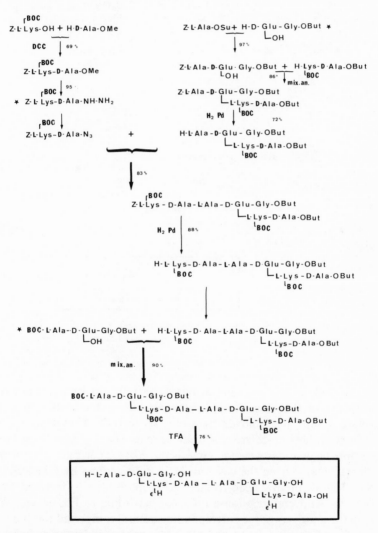

Fig. 18. Steps of the synthesis of the decapeptide representing the *M. lyso-deikticus* peptidoglycan "head-to-tail" dimer of two pentapeptide subunits.

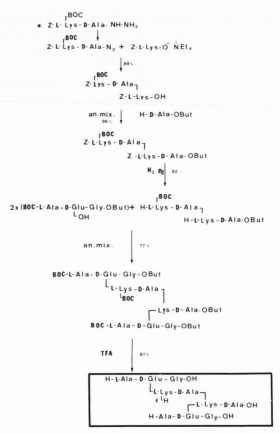

Fig. 19. Steps of the synthesis of a decapeptide representing the *M. lyso-deikticus* peptidoglycan ε-(D-Ala-L-Lys) cross-linked dimer of two penta-peptide subunits.

yield in one step by the carbonic-carboxylic mixed anhydride method. As in the synthesis of the "head-to-tail" decapeptide, in the final step all the protecting groups are removed by acidolysis with CF_3COOH.

The R_f values in different solvent systems and the migration rate of these synthetic peptides were the same as those of the corresponding peptide fragments isolated from the cell wall of *M. lysodeikticus* by Ghuysen *et al.* (1968a). The synthetic "head-to-tail" decapeptide was hydrolyzed by the *Myxobacter* AL.I endopeptidase prepared according to Ensign and Wolfe 1966), in the same way as the corresponding natural compound was (Ghuysen *et al.*, 1968a). In both cases, the -D-Ala-L-Ala- peptide linkage was selectively hydrolyzed, forming the monomer corresponding to the structure of L-Ala-

D-Glu-(α-Gly)-Lys-D-Ala. In contrast, the ε-(D-Ala-Lys) cross-linked synthetic decapeptide was resistant to the action of this enzyme.

Thus the results concerning the enzymatic degradation of the natural and the synthetic peptide fragments (unpublished results of Mulliez *et al.*) together with those concerning the structural work (Ghuysen *et al.*, 1968*a*; Schleifer and Kandler, 1967) confirm the presence of a unique type of double cross-linking in this peptidoglycan. These results also explain the "anomalous" value of the ratio between the number of the muramic acid and glutamic acid residues, which is equal to 1 despite the presence of unsubstituted muramic acid residues. The existence of this loose network in this peptidoglycan is probably related to the rapid lysis of the cell wall of *M. lysodeikticus* by lysozyme.

VI. CONCLUSIONS

The chemistry of bacterial cell wall peptidoglycans represents a relatively recent achievement in the field of naturally occurring peptides. It is noteworthy that in the general review on natural peptides written in 1953 (Bricas and Fromageot, 1953) only one peptide related to bacterial cell wall, the Park nucleotide peptide, was described.

In the following years, the chemistry of the bacterial cell walls was developed rapidly. Starting from the analytical determination of the chemical compounds of the cell walls and those of the peptidoglycans, these studies were further extended to the elucidation of the detailed structure of these peptidoglycans.

In order to prove the proposed structures and also to prepare model compounds related to the peptide part of the bacterial peptidoglycans, the synthetic approach has been the next step in the development of this field; considering the structural and configurational peculiarities of the peptide moiety of the bacterial peptidoglycans, stereochemical problems usually not encountered during the synthesis of "normal" peptide chains have had to be taken into account. The stereochemical preparation of synthetic peptides containing meso-DAP residues was one of the main problems which had to be solved first. Associating usual synthetic procedures with stereoselective enzymatic hydrolysis of meso-DAP symmetrical derivatives, we have established general methods for preparing various unsymmetrical meso-DAP compounds which can be conveniently used for the stereospecific synthesis of these bacterial peptides.

On the other hand, the synthesis of ω-branched peptides present in the lysine-containing peptidoglycans requires adequate choice of a variety of protecting groups which can be eliminated independently of each other.

The presence of alternating L-D-L-D stereochemical sequences in these cell wall peptides makes it necessary in these syntheses to use procedures excluding any racemization during the coupling steps.

Another aim of the synthetic approach in this field is the preparation by synthesis of model substrates which reproduce the peculiar peptide bonds encountered in bacterial peptidoglycans. Synthetic substrates have been particularly useful in investigating various lytic enzymes which selectively split these linkages.

According to recent studies, various amidases and peptidases (carboxy-, amino-, endo-, and transpeptidases) appear to be very important factors in the continuous makeup of the peptidoglycan during growth and division of the bacterial cell.

In connection with this work, the semi-chemical and semi-enzymatic synthesis of various labeled uridine diphosphate N-acetyl muramyl peptide fragments used as substrates in studies of the specificities of these enzymes (van Heijenoort and van Heijenoort 1971) is worthy of note.

The specific action of several antibiotics (penicillin, cycloserine, van-comycin, ristomycin, bacitracin, etc.) in different steps of the cell wall pepti-doglycan biosynthesis offers a new tool for investigations concerning the interaction between these unusual peptidases and their substrates and inhibitors. The synthetic approach to these problems is very promising.

On the other hand, the progress made in the synthesis of the peptide part of these peptidoglycans has presented the problem of the preparation by chemical synthesis of short peptidoglycan fragments containing di- or oligosaccharide peptides. Several successful attempts to synthesize compounds containing a N-acetylmuramic acid residue bound to peptide fragments have been made by Lanzilotti *et al.* (1964), Chaturvedi *et al.* (1966), and more recently Seymour *et al.* (1972).

However, the major difficulty in this work is the chemical synthesis of appropriate derivatives of the β-1-4-N-acetylglucosaminyl-N-acetylmuramic acid disaccharide necessary for these syntheses.

ACKNOWLEDGMENTS

In the studies on the chemistry of bacterial cell wall peptides conducted in my laboratory, I have enjoyed the collaboration of Cl. Nicot, J. van Heijenoort, P. Lefrancier, P. Dezélée, and M. Mulliez, to all of whom my thanks are due. I am also grateful to Dr. J. van Heijenoort for fruitful discussions concerning this paper and for careful reading of the manuscript.

REFERENCES

Bogdanovsky, D., Bricas, E., and Dezélée, P., 1969, Sur l'identité de la "mucoendopeptidase" et de la "carboxypeptidase I" d'*Escherichia coli,* enzymes hydrolysant des liaisons de configuration D-D et inhibées par la pénicilline, *Compt. Rend.* **269**:390.

Bricas, E., and Fromageot, C., 1953, Naturally occurring peptides, *Advan. Protein Chem.* **8**:1–125.

Bricas, E., and Nicot, C., 1965, Synthèse de peptides non symétriques de l'acide α,α'-meso-diaminopimélique, *in* "Peptides: Proceedings of the Sixth European Peptide Symposium, Athens, 1963" (L. Zervas, ed.), pp. 329–336, Pergamon Press, Oxford.

Bricas, E., Nicot, C., and Lederer, E., 1962, Action de la carboxypeptidase A sur des peptides synthétiques de l'acide meso-α,α'-diaminopimélique; Préparation stéréospécifique d'un dipeptide meso-DAP-L-Ala, *Bull. Soc. Chim. Biol.* **44**:1115.

Bricas, E., Dezélée, P., Gansser, C., Lefrancier, P., Nicot, C., and van Heijenoort, J., 1965, Dérivés symétriques de l'acide meso-α,α'-diaminopimélique, *Bull. Soc. Chim. France* **1965**:1813.

Bricas, E., Ghuysen, J. M., and Dezélée, P., 1967a, The cell wall peptidoglycan of *Bacillus megaterium* KM. I. Studies on the stereochemistry of α,α'-diaminopimelic acid, *Biochemistry* **6**:2598.

Bricas, E., Lefrancier, P., Ghuysen, J. M., Munoz, E., Leyh-Bouille, M., Petit, J. F., and Heymann, H., 1967b, Structure et synthèse de la subunité peptidique de la paroi des trois bacteries Gram-positif, *in* "Peptides: Proceedings of the Eighth European Peptide Symposium," pp. 286–292, North-Holland, Amsterdam.

Chaturvedi, N. C., Khosla, M. C., and Anand, N., 1966, Synthesis of analogs of bacterial cell wall glucopeptides, *J. Med. Chem.* **9**:971.

Dezélée, P., and Bricas, E., 1967, Nouvelle méthode de synthèse stéréospécifique de peptides de l'acide meso-α,α'-diaminopimélique. I. Synthèse du dipeptide meso-diaminopimélyl. (L)-D-alanine, *Bull. Soc. Chim. Biol.* **49**:1579.

Dezélée, P., and Bricas, E., 1968, Synthèse des α- et γ-isomères du tripeptide non symétrique de la paroi d'*Escherichia coli* et de *Bacillus megaterium*: L-Alanyl-D-glutamyl (L)-meso-diaminopimélique, *in* "Peptides 1968. Proceedings of the Ninth European Peptide Symposium (E. Bricas, ed.), p. 299–304, North-Holland, Amsterdam.

Dezélée, P., and Bricas, E., 1970, Structure of the peptidoglycan in *E. coli* B and *B. megaterium* KM. Stereospecific synthesis of two meso-diaminopimelic peptides isomeric with the tetrapeptides subunit of bacterial cell wall peptidoglycan, *Biochemistry* **9**:823.

Dezélée, P., and Bricas, E., 1971, Stereospecific synthesis of two meso-diaminopimelic acid peptides, isomeric with the tetrapeptide subunit of the bacterial cell wall peptidoglycan, *in* "Peptides 1969: Proceedings of the Tenth European Peptide Symposium" (E. Scoffone, ed.), pp. 347–355, North-Holland, Amsterdam.

Diringer, H., 1968, Über die Bindung der D-Glutaminsäure im Murein von *E. coli, Z. Naturforsch.* **23b**:883.

Diringer, H., and Jusic, D. Z., 1966, Über die Bindung der meso-Diaminopimelinsäure im Murein von *E. coli, Z. Naturforsch.* **21b**:603.

Ensign, J. C., and Wolfe, R. S., 1966, Characterization of a small proteolytic enzyme which lyses bacterial cell walls, *J. Bacteriol.* **91**:524.

Garg, H. G., Khosla, M. C., and Anand, N., 1962, Synthesis of completely protected L-alanyl-L(and D)-glutamyl-L-lysyl-D-alanyl-D-alanine, *J. Sci. Ind. Res.* **21B**:286.

Ghuysen, J. M., 1968, Use of bacteriolytic enzymes in determination of wall structure and their role in cell metabolism, *Bacteriol. Rev.* **32**:425–464.

Ghuysen, J. M., Bricas, E., Leyh-Bouille, M., Lache, M., and Shockman, G. D., 1967, The peptide Nα-(L-alanyl-D-isoglutaminyl)Nε-(D-isoasparaginyl)-L-lysyl-D-alanine in cell wall peptidoglycan of St. faecalis, Biochemistry 6:2607.

Ghuysen, J. M., Bricas, E., Lache, M., and Leyh-Bouille, M., 1968a, Structure of the cell wall of Micrococcus lysodeikticus. III. Isolation of a new peptide dimer, Biochemistry 7:1450.

Ghuysen, J. M., Strominger, J. L., and Tipper, D. J., 1968b, Bacterial cell walls, in "Comprehensive Biochemistry," Vol. 26A (M. Florkin and E. H. Stotz, eds.), pp. 53–104, Elsevier, Amsterdam.

Honzl, J., and Rudinger, J., 1961, Nitrosylchloride and butylnitrile as reagents in peptide synthesis by the azide method: Suppression of amide formation, Coll. Czech. Chem. Commun. 26:2333.

Ito, E., and Strominger, J. L., 1964, Enzymatic synthesis of the peptide in bacterial uridine nucleotides, J. Biol. Chem. 239:210.

Jarvis, D., and Strominger, J. L., 1967. Structure of the cell wall of Staphylococcus aureus. VIII. Structure and chemical synthesis of the basic peptides released by the Myxobacterium enzyme, Biochemistry 6:2591.

Kelly, R. B., 1963, Phenylhydrazide as a protective group in peptide synthesis. The oxidation of γ-phenylhydrazides of N-carbobenzoxy α-L-glutamyl-amino acid esters with manganese oxide, J. Org. Chem. 28:453.

Khosla, M. C., Chaturvedi, N. C., Garg, H. G., and Anand, N., 1965, Peptides related to bacterial cell wall components: Synthesis of L-alanyl-D-(α and γ)glutamyl-L-lysyl-D-alanyl-D-alanine and partially protected L-glutamyl isomer, Indian J. Chem. 3:111.

Lanzilotti, A. E., Benz, E., and Goldman, L., 1964, Total synthesis of Nα[1-(2 acetamino-3.0-D-glycosyl)-D-propionyl-L-alanyl-D-α and γ-glutamyl]-L-lysyl-D-alanyl-D-alanine, and the identity of the γ-glutamyl isomer with the glycopeptide of a bacterial cell wall precursor, J. Am. Chem. Soc. 86:1880.

Lefrancier, P., and Bricas, E., 1967, Synthèse de la subunité peptidique du peptidoglycane de la paroi de trois bacteries Gram positif et de peptides de structure analogue, Bull. Soc. Chim. Biol. 49:1257.

Lefrancier, P., and Bricas, E., 1968, Synthèse d'un tridécapeptide de la paroi de Staphylococcus aureus de structure ramifiée, in "Peptides 1968: Proceedings of the Ninth European Peptides Symposium" (E. Bricas, ed.), pp. 293–298, North-Holland, Amsterdam.

Lefrancier, P., and Bricas, E., 1969, Synthèse d'un tridécapeptide de la paroi de Staphylococcus aureus de structure ramifiée, Bull. Soc. Chim. France 1969:3561.

Martin, H. H., 1966, Biochemistry of bacterial cell walls, Ann. Rev. Biochem. 35 II:457–484.

Mulliez, M., Šavrda, J., and Bricas, E., 1973, Synthesis of two decapeptides and one pentapeptide of the cell wall pentidoglycan of Micrococcus lysodeikticus, in "Peptides 1971" Proceedings of the Eleventh European Peptide Symposium (H. Nesvadba, ed.), pp. 281–285, North Holland, Amsterdam.

Munoz, E. Ghuysen, J. M., Ley-Bouille, M., Petit, J. F., Heymann, H., Bricas, E., and Lefrancier, P., 1966, The peptide subunit Nα-(L-alanyl-D-isoglutaminyl-L-lysyl-D-alanine) in cell wall peptidoglycans of Staphylococcus aureus, Biochemistry 5:3748.

Nefkens, G. H. L., and Nivard, R. J. F., 1965, Synthesis of α-esters of N-acyl-aspartic acid, Rec. Trav. Chim. 84:1315.

Nicot, C., and Bricas, E., 1963a, Action de la leucine aminopeptidase sur des peptides symétriques de l'acide meso-diaminopimélique, Compt. Rend. 256:1391.

Nicot, C., and Bricas, E., 1963b, Action de la leucine-aminopeptidase et de l'aminoacide-amidase sur quelques nouveaux substrats, Bull. Soc. Chim. Biol. 45:455.

Nicot, C., and Bricas, E., 1965, Synthèse stéréospécifique d'un dérivé mono-N-acylé de l'acide meso-α,α'-diaminopimélique et obtention par son intermédiaire d'un tripeptide non symétrique, Acta Chim. Hung. 44:229.

Nicot, C., van Heijenoort, J., Lefrancier, P., and Bricas, E., 1965, α,α'-Diaminopimélic acid peptides. VIII. Synthesis of symmetrical peptides containing meso-α,α'-diaminopimelic acid, D- or L-alanine and L-glutamic acid, *J. Org. Chem.* **30**:3746.

Park, J. T., 1952, Uridine-5-pyrophosphate derivatives. III. Amino acid containing derivatives, *J. Biol. Chem.* **194**:897.

Petit, J. F., Munoz, E., and Ghuysen, J. M., 1966, Peptide cross-links in bacterial cell wall peptidoglycans studied with specific endopeptidase from *Streptomyces albus* G, *Biochemistry* **5**:2764.

Ressler, C., and Kashelikar, D. V., 1966, Identification of asparaginyl and glutaminyl residues in endo position in peptides by dehydration-reduction, *J. Am. Chem. Soc.* **88**:2025.

Rogers, H. J., and Perkins, H. R., 1968, "Cell Walls and Membranes," E. and F. N. Spon, London.

Salton, M. R. J., 1964, "The Bacterial Cell Wall," Elsevier, Amsterdam.

Schleifer, K. H., and Kandler, O., 1967, *Micrococcus lysodeikticus*: A new type of crosslinkage of the murein, *Biochem. Biophys. Res. Commun.* **28**:965.

Schleifer, K. L., and Kandler, O., 1972, Peptidoglycan types of bacterial cell walls and their taxonomic implications, *Bacteriol. Rev.* **36**:407–477.

Seymour, S. F., Lerique, F., David, S., and Bricas, E., 1972, Unpublished results.

Strominger, J. L., 1970, Penicillin sensitive enzymatic reactions in bacterial cell wall synthesis, *Harvey Lectures* **64**:179–213.

Tesser, G. I., and Nivard, R. J. F., 1964, Synthesis of completely protected pentapeptide found in bacterial cell walls, *Rec. Trav. Chim.* **83**:53.

Tipper, D. J., Katz, W., Strominger, J. L., and Ghuysen, J. M., 1967, Substituents on α-carboxyl groups of D-glutamic acid in the peptidoglycan of several bacterial cell walls, *Biochemistry* **6**:921.

Van Heijenoort, Y., and van Heijenoort, J., 1971, Study of the *N*-acetylmuramyl-L-alanine amidase activity in *Escherichia coli*, *FEBS Letters* **15**:137.

Van Heijenoort, J., Bricas, E., and Nicot, C., 1969a, Synthèse stéréospécifique de dérivés non symétriques de l'acide meso-α,α'-diaminopimélique, *Bull. Soc. Chim. France* **1969**:2743.

Van Heijenoort, J., Elbaz, L., Dezélée, P., Petit, J. F., Bricas, E., and Ghuysen, J. M., 1969b, Structure of the meso-diaminopimelic acid containing peptidoglycans in *E. coli* B and *B. megaterium* KM, *Biochemistry* **8**:207.

Weidel, W., and Pelzer, H., 1964, Bagshaped macromolecules: A new outlook on bacterial cell walls, *Advan. Enzymol.* **26**:193–232.

Wergin, A., 1965, Hydrolyse von Hydrazidbindungen durch Leucinaminopeptidase, *Naturwissenschaften* **52**:34.

Whittle, P. J., and Anderson, J. C., 1969, The linkage of diaminopimelic acid in the peptidoglycan of *B. megaterium* NCTC 7581, *Biochim. Biophys. Acta* **192**:165.

Work, E., Birnbaum, S. M., Winitz, M., and Greenstein, J., 1955, Separation of the three isomeric components of synthetic α,ε-diaminopimelic acid, *J. Am. Chem. Soc.* **77**:1916.

CHAPTER 11

INTRACELLULAR PROTEOLYSIS AND ITS DEMONSTRATION WITH SYNTHETIC AND NATURAL PEPTIDES AND PROTEINS AS SUBSTRATES

Horst Hanson

Physiologisch-Chemisches Institut
Martin-Luther-Universität
Halle-Wittenberg, G.D.R.

I. INTRODUCTION

It is known that in multicellular organisms the life span of the cells, their organelles, and other constituents, including proteins, is limited and varies according to the organ, the species, the cell organelle, and the type of substance. In each living cell, exchange and metabolic reactions are taking place in processes in which the structural elements of the cell are also involved. Accordingly, the cell components, including the proteins of the organelles, have a much shorter life span than the cell itself. An expression of the difference in the life span of various proteins of the cell is given by their half-lives, which lie between $t/2 = 10$ days for the cytochromes of the mitochondria and $t/2 = 0.1$ day for tryptophan pyrrolase (see Table I). Since, in the living cell, the relative amounts of proteins remain constant under a given set of cellular functions, the levels of the proteins of the cell must be directed not only by their rates of synthesis but also by their rates of degradation. Thus a protein needed in a larger amount for a specific function in the cell, such as an enzyme protein, may be increased by enhancing its rate of synthesis and/or by diminishing its rate of degradation. Similarly, a protein that is needed in smaller amounts by the cell may be lowered by decreasing its rate of synthesis and/or increasing its rate of degradation. Schimke (1970) has developed mathematical formulations of the steady-state relations

Table I. Half-Life *in Vivo* of Rat Liver
Proteins

Protein	Half-life (days)
Collagen	20
Histone	18
Cytochrome	6
Arginase	4
Alanine aminotransferase	3.5
Ferritin	3.0
Aspartate transcarbamylase	1.0
Catalase	1.0
Ornithine aminotransferase	0.9
Alanine aminotransferase	0.9
Amylase	0.18
Thymidine kinase	0.11
Tryptophan pyrrolase	0.08
Tyrosine aminotransferase	0.06
σ-Aminolevulinic acid	0.04

From Bohley *et al.* (1971).

between synthesis and degradation of intracellular proteins using enzyme proteins as examples.

II. THE CONCEPT OF PROTEOLYSIS

For an understanding of the intracellular turnover of proteins and its regulation, analysis of both the biosynthesis of proteins and their intracellular proteolysis is important. The experiments performed in the past years by our group were pertinent to the latter. Two major problems were mainly considered:

1. Determination of intracellular proteolysis and examination of the differences between organs and between the organelles of the same cell type.
2. Determination of, and attempts to differentiate between, the enzymes taking part in intracellular proteolysis.

The experiments with regard to both questions were performed by use of a large number of peptides and proteins and their derivatives. Intracellular proteolysis was primarily assessed by the *in vitro* autolysis of parts of organs, of homogenates of the organs, and of subcellular fractions (organelles), which were obtained by differential centrifugation. The enzymatic processes functioning during intracellular proteolysis were demonstrated by use of synthetic peptide substrates, as well as by use of natural or chemically modified proteins. It quickly became evident from these investigations that for the classification and characterization of the enzymes taking part in intracellular proteolysis the usual designations given to the proteolytic enzymes of the gastrointestinal tract are not sufficient. The all-inclusive old classical name for the intracellular proteases, "cathepsins," is not justified, since the enzymes that participate in the degradation of proteins are widely distributed in the cell and show different specificities. The use of this designation for these enzymes causes one to think only of the hydrolytic splitting of peptide bonds and does not take into account the enzymatic reduction of inter- or intra-chain disulfide bridges which can also participate in intracellular protein degradation processes.

III. SYNTHETIC SUBSTRATES FOR PROTEASE DETERMINATIONS

There is no doubt that the application of synthetic peptide substrates was a significant step in the determination of the substrate specificity of the intracellular enzymes participating in the degradation of proteins, a procedure similar to that used with the proteolytic enzymes of the gastrointestinal tract. The first conditions required for that were met by the school of Emil Fischer with the chemical synthesis of peptides. This orientation obtained a significant new lift in about 1930 from the outstanding research of Max Bergmann and his collaborators in methods for peptide synthesis and the use of these substances to differentiate among the endo- and exopeptidases of the digestive tract, animal organs, yeast, etc. Even today the synthetic peptide substrates in the large group of the so-called chromogen substrates such as the amino acid- and peptide-β-naphthylamides and p-nitroanilides are widely used for the identification and specification of the intracellular peptidases with pepsin-, trypsin-, chymotrypsin-, carboxypeptidase-, and aminopeptidase-like action. At the same time, it is recognized that there exist intracellular peptidases, such as cathepsins D and E, the specificity of which cannot be identified with any of the abovementioned substrates. It should be added that the synthetic model substrates have a drawback for the clarification of the intracellular degradation of proteins under *in vivo* conditions, since they are not the natural substrates for the intracellular proteases. These substrates are the proteins of the cells in which the proteases act, proteins entering the cell from the outside, such as proteohormones (insulin, etc.), and peptides which result from the action of intracellular endopeptidases and which then by the action of intracellular exopeptidases can be degraded to free amino acids.

Model peptides with protected NH_2 and/or COOH groups have only a limited value for indicating the specificity of intracellular proteases and

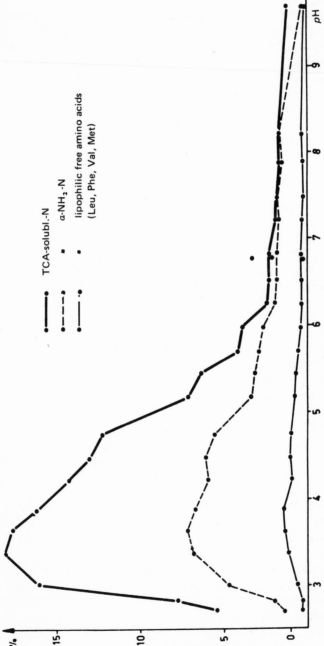

Fig. 1. Autoproteolysis in the homogenate of rat liver in relationship to pH. Homogenization in 0.25 M sucrose; incubation in 0.2% Triton X100 in 0.05 M K–citrate–phosphate buffer or without addition of buffer (single point at pH 6.8); precipitation with an equal volume 10% TCA before and after a 2-hr incubation at 37.5°C; this followed by 30 min heating at 70°C; centrifugation (10 min at 3000 × g); separation of the supernatant and determination of the residual N, α-amino-N, and lipophilic free amino acids (Leu, Phe, Val, Met) (from Hanson, 1969).

for comparing the activities of various organs or subcellular fractions. The same holds for the protein substrates which are still frequently used today for the demonstration of intracellular proteolytic activities, for example, casein. These statements are not intended to diminish the value of these investigations but only to indicate the limitations of the results obtained with these substrates.

IV. AUTOPROTEOLYSIS OF ORGANS

To demonstrate the intrinsic proteolytic potential of an organ, the process of autoproteolysis is still a very appropriate method, as demonstrated by the autolysis curves in Figs. 1, 2, and 3 for homogenates of rat liver, beef lens, and parts of a 10-day-old chick embryo. The liver, after 2 hr autolysis at the optimal pH 4.0, shows a threefold increase in trichloroacetic acid soluble nitrogen over that of the lens homogenate after 10 days autolysis at 37°C at the optimal pH 6.0. It is noteworthy that the autolysis of the lens at acid pH values, at which the liver shows optimal autoproteolysis, drops to zero and that the lens displays an optimal pH depending on the temperature of the autolysis: pH 6.0 at 37°C and pH 7.5 at 56°C. In this latter curve, the optimum at 37°C shows up as a shoulder. We see in this behavior of the autolysis of the lens, and in the decrease of the autoproteolysis of the liver

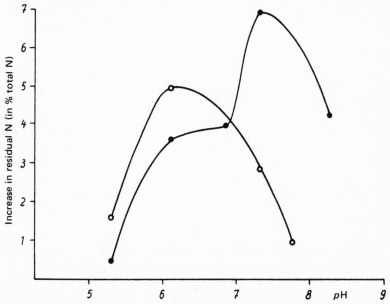

Fig. 2. pH dependence of the autoproteolysis of beef lens homogenate (from Methfessel *et al.*, 1963).

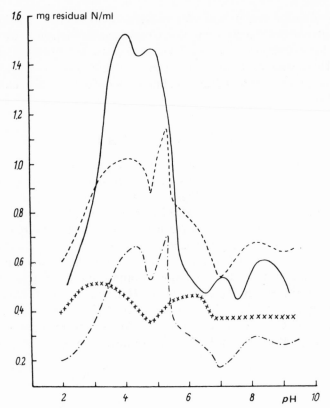

Fig. 3. pH dependence of proteolysis of egg yolk sac and egg yolk.
Effect of egg yolk sac on egg yolk of 10-day incubated chick embryo
and determination of the increase of residual N after 9 hr at 30°C
under sterile conditions (from Hanson, 1963).

toward the alkaline side of the pH values, and furthermore in the course of
the proteolysis of chick embryo with the different pH optima the expression
of the existence of different intracellular proteases with distinguishable pH
optima. The so-called acid proteases (cathepsins at pH 4.0) dominate in the
liver, and the neutral endo- and exopeptidases with optimal pH values
around pH 6.0–8.0 are prevalent in the eye lens.

V. QUALITATIVE AND QUANTITATIVE
DIFFERENTIATION OF AUTOPROTEOLYSIS
BY USE OF SYNTHETIC SUBSTRATES

Data bearing on the differences in autoproteolysis in the eye lens and in
the liver follow. In the eye lens, there was no, or barely any, activity toward
typical endopeptidase substrates, such as synthetic substrates for the

demonstration of tryptic and chymotryptic specificity, as well as toward carboxypeptidase substrates. Significant activity, however, was found toward synthetic substrates used for the demonstration of the presence of di- and aminopeptidases (see Tables II and III), such as leucine aminopeptidase. The last enzyme was obtained in crystalline form from the lens and is initially optimally active at 56°C (Table IV).

Table II. Relative Average Peptidase, Esterase, and Amidase Activities of Homogenates of Six Different Rat Tissues[a]

Substrate	pH	Kidney	Liver	Lung	Spleen	Heart	Muscle
Leu-Gly-Gly	7.8	150 = 100	86.6	143.0	102.0	18.1	11.8
Leu-Gly	8.0	170 = 100	51.0	136.0	40.7	14.6	7.1
Leu-amide	8.4	49 = 100	65.6	29.8	19.7	9.2	6.7
Gly-Leu	8.1	247 = 100	52·6	202.0	96.8	29.6	10.4
Gly-Gly-Gly	7.8	232 = 100	20.2	66.5	45.2	5.9	5.6
Gly-Gly	8.1	248 = 100	8.9	31.2	8.4	3.6	1.4
Cl-acetyl-Tyr	7.8	47.6 = 100	8.4	9.3	6.7	2.9	2.9
Z-Gly-Phe	5.1	38.3 = 100	16.8	17.5	3.10	0	0
Tyr-AE[b]	8.1	293 = 100	119.0	33.2	2.8	2.9	1.5
Gly-Tyr-AE	7.3	0	65.7 = 100	43.6	69.0	32.8	44.5
Acetyl-Tyr-AE	8 or 6[c]	116 = 100	12.3	19.4	16.0	15.4	9.4
Tyr-amide	8.1	35.8 = 100	26.3	15.0	10.0	16.8	3.5
Gly-Tyr-amide	8.1[d]	359 = 100	6.3	9.1	3.6	1.9	1.8

From Hanson (1969).
[a]Rate of hydrolysis in kidney is arbitrary given as 100; activities were determined at the pH optima of the particular enzymes (see left column).
[b]Esterase activity.
[c]Kidney ATEEase (at pH 8 = 0).
[d]Hydolysis by aminopeptidases.

Table III. Rate of Hydrolysis[a] of Synthetic Substrates Under the Influence of Beef Lens Extract, Serum, and Homogenates of Liver and Kidney

Substrate	Enzyme preparation			
	Lens	Liver	Kidney	Serum
L-Leucinamide	30.8 ± 2.5	20.5 ± 2.1	26.7 ± 2.5	1.45 ± 0.3
L-Prolylglycine	1·0 ± 0.13	12.2 ± 0.7	34.5 ± 1.5	0.32 ± 0.06
Glycyl-L-proline	0.5 ± 0.1	5.3 ± 1.1	28.4 ± 2.2	—
L-Prolylglycylglycine	4.1 ± 0.9	60.2 ± 2.1	59.7 ± 3.1	4.5 ± 0.27
Glycyl-L-prolylglycine	0.1 ± 0.1	3.3 ± 0.21	29.5 ± 2.3	0.06 ± 0.01
Glycylglycyl-L-proline	0.4 ± 0.1	5.9 ± 1.2	31.7 ± 1.7	—

From Hanson (1969).
[a]Rate of hydrolysis equals μmoles of proline or NH_3 liberated per mg protein N of the enzyme preparation per 60 min at pH 7.9–8.0. Since the course of hydrolysis was linear up to 140 min under the selected conditions, the values obtained from incubations at 30 and 120 min were extrapolated to 60 min.

Table IV. Proteolytic Activity of Beef Lens

Substrate	pH	Rate of hydrolysis
DL-Ala-Gly	8.0	21.4
Gly-Gly	8.0	10.7
Gly-Gly + Co^{2+}	8.0	44.5
Gly-D-Leu + Mn^{2+} + cysteine	8.0	0
Gly-L-Leu	8.0	30.2
L-Leu-Gly-Gly	8.0	80.0
L-Leu-NH$_2$	8.0	35.1
L-Leu-NH$_2$ + Mn^{2+}	8.5	130.0
Gly-Gly-Gly	8.0	0
L-Pro-Gly	8.3	1.0
L-Pro-Gly-Gly	8.0	4.1
Gly-L-Pro-Gly	8.0	0.1
Gly-L-Pro	8.0	0.5
Z-Gly-L-Phe	7.3	0
Cl-Ac-L-Tyr	8.0	0
Casein	7.3	0
Hemoglobin	4.6	0
Bz-L-Arg-NH$_2$	4.6	0
Gly-DL-Phe-NH$_2$ + cysteine	4.6	0
Z-L-Glu-L-Tyr	4.0	0
L-Leu-OET	8.0	4.6
L-Leu-OET + Mn^{2+}	8.5	10.6
L-Tyr-OET	8.3	0.8
Ac-L-Tyr-OET	8.0	0
Gly-L-Tyr-OET	7.2	(+)

From Hanson (1969).

On the other hand, there are obvious differences in the activity and specificity of the proteolytic enzyme spectrum of the homogenates of rat organs and the relations of the enzymatic activities among them. In both Table II and Table III, values are given for such characteristic synthetic substrates, which are mainly used for the estimation of exopeptidases and neutral chymotrypsin-like activities. However, one can see clearly by comparing these incomplete values to measurable enzymatic activities on the whole that organs such as liver and kidney show marked differences from each other, as well as compared to other organs. In this respect, the special position of the eye lens with its extraordinary aminopeptidase activity (LAP) is especially remarkable. The question of the acid and neutral intracellular proteases and exopeptidases will be discussed later with respect to the proteolytic activity of the components of the cell.

Table V. Rates of Hydrolysis[a]

	HP	NS	PS	RK	RL	RU	RD	RP	BEL
L-Leucinamide	50.0	1.02	2.44	235	82	167	57	40	60.5
L-Leucylglycine	89.0	0.36	0.87	158	92	—	—	—	72
S-Benzyl-L-cysteinamide	74.0	1.07	2.6	215	258	154	63	60.5	63
S-Benzyl-L-cysteinylglycinamide	191.0	1.76	2.36	265	415	353	129	118	131
L-Cystinyl-di-L-tyrosinamide	51.5	1.25	2.14	330	242	275	123	50	60.5
S-Methyl-L-cysteinyl-L-tyrosinamide	720.0	3.67	4.4	1300	1620	1690	670	478	560
S-Benzyl-L-cysteinyl-L-tyrosinamide	264.0	2.5	6.6	720	910	860	230	232	306
L-Tyrosinamide	17.5	0.32	1.15	165	36	117	41	34.5	23.3
L-Leucine-β-naphthylamide	41.5	1.35	5.1	154	20	36	13	34	3.5
L-Cystine-di-β-naphthylamide	4.6	0.115	0.6	5.1	0.93	5.8	1.75	3.8	0.55
S-Benzyl-L-cysteine-β-naphthylamide	13.3	0.032	2.74	11.6	4.3	22.4	11.2	2.0	0.53
L-Leucine-p-nitroanilide	177.0	1.01	10.3	127	61	229	67	52	1.51
S-Benzyl-L-cysteine-p-nitroanilide	21.5	0.12	3.35	29.5	9.2	36.5	8.3	10.2	0.46

From Hanson and Mannsfeldt (1971).

[a] In human placenta (HP), normal human female serum (NS), pregnant female serum (PS), rat tissues (RK, rat kidney; RL, rat liver; RU, rat uterus; RD, rat diaphragm; RP, rat placenta), beef eye lenses (BEL), pH 7.5, 0.065 N veronal-Na–HCl buffer or triethanolamine–HCl buffer. The assays with S-benzyl compounds contained 15 % dimethylformamide.

As already mentioned, intracellular proteolytic enzymes can attack not only proteins and polypeptides made in the cell but also, for example, polypeptide hormones that are brought into the cell. It appears that for these substances there is more of a quantitative than qualitative difference between the various organs. Under certain conditions (morphological organ structure, blood flux), enzymes of very well-defined, albeit not absolute, specificity are transported into the blood, in which they may be easily demonstrated. An example of this is the oxytocin-splitting activities, which according to the investigations of Fekete (1930) are present in the serum of pregnant women, and which, with the application of synthetic substrates, were demonstrated to have the character of an aminopeptidase by Werle (1941, 1956), Page (1946, 1947), and especially by Tuppy et al. (1957), and Hanson and Mannsfeldt (1971). It turns out that even substrates to which a relative specificity for oxytocinase was ascribed, such as S-benzyl-L-Cys-L-Tyr-amide,   L-cystinyl-di-β-naphthylamide,   S-benzyl-L-Cys-β-naphthylamide, and S-benzyl-L-Cys-p-nitroanilide are split by all the organs so far investigated, but at different rates. In addition, the serum of pregnant women but not the serum of pregnant rats shows a very high ratio (between 30 and 80) compared to the serum of nonpregnant women for two substrates (S-benzyl-L-Cys-β-naphthylamide and p-nitroanilide). All other synthetic substrates are attacked by aminopeptidase-like enzymes by all of the organ homogenates investigated. The eye lens shows again a special position; obviously, because of the lack of other aminopeptidases except LAP, it splits both the abovementioned S-benzyl compounds very slightly (Table V).

VI. AUTOPROTEOLYSIS OF THE CELL COMPONENTS AND PROTEOLYTIC INTERACTIONS BETWEEN THEM

For an understanding of the intracellular degradation of proteins, in addition to knowledge of the enzymes which participate in this process, their intracellular localization and the interaction of the cell components during proteolysis are of great significance. Our experiments performed with rat liver showed that autoproteolysis in the subcellular fractions obtained by differential centrifugation (nuclei, mitochondria, lysosomes, microsomes, and cytosol) is not equal, but shows great differences in the various fractions when expressed per milligram protein nitrogen (Figures 4, 5, 6, and 7). The experiments were performed as in Fig. 1. The optimum of the pH activity curve in all preparations is in the region between 3 and 4. It is noteworthy that the hyaloplasma (cytosol) shows only a fraction of the autoproteolysis of the whole homogenate (Fig. 1), and in comparison with the lysosomal fraction it has only a very minimal activity. It is also noteworthy that in the

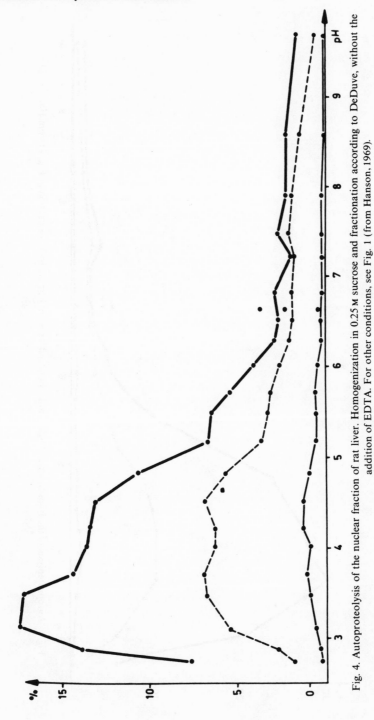

Fig. 4. Autoproteolysis of the nuclear fraction of rat liver. Homogenization in 0.25 M sucrose and fractionation according to DeDuve, without the addition of EDTA. For other conditions, see Fig. 1 (from Hanson, 1969).

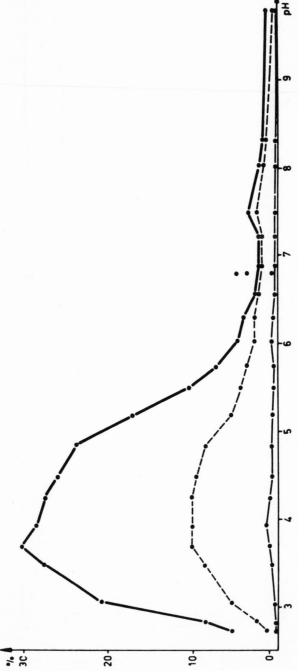

Fig. 5. Autoproteolysis of the lysosome-rich "light" mitochondrial fraction. For method, see caption of Fig. 4 (from Hanson, 1969).

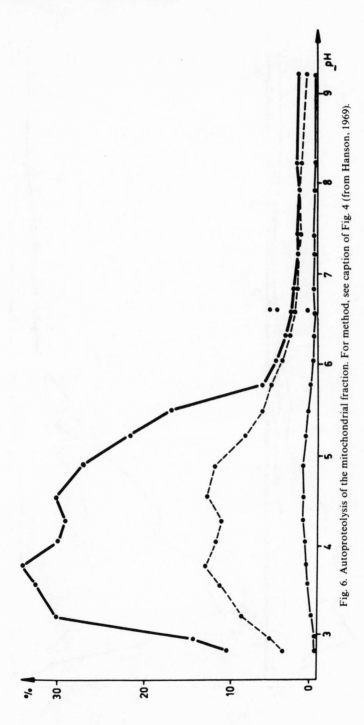

Fig. 6. Autoproteolysis of the mitochondrial fraction. For method, see caption of Fig. 4 (from Hanson, 1969).

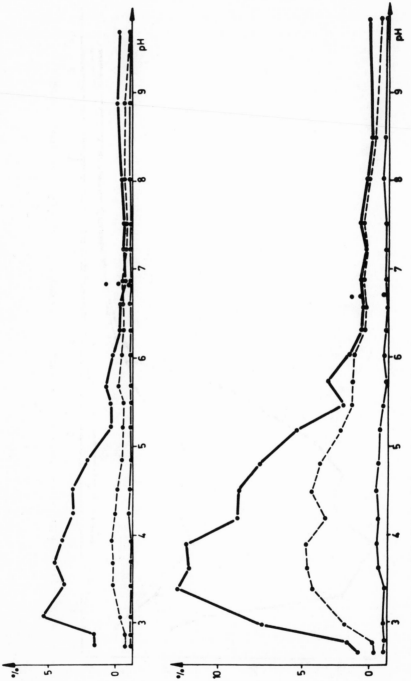

Fig. 7. Autoproteolysis of the hyaloplasma (top) and the microsomes (bottom). For method, see caption of Fig. 4 (from Hanson, 1969).

lysosomal fraction there is a small, but always reproducible, optimum of autoproteolysis around pH 7–8. Such *in vitro* experiments, i.e., with homogenized and then fractionated liver, at pH 6.8 show significant and characteristic differences from *in vivo* protein degradation of the liver as taken from data in the literature (Bohley, 1968) (Table VI). For example, the lysosomal

Table VI. Turnover Rates of Proteins of Subcellular Fractions of Rat Liver *in Vitro* and *in Vivo* Expressed as Percent Protein Degradation per Hour[a]

	In vitro	In vivo
Total liver proteins	1.6	1.0
Nuclear proteins	1.3	(0.1)
Mitochondrial proteins	4.4	0.2
Lysosomes	5.9	<0.2
Microsomal proteins	1.1	0.5
Hyaloplasmatic proteins	0.6	>1.5

From Bohley (1968).
[a]All the data for protein degradation *in vitro* were obtained by determinations of nonprotein N. Incubations were carried out in 0.2% Triton × 100 at pH 6.8 without addition of buffer. For nuclear and hyaloplasmatic proteins of rat liver, the turnover rates *in vivo* were not determined directly until now. Therefore, for nuclei only the value for histones can be given in parentheses. The *in vivo* turnover rate for hyaloplasmatic proteins was calculated from the turnover rates of total liver proteins and of the proteins of the particle fractions (hyaloplasmatic proteins equal 40% of total liver proteins).

proteins show the highest autoproteolysis *in vitro* among all cellular components, while *in vivo* they show the lowest. In contrast, the proteins of the cytosol show the lower *in vitro* and the highest *in vivo*. The latter are the only cellular proteins that have a higher protein turnover *in vivo* than *in vitro*. Taking into consideration the conditions of the *in vitro* and *in vivo* experiments, these differences seem to be understandable. In the *in vitro* experiments, after homogenization the cell components are rapidly separated from each other in the cold by centrifugation. The enzymes of the various cell fractions cannot, or can only in a very limited way, act on each other, and only under conditions of autolysis (37°C) can the proteolysis of their own proteins begin. In the intact cell under *in vivo* conditions, the situation is quite different since the enzymes in the various compartments can be coordinated and regulated to act on the available protein. Accordingly, the well-separated proteins of the particles are only slightly degraded. The cytosol, as mentioned above (Fig. 7), has the lowest autoproteolytic capacity. The high *in vivo* turnover rates of the proteins in the cytosol are the result of their interactions with proteolytic enzymes, primarily of the lysosomes.

According to these assumptions and interpretations, it was possible in the course of further investigations to demonstrate the following:

1. The lysosomes possess a great multiplicity and high activity in their proteolytic enzymes.
2. The lysosomes are able to attack *in vitro* the cytosol proteins from the same cell from which they are obtained in conditions that are comparable to *in vivo* conditions.

For this purpose, fractionations of rat liver were performed according to DeDuve *et al.* (1955) in order to allow lysosomal and cytosol fractions to act separately and in combination on each other. In addition, RNA-free cytosol proteins of molecular weight 70,000–15,000 (CP_1) were prepared by gel filtration (Sephadex G-75) in order to have a better-defined mixture of cytosol proteins as substrate (Bohley *et al.*, 1971).

At *p*H 6.9, this CP_1 fraction was degraded to peptides and amino acids by the lysosomal fraction which had the highest specific activity of all subcellular fractions and by the microsomal fraction practically up to only the liberation of peptides. The mitochondrial and nuclear fractions and the cytosol had significantly lower, or very low, degradative action on CP_1.

VII. PROTEOLYTIC ENZYME ACTIVITY OF LYSOSOMES TOWARD THE CELL'S OWN CYTOSOL PROTEINS IN COMPARISON TO FOREIGN PROTEINS AND SYNTHETIC PEPTIDES

The finding that the proteolytic activity of a lysosomal fraction of rat liver toward its own cytosol proteins (CP_1) was higher at *p*H 6.9 than that toward most usual protease substrates, i.e., hemoglobin, led us to investigate how one can define the proteolytic activities present in the lysosomes in terms of enzymes with known specificities. For this purpose, the water-soluble part of the lysosomal fraction was separated by high-speed centrifugation from the insoluble part (membranes, mitochondria, and microsomal contaminants) and subjected to several fractionation steps. The water extract of the lysosomes (LS) was subjected to gel filtration of Sephadex G-75 (three fractions: L120, L60, and L20); this was followed by an additional fractionation on an ion exchanger CM-Sephadex C-50 (fractions L20, C1–5 and L60, C1–7). The individual fractions were tested for their specific proteolytic activity toward synthetic and natural protease substrates (see Table VII).

Fraction L20 (mol. wt. 10,000–30,000) showed the highest proteolytic activity towards the cytosol proteins (CP_1). The lowest proteolytic activity was shown by fraction L120 (mol. wt. 80,000), which, similar to fraction L60 (mol. wt. 40,000–80,000), showed very little activity toward CP_1 and azo-casein. With the CM-Sephadex C-50 fractionation, five fractions were obtained which were active toward azocasein, and in L20C5 only one band could be demonstrated with disc electrophoresis. In the other fractions, there were two or more bands, among them some with endopeptidase activity. These, on the basis of their behavior toward glutathione, EDTA, and p-chloromercuribenzoate, could not be identified with the so-called cathepsins A, B, C, and D. CP_1, glucagon, and ribonuclease were well attacked by the endopeptidases of L20. Only a certain fraction of hemoglobin hydrolysis at pH 4.0 can be attributed to cathepsin D, since proteases of a molecular weight below 50,000 are much more able to hydrolyze Hb due to conformational changes brought about at pH 4.0. (For further details with discussion, see Bohley et al., 1971.)

The endopeptidases may have a special significance as limiting enzymes in the intracellular degradation of proteins. In order to clarify this question, the relationship of the NH_2 nitrogen, determined in the TCA-soluble super-natant after enzyme incubation, was compared with that obtained after hydrolysis with 6 N HCl. In order to inactivate the exopeptidase present in CP_1, which could further degrade the peptides formed from CP_1, CP_1 was heated prior to the incubation experiments. It was demonstrated in some fractions of LS that with the exclusion of exopeptidase activity the ratio of NH_2 nitrogen after hydrolysis to NH_2 nitrogen before hydrolysis was 5; i.e., peptides with an average of four to six amino acids per molecule were present.

From the described experiments on the fractionation of lysosomes and by use of synthetic and natural substrates as shown in Table VII, it is clearly demonstrated that there are a large number of endo- and exopeptidases in the lysosomes, which even at a neutral pH are able to degrade the cytosol proteins of their own cells all the way down to amino acids, and these are then available to the cell for further utilization.

VIII. SIGNIFICANCE OF ENZYMATIC REDUCTIVE CLEAVAGE OF DISULFIDE BRIDGES FOR INTRACELLULAR PROTEOLYSIS

During intracellular degradation of proteins, not only peptide linkages are split hydrolytically by the proteases. Also to be taken into considera-tion is the reductive cleavage of the disulfide bridges, present in many proteins, in order to make these proteins accessible to the proteases. Without wishing to generalize, we could show that insulin, which has three disulfide bridges, is degraded by a different molecular mechanism than is glucagon, which has no disulfide bridges (Bohley et al., 1971; Ansorge et al., 1971, 1972). It seems

Table VII. Specific Activities of Liver Homogenates, Lysosomal Extract, and Lysosomal

					Fractions of		
	G^b	LS^c	L20	L20C1	L20C2	L20C3	L20C5
A. pH 7.0							
Cytosol proteins^d							
(CP$_1$)	5	130	400	280	270	270	4200
Glucagon	20	280	490	290	330	520	1590
Ribonuclease^d	5	150	160	160	160	210	800
Polysyline^d			310	30	160	30	740
Hemoglobin^d	5	230	170	50	30	80	570
Insulin^{df}	5	40	10	20	0	30	0
Insulin B chain^d (oxid.)^f	7		340	30	60	140	570
Z-Glu-Tyr, pH 5			30	30	0	0	0
Leu-β-Na^{+d}	2	9	650	28	95	680	0
Bz-Arg-β-Na^e	1	4	450	20	69	240	0
Bz-Arg-NA^e	1	1	6	0	1	4	0
Bz-Arg-NH$_2{}^e$							
Gly-Phe-NH$_2{}^e$							
Leu-Gly-NA^e	1	1	34	0	0	14	0
Leu-Gly-Gly^e				16	65	220	0
Leu-Phe^e				14	58	220	0
Leu-Phe							
Z-Gly-Phe, pH 5							
B. pH 4.0							
Ribonuclease^d		440	400	100	0	50	1880
Hemoglobin^d	52	2190	1360	1000	270	310	1940
Leu-Gly-Gly^e				8	39	270	0
Phe-Phe^e				9	35	210	0

^{a}The specific activities of the lysosomal preparations tested for peptidases are expressed as the increase in α-NH$_2$ nitrogen (μg) of the TCA supernatant for proteins and as μmoles of substrate cleaved for peptide substrates after incubation time of 1 hr at 37°C per mg protein N of the enzyme preparation. Assays at pH 7.0, 0.01 M potassium phosphate buffer; assays at pH 4.0, 0.2 M potassium acetate buffer. Substrate concentrations: proteins 0.6% (w/v), peptides and peptide derivatives 10^{-2} to 2×10^{-3} M, β-naphthylamide and p-nitroanilide 10^{-3} M. Some other substrates tested at pH 7.0 and with L fractions give results similar to the specific activities listed in the table: myoglobin and catalase similar to Hb; γ-globulin and edestine similar to insulin; Phe-Phe similar to Leu-Phe; and Leu-Gly-Gly-Gly, Leu-Leu-p-NA, and Gly-Gly-p-NA similar to Leu-Gly-p-NA.
^{b}G, homogenate.
^{c}LS, lysosomal extract.
^{d}10^{-3} M glutathione.
^{e}10^{-3} M dithiothreitol.
^{f}10^{-3} M EDTA.

Enzyme Fractions Toward Various Substrates[a]

lysosomal extracts

L60	L60C1	L60C2	L60C3	L60C4	L60C5	L60C6	L60C7	L120	Trypsin
7	60	75	65	330	165	120	390	25	3510
									180
								10	165,000
20	7	0	0	25	125	0	0		330
									80
									50
									90
1	5	3	0	0	0	0	0		
30	15	35	50	220	40	15	0	95	
1330	35	55	205	2860	650	295	70		
2370				6200	1000			1720	
	50	10	0	0	5	0	0		
925	1460	250	220	190	1830	600	290		

very likely that insulin is first degraded by a transhydrogenolytic step, and then by proteolytic–hydrolytic reactions; this is brought about probably with the participation of the proteolytic lysosomal enzymes. Contrary to the findings of Mirsky (1964), who found a maximal rate of degradation of about 20 μg insulin/g liver/min, we established a ten times higher degradation rate of insulin by liver, homogenate or subcellular fractions, when gluathione (GSH, 10^{-3} M) and EDTA (5×10^{-3} M) were added and the experiments performed with ^{125}I- or ^{131}I-labeled insulin (225 ± 83 μg insulin degraded/g liver/min). We have found values between 10 and 20 μg of insulin degradation in homogenates without GSH and EDTA activation, and in the cytosol fractions with and without these additions. It is to be concluded from this that in addition to the enzymatic systems of the cytosol with only a weak degradative activity toward insulin that have so far been mainly recognized, the liver possesses other, strongly active, GSH-dependent insulin-degrading enzyme systems. It was possible to demonstrate conclusively that when GSH and EDTA were added the microsomal fraction had by far the highest capacity for proteolytic degradation of insulin, and it seems likely that the degradation is started by the hydrogenolytic systems demonstrated by Narahara and Williams (1959) and Tomizawa and Halsey (1959) in both the microsomal and cytosol fractions. In both insulin-degrading fractions, a transhydrogenolytic activity was demonstrated (Ansorge et al., 1971). At acid pH value (4.0), only lysosomes can degrade insulin (and glucagon as well), and this is also potentiated by the addition of SH reagents and EDTA. The strongest SH-reagent-independent ^{125}I-glucagon degradation at neutral pH is in the cytosol, whereas in the lysosomes at that pH one finds only 10–15% of the value of the cytosol. The degradation of glucagon by the cytosol is, moreover, inhibited by GSH and EDTA. With regard to insulin, we must consider that more than one enzyme attacks this molecule, i.e., both the hydrolytic peptide-splitting and the reductive transhydrogenolytic disulfide-cleaving mechanisms.

IX. CONCLUSIONS

Experiments aimed at clarifying the mechanisms of intracellular proteolysis must take into consideration the fact that each cell contains several thousand structural and functional proteins, including the numerous enzymes with individually distinguishable turnover rates. This does not mean anything else but that there are a great number of protein substrates present. At the present time, it is not possible to make any statements as to which properties and conditions are decisive for the susceptibility of cell proteins to a particular endopeptidase present among the large number of intracellular, primarily intralysosomal, proteases, especially the endopeptidases. At any rate, it is certain that the low molecular weight substrates synthesized in an ever increasing number since the investigations of Max Bergmann do not suffice for the complete elucidation and characterization of the intracellular endopeptidases, although they have been, and still are, invaluable for rapid differentiation of proteases. One needs only remember that cathepsin A was demonstrated to be a carboxypeptidase by the use of Glu-Tyr (and of glucagon) (Jodice, 1967). Cathepsin C was also characterized by the use of synthetic substrates as being a dipeptidyl transferase, in addition to having hydrolytic properties, and today it is considered as an exopeptidase with broad specificity for the splitting of N-terminal dipeptides from polypeptides and from dipeptide naphthylamides (dipeptidylaminopeptidase I). It can be differentiated specifically from other dipeptidylaminopeptidases (II–IV) by the use of synthetic substrates (McDonald et al., 1971). Similarly, this holds for cathepsins B_1 and B_2. Even cathepsin D, for which up to now no typical synthetic substrates were known, can possibly (according to the investigation of Keilova, 1971) be characterized with low molecular weight peptides for substrate, such as Gly-Phe-Leu-Gly-Phe or Gly-Phe-Leu-Gly-Phe-Leu, as splitting N-terminal dipeptides, with the carboxyl end of phenylalanine as the preferred splitting site. In all investigations on substrate specificity in the last years, substrates with natural sequences, such as glucagon, angiotensin, and ACTH and primarily the oxidized B chain of insulin, have been increasingly utilized. On the basis of the findings with the last substrate, we have made an attempt to derive some regularities for the specificity of endopeptidases (see Bohley, 1968; Bohley et al., 1971; Keilova, 1971). In general, one can conclude that intracellular peptidases preferentially split peptide linkages which consist of hydrophobic amino acids and that in the oxidized B chain of insulin the most frequently and rapidly hydrolyzed peptide bonds are the ones between amino acids 11–18 and 23–27.

REFERENCES

Ansorge, S., Bohley, P., Kirschke, H., Langner, J., and Hanson, H., 1971, *Europ. J. Biochem.* **19**:283.

Ansorge, S., Bohley, P., Kirschke, H., Langner, J., Wiederanders, B., and Hanson, H., 1972, in press.

Bergmann, M., 1942, A classification of proteolytic enzymes, in "Advances in Enzymology," Vol. II (F. F. Nord and C. H. Werkman, eds.), p. 49, Interscience, New York.

Bergmann, M., and Zervas, L., 1932, *Ber. Deutsch. Chem. Ges.* **65**:1192.

Bergmann, M., Zervas, L., Schleich, H., and Leinert, F., 1932, *Hoppe-Seyler's Z. Physiol. Chem.* **212**:72.

Bohley, P., 1968, *Naturwissenschaften* **55**:211.

Bohley, P., Kirschke, H., Langner, J., Ansorge, S., Wiederanders, B., and Hanson, H., 1971, Intracellular protein breakdown, in "Tissue Proteinases" (A. J. Barret and J. T. Dingle, eds.), p. 187, North-Holland, Amsterdam.

DeDuve, C., Pressman, B. C., Gianetto, R., Wattiaux, R., and Applemans, F., 1955, *Biochem. J.* **60**:604.

Hanson, H., 1963, *Abhandl. Sächsisch. Akad. Wiss., Math.-Naturwiss. Klasse* **48**:1.

Hanson, H., 1969, Zur Kennzeichnung und Differenzierung proteolytischer Gewebsenzyme, *in* "Stoffwechselvorgänge und Enzym-mechanismen in Zellen und Geweben," Vol. 23 (K. Lohmann *et al.*, eds.), p. 52, Verhandlungen der Deutschen Gesellschaft für Experimentelle Medizin.

Hanson, H., and Mannsfeldt, H.-G., 1971, Biochemische Beiträge zur Kenntnis der Serumoxytocinase, *Nova Acta Leopoldina* **36**:1 (N.F., No. 200).

Jodice, A. A., 1967, *Arch. Biochem.* **121**:241.

Keilova, H., 1971, On the specificity and inhibition of cathepsins D and B, *in* "Tissue Proteinases" (A. J. Barret and J. T. Dingle, eds.), p. 45, North-Holland, Amsterdam.

McDonald, J. K., Callahan, P. X., Ellis, S., and Smith, R. E., 1971, Polypeptide degradation by dipeptidyl aminopeptidase I (cathepsin C) and related peptidases, *in* "Tissue Proteinases" (A. J. Barret and J. T. Dingle, eds.), p. 69, North-Holland, Amsterdam.

Methfessel, J., Langner, J., and Hanson, H., 1963, *v. Graefes Arch. Ophthalmol.* **165**:529.

Metrione, R. M., Neves, A. G., and Fruton, J. S., 1966, *Biochemistry* **5**:1597.

Mirsky, J. A., 1964, The metabolism of insulin, *Diabetes* **13**:225.

Narahara, H. T., and Williams, R. H., 1959, *J. Biol. Chem.* **234**:71.

Schimke, R. T., 1969, On the roles of synthesis and degradation in regulation of enzyme levels in mammalian tissues, *in* "Current Topics in Cellular Regulation," Vol. 1 (B. L. Horecker and E. R. Stadtman, eds.), p. 77, Academic Press, New York.

Schimke, R. T., 1970, Regulation of protein degradation in mammalian tissues, *in* "Mammalian Protein Metabolism," Vol. IV (H. N. Munro, ed.), p. 177, Academic Press, New York.

Schimke, R. T., and Dehlinger, D. J., 1971, Turnover of protein constituents of rat liver membranes, *in* "Drugs and Cell Regulation" (E. Mikisch, ed.), p. 121, Academic Press, New York.

Schimke, R. T., and Doyle, P., 1970, Control of enzyme levels in animal tissues, *in* "Annual Review of Biochemistry," Vol. 39 (E. E. Snell, ed.), p. 929, Ann. Rev., Inc., Palo Alto, Calif.

Tomizawa, H. H., and Halsey, Y. D., 1959, *J. Biol. Chem.* **234**:307.

CHAPTER 12

SYNTHESIS OF HUMAN ACTH AND ITS BIOLOGICALLY ACTIVE FRAGMENTS

Kálmán Medzihradszky

Institute of Organic Chemistry
Eötvös L. University
Budapest, Hungary

I. INTRODUCTION

Of the numerous biologically active polypeptides of the anterior pituitary, the adrenocorticotropic hormone (ACTH) has been the most thoroughly investigated. The corticotropins have been isolated in chemically pure form from four different species, and their primary structures were established by several research groups. Thus the porcine hormone (Bell, 1954; Bell et al., 1956; Shepherd et al., 1956a,b; White and Landmann, 1955), the sheep hormone (Li et al., 1955), the bovine hormone (Li et al., 1958), and the human hormone (Lee et al., 1961) all proved to be single polypeptide chains consisting of 39 amino acid residues with an essentially identical sequence, which is different and species-characteristic at the 25–33 positions only (Fig. 1).

According to recent findings (Gráf et al., 1971; Riniker et al., 1972), it is very likely that in the case of human and porcine corticotropins the seemingly well-established structures, supported by synthesis, need revision. As indicated by these investigations, both hormones contain asparagine instead of aspartic acid at position 25 and glutamic acid instead of glutamine at position 30. Furthermore, the sequence of the human hormone is reversed at positions 26–27; instead of Ala-Gly, the correct structure is Gly-Ala. Correspondingly, the amino acid sequence in the species-specific part is

	25								33
Porcine:	-Asn-Gly-Ala-Glu-Asp-Glu-Leu-Ala-Glu-
Human:	-Asn-Gly-Ala-Glu-Asp-Glu-Ser-Ala-Glu-

259

1   2   3   4   5   6   7   8   9 10 11 12 13 14 15 16 17 18   19

Ser-Tyr-Ser-Met-Glu-His-Phe-Arg-Trp-Gly-Lys-Pro-Val-Gly-Lys-Lys-Arg-Arg-Pro-

20 21 22 23 24

Val-Lys-Val-Tyr-Pro-

	25	26	27	28	29	30	31	32	33

α_p-ACTH   -Asp-Gly-Ala-Glu-Asp-Gln-Leu-Ala-Glu-

α_s-ACTH   -Ala-Gly-Glu-Asp-Asp-Glu-Ala-Ser-Gln-

α_b-ACTH   -Asp-Gly-Glu-Ala-Glu-Asp-Ser-Ala-Gln-

α_h-ACTH   -Asp-Ala-Gly-Glu-Asp-Gln-Ser-Ala-Glu-

34 35 36 37 38 39

-Ala-Phe-Pro-Leu-Glu-Phe

Fig. 1. Primary structures of corticotropins.

The observations that the whole molecule (containing all of the 39 amino acid residues) is not necessary for biological activity of the adreno-corticotropic hormone and that the smaller fragments containing the sequences 1–28 and 1–24 possess the full biological activity of the hormone greatly promoted the efforts to synthesize peptides with corticotropic activity.

The first publication on the synthesis of a biologically active ACTH fragment containing the first 20 amino acid residues appeared in 1956 (Boissonnas *et al.*, 1956). During the following decade, this pioneering work has been followed by the synthesis of a great number of different corticotropin fragments and, eventually, by the successful synthesis of the complete molecule of porcine and human corticotropins. Table I (p. 269) lists synthetic fragments described in the literature containing the first ten or more amino acid residues.

II. SYNTHESIS OF HUMAN ACTH

The Hungarian peptide chemists entered the ACTH field relatively early, first aiming at the synthesis of the *N*-terminal 1–28 octacosapeptide fragment of porcine corticotropin. This was the smallest isolated and characterized fragment, from the partial enzymatic degradation experiments, with full biological activity (Bell *et al.*, 1956). At that time, there was only indirect evidence for the existence of a fully active 1–24 tetracosapeptide fragment.

Although in this chapter I wish to describe the synthesis of human corticotropin, it is essential to briefly consider our earlier work on this topic, since it served as a basis for our later synthetic studies.

In our first experiments, the octacosapeptide was constructed by coupling three peptide fragments with the following structures:

Z-Ser-Tyr-Ser-Met-Glu-His-Phe-Arg-Trp-N₂H₃ (sequence 1–9)

Z-Gly-Lys(Tos)-Pro-Val-Gly-Lys(Tos)-Lys(Tos)-Arg-Arg-Pro-Val-Lys(Tos)-OH (sequence 10–21)

H-Val-Tyr-Pro-Asp(OEt)-Gly-Ala-Glu(OEt)-OEt (sequence 22–28)

Tosyl protection of the ε-amino group of lysine permits the selective removal of the terminal carbobenzoxy groups, which in turn can be split off by sodium–ammonia reduction simultaneously with the tosyl groups in the last step of the synthesis. It might be pointed out that although use of alkyl esters is not advantageous, we have employed these protecting groups in place of benzyl esters in order to avoid the unwanted partial splitting of the latter derivatives during the hydrogen bromide–acetic acid cleavage of the carbobenzoxy groups (Bayer et al., 1961).

Synthesis of the N-terminal nonapeptide hydrazide (Medzihradszky et al., 1962a,b) was performed according to the schema in Fig. 2.

The protected tetrapeptide hydrazide Z-Ser-Tyr-Ser-Met-N₂H₃ was known (Hofmann et al., 1957) and was later also described by Li et al. (1961). In the synthesis of this compound, removal of the carbobenzoxy group from Z-Ser-Met-OMe was accomplished by catalytic hydrogenolysis in the presence of hydrochloric acid. Hydrogenolysis in the presence of bases (e.g., triethylamine), which proved to be useful for the decarbobenzoxylation of methionine derivatives (Medzihradszky and Medzihradszky-Schweiger, 1965; Medzihradszky-Schweiger and Medzihradszky, 1966), could not be applied because of the possibility of diketopiperazine ring formation. As can be seen from Fig. 2, coupling of the fragments was carried out via the azide route to diminish the danger of racemization. Peptide azides were generally prepared in dimethylformamide solution and were not isolated before

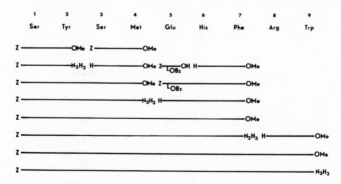

Fig. 2. Synthesis of the N-terminal nonapeptide hydrazide.

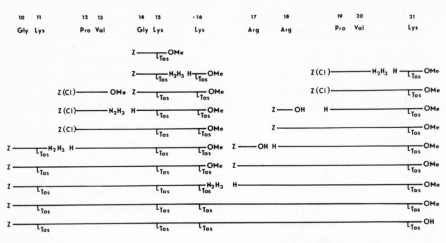

Fig. 3. Synthesis of the dodecapeptide containing the 10–21 amino acids.

coupling. The protected nonapeptide hydrazide was crystallized from a dioxane–water mixture.

As shown in Figs. 3 and 4, the dodecapeptide derivative containing the sequence 10–21 was prepared by the coupling of two fragments, the heptapeptide 10–16 and the pentapeptide 17–21. The latter fragment was synthesized by two different routes using protected (nitro group) or unprotected arginine (Bajusz and Lénárd, 1962; Bajusz et al., 1962).

In this synthesis, the azide method was employed for fragment condensation and the p-chlorocarbobenzoxy group as the amino protector (Kisfaludy and Dualszky, 1960). Purification of the nitro-substituted arginine compounds was easier than that of the unprotected peptides.

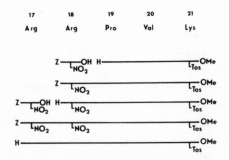

Fig. 4. Alternative route for the 17–21 penta-peptide synthesis.

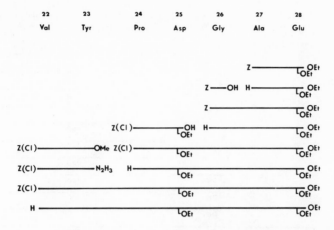

Fig. 5. Synthesis of the *C*-terminal heptapeptide.

The synthesis of the *C*-terminal heptapeptide derivative (Kisfaludy and Dualszky, 1962; Kisfaludy *et al.*, 1962) is depicted in Fig. 5. As was the case in the synthesis of the dodecapeptide derivative (Fig. 3), the use of the *p*-chlorocarbobenzoxy group as the amino-protecting function in the synthesis of the *C*-terminal heptapeptide derivative permitted the isolation of several intermediates, as well as of the final product in crystalline form.

Coupling of the three fragments was accomplished by several different techniques. The protected octacosapeptide was obtained in chromatographically pure state by dissolving in hot methanol and cooling to room temperature (Bruckner *et al.*, 1962). After deblocking in sodium–liquid ammonia, however, the biologically active free octacosapeptide could be isolated in homogeneous form only by preparative electrophoresis. Although the intensive study of numerous model compounds showed that under strictly anhydrous conditions the bond splitting known from the literature (Hofmann *et al.*, 1960) does not take place (Bajusz and Medzihradszky, 1963), it was clear that this degradation could not be fully avoided in the case of the rather complex octacosapeptide containing many tosyl groups and certainly some water as well.

These difficulties in the final step of the octacosapeptide synthesis forced us to look for protecting groups which would give unequivocal results in the deprotection procedure. This we found in the combination of the *tert*-butyloxycarbonyl and *tert*-butyl ester groups, elaborated by Schwyzer and his coworkers. Furthermore, aiming at the synthesis of a series of sequential homologues for studying relationships between biological activity and the growing number of amino acid residues in the peptide chain, we sought for

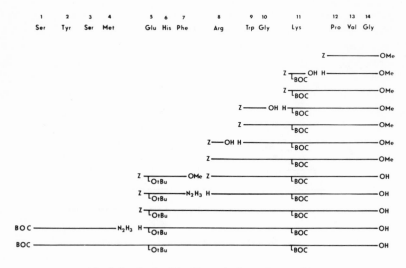

Fig. 6. Synthesis of the *N*-terminal tetradecapeptide.

a key compound which could serve as a preferable starting material for the synthesis of all the larger fragments. For this purpose, the partially protected *N*-terminal tetradecapeptide of the corticotropin was selected. The synthesis of this intermediate is outlined in Fig. 6.

Synthesis of the protected *N*-terminal tetrapeptide hydrazide has been described by Iselin and Schwyzer (1961), and we changed only some of the experimental procedures. Building up the tripeptide sequence Glu-His-Phe similarly to the schema shown for the previous synthesis in Fig. 2, we found a simple process with excellent yield for the preparation of Z-Glu(OtBu)-OH (Medzihradszky *et al.*, 1967*b*). The required tripeptide hydrazide was prepared in crystalline form from the tripeptide diester by selective hydrazinolysis. The *C*-terminal half of the tetradecapeptide was synthesized essentially by the stepwise method. After saponification and hydrogenolysis, compound H-Arg-Trp-Gly-Lys(BOC)-Pro-Val-Gly-OH was acylated with the Z-Glu(OtBu)-His-Phe-N_2H_3 tripeptide hydrazide without isolating the intermediate azide. Repeated hydrogenolysis and coupling with the *N*-terminal tetrapeptide azide led to the desired protected tetradecapeptide, which was obtained in pure state after column chromatography on silica gel and was converted to the hydrochloride (Bruckner *et al.*, 1966; Bajusz and Medzihradszky, 1967; Medzihradszky *et al.*, 1973*a*).

The synthetic steps leading to the *C*-terminal tetradecapeptide are shown in Fig. 7. The intermediate peptides 20–24 and 25–28 were first coupled, and the resulting *C*-terminal nonapeptide (20–28) was hydrogenated and acylated with the Lys-Lys-Arg-Arg-Pro pentapeptide (sequence 15–19).

The presence of five basic residues made this synthesis rather laborious, whereas by the advantageous distribution of the proline residues the danger of racemization could be avoided. In addition to the *tert*-butyl ester and *tert*-butyloxycarbonyl protection, the tyrosine hydroxyl function was also blocked by conversion into *tert*-butyl ether. By building up the C-terminal fragment, stepwise synthesis, using N-hydroxysuccinimide esters, was successfully applied (Löw and Kisfaludy, 1965).

The combination of the two tetradecapeptides was the object of an intensive study (Medzihradszky and Bajusz, 1968). Although, in principle, a great number of coupling procedures can be used for acylation with a glycine-terminating peptide, mixed anhydride synthesis always gave unsatisfactory results. We believe that either the formation of the anhydride was hindered, or, once formed, it decomposed during the slow acylation process between the relatively large molecules. With the *p*-nitrophenyl ester method, the yield of the ester prepared from *p*-nitrophenyl sulfite was very low, while the dicyclohexylcarbodiimide procedure led to the formation of the dicyclohexyl urea derivative of the BOC-tetradecapeptide. Direct coupling in the presence of DCCI also gave significant amounts of urea derivative, which could be separated by silica gel column chromatography only with great losses of the desired octacosapeptide. We obtained, however, a very good conversion with DCCI in the presence of pentachlorophenol using these reagents either directly (Bajusz and Medzihradszky, 1967) or in the form of their previously prepared complex (Kovács *et al.*, 1967).

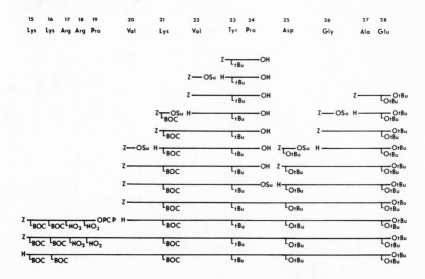

Fig. 7. Synthesis of the C-terminal tetradecapeptide.

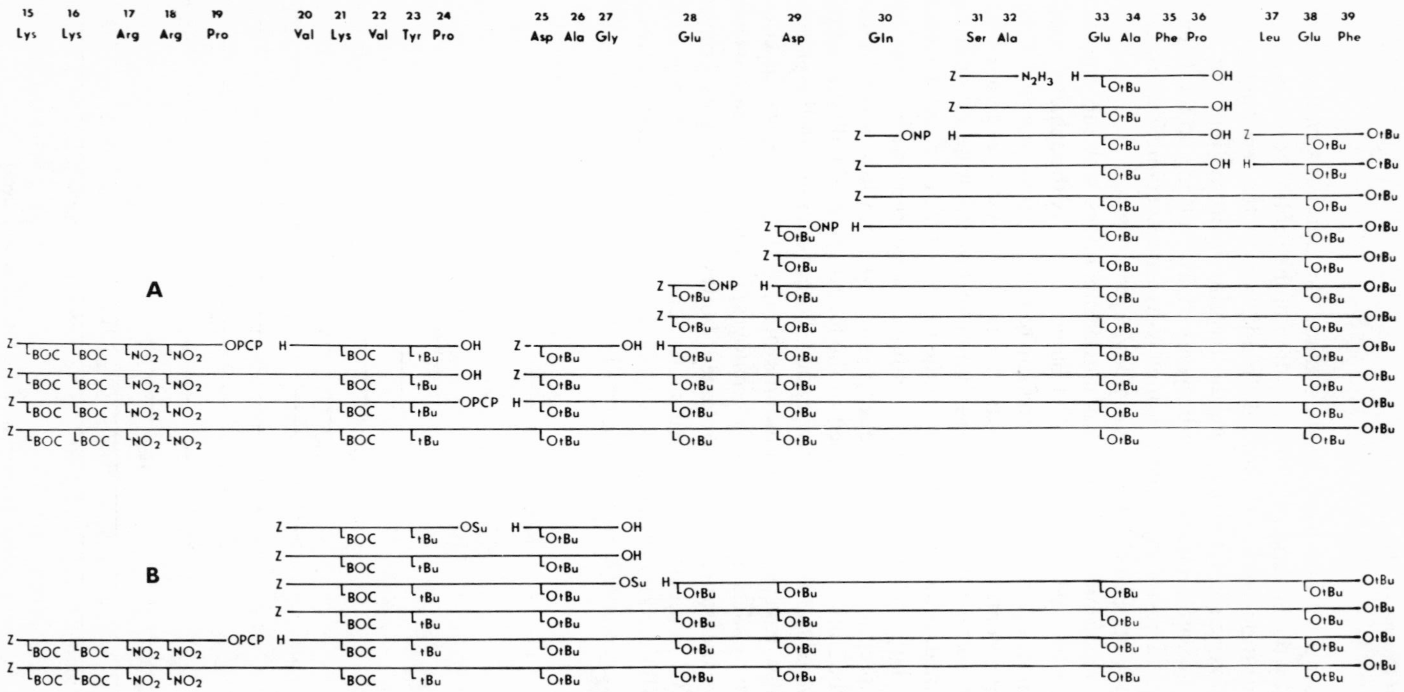

Fig. 8. Synthesis of the C-terminal pentacosapeptide of human corticotropin.

The protected octacosapeptide was purified on silica gel column, deblocked by dissolving in trifluoroacetic acid containing some mercaptoethanol, and converted into the acetate salt. Pure product could be obtained by deblocking the crude protected octacosapeptide in the same way, followed by an ammonium acetate buffer gradient elution on carboxymethylcellulose column. The free peptide possessed essentially the same biological activity as the natural corticotropins (Bruckner *et al.*, 1966; Bajusz and Medzihradszky, 1967; Medzihradszky *et al.*, 1967*a*; Bajusz *et al.*, 1973*a*).

Having in our possession the octacosapeptide intermediates, it did not seem too difficult a task to synthesize the complete sequence of human corticotropin, whose structure had been established by Lee *et al.* (1961).

This synthesis was justified from several points of view. Although partial hydrolysis and synthesis of fragments proved that the N-terminal half of the corticotropins possesses the characteristic biological activity, it was evident that immunological activity required the presence of the species-specific C-terminal part as well. Moreover, one could speculate that the intact molecule's lifetime is longer, its resistance against proteolytic enzymes is greater, and possibly the transport properties and perhaps the spectrum of steroids produced under the stimulatory influence of the hormone are different from those of the smaller fragments. Mainly because of the above-mentioned properties, the molar biological activity of the intact hormone is about twice as high as that of the N-terminal half, whereas the activities on a weight basis are similar. Since the natural hormone is not easily accessible, closer investigation of these phenomena can be attained with synthetic product only.

For the synthesis of human adrenocorticotropic hormone (Bajusz *et al.*, 1967, 1968), the 1–14, 15–19, and 20–24 fragments of the octacosapeptide synthesis could be used without any modifications. Quite new synthetic work was necessary, however, to build the C-terminal 25–39 sequence, although some of the intermediates had been prepared previously (Bajusz and Lázár, 1966). Using these smaller peptides, the pentacosapeptide consisting of the 15–39 amino acid residues was synthesized by two alternative routes (Fig. 8). In route A, two pairs of four polypeptides were coupled to a decapeptide and to a pentadecapeptide, respectively. The decapeptide 15–24 obtained in this way was converted to the pentachlorophenyl ester derivative, and this in turn was coupled to the C-terminal pentadecapeptide ester. The first step in route B was the synthesis of the centrally positioned, 20–27, octapeptide, followed by the acylation of the C-terminal dodecapeptide with this product. The resulting icosapeptide ester was lengthened at the N-terminus with the 15–19 pentapeptide, containing the basic core of ACTH. The pentacosapeptides obtained by the two different procedures were identical in every respect.

Coupling the 1–14 and 15–39 components to produce the protected human adrenocorticotropin was performed with the aid of the dicyclohexylcarbodiimide–pentachlorophenol method, which we used previously in the synthesis of the octacosapeptide. The pure, homogeneous product (thin layer chromatography, paper and gel electrophoresis) proved to be identical to the natural human ACTH in all biological activities, including radiobioassay*. It is noteworthy to mention that clinical trials have shown that the synthetic hormone could be given to patients having an anaphylactic response to highly purified porcine ACTH (Kovács et al., 1968) without any side-effects.

Having synthesized human corticotropin according to the structure proposed by Lee et al. (1961), we were able not only to determine its biological activities but also to compare its chemical properties with those of the natural hormone. To our surprise, the synthetic polypeptide did not show a tendency for deamination at pH 10, a property exhibited by a fragment containing the 22–39 sequence which was obtained from papain-catalyzed partial hydrolysis of natural human ACTH (Gráf et al., 1971). This deamination is also shown by natural porcine ACTH and can be explained by the presence of an Asn-Gly bond at positions 25–26 rather than Asp-Gly as was originally proposed (Shepherd et al., 1956b). We assume that human corticotropin also contains a similar sequence at this point. The same suggestion was made by Riniker et al. (1972) based on similar observations. Thus the structures of porcine and human corticotropins differ only at position 30 (Leu in porcine and Ser in the human hormone), a fact that makes the interpretation of their immunological specificity more difficult. In view of our experience with our previous synthesis, we were able to synthesize human corticotropin with the proposed new structure with only minor changes in the synthetic pathway described above (Bajusz et al., 1973b).

III. SYNTHESIS OF SEQUENCE HOMOLOGS OF CORTICOTROPINS

Once the synthesis of a natural product such as a polypeptide is completed, it is most interesting to study the relationship between structure and biological activity. One approach toward that end is the synthesis of fragments of the original molecule with different chain lengths. The polypeptides listed in Table I have been prepared mainly for this purpose. Investigations of the biological properties of these compounds have contributed considerably to the determination of the active center of this hormone and identification of the structural requirements for the different biological activities of the molecule.

*Personal communication from Dr. R. S. Yalow, Veterans Administration Hospital, Bronx, N.Y.

Table I. Synthetic ACTH Fragments

Fragment	Reference
ACTH-(1–10)-decapeptide	Hofmann and Yajima (1961)
	Otsuka et al. (1965)
	Li et al. (1964b)
	Schwyzer and Kappeler (1961)
ACTH-(1–11)-undecapeptide amide	Medzihradszky and Pongrácz (1973)
ACTH-(1–12)-dodecapeptide amide	Medzihradszky and Pongrácz (1973)
ACTH-(1–13)-tridecapeptide amide	Hofmann and Yajima (1961)
	Guttmann and Boissonnas (1961)
ACTH-(1–14)-tetradecapeptide amide	Bajusz and Medzihradszky (1967)
ACTH-(1–15)-pentadecapeptide amide	Bajusz and Medzihradszky (1967)
ACTH-(1–16)-hexadecapeptide	Hofmann et al. (1962a)
ACTH-(1–16)-hexadecapeptide methyl ester	Schwyzer et al. (1962)
ACTH-(1–16)-hexadecapeptide amide	Bajusz and Medzihradszky (1967)
ACTH-(1–17)-heptadecapeptide	Li et al. (1964b)
ACTH-(1–17)-heptadecapeptide amide	Ramachandran et al. (1965)
	Bajusz and Medzihradszky (1967)
ACTH-(1–18)-octadecapeptide	Otsuka et al. (1965)
ACTH-(1–18)-octadecapeptide amide	Otsuka et al. (1965)
	Ramachandran et al. (1965)
	Bajusz and Medzihradszky (1967)
ACTH-(1–19)-nonadecapeptide	Li et al. (1961)
	Hofmann et al. (1962d)
	Li et al. (1964a)
ACTH-(1–19)-nonadecapeptide amide	Ramachandran et al. (1965)
	Bajusz and Medzihradszky (1967)
ACTH-(1–20)-icosapeptide methyl ester	Boissonnas et al. (1956)
ACTH-(1–20)-icosapeptide amide	Hofmann et al. (1962c)
	Bajusz and Medzihradszky (1967)
ACTH-(1–21)-henicosapeptide amide	Geiger et al. (1964)
	Bajusz and Medzihradszky (1967)
ACTH-(1–22)-docosapeptide amide	Geiger et al. (1964)
ACTH-(1–23)-tricosapeptide	Hofmann et al. (1962b)
ACTH-(1–23)-tricosapeptide amide	Geiger et al. (1964)
	Gelger et al. (1969)
ACTH-(1–24)-tetracosapeptide	Schwyzer and Kappeler (1963)
α_{bp}-ACTH-(1–26)-hexacosapeptide	Ramachandran and Li (1965)
α_p-ACTH-(1–28)-octacosapeptide	Bruckner et al. (1962)
	Bruckner et al. (1966)
α_h-ACTH-(1–28)-octacosapeptide	Kisfaludy and Löw (1968)
α_h-ACTH-(1–32)-dotriacontapeptide	Kisfaludy and Löw (1968)

As mentioned previously, preparation of the N-terminal protected tetradecapeptide was undertaken partly in order to synthesize sequential homologs containing more than the terminal 14 amino acid residues. Therefore, a series of peptides, listed in Table II, was synthesized and combined with the N-terminal tetradecapeptide to produce the 1–15, 1–16, 1–17, 1–18, 1–19, 1–20, and 1–21 polypeptide amides (Bajusz and Medzihradszky, 1967). The amides were selected because they show a higher biological activity compared to the free carboxylic acids with the same amino acid sequence.

Table II. Structures of Amino Components for the Synthesis of ACTH Fragments

Product	Amino component
ACTH-(1–15)-pentadecapeptide amide	H-Lys(Boc)-NH$_2$
ACTH-(1–16)-hexadecapeptide amide	H-Lys(Boc)-Lys(Boc)-NH$_2$
ACTH-(1–17)-heptadecapeptide amide	H-Lys(Boc)-Lys(Boc)-Arg-NH$_2$
ACTH-(1–18)-octadecapeptide amide	H-Lys(Boc)-Lys(Boc)-Arg-Arg-NH$_2$
ACTH-(1–19)-nonadecapeptide amide	H-Lys(Boc)-Lys(Boc)-Arg-Arg-Pro-NH$_2$
ACTH-(1–20)-icosapeptide amide	H-Lys(Boc)-Lys(Boc)-Arg-Arg-Pro-Val-NH$_2$
ACTH-(1–21)-henicosapeptide amide	H-Lys(Boc)-Lys(Boc)-Arg-Arg-Pro-Val-Lys(Boc)-NH$_2$

Synthesis of the amino components recorded in Table II is outlined in Fig. 9. For the preparation of these peptide derivatives, Z-Lys(BOC)-Lys(BOC)-N$_2$H$_3$ served as the starting material; coupling with nitroarginine amide or with nitroarginyl-nitroarginine amide led to the synthesis of the corresponding tri- and tetra-peptides, the C-terminal moieties of the 1–17 heptadecapeptide and the 1–18 octadecapeptide, respectively. To obtain the 15–19 fragment, the 17–19 tripeptide containing the Arg-Arg-Pro sequence was prepared, followed by its acylation with Z-Lys(BOC)-Lys(BOC)-N$_3$. Activated ester derivatives of the resulting pentapeptide were suitable intermediates in the synthesis of the 15–20 and 15–21 protected peptides. The coupling of these derivatives with the N-terminal tetradecapeptide and the deblocking and purification procedures for obtaining the final product were similar to those mentioned earlier. As these sequence homologs are found in the nonspecific part of the molecule of the adrenocorticotropic hormone, they can be regarded as fragments of all the known corticotropins. Their biological activities will be discussed later.

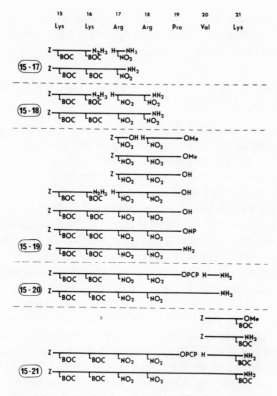

Fig. 9. Synthesis of amino components for the preparation
of corticotropin fragments.

For the synthesis of corticotropin fragments, which contain the species-specific sequence of the human hormone, we have prepared the 1–28 polypeptide according to the scheme employed in the synthesis of the porcine octacosapeptide, and subsequently the 1–32 fragment (Kisfaludy and Löw, 1968). In these syntheses, the amino acid sequence at 25–28 was changed from Asp-Gly-Ala to Asp-Ala-Gly.

For the preparation of the 1–32 fragment, the octapeptide N-hydroxy-succinimide ester containing the 20–27 amino acid sequence (used in the human ACTH synthesis) was employed for the acylation of the pentapeptide derivative H-Glu(OtBu)-Asp(OtBu)-Gln-Ser-Ala-OtBu (sequence 28–32). The latter compound was synthesized by the stepwise procedure. The remaining steps for the completion of the 1–28 and 1–32 syntheses were similar to those used in the preparation of the porcine octacosapeptide.

After the revision of the primary structure of the human ACTH, the abovementioned peptide fragments were prepared according to the new sequence as well, with Asn in place of Asp at position 25, Glu in place of Gln at position 30, and the change of Ala-Gly to Gly-Ala in the 26–27 sequence (Kisfaludy and Löw, 1973).

IV. BIOLOGICAL ACTIVITY OF THE SYNTHETIC PEPTIDE FRAGMENTS

In its direct or indirect biological effects, the adrenocorticotropic hormone is one of the most many-sided active substances in the living organism. An excellent compilation of these effects known until 1961 is given in Li (1962). The most characteristic properties of ACTH are undoubtedly the stimulatory effects on the adrenal cortex and the adipokinetic and melanocyte-stimulating activities. Biological evaluation is carried out by measuring stimulation of the adrenal cortex by one of the following tests: depletion of ascorbic acid *in vivo* (Sayers *et al.*, 1948) in hypophysectomized rats or animals with chemically blocked hypophysis function; measurement of the produced steroids *in vitro* by the Saffran and Schally (1955) technique, whose *in vivo* modification was worked out by Guillemin *et al.* (1959); and direct isolation of the steroids from adrenal venous blood, as proposed by Lipscomb and Nelson (1962).

The adrenal cortex stimulating activity of our synthetic products was determined mainly with the Sayers test* and their adipokinetic activity by the procedure described by White and Engel (1958).† Table III shows the biological activities of the synthesized peptides; corticotropic activity is expressed as International Units per milligram and adipokinetic activity in the minimal effective dose (micrograms per milliliter). Some of these biological data were published in Bajusz and Medzihradszky (1967), Medzihradszky and Bajusz (1968), and Szporny *et al.* (1968a,b).

V. CONCLUSIONS

On the basis of these data, the following conclusions can be drawn regarding the relationship between biological function and chain length of corticotropin fragments. The adrenal cortex stimulating activity seems to appear at the 1–13 and 1–14 oligopeptides, although a significant level of activity appears with the 1–15 and 1–16 peptide amides. From then on, the activity increases gradually, reaching the value possessed by the intact corticotropin molecule at the nonadecapeptide amide level when it is

---

*These measurements were kindly performed in the Pharmacological Laboratory of the Chemical Works of Gedeon Richter (leader Dr. L. Szporny).

†Dr. Gy. Cseh cooperated in this determination (Research Institute for Pharmaceutical Chemistry).

Table III. Biological Activity of Synthetic Corticotropins and Corticotropin Fragments

	Steroidogenesis (IU/mg)	Lipid mobilization, minimum effective dose, (μg/ml)
ACTH-(1–15)-pentadecapeptide amide	0.2–0.5	0.15
ACTH-(1–16)-hexadecapeptide amide	1.4	0.027
ACTH-(1–17)-heptadecapeptide amide	10.0	0.0033
ACTH-(1–18)-octadecapeptide amide	27.5	0.0038
ACTH-(1–19)-nonadecapeptide amide	~100	0.0060
ACTH-(1–20)-icosapeptide amide	~100	0.0027
ACTH-(1–21)-henicosapeptide amide	~100	0.0055
$\alpha_{..p..}$-ACTH-(1–28)-octacosapeptide	107–147	0.0020
$\alpha_{..h..}$-ACTH-(1–28)-octacosapeptide	69–115	—
$\alpha_{..h..}$-ACTH-(1–32)-dotriacontapeptide	70–129	—
α_h-ACTH-(1–32)-dotriacontapeptide	130	—
$\alpha_{..h..}$-ACTH	78–153	0.0057
α_h-ACTH	130	—

expressed in units per milligram. The role of the basic amino acids (sequence 15–18) in increasing the biological activity of the ACTH fragments is clearly seen, and this activity–basicity relationship is also reflected in the higher potency of the amides compared to the corresponding derivatives having a free carboxyl terminus. Since small but not negligible activity is shown by the shorter N-terminal sequences, this part of the molecule can be regarded as the active center. There are suggestions that the basic core serves as the binding site to the hormone receptor (Schwyzer, 1968). It must be emphasized that reliable biological evaluations can be best made when the same laboratory under standard conditions carries out the comparisons of the various substances. This has been done in the investigations carried out, for example, by Ney et al. (1964) and in the collected results given in Table III.

The adipokinetic activity of the ACTH fragments also varies greatly with the length of the peptide chain. Activity appears in the N-terminal tetradecapeptide sequence and increases as the length of the peptide chain increases. The maximal activity is reached at the heptadecapeptide amide level, and it is not changed by further lengthening of the molecule at the carboxyl terminus.

It is interesting to note that the octacosapeptide molecule is more active than the intact hormone in a biological effect that can be both adrenal and extra-adrenal in nature: the increase of blood flow in the adrenal cortex and in the ovaries (Stark et al., 1970). In this function, even the 1–14 fragment possesses significant activity. Shortening of the peptide sequence causes a separation of the adrenal and extra-adrenal effects.

Besides studying the relationship between chemical structure and biological activity, our laboratory is now involved with problems related to the mechanism of action of the corticotropin. In the course of these investigations, and applying slightly modified synthetic routes, we have prepared the ^{14}C- and tritium labeled 1–10 peptide fragment (Medzihradszky et al., 1970, 1973b) and furthermore the tritium- and ^{14}C-labeled N-terminal heptadecapeptides (Medzihradszky et al., 1973c). Synthesis of other labeled hormone fragments as well as their use in the study of the mechanism of action of ACTH is in progress.

ACKNOWLEDGMENT

All work reported in this chapter has been carried out cooperatively with or independently from the author's laboratory by research groups in the Research Institute for Pharmaceutical Chemistry, Budapest (Dr. S. Bajusz), and in the Chemical Works of Gedeon Richter, Budapest (Dr. L. Kisfaludy).

REFERENCES

Bajusz, S., and Lázár, T., 1966, Synthesis of peptides related to the C-terminal 25–39 sequences of corticotropins, Acta Chim. Acad. Sci. Hung. 48:111.

Bajusz, S., and Lénárd, K., 1962, Synthesis of the dodecapeptide sequence 10–21 of β-corticotropin, Coll. Czech. Chem. Commun. 27:2257.

Bajusz, S., and Medzihradszky, K., 1963, Erfahrungen über das Verhalten geschützter Peptide bei der Reaktion mit Natrium in flüssigem Ammoniak, in "Peptides: Proceedings of the Fifth European Peptide Symposium, Oxford, 1962" (G. T. Young, ed.), pp. 49–52, Pergamon Press, Oxford.

Bajusz, S., and Medzihradszky, K., 1967, Synthesis and biological activity of corticotropin fragments, in "Peptides: Proceedings of the Eighth European Peptide Symposium, Noordwijk, 1966" (H. C. Beyerman, A. van de Linde, and W. Maassen van den Brink, eds.), pp. 209–213, North-Holland, Amsterdam.

Bajusz, S., Lénárd, K., Kisfaludy, L., Medzihradszky, K., and Bruckner, V., 1962, Synthese eines Dodekapeptid-Derivats für den Aufbau corticotrop wirksamer Polypeptide, Acta Chim. Acad. Sci. Hung. 30:239.

Bajusz, S., Medzihradszky, K., Paulay, Z., and Láng, Zs., 1967, Totalsynthese des menschlichen Corticotropins (α_h-ACTH), Acta Chim. Acad. Sci. Hung. 52:335.

Bajusz, S., Paulay, Z., Láng, Zs., Medzihradszky, K., Kisfaludy, L., and Löw, M., 1968, Synthesis and biological properties of human corticotropin and its fragments, in "Peptides: Proceedings of the Ninth European Peptide Symposium, Paris, 1968" (E. Bricas, ed.), pp. 237–242, North-Holland, Amsterdam.

Bajusz, S., Kisfaludy, L., Medzihradszky, K., Paulay, Z., and Löw, M., 1973a, Synthesis of the N-terminal octacosapeptide of porcine corticotropin, Acta Chim. Acad. Sci. Hung. (in press).

Bajusz, S., Kisfaludy, L., and Medzihradszky, K., 1973b, Synthesis of human ACTH with revised structure, In preparation.

Bayer, J., Dualszky, S., and Kisfaludy, L., 1961, Decomposition of amino acid esters by treatment with hydrobromic acid in acetic acid, *J. Chromatog.* **6**:155.

Bell, P. H., 1954, Purification and structure of β-corticotropin, *J. Am. Chem. Soc.* **76**:5565.

Bell, P. H., Howard, K. S., Shepherd, R. G., Finn, B. M., and Meisenhelder, J. H., 1956, Studies with corticotropin. II. Pepsin degradation of β-corticotropin, *J. Am. Chem. Soc.* **78**:5059.

Boissonnas, R. A., Guttmann, St., Waller, J.-P., and Jaquenoud, P.-A., 1956, Synthesis of a polypeptide with ACTH-like activity, *Experientia* **12**:446.

Bruckner, V., Medzihradszky, K., Bajusz, S., and Kisfaludy, L., 1962, Biologically active polypeptides, Hung. Pat. 151,214.

Bruckner, V., Bajusz, S., Kisfaludy, L., Medzihradszky, K., Löw, M., Paulay, Z., Szporny, L., and Hajós, Gy., 1966, Biologically active polypeptides, Hung. Pat. 155,880.

Geiger, R., Sturm, K., and Siedel, W., 1964, Synthese eines biologisch aktiven Tricosapeptidamids mit der Aminosäure-sequenz 1–23 des Corticotropins (ACTH), *Chem. Ber.* **97**:1207.

Geiger, R., Schröder, H.-G., and Siedel, W., 1969, Synthetische Analoga des Corticotropins, *Liebigs Ann. Chem..* **726**:177.

Gráf, L., Bajusz, S., Patthy, A., Barát, E., and Cseh, G., 1971, Revised amide location for porcine and human adrenocorticotropic hormone, *Acta Biochim. Biophys. Acad. Sci. Hung.* **6**:415.

Guillemin, R., Clayton, G. W., Lipscomb, H. S., and Smith, J. D., 1959, Fluorometric measurement of rat plasma and adrenal corticosterone concentration, *J. Lab. Clin. Med.* **53**:830.

Guttmann, St., and Boissonnas, R. A., 1961, Influence of the structure of the N-terminal extremity of α-MSH on the melanophore stimulating activity of this hormone, *Experientia* **17**:265.

Hofmann, K., and Yajima, H., 1961, Studies on polypeptides. XX. Synthesis and corticotropic activity of a peptide amide corresponding to the N-terminal tridecapeptide sequence of the corticotropins, *J. Am. Chem. Soc.* **83**:2289.

Hofmann, K., Jöhl, A., Furlenmeier, A. E., and Kappeler, H., 1957, Studies on polypeptides. VIII. Synthesis of peptides related to corticotropin, *J. Am. Chem. Soc.* **79**:1636.

Hofmann, K., Stutz, E., Spühler, G., Yajima, H., and Schwartz, E. T., 1960, Studies on polypeptides. XVI. The preparation of N^ε-formyl-L-lysine and its application to the synthesis of peptides, *J. Am. Chem. Soc.* **82**:3727.

Hofmann, K., Yanaihara, N., Lande, S., and Yajima, H., 1962a, Studies on polypeptides. XXIII. Synthesis and biological activity of a hexadecapeptide corresponding to the N-terminal sequence of the corticotropins, *J. Am. Chem. Soc.* **84**:4470.

Hofmann, K., Yajima, H., Liu, T.-Y., and Yanaihara, N., 1962b, Studies on polypeptides. XXIV. Synthesis and biological evaluation of a tricosapeptide possessing essentially the full biological activity of ACTH, *J. Am. Chem. Soc.* **84**:4475.

Hofmann, K., Yajima, H., Liu, T.-Y., Yanaihara, N., Yanaihara, C., and Humes, J. L., 1962c, Studies on polypeptides. XXV. The adrenocorticotropic potency of an eicosapeptide amide corresponding to the N-terminal portion of the ACTH molecule; contribution to the relation between peptide chain-length and biological activity, *J. Am. Chem. Soc.* **84**:4481.

Hofmann, K., Liu, T.-Y., Yajima, H., Yanaihara, N., Yanaihara, C., and Humes, J. L., 1962d, Studies on polypeptides. XXII. High adrenocorticotropic activity in the rat and in man of a synthetic eicosapeptide amide, *J. Am. Chem. Soc.* **84**:1054.

Iselin, B., and Schwyzer, R., 1961, Synthese von Peptid-Zwischenprodukten für den Aufbau eines corticotrop wirksamen Nonadecapeptids. II. Derivate des L-Seryl-L-tyrosyl-L-serins, *Helv. Chim. Acta* **44**:169.

Kisfaludy, L., and Dualszky, S., 1960, p-Chlorocarbobenzoxy-Aminosäuren und -Peptide. I, *Acta Chim. Acad. Sci. Hung.* **24**:301.

Kisfaludy, L., and Dualszky, S., 1962, Synthese der Heptapeptidsequenz 22-28 des β-Corticotropins, *Coll. Czech. Chem. Commun.* **27**:2258.

Kisfaludy, L., and Löw, M., 1968, Synthesis of the 1-28 and 1-32 fragments of human corticotropin, *Acta Chim. Acad. Sci. Hung.* **58**:231.

Kisfaludy, L., and Löw, M., 1973, Synthesis of the 1-31 and 1-32 fragments of human ACTH with revised structure, In preparation.

Kisfaludy, L., Dualszky, S., Medzihradszky, K., Bajusz, S., and Bruckner, V., 1962, Synthese eines Heptapeptid-Derivats für den Aufbau corticotrop wirksamer Polypeptide, *Acta Chim. Acad. Sci. Hung.* **30**:473.

Kovács, J., Kisfaludy, L., and Ceprini, M. Q., 1967, On the optical purity of peptide active esters prepared by N,N'-dicyclohexylcarbodiimide and complexes of N,N'-dicyclohexylcarbodiimide-pentachlorophenol and N,N'-dicyclohexylcarbodiimide-pentafluorophenol, *J. Am. Chem. Soc.* **89**:183.

Kovács, K., László, F. A., Durszt, F., Szijj, I., Faredin, I., Tóth, I., Czakó, L., Biró, A., and Julesz, M., 1968, Effects of synthetic human corticotropin, *Lancet* **1968**:698.

Lee, T. H., Lerner, A. B., and Buettner-Janusch, V., 1961, On the structure of human corticotropin (adrenocorticotropic hormone), *J. Biol. Chem.* **236**:2970.

Li, C. H., 1962, Synthesis and biological properties of adrenocorticotropin peptides, *in* "Recent Progress of Hormone Research" (G. Pincus, ed.), pp. 1-40, Academic Press, New York.

Li, C. H., Geschwind, I. I., Cole, R. D., Raacke, I. D., Harris, J. I., and Dixon, J. S., 1955, Amino acid sequence of alpha-corticotropin, *Nature* **176**:687.

Li, C. H., Dixon, J. S., and Chung, D., 1958, The structure of bovine corticotropin, *J. Am. Chem. Soc.* **80**:2587.

Li, C. H., Meienhofer, J., Schnabel, E., Chung, D., Lo, T.-B, and Ramachandran, J., 1961, Synthesis of a biologically active nonadecapeptide corresponding to the first nineteen amino acid residues of adrenocorticotropins, *J. Am. Chem. Soc.* **83**:4449.

Li, C. H., Chung, D., and Ramachandran, J., 1964a, A new synthesis of a biologically active nonadecapeptide corresponding to the first nineteen amino acid residues of adrenocorticotropins. *J. Am. Chem. Soc.* **86**:2715.

Li, C. H., Ramachandran, J., Chung, D., and Gorup, B., 1964b, Synthesis of a biologically active heptadecapeptide related to adrenocorticotropin, *J. Am. Chem. Soc.* **86**:2703.

Lipscomb, H. S., and Nelson, D. H., 1962, A sensitive biologic assay for adrenocorticotropin (ACTH), *Endocrinology* **71**:13.

Löw, M., and Kisfaludy, L., 1965, Some observations with N-hydroxysuccinimide esters, *Acta Chim. Acad. Sci. Hung.* **44**:61.

Medzihradszky, K., and Bajusz, S., 1968, Synthesis of biologically active ACTH fragments (in Hungarian), *Kémiai Közl.* **29**:379.

Medzihradszky, K., and Medzihradszky-Schweiger, H., 1965, Über die katalytische Hydrierung schwefelhaltiger Peptid-derivate, *Acta Chim. Acad. Sci. Hung.* **44**:15.

Medzihradszky, K., and Pongrácz, K., 1973, Synthesis of an undecapeptide amide and a dodecapeptide amide with the N-terminal sequence of corticotropins, *Acta Chim. Acad. Sci. Hung.* (submitted for publication).

Medzihradszky, K., Kajtár, M., and Löw, M., 1962a, Synthese des Nonapeptids der Sequenz 1-9 des β-Corticotropins, *Coll. Czech. Chem. Commun.* **27**:2256.

Medzihradszky, K., Bruckner, V., Kajtár, M., Löw, M., Bajusz, S., and Kisfaludy, L., 1962b, Synthese eines Nonapeptid-Derivats für den Aufbau corticotrop wirksamer Polypeptide, *Acta Chim. Acad. Sci. Hung.* **30**:105.

Medzihradszky, K., Bajusz, S., and Kisfaludy, L., 1967a, Total synthesis of the human corticotropin and of polypeptides with adrenocorticotrop activity (in Hungarian), *Kémiai Közl.* **28**:219.

Medzihradszky, K., Kótai, A., Kajtár, M., Szókán, Gy., and Vajda, T., 1967b, Beiträge zur Herstellung einiger tert.-Butyloxycarbonyl-Aminosäurederivate, Annal. Univ. Rol. Eötvös Nom. Sect. Chim. 9:71.

Medzihradszky, K., S.-Vargha, H., Fittkau, S., and Marquardt, I., 1970, Labelled polypeptides, I. Synthesis of the N-terminal decapeptide of ACTH labelled with ^{14}C on the glycine residue, Acta Chim. Acad. Sci. Hung. 65:449.

Medzihradszky, K., Bruckner, V., Bajusz, S., Kisfaludy, L., Paulay, Z., and Löw, M., 1973a, Synthesis of the N-terminal tetradecapeptide of corticotropins, Acta Chim. Acad. Sci. Hung. (in press).

Medzihradszky, K., Nikolics, K., and Seprödi, J., 1973b, Labelled polypeptides. II. Synthesis of the N-terminal decapeptide of ACTH labelled with tritium on the tyrosine residue, Ann. Univ.Rol. Eötvös Nom. Sect. Chim. (in press).

Medzihradszky, K., S.-Vargha, H., Fittkau, S., and Teplán, I., 1973c, Synthesis of polypeptide hormones labelled with carbon 14 and tritium, in "Peptides: Proceedings of the Eleventh European Peptide Symposium, Vienna, 1971" (H. Nesvadba, ed.), pp. 306–309, North-Holland, Amsterdam.

Medzihradszky-Schweiger, H., and Medzihradszky, K., 1966, Über die hydrogenolytische Abspaltung der Carbobenzoxy-Schutzgruppe von Methioninpeptiden, Acta Chim. Acad. Sci. Hung. 50:339.

Ney, R. L., Ogata, E., Shimizu, N., Nicholson, W. E., and Liddle, G. W., 1964, Structure–function relationships of ACTH and MSH analogues, in "Proceedings of the Second International Congress of Endocrinology, London," p. 1184.

Otsuka, H., Inouye, K., Shinozaki, F., and Kanayama, M., 1965, Synthesis of an octadecapeptide and its 18-amide analog corresponding to the first eighteen amino acid residues of corticotropin, and their biological activities, J. Biochem. (Tokyo) 58:512.

Ramachandran, J., and Li, C. H., 1965, Adrenocorticotropins. XXXIII. Synthesis of a biologically active hexacosapeptide corresponding to the first 25 residues of bovine ACTH, J. Am. Chem. Soc. 87:2691.

Ramachandran, J., Chung, D., and Li, C. H., 1965, Adrenocorticotropins. XXXIV. Aspects of structure–activity relationships of the ACTH-molecule. Synthesis of a heptadecapeptide amide, an octadecapeptide amide, and a nonadecapeptide amide possessing high biological activities, J. Am. Chem. Soc. 87:2696.

Riniker, B., Sieber, P., Rittel, W., and Zuber, H., 1972, Revised amino acid sequences for porcine and human adrenocorticotropic hormone, Nature New Biol. 235:115.

Saffran, M., and Schally, A. V., 1955, In vitro bioassay of corticotropin: Modification and statistical treatment, Endocrinology 56:523.

Sayers, M. A., Sayers, G., and Woodbury, L. A., 1948, The assay of adrenocorticotropic hormone by the adrenal ascorbic acid depletion method, Endocrinology 42:379.

Schwyzer, R., 1968, Hormones with polypeptide structure, J. Mondial Pharm. 3:254.

Schwyzer, R., and Kappeler, H., 1961, Synthese von Zwischenprodukten für den Aufbau corticotrop wirksamer Polypeptide. III. Das Decapeptid H-Ser-Tyr-Ser-Met-Glu-His-Phe-Arg-Try-Gly-OH und einige seiner Derivate, Helv. Chim. Acta 44:1991.

Schwyzer, R., and Kappeler, H., 1963, Synthese eines Tetracosapeptids mit hoher corticotroper Wirksamkeit: β^{1-24}-Corticotropin, Helv. Chim. Acta 46:1550.

Schwyzer, R., Rittel, W., and Costopanagiotis, A., 1962, β^{1-16}-Corticotropin-methylester, Helv. Chim. Acta 45:2473.

Shepherd, R. G., Howard, K. S., Bell, P. H., Cacciola, A. R., Child, R. G., Davies, M. C., English, J. P., Finn, B. M., Meisenhelder, J. H., Moyer, A. W., and van der Scheer, J., 1956a, Studies with corticotropin. I. Isolation, purification and properties of β-corticotropin, J. Am. Chem. Soc. 78:5051.

Shepherd, R. G., Willson, S. D., Howard, K. S., Bell, P. H., Davies, D. S., Davis, S. B., Eigner, E. A., and Shakespeare, N. E., 1956b, Studies with corticotropin. III. Determination of the structure of β-corticotropin and its active degradation products, *J. Am. Chem. Soc.* **78**:5067.

Stark, E., Varga, B., Medzihradszky, K., Bajusz, S., and Hajtman, B., 1970, Relationship between the structure and the adrenal and extra-adrenal effects of ACTH fragments, *Acta Physiol. Acad. Sci. Hung.* **38**:193.

Szporny, L., Hajós, Gy. T., Szeberényi, Sz., and Fekete, Gy., 1968a, Pharmacology of synthetic ACTH peptides of different chain lengths, *in* "Pharmacology of Hormonal Polypeptides and Proteins, Proceedings of the International Symposium, Milan, 1967" (N. Back, L. Martini, and R. Paoletti, eds.), pp. 196–202, Plenum Press, New York.

Szporny, L., Hajós, Gy. T., Szeberényi, Sz., and Fekete, Gy., 1968b, Biological activity of synthetic 1–39 human corticotropin, *Nature* **218**:1109.

White, J. E., and Engel, F. L., 1958, Lipolytic action of corticotropin on rat adipose tissue *in vitro*, *J. Clin. Invest.* **37**:1556.

White, W. F., and Landmann, W. A., 1955, Studies on adrenocorticotropin. XI. A preliminary comparison of corticotropin-A with β-corticotropin, *J. Am. Chem. Soc.* **77**:1711.

CHAPTER 13

SYNTHESIS OF PURE POLYPEPTIDE HORMONES WITH FULL BIOLOGICAL ACTIVITY

Erich Wünsch

Max-Planck-Institut für Biochemie
Abteilung für Peptidchemie
München, Germany

I. INTRODUCTION

In the synthesis and isolation of naturally occurring macromolecules, especially polypeptides, the two major methods of purification employed in organic synthesis, namely, recrystallization and fractional distillation, are usually not applicable. Only in the so-called lower region of smaller peptides or in a very few special cases is it possible to obtain pure materials by crystallization; in the so-called upper region of larger peptides, crystallization is successful only if preceded by other methods of purification. The present situation of research into the purification of polypeptide hormones can be comprehensively considered using as examples the total syntheses of the hormones glucagon, secretin, and [leucine-15]-human gastrin I, an analogue of the natural product.

II. GLUCAGON

A. Pathway of Synthesis

After several years of work, the first synthesis of glucagon was successfully carried out by my research group (Wünsch, 1967, 1968; Wünsch and Wendlberger, 1968a) by fragment condensation of the five synthetic fragments

> H-Met-Asn-Thr(tBu)-OtBu (Wünsch *et al.*, 1965)
>
> NPS-Phe-Val-Gln-Trp-Leu-OH (Wünsch and Drees, 1967)
>
> NPS-Ser(tBu)-Arg(HBr)-Arg-Ala-Gln-Asp(OtBu)-OH (Wünsch and Wendlberger, 1968*b*)
>
> NPS-Thr(tBu)-Ser(tBu)-Asp(OtBu)-Tyr(tBu)-Ser(tBu)-Lys(BOC)-Tyr(tBu)-Leu-Asp(OtBu)-OH (Wünsch *et al.*, 1968*a*)
>
> AdOC-His(AdOC)-Ser(tBu)-Gln-Gly-Thr(tBu)-Phe-OH (Wünsch *et al.*, 1968*b*)

according to the scheme shown in Fig. 1.

Without discussing details of the synthesis, I would like to mention that all polyfunctional amino acid residues were applied with masked third functions: the hydroxyamino acids as *tert*-butyl ethers, the aminodicarboxylic acids as *ω-tert*-butyl esters, lysine as *ε-tert*-butyloxycarbonyl derivative, arginine in its guanido protonated form, histidine as N^{im}-adamantyloxycarbonyl derivative and the *C*-terminal amino acid as *α-tert*-butyl ester. In

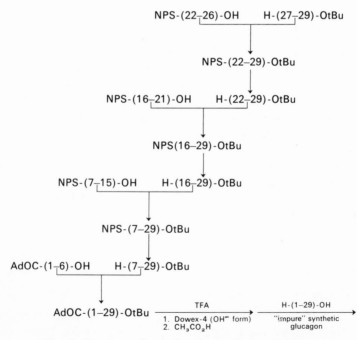

Fig. 1. Reaction scheme for the synthesis of glucagon.

other words, only those protective groups were used which can finally be removed simultaneously by the trifluoroacetic acid treatment.

B. Purification (Wünsch et al., 1968c)

The impure glucagon obtained by the aforementioned method could easily be separated into two major fractions, A and B, by gel filtration on Sephadex G-50 (Fig. 2). Fraction B proved to be the desired nonacosapeptide with the characteristics of the hormone, as could be seen by a thin layer chromatogram comparing it to the natural product. Under conditions employed for the natural hormone, the purified synthetic material crystallized in the shape of beautiful rhombic dodecahedra (Fig. 3). By this crystallization, a further purification was possible, representing one of the few cases in the chemistry of naturally occurring peptides where this can be done.

The decisive tests of purity, which at the same time represent proofs of identity between the natural product and the synthetic material, were obtained by the following analytical determinations:

1. Comparison of the amino acid analyses of "impure glucagon," fractions A and B, and natural glucagon with respect to a quantitative amino acid determination. The values for the synthetic and for the natural product were 85.5 ± 0.5 % within the range of error.
2. Comparison of the ultraviolet absorption curves of synthetic and natural glucagon. They proved to be almost identical to each other and different from the curve obtained with "impure glucagon."

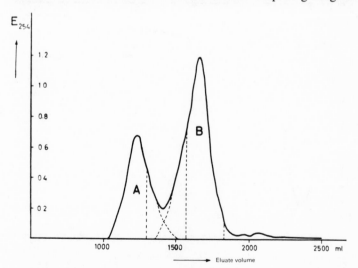

Fig. 2. Gel filtration of synthetic glucagon. Sephadex G-50 column; elution by 0.5 % acetic acid.

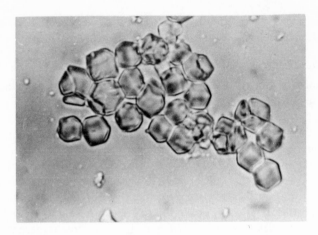

Fig. 3. Photomicrograph of synthetic crystalline glucagon.

3. Optical rotatory dispersion measurements of natural and synthetic glucagon. The course of the two curves (Fig. 4) shows almost complete identity, including the Cotton effect with a trough at 233 nm. This Cotton effect proves the existence of a partial α-helical structure, and it is characteristic for this peptide hormone.

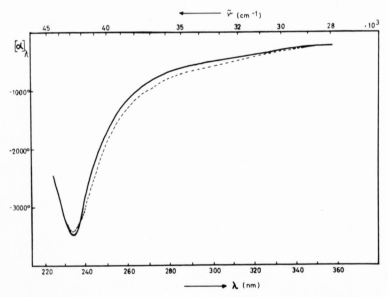

Fig. 4. Optical rotatory dispersion of natural crystalline glucagon (solid line) and synthetic crystalline glucagon (dashed line) ($c = 0.02$ in 0.02 % acetic acid).

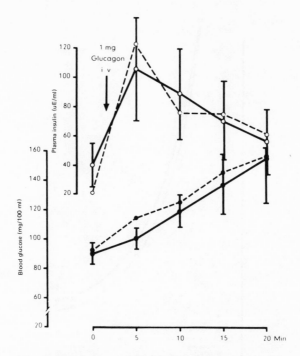

Fig. 5. Blood glucose and plasma insulin response to intravenous injection of 1 mg synthetic (dashed line) and natural (solid line) glucagon into human subjects.

4. Comparison of the biological activities of synthetic and natural peptides by measuring the levels of blood glucose and plasma insulin subsequent to intravenous injection (Fig. 5). These experiments, carried out by Weinges (1968) at the University of Homburg/Saar, showed that the synthetic material is somewhat more active than natural glucagon.

5. Comparison of the immunological activity of the synthetic and natural nonacosapeptides (Fig. 6), carried out by Weinges et al. (1969).

6. Behavior of synthetic and natural glucagon on polyacrylamide gel electrophoresis. It was found (Markussen, 1972) that pure natural crystalline Novo-glucagon shows a minor impurity, whereas the synthetic product contains a minor side-fraction which we have identified as fraction A.

The fraction A has the same amino acid composition as the synthetic hormone; it does, however, show double the molecular weight in the ultra-centrifuge compared to glucagon. Fraction A, tentatively called "glucagon

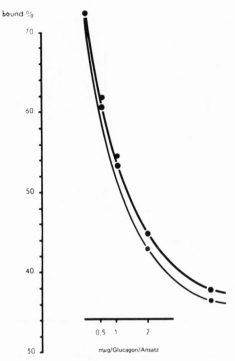

Fig. 6. Comparison of immunoreactivity of
synthetic and natural glucagon with anti–pork
glucagon serum from rats.

dimer," is biologically inactive but has the full immunological activity of
glucagon. This compound is possibly identical to fraction I of the "glucagon-
like-hormone," which has been isolated from the intestine (Valverde *et al.*,
1970). The "glucagon dimer" displays a characteristic pattern when subjected to enzymatic degradation: carboxypeptidase A degrades, from the
carboxylic end, in the correct manner until the tryptophan residue in position 25 is reached; the enzymatic attack by aminopeptidase M is suppressed
completely. This enzymatic behavior implies that tryptophan as well as
histidine or serine, its neighbor in the sequence, might be involved in reactions
which lead to formation of the dimer.

At the European Peptide Symposium in Paris in 1968 (Wünsch *et al.*,
1968*d*), colleagues of Ciba AG, Basel, pointed out that the demasking of
"*tert*-butyl-protected" natural polypeptides by trifluoroacetic acid–water
or trifluoroacetic acid–aqueous hydrochloric acid mixtures is more advantageous than using anhydrous trifluoroacetic acid. This method proved,
however, to be a disadvantage in the case of glucagon, since it leads

to the formation of a byproduct in relatively high yield. This byproduct was located between fractions A and B during gel filtration and rendered the isolation of pure fraction B more difficult. We believe that this fraction is a glutamine-desamidated glucagon and is probably an impurity of natural glucagon as well. Due to the presence of this material, however, it was possible to obtain pure fraction B if gel filtration was followed by an additional electrophoretic purification step.

III. [15-LEUCINE]-HUMAN GASTRIN I

A. Synthetic Route

The two polypeptide hormones gastrin I and II, which stimulate gastric secretion, were isolated by Gregory and Tracy (1964) and Gregory et al. (1964). A subsequent structural determination showed that both substances are linear polypeptides composed of 17 amino acid residues, two of which are methionine residues in positions 5 and 15. The only difference between the two peptides is that in gastrin II the phenylhydroxyl group of tyrosin is masked by esterification with sulfuric acid. Human gastrin I,* which was discovered somewhat later (Gregory et al., 1966), shows an amino acid substitution in position 5, where the methionine is replaced by leucine. Our British colleagues (Morley and Smith, 1968) found that a replacement of the second methionine in position 15 by leucine does not alter the biological activity of the molecule. This is of particular importance because oxidation of methionine at position 15 to methionine sulfoxide causes an almost complete loss of biological activity (Morley et al., 1965).

With the knowledge of these facts, my research group synthesized [15-leucine]-human gastrin, expecting to obtain a more stable compound than the natural gastrin. The construction of this heptacosapeptide amide (Wünsch and Deimer, 1972a) was carried out by fragment condensation of the three synthetic fragments (Wünsch and Deimer, 1972b)

H-Trp-Leu-Asp(OtBu)-Phe-NH$_2$

Z-Glu(OtBu)-Glu(OtBu)-Glu(OtBu)-Glu(OtBu)-Glu(OtBu)-Ala-
  Tyr(tBu)-Gly-OH

Pyr-Gly-Pro-Trp-Leu-OH

according to the reaction scheme shown in Fig. 7 using the racemization-free Wünsch–Weygand carbodiimide–hydroxysuccinimide technique. Again all polyfunctional amino acids were protected: the hydroxyamino acids as tert-butyl ethers, the aminodicarboxylic acids as tert-butyl esters.

---

*I will refer simply to "gastrin" throughout this paper, although, as in the case of the porcine hormone, there is also a human gastrin II.

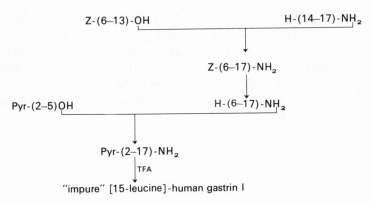

Fig. 7. Reaction scheme for the synthesis of [15-leucine]–human gastrin I.

B. Methods of Purification (Wünsch et al., 1972a)

The "impure gastrin" obtained after removal of the protective groups by means of trifluoroacetic acid was submitted to purification by gel filtration on Sephadex G-25, in analogy to the work of our British colleagues (Kenner et al., 1968), but without success. The use of Sephadex G-15 gave better results; the elution curve showed the presence of a fraction of unknown composition, of another fraction (mainly a "failure sequence" 1–13), and of a third fraction which contained several impurities besides [15-leucine]–human gastrin as became obvious by thin layer chromatography. A further purification of the third Sephadex fraction was possible by application of continuous carrier-free electrophoresis. Thin layer chromatography of the material thus obtained indicated that it was a mixture of two components. In fact, we could separate the product into two compounds by thin layer electrophoresis or, even better, by countercurrent distribution in the two-phase system sec-butanol–aqueous ammonia–formic acid.

Later on, we found that countercurrent distribution (Fig. 8) is sufficient for purification of [15-leucine]–human gastrin in one step: fraction A contains a chromatographically and electrophoretically pure heptadeca-peptide amide (Fig. 9) and is well separated from the byproduct, fraction B.

C. Biological and Antigenic Activity

Investigations by M. Grossman (private communication) with purified [15-leucine]–human gastrin have demonstrated that it has the full biological activity of natural [15-methionine]–human gastrin I. Similarly, an intra-venous infusion of 1 μg of [15-leucine]–human gastrin per kilogram body weight per hour causes a maximal acid output in humans, comparable to that of the natural hormone (Konz et al., 1971; Wünsch et al., 1971a). With

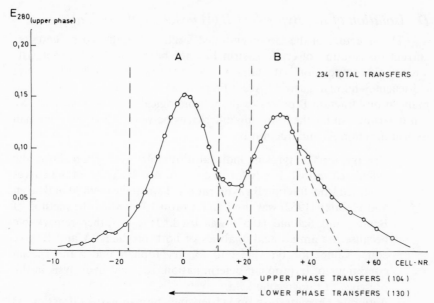

Fig. 8. Countercurrent distribution of crude [15-leucine]–human gastrin I in the system *sec*-butanol–0.075 % aqueous ammonia–formic acid (5:5:0.009, v/v), pH 5.

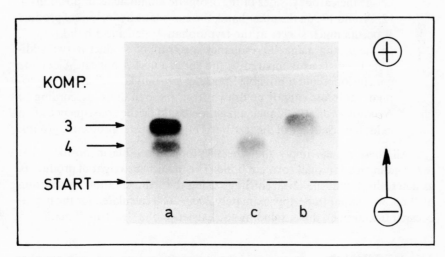

Fig. 9. Thin layer electrophoresis of [15-leucine]–human gastrin I in pyridine–formic acid–water (10:1:89, v/v), pH 6.5. (a) Crude product, (b) pure component A, (c) pure component B.

the same synthetic heptadecapeptide amide, Feurle *et al.* (1972) produced an excellent radioimmunoassay for gastrin. Iodinated [leucine-15]–human gastrin, necessary for this assay, was produced from the synthetic material.

D. Isolation of the Byproduct B (Wünsch et al., 1972a)

The structure of the aforementioned fraction B obtained by counter-current distribution of crude gastrin has not been elucidated as yet. The amino acid composition is the same as that of fraction A, namely, of the pure [15-leucine]–human gastrin. According to M. Grossman (private communication), fraction B possesses the full biological activity of gastrin. We can determine so far the following differences between [15-leucine]–human gastrin (fraction A) and fraction B:

1. The tryptophan–tyrosine induced ultraviolet absorption shows the calculated ratio 2:1 only for fraction A. A calculation of the curves obtained from fraction B gave a ratio 1:1 when the method of Beaven and Holiday (1952) was used and a ratio 1.7:1 when the method of Bencze and Schmid (1957) was used. This is rather remarkable because the amino acid analyses of both products, A and B, give equal values for tyrosine and for tryptophan under the special conditions of tryptophan determination, i.e., acid hydrolysis in the presence of thioglycolic acid.
2. Action of chymotrypsin on [15-leucine]–human gastrin (fraction A) results in a unequivocal cleavage at the positions to be expected, i.e., at the carboxyl sides of the aromatic amino acids in positions 4, 12, 14, and 17; on the other hand, the enzymatic hydrolysis of product B occurs much slower at the tryptophan-4—leucine-5 bond.
3. In the proton magnetic resonance spectrum of product B, two additional signals as compared to the spectrum of A appear. A definite conclusion about a possible "nucleus-*tert*-butylation" at the tryptophan molecule cannot be drawn from these findings. According to Alakhov *et al.* (1970), such a reaction could be the consequence of an acidolytic cleavage of the *tert*-butyl ether and ester protective groups.

All these results imply that the tryptophan residue in position 4 is involved in an additional covalent bond. The molecular weight of product B, as determined with the ultracentrifuge using the sedimentation equilibrium method, was found to be approximately 2000 (2080 calculated for the heptadecapeptide amide), thus excluding the existence of a "gastrin dimer."

IV. SECRETIN

A. Synthetic Route

The pancreas-stimulating heptacosapeptide amide secretin, which was first isolated by Jorpes and Mutt (1959, 1961), who also determined the primary structure, is very similar to glucagon in several regions of the sequence. Our synthesis of secretin was based mainly on the experiences which we had gained previously in the total synthesis of glucagon. Thus the synthesis of this polypeptide involved three major sections:

1. Preparation of six fragments suitable for further coupling, i.e.,

H-Arg(HBr)-Leu-Leu-Gln-Gly-Leu-Val-NH$_2$ Wünsch *et al.*, 1971*b*)

Z-Arg(Z, Z)-Leu-Gln-OH (Wünsch *et al.*, 1971*b*)

Z-Arg(Z, Z)-Asp(OtBu)-Ser(tBu)-Ala-OH (Wünsch *et al.*, 1971*c*)

Z-Arg(Z, Z)-Leu-OH (Wünsch *et al.*, 1971*c*)

Z-Thr(tBu)-Ser(tBu)-Glu(OtBu)-Leu-Ser-(tBu)-OH (Wünsch and Thamm, 1971)

AdOC-His(AdOC)-Ser(tBu)-Asp(OtBu)-Gly-Thr(tBu)-Phe-OH (Wünsch *et al.*, 1971*d*)

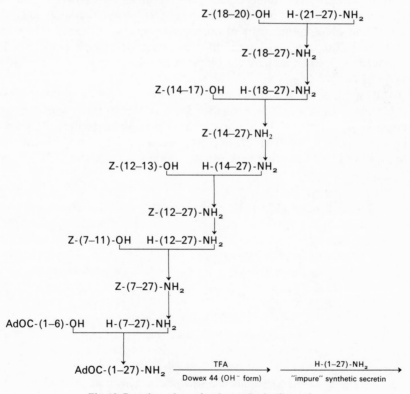

Fig. 10. Reaction scheme for the synthesis of secretin.

according to the reaction shown in Fig. 10. All side-chain functions of the trifunctional amino acids were masked by *tert*-butyl protective groups, which are easily cleaved by acidolysis; the complex function of arginine was protected either by N^ω,N^δ-bisacylation or by protonation (salt formation with hydrogen bromide).

2. Condensation of these fragments by the use of the racemization-free Wünsch–Weygand coupling procedure or the appropriate König–Geiger modification to yield the overall protected sequence 1–27 (Wendlberger *et al.*, 1971).

3. Purification of the synthetic hormone subsequent to the cleavage of all protective groups (Jaeger *et al.*, 1971; Wünsch *et al.*, 1972*b*).

B. Attempts to Purify the Product of a Preliminary Synthesis (*Jaeger et al., 1971, Wünsch et al., 1972b*)

In our first synthesis of "impure secretin," gel filtration on Sephadex G-50 was used as a method of purification as we had already done successfully in purification of "impure glucagon." As can be seen, however, from the elution curve (Fig. 11) as well as by thin layer chromatographic analysis, separation of all components of the mixture was not achieved. The main fraction B still contained four different impurities which could not be removed by repeated gel filtration on the same gel type or by use of Sephadex G-25 or G-100.

By application of a carrier-free electrophoretic method (Hannig, 1964), a further separation of the Sephadex fraction B into three new fractions could be accomplished. In contrast to the two major impurities, the desired heptacosapeptide amide (1–27) migrated much further toward the cathode in the electric field together with another unknown component.

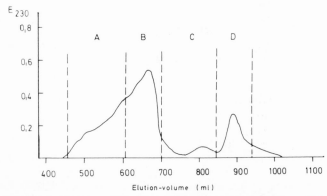

Fig. 11. Gel filtration of synthetic secretin (product of preliminary synthesis). Sephadex G-50 column; elution by 0.5 % acetic acid.

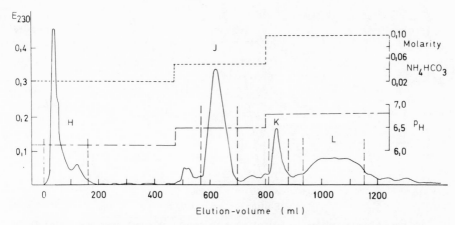

Fig. 12. Ion exchange chromatography of synthetic secretin (product of preliminary synthesis on SP-Sephadex C-25. Initial buffer: $0.02\,M$ NH_4HCO_3 solution adjusted to pH 6.1 with CO_2.

Only by means of a third method of purification, namely, ion exchange chromatography on the SP-Sephadex ion exchanger C-25 (Fig. 12), did it become possible to obtain the major component of the main electrophoretic fraction (fraction G) free of contaminants. The material isolated from the final fraction J (Fig. 12) was proved to be homogeneous by the thin layer chromatographic test as well as by polyacrylamide gel electrophoresis. Using a method described by Lehnert et al. (1969), the biological activity of this product was 75% in comparison to pure natural secretin. A comparison of the amino acid analyses of the main fractions B, G, and J showed increasing similarity to the theoretical values parallel to ongoing purification. An increase of the arginine values (and a simultaneous decrease of ornithine values) gave an especially clear indication of the successive removal of these side-products (they were found partially in the Sephadex fraction A), which gave ornithine values in the final analytical result due to a so far unknown change at the guanido function of one or several arginine residues.

During the course of our endeavor to detect possible failures of our secretin synthesis, we were trying to isolate several impurities in a rather pure form and to elucidate their structure at least partially. Especially the proof of existence of the two failure sequences 1–6/12–27 and 1–6/(7–11)$_2$/ 12–27 was possible. (The latter showed rather high ornithine values). This enabled us to draw conclusions that led to decisive improvements during a second analysis. This repeated synthesis was watched more carefully and was also changed at some steps. Our endeavor—the final synthetic route demonstrated in Fig. 10—was finally rewarded by success.

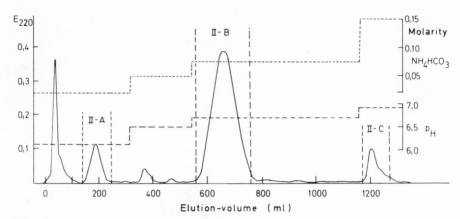

Fig. 13. Ion exchange chromatography of synthetic secretin (product of second synthesis).
Conditions as in Fig. 12.

C. Preparation of Pure Synthetic Secretin (*Jaeger et al., 1971; Wümsch et al., 1972b*)

Subsequent to the cleavage of all protective groups from the masked heptadecapeptide amide derivative resulting from the "corrected" synthesis, we could isolate an "impure material II," which displayed a biological activity of about 50% that of the natural hormone. A simple ion exchange chromatography of this material on SP-Sephadex C-25 (Fig. 13) led to the separation of two side-products and enabled us to obtain pure synthetic secretin (fraction II-B) in about 45% yield (see Table I for amino acid

Table I. Amino Acid Analysis of Synthetic Secretin

	Crude Secretin II	Fractions from ion exchange chromatography			Calculated
		II-A	Pure secretin	II-C	
His	0.72	0.07	0.98	0.43	1
Arg	3.93	3.81	3.98	3.91	4
Asp	1.80	1.17	1.99	1.47	2
Thr	1.78	1.10	1.93	1.40	2
Ser	3.84	3.24	4.04	3.45	4
Glu	3.02	3.00	2.99	2.96	3
Gly	1.87	1.20	2.00	1.59	2
Ala	1.01	1.00	1.00	1.00	1
Val	0.98	0.95	0.99	1.01	1
Leu	6.00	5.96	6.02	6.10	6
Phe	0.83	0.15	0.98	0.50	1

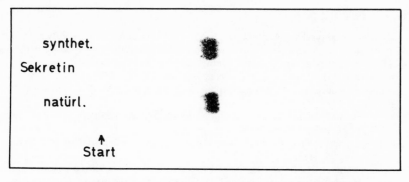

Fig. 14. Comparison of natural and synthetic secretin by thin layer chromatography on silica gel G; *n*-butanol–acetic acid–water–pyridine (60:6:24:20).

analysis). This synthetic product behaved like natural secretin on thin layer chromatography (Fig. 14) and disc electrophoresis; the biological activity of the synthetic material was found to be $100 \pm 4\%$ (Lehnert *et al.*, 1973) when compared with purest "Jorpes–Mutt secretin." A tryptic digest* of the synthetic heptacosapetide amide gave the characteristic pattern obtained when natural secretin was subjected to the same procedure (Mutt *et al.*, 1966).

V. CONCLUDING REMARKS

From the data presented in this chapter, some general conclusions may be drawn regarding the synthesis of polypeptides:

1. The planning of synthesis of long-chain polypeptides (with more than 15 amino acid residues) does not always give an absolute guarantee of success, even when it is performed with the most recent and modern scientific techniques. In most cases, experience gained in a "preliminary synthesis" is required to formulate the "specific" strategy which leads to a successful synthesis.
2. The isolation and structural elucidation of byproducts which appear during the course of purification can be of tremendous value. From such data conclusions may be drawn regarding the changes necessary for an improved synthesis. It may also be possible to gain information concerning the roles which single amino acid residues or certain amino acid sequences might play in biochemical reactions, such as biological or antigenic action mechanisms or transport mechanisms.

*We are very much indebted to Professor V. Mutt for carrying out these experiments.

REFERENCES

Alakhov, Yu. B., Kiryushkin, A. A., Lipkin, V. M., and Milne, G. W. A., 1970, Butylation of the tryptophan indole ring: A side reaction during the removal of *t*-butyloxycarbonyl and *t*-butyl protecting groups in peptide synthesis, *Chem. Commun.* **1970**:406.

Assan, R., Drouet, J., Rosselin, G., Wünsch, E., and Schröder, E., 1969, Étude radio-immunologique de glucagons naturel et synthétique et de peptides synthétiques apparentés, *Path. Biol.* **17**:757.

Beaven, G. H., and Holiday, E. R., 1952, Ultraviolet absorption spectra of proteins and amino acids, *Advan. Protein Chem.* **7**:319.

Bencze, W. L., and Schmid, K., 1957, Determination of tyrosine and tryptophan in proteins, *Anal. Chem.* **29**:1193.

Feurle, G., Ketterer, H., Becker, H. D., and Creutzfeldt, W., 1972, Circadian serum gastrin concentration in control persons and in patients with ulcer disease, *Gastroenterologie* **7**:177.

Gregory, R. A., and Tracy, H. J., 1964*b*, Constitution and properties of two gastrins extracted from hog antral mucosa, *Gut* **5**:103.

Gregory, R. A., Hardy, P. M., Jones, D. S., Kenner, G. W., and Sheppard, R. C., 1964, The antral hormone gastrin, *Nature* **204**:931.

Gregory, R. A., Tracy, H. J., and Grossman, M. I., 1966, Human gastrin: Isolation, structure and synthesis, *Nature* **209**:583.

Hannig, K., 1964, Eine Neuentwicklung der trägerfreien kontinuierlichen Elektrophorese, *Hoppe-Seyler's Z. Physiol. Chem.* **338**:211.

Jaeger, E., Deffner, M., Deimer, K.-H., Scharf, R., and Wünsch, E., 1971, Zur Reindarstellung der synthetischen Peptidwirkstoffe Sekretin, [15-Leucin]-Human-Gastrin I und Glucagon, *in* "Proceedings of the Eleventh European Peptide Symposium, Wien, 1971," North Holland Publ. Co. Amsterdam (1973), p. 237.

Jorpes, J. E., and Mutt, V., 1959, Secretin, pancreozymin, and cholecystokinin: Their preparation and properties, *Gastroenterologie* **36**:377.

Jorpes, J. E., and Mutt, V., 1961, On the biological activity and amino acid composition of secretin, *Acta Chem. Scand.* **15**:1790.

Kenner, G. W., Mendive, J. J., and Sheppard, R. C., 1968, Peptides. Part XXVI. Analogues of gastrin containing leucine in place of methionine, *J. Chem. Soc.* (*C*) **1968**:761.

Konz, B., Holle, F., Wünsch, E., Kissler, K., and Leimer, E., 1971, Magensekretion nach Stimulation mit [15-Leucin]-Human-Gastrin I, *Gastroenterologie* **9**:413.

Lehnert, P., Stahlheber, H., Forell, M. M., Dost, F. H., Fritz, H., Hutzel, M., and Werle, E., 1969, Bestimmung der Halbwertzeit von Sekretin, *Klin. Wschr.* **47**:1200.

Lehnert, P., Stahlheber, H., Roder, O., Zoelch, M., Forell, M. M., Wünsch, E., Jaeger, E., and Martens, H. L., 1973, Bestimmung der biologischen Aktivität von synthetischem Sekretin, *Klin. Wschr.* **51**:44.

Markussen, J., 1972, Private communication.

Morley, J. S., and Smith, J. M., 1968, Variations of the methionyl position in the *C*-terminal tetrapeptide amide sequence of the gastrins, *J. Chem. Soc.* (*C*) **1968**:726.

Morley, J. S., Tracy, H. J., and Gregory, R. A., 1965, Structure function relationships in the active *C*-terminal tetrapeptide sequence of gastrin, *Nature* **207**:1356.

Mutt, V., Magnusson, S., Jorpes, J. E., and Dahl, E., 1966, Structure of porcine secretin. I. Degradation with trypsin and thrombin. Sequence of the tryptic peptides. The *C*-terminal residue, *Biochemistry* **4**:2358.

Valverde, I., Rigopoulou, D., Marco, J., Faloona, G. R., and Unger, R. H., 1970, Molecular size of extractable glucagon and glucagon-like immunoreactivity (GLI) in plasma, *Diabetes* **19**:624.

Weinges, K. F., 1968, Untersuchungen der biologischen Aktivität und immunologischen Reaktionsfähigkeit von synthetischem Glucagon, in "Fourteenth Symp. Deutsch. Ges. Endokrinol.," p. 219, Springer-Verlag, Berlin, Heidelberg, New York.

Weinges, K. F., Wünsch, E., Biro, G., Kettl, H., and Mitzuno, M., 1969, The immunological reactivity and biological activity of synthetic glucagon, Diabetologia 5:97.

Wendlberger, G., Högel, A., Thamm, P., Spangenberg, R., and Wünsch, E., 1971, Eine neue Totalsynthese des Sekretins, in Proceedings of the Eleventh European Peptide Symposium, Wien, 1971, North Holland Publ. Co. Amsterdam (1973), p. 96.

Wünsch, E., 1967, Die Totalsynthese des Pankreas-Hormons Glucagon, Z. Naturforsch. 22b:1269.

Wünsch, E., 1968, Die Biochemie der Polypeptid-Naturstoffe: Totalsynthese des biologisch-wirksamen Glucagons, in "Fourteenth Symp. Deutsch. Ges. Endokrinol.," p. 206, Springer-Verlag, Berlin, Heidelberg, New York.

Wünsch, E., and Deimer, K.-H., 1972a, Zur Synthese des [15-Leucin]-Human-Gastrins I, II. Mitteilung. Erstellung der Gesamtsequenz, Hoppe-Seyler's Z. Physiol. Chem. 353:1255.

Wünsch, E., and Deimer, K.-H., 1972b, Zur Synthese des [15-Leucin]-Human-Gastrins I, I. Mitteilung. Erstellung der Teilsequenzen 1–5, 6–13 und 14–17, Hoppe-Seyler's Z. Physiol. Chem. 353:1246.

Wünsch, E., and Drees, F., 1967, Zur Synthese des Glucagons. XIII. Darstellung der Sequenz 22–29 (neuer Weg), Chem. Ber. 100:816.

Wünsch, E., and Thamm, P., 1971, Zur Synthese des Sekretins. III. Darstellung der Sequenz 7–11, Chem. Ber. 104:2454.

Wünsch, E., and Wendlberger, G., 1968a, Zur Synthese des Glucagons. XVIII. Darstellung der Gesamtsequenz, Chem. Ber. 101:3659.

Wünsch, E., and Wendlberger, G., 1968b, Zur Synthese des Glucagons. XVII. Darstellung der Sequenz 7–29, Chem. Ber. 101:341.

Wünsch, E., Drees, F., and Jentsch, J., 1965, Zur Synthese des Glucagons. VII. Darstellung der Sequenz 24–29, Chem. Ber. 98:803.

Wünsch, E., Zwick, A., and Fontana, A., 1968a, Zur Synthese des Glucagons. XV. Darstellung der Sequenzen 7–15 und 7–23, Chem. Ber. 101:326.

Wünsch, E., Zwick, A., and Jaeger, E., 1968b, Zur Synthese des Glucagons. XVI. Darstellung der Sequenz 1–6, Chem. Ber. 101:336.

Wünsch, E., Jaeger, E., and Scharf, R., 1968c, Zur Synthese des Glucagons. XIX. Reindarstellung des synthetischen Glucagons, Chem. Ber. 101:3664.

Wünsch, E., Wendlberger, G., Jaeger, E., and Scharf, R., 1968d, Zur Totalsynthese des biologisch-aktiven Pankreas-Hormons Glucagon, in "Peptides—1968: Proceedings of the Ninth European Peptide Symposium," p. 229, North-Holland, Amsterdam.

Wünsch, E., Konz, B., and Holle, F., 1971a, Gastric secretion in man after stimulation with a new human-gastrin I, in "Proceedings of the International Union of Physiological Sciences," Vol. IX, p. 610, Munich.

Wünsch, E., Wendlberger, G., and Högel, A., 1971b, Zur Synthese des Sekretins. I. Darstellung der Sequenz 18–27, Chem. Ber. 104:2430.

Wünsch, E., Wendlberger, G., and Thamm, P., 1971c, Zur Synthese des Sekretins. II. Darstellung der Sequenz 12–27, Chem. Ber. 104:2445.

Wünsch, E., Wendlberger, G., and Spangenberg, R., 1971d, Zur Synthese des Sekretins. IV. Darstellung der Sequenz 1–6, Chem. Ber. 104:3854.

Wünsch, E., Jaeger, E., Deffner, M., and Scharf, R., 1972a, Zur Synthese des [15-Leucin]-Human-Gastrins I. III. Mitteilung. Reindarstellung des synthetischen Hormons, Hoppe-Seyler's Z. Physiol. Chem. 353:1716.

Wünsch, E., Jaeger, E., Deffner, M., Scharf, R., and Lehnert, P., 1972b, Zur Synthese des Sekretins. VI. Reindarstellung des synthetischen Sekretins, Chem. Ber. 105:2515.

CHAPTER 14

NEUROHYPOPHYSEAL HORMONES: OLD AND NEW SLANTS ON THE RELATIONSHIP OF CHEMICAL STRUCTURE TO BIOLOGICAL ACTIVITY

Irving L. Schwartz

Department of Physiology and Biophysics
Mount Sinai School of Medicine of the City University of New York
and
Medical Research Center
Brookhaven National Laboratory
New York, N.Y.

Study of the relationship between the chemical structure and the biological activity of hormones has long been recognized as a valid approach to the elucidation of that initial chemical interplay between hormone and responsive cell which constitutes the primary step in hormone action. It would of course be desirable also to know the sequence and detailed nature of all other events leading to and including the final effector process for each of the final physiological effects of a given hormone on each of its target tissues. However, a thorough understanding of even the initial event in hormone action requires that structure–activity analysis be carried to the conformational level, and, indeed, work directed to this goal is under way (Walter *et al.*, 1971). Nevertheless, the approach to the study of peptide hormone–receptor interaction at the three-dimensional (topochemical) level of structure must be regarded as derivative from antecedent classical structure–activity studies—for it is the primary structure of both hormone and receptor that determines the conformation of the active hormone–receptor complex.

Classical structure-activity studies have entailed the synthesis of hormone analogs and, to a lesser extent, the chemical modification of purified natural hormones and the observation of the structural modifications that have occurred by natural selection during the course of evolution.

In considering the relationship between hormone structure and biological activity, it is important to remember that data obtained from *in vivo*
assays or from highly organized *in vitro* systems require evaluation, for each
hormone and analog of many factors (transport in the bloodstream, diffusion through cellular barriers to the receptor sites, inhibitory and degradation mechanisms, etc.) which have been detailed adequately elsewhere
(Rudinger and Jost, 1964*a*; Schwartz and Livingston, 1964). In addition, a
number of purely chemical considerations must be resolved before structure–
activity data can be useful in the analysis of hormone–receptor interactions.
For example, there is in all probability a multipoint fixation of hormone to
receptor, although the critical (catalytic) reaction which initiates the chain
of events culminating in the final biological effect may involve a limited
region on the hormone molecule. In a given analog, this region could be
intact but nonfunctional biologically because of inappropriate orientation
to its complementary region on the receptor. It is also possible for a given
analog to induce a change in the receptor at a point removed from the site
of catalytic interaction which is capable of increasing or decreasing the
potency of the analog (allosteric effect). Thus a simple single structural alteration in the hormone may produce mild, moderate, or extreme changes in
the conformation of the hormone and of the hormone–receptor complex.
In fact, the very concept of "a catalytic center" or "active site" is not applicable to either hormone or receptor if the formation of the hormone–
receptor complex involves an overall change in conformation on which
biological activity (signal generation) depends. Despite these considerations, a number of important structure–function correlations, if not principles,
have been established and have led to (1) the production of analogs which are
more potent than the natural (parent) hormones or in which therapeutically
valuable pharmacological properties have been selectively enhanced, (2) the
development of partial agonists and antagonists of the natural hormones,
(3) the analysis of the enzymatic degradation of oxytocin and vasopressins,
(4) advances in knowledge of genetic alteration of neurohypophyseal hormones in phylogeny, and (5) an analysis of hormone action by comparison
of the structure–activity aspects of early and late events in the sequence of
reactions initiated by hormone–receptor interactions and leading to the final
physiological effect (see below).

Approximately 50 years ago, the biological effects of mammalian posterior pituitary extracts were separated chemically (Dudley, 1923; Schlapp,
1925; Draper, 1927; Kamm *et al.*, 1928) into two fractions, one containing
predominantly pressor and antidiuretic activities and the other containing
predominantly oxytocic (uterotonic, avian depressor, and milk-ejecting)
activities. However, almost five decades were to pass before oxytocin was to
be isolated in highly purified form (Livermore and du Vigneaud, 1949),

synthesized (du Vigneaud *et al.*, 1953), and characterized conformationally (Urry and Walter, 1971; Walter, 1971). Despite this span of time, oxytocin has become the first peptide hormone to yield to "sequencing," synthesis, and elucidation of peptide backbone conformation in solution.

The primary structure of oxytocin consists of a ring containing cystine (Cys), tyrosine (Tyr), isoleucine (Ile), glutamine ($GluNH_2$), and asparagine ($AspNH_2$) residues with one half-cystine residue as N-terminus (and therefore possessing a free amino group) and the other half-cystine residue joined in peptide linkage through its amino group to the ring and through its carboxyl group to the carboxyl-terminal tripeptide sequence, prolyl-leucyl-glycinamide (Pro-Leu-Gly-NH_2). In the mid-1950s, it was shown that the vasopressins had the same essential structure (du Vigneaud, 1956a,b), and the first synthetic peptide hormone analogs ([Leu^3]-oxytocin, [Val^3]-oxytocin, and [Phe^3]-oxytocin) were prepared (Rudinger *et al.*, 1956; Boissonnas *et al.*, 1956; Katsoyannis, 1957). During the past 15 years, more than 300 additional oxytocin and vasopressin analogs have been prepared; consequently, the neurohypophyseal hormone series has become the most extensive material available for the analysis of peptide hormone structure–function relationships.

The relationship between covalent structure and biological activities of natural and synthetic neurohypophyseal peptides has been reviewed extensively elsewhere (Sawyer, 1961, 1965; Rudinger and Jost, 1964a; Schwartz and Livingston, 1964; Schwartz *et al.*, 1964a; Walter *et al.*, 1967; Berde and Boissonnas, 1968; Rudinger, 1968; Pickering, 1970). Therefore, we will consider there only some highlights of this relationship, emphasizing those structure–activity correlates which are concerned with the development of highly selective changes in the biological activity spectrum of the neurohypophyseal hormones and which may provide insight into the mechanism of action of these peptides (especially in relation to antidiuresis in mammals and hydro-osmotic effects in amphibian membranes).

The mammalian neurohypophyseal hormones—oxytocin and lysine- and arginine-vasopressin—differ only in the two amino acid residues in positions 3 and 8; the vasopressins differ from each other only in the amino acid residue in position 8. Arginine-vasopressin (the antidiuretic hormone of most mammals) has an arginine residue in position 8, whereas lysine-vasopressin (the antidiuretic hormone of pigs and a few related species) has a lysine residue in this position. Oxytocin, unlike either of the vasopressins, has an isoleucine residue in position 3 and a leucine residue in position 8. These differences served as the initial focus for much of the synthetic work following the characterization of the vasopressins and oxytocin and led to the felicitous synthesis of arginine-vasotocin (Katsoyannis and du Vigneaud, 1958), before it was recognized as a natural hormone.

This interesting historical sequence of synthesis followed by discovery in nature has been reproduced again in the case of mesotocin (Berde and Konzett, 1960; Jaquenoud and Boissonnas, 1961) and Pro-Leu-Gly(NH$_2$), the recently proposed structure for the factor inhibiting the release of melano-cyte-stimulating hormone (Celis *et al.*, 1971*a,b*).

It is known from studies of a variety of analogs—with single and multiple replacements—that the biological activities of neurohypophyseal hormones are influenced substantially by the presence (closure) and size of the pentapeptide ring, the presence of the free amino group at position 1, the nature (bulk, shape, polarity) of the side-chains of positions 2 and 3, the presence of the carboxamide moiety at position 5, the basicity of the side-chain at position 8, and the omission, substitution, or displacement of the glycinamide residue from position 9. The side-chains of the amino acid residues in positions 3, 4, 7, and 8 are free to engage in intermolecular interactions, but they affect the peptide backbone conformation only to the (limited) extent to which their bulk helps to stabilize the two β turns (Walter *et al.*, 1971). Thus these side-chains would appear to be the primary determinants of differential potency (affinity) in various neurohypophyseal hormone–receptor interactions.

The natural mammalian hormones provided the earliest structure–activity correlation in the neurohypophyseal peptide series, namely, the shift from a predominantly "oxytocic" (uterotonic, avian depressor, and milk-ejecting) to a predominantly antidiuretic–pressor spectrum of biological activities when isoleucine in position 3 and leucine in position 8 are replaced by phenylalanine and arginine (or lysine), respectively.

Uterotonic activity is quite sensitive to structural alterations on the side-chain of the isoleucine residue in position 3, the terminal methyl group of which appears to be involved in lipophilic binding to the uterine receptor (Rudinger and Krejci, 1962; Nesvadba *et al.*, 1963): this conclusion is supported by the fact that the activity of [3-*O*-methylthreonine]-oxytocin resembles that of [3-valine]-oxytocin (an analog in which the side-chain in position 3 is shortened) rather than the approximately isosteric parent hormone (Chimiak and Rudinger, 1965).

Recent evidence shows that the side-chain of position 3 also may influence the catalytic function of the hormone–receptor complex (Walter *et al.*, 1968, 1969), indicating that this side-chain may be concerned with intrinsic hormonal activity as well as with hormone–receptor affinity.

Deletion of nonfunctional groups, such as a single methylene moiety from the isoleucine side-chain in position 3 of oxytocin, may result in a profound reduction in potency, whereas more extensive deletions of non-functional groups at other loci, for example, in positions 7 and 8, may be associated with substantial retention of hormonal potency (Jaquenoud, 1965; Walter and du Vigneaud, 1966; Bespalova *et al.*, 1968).

Synthetic analogs with replacements at position 3 in which a methylene group occupies the same position as the terminal methyl group of isoleucine, *viz.*, [3-β-diethylalanine]-oxytocin and [3-1-cyclopentylglycine]-oxytocin (Eisler *et al.*, 1966) retain substantial uterotonic, avian depressor, and milk-ejecting potencies; however, numerous other 3-substituted analogs, e.g., those with alloisoleucine, leucine, norleucine, tyrosine, or phenylalanine (Kaysoyannis, 1957; Boissonnas and Guttman, 1960; Nesvadba *et al.*, 1963), all exhibit gross reduction in these "oxytocic" activities. Substitutions at position 8 are generally associated with less reduction in uterotonic and still less reduction in avian depressor and milk-ejecting activities (particularly in the presence of magnesium ions), suggesting that this position is less significant than position 3 for the reactions between hormone and uterine, avian arteriolar, and mammary receptors. It is the introduction at position 8 of a basic amino acid residue (arginine, lysine, or even an "unnatural" residue such as ornithine) which contributes mostly to the shift from the "oxytocic" to the antidiuretic–pressor spectrum. However, within the latter spectrum, phenylalanine at position 3 enhances antidiuretic activity more than pressor activity, and the substitution of lysine for arginine at position 8 reduces the antidiuretic and pressor potencies of the molecules by approximately one-half, except in pigs where lysine-vasopressin is the naturally occurring neurohypophyseal hormone. The finding that lysine-vasopressin is more active than arginine-vasopressin in one species and less active in others indicates that the basicity of the side-chain in position 8 can be overcome by other factors that contribute to antidiuretic (and pressor) potency. This phenomenon is also exemplified by (1) the relatively low antidiuretic activity of [8-ornithine]-vasopressin (Berde *et al.*, 1964) and [8-D-lysine]-vasopressin (Zaoral *et al.*, 1967a), (2) the high antidiuretic activity of [8-thialysine]-vasopressin (Hermann and Zaoral, 1965), (3) the relatively high antidiuretic activity of [8-citrulline]-vasopressin (Van Dyke *et al.*, 1963), and (4) the low mammalian pressor activities of analogs in which basic L-amino acid residues have been replaced by basic D-amino acid residues (Zaoral *et al.*, 1967a,b). Other exceptions to the principle which relates basicity to pressor–antidiuretic activity are exemplified at the level of the molecule as a whole—by the fact that *deletion* of the N-terminal amino group of the vasopressins is associated with *enhancement* of antidiuretic potency (Kimbrough *et al.*, 1963; Huguenin and Boissonnas, 1966).

The studies of various replacements in position 8 have yielded a number of analogs with ratios of pressor potency to antidiuretic potency that differ strikingly from that of the natural hormones (1.0 for arginine-vasopressin, 1.2 for lysine-vasopressin). The high selectivity of the pressor

receptor and the lower (and more variable) selectivity of the antidiuretic
receptor for L-amino acid residues at position 8 make it possible to develop
analogs with low pressor/antidiuretic ratios, e.g., 0.04 in the case of [8-D-
arginine]-vasopressin (Zaoral et al., 1967a,b). However, the specificity for the
pressor and antidiuretic receptors is reversed in the case of other analogs,
such as [8-ornithine]-vasopressin (Berde and Boissonnas, 1966, 1968) which
exhibits a pressor/antidiuretic ratio of 4. If, in addition, the phenolic
hydroxyl group in position 2 of this analog is replaced by hydrogen, the
modified analog [2-phenylalanine, 8-ornithine]-oxytocin, exhibits a further
dissociation of pressor and antidiuretic activities, the ratio having increased
to 10. These activities are still further differentiated if the phenylalanine
residue in position 3 is replaced by an isoleucine residue, the pressor/anti-
diuretic ratio of the resulting analog, [2-phenylalanine, 8-ornithine]-oxytocin,
having in this case increased to 220 (Berde et al., 1964).

The high antidiuretic potency of lysine-vasopressin in pigs has previously invited attention to the question of whether there has been a
genetically determined alteration in the structure of the porcine renal
receptor as well as in the porcine hormone (Schwartz et al., 1964b) or whether
the target cell has adapted to the mutant hormone by modification of an
"amplification" component of some step subsequent to hormone–receptor
interaction. We have recently explored this question by comparing the
effects of lysine-vasopressin (and its congeners) on the earliest measurable
event in neurohypophyseal hormone action (activation of adenylate cyclase)
and on a final effect (antidiuresis) in several mammalian species including
the pig. The results of this study (Dousa et al., 1970a, 1971) indicate that the
renal receptor for neurohypophyseal hormones in a particular species
exhibits the highest affinity for the specific antidiuretic hormone which
occurs naturally in that species. Accordingly, the target cell in the porcine
distal nephron appears to have preserved a homeostatically efficient antidiuretic response via an adaptive structure modification at the receptor level.

When lysine is substituted for arginine in position 8 of arginine-vasotocin, thus converting the natural hormone to the analog, lysine-vasotocin,
the antidiuretic and pressor activities of the molecule fall from 195 to 24
units/mg and from 130 to 24 units/mg, respectively (Sawyer, 1965; Berde
and Boissonnas, 1968); the corresponding specific antidiuretic and pressor
activities for arginine-vasopressin, lysine-vasopressin, and [8-leucine]-
vasopressin (the analog of oxytocin in which phenylalanine replaces isoleucine in position 3) are 400 and 400 units/mg, 165 and 270 units/mg, and
22 and 4.2 units/mg, respectively (Schwartz and Livingston, 1964; Sawyer,
1965; Pickering, 1970). Therefore, the single change in the oxytocin molecule
from leucine to lysine in position 8 has about the same effect on the renal
response to the hormone as the single change from isoleucine to phenylalanine in position 3, the antidiuretic activity increasing from 1 to 24 units/mg

in the former case and from 1 to 22 units/mg in the latter case. However, when lysine already occupies position 8, as it does in lysine-vasotocin, then the same single change in position 3 (isoleucine to phenylalanine) brings about an increase in antidiuretic activity from 24 to 165 units/mg. Thus it would appear that for lysine-vasopressin, as for arginine-vasopressin, it is necessary to have the participation of the side-chains in both positions 3 and 8 in order to achieve a physiologically effective hormone–receptor interaction—the requirement for the participation of position 3 in the hormone–receptor interaction being masked when leucine occupies position 8 instead of lysine. These findings indicate that the reaction of the antidiuretic receptor with lysine at position 8 brings about a change in the hormone, receptor, or both which then facilitates the reaction of the receptor with phenylalanine (but not isoleucine) at position 3. The possibility that folding of the hormone molecule has resulted in the juxtaposition of positions 3 and 8 to form a single topochemical locus seems unlikely because these positions appear to be substantially separated in the three-dimensional structure of the hormone; thus we have here an illustration of a biofunctional multipoint hormone–receptor interaction.

In rats, mice, cows, humans, and probably all other mammals in which AVP is the natural antidiuretic hormone, the decreased potency of the 8-lysine peptides relative to the 8-arginine peptides involves the duration as well as the intensity of the response despite the fact that both hormones have a similar circulating half-life (1 min). This difference in the antidiuretic potencies of arginine-vasopressin and lysine-vasopressin is reflected quantitatively at the adenyl cyclase step in the action of these hormones (Dousa et al., 1970a, 1971), a finding which, in conjunction with related studies (see below), shows potency differences among neurohypophyseal peptides to be determined by the initial events in hormone action (namely, the hormone–receptor interaction per se and/or its coupling to adenyl cyclase activation) and not by subsequent events in the chain of reactions that leads to the final homonal effect. A similar parallelism between an effect at an "early" step in hormone action (adenyl cyclase activation) and a final effector event (the hydro-osmotic response of a vasopressin-sensitive amphibian epithelial membrane) has been noted in the case of the action on the toad bladder of arginine-vasotocin, the natural amphibian hormone. Arginine-vasotocin is the most potent of all neurohypophyseal peptides in eliciting the characteristic hydro-osmotic response of the toad bladder, and it also has been found to be the most potent of all hormonal agents capable of activating cell-free adenyl cyclase preparations derived from the toad bladder epithelium (Kirchberger et al., 1972).

In order to approach the question of how and to what degree various parts of the oxytocin molecule contribute to biological activity, a number of laboratories, particularly that of du Vigneaud (1964), have synthesized analogs in which hydrogen replaces the terminal free amino group in position 1, the phenolic hydroxyl group in position 2, the carboxamide groups in positions 4, 5, and 9, and the disulfide bridge spanning positions 1 and 6. Pharmacological evaluation of such "deletion analogs" (see below) showed that, of these chemically functional groups, *only* the carboxamide group in position 5 can be considered essential for the biological activities of oxytocin but that *all* of these groups—with the exception of the terminal amino group —increase the affinity of the hormone for its receptor and thus contribute to hormonal potency. The presence of the terminal amino group in fact serves to diminish hormonal potency, as illustrated by the finding that the deamino analog of oxytocin has considerably greater uterotonic and avian depressor potencies than oxytocin itself and, similarly, that the deamino analogs of lysine- and, particularly, arginine-vasopressin have very much greater antidiuretic potency than the parent natural hormones—a generalization which does not include the mammalian pressor receptor systems, which are somewhat less responsive to deamino analogs than to the parent hormones (Walter *et al.*, 1967). The addition of a substituent to the terminal amino group of oxytocin, or the vasopressins, may lead to analogs with very low activity, e.g., 1-N-acetyloxytocin (Boissonnas *et al.*, 1961), 1-N-carbamyloxytocin (Smyth, 1967), 1-N-acetyl-[8-arginine]-vasopressin (Studer and Cash, 1963), or, if the added substituent is an amino acid or peptide, analogs (hormonogens) which in effect have protracted activity because they serve as pharmacological "precursors" or "reservoirs" from which native hormone is more or less gradually released by aminopeptidase action (Rychlik, 1964; Berankova-Ksandrova *et al.*, 1966). The "supranormal" activity of deamino-oxytocin, contrasted with the very low activity of amino-substituted analogs such as N-acetyloxytocin, shows that the free amino group does not facilitate and, in fact, operates to diminish binding of the native hormone to the receptor, and, furthermore, it reveals steric hindrance rather than altered chemical functionality as the basis for the ineffective hormone–receptor interaction of analogs with "blocking" substituents in position 1.

Deletions or substitutions in position 2 of oxytocin and vasopressins can lead to quantitative effects, i.e., changes in potency (affinity), and also to qualitative effects, i.e., changes in intrinsic activity (mode of induction of response following hormone–receptor interaction). Replacement of the tyrosyl hydroxyl group of oxytocin with hydrogen yields a 2-phenylalanine analog which acts as a partial agonist (depressed intrinsic activity); replacements at this locus with p-substituents of increasing size yields analogs (Rudinger and Krejci, 1968) which act as partial agonists or inhibitors, indicating that they can occupy the receptor in a *nonfunctional* complex which serves to prevent the parent hormone or related agonists from forming a *functional* complex. Accordingly, 2-O-methyltyrosine analogs of oxytocin and vasopressins act as partial agonists or antagonists in various hormone-responsive systems, and the inhibitory properties of these analogs can be accentuated by changes at other positions in the molecule, e.g., deletion of or substitution of a carbamyl moiety on the terminal amino group (Smyth, 1967, 1970; Rudinger, 1969). Such inhibitory phenomena have heuristic value and, in addition, may lead to important generalizations. Indeed, the recognition of the broad inhibitory spectrum of N-carbamyl-O-methyloxytocin—an analog which antagonizes the uterotonic, galactobolic, pressor, and antidiuretic activities of oxytocin and the vasopressins—has led to the suggestion that there may be a common denominator in the fundamental mechanisms by which the neurohypophyseal hormones initiate their varied biological effects (Rudinger, 1969).

We have already noted that the side-chain of the amino acid residue in position 3 can influence processes concerned with intrinsic hormonal activity (Walter et al., 1968) as well as in interactions concerned with hormone–receptor affinity (Rudinger and Krejci, 1962; Nesvadba et al., 1963; Rudinger, 1969) and, indeed, it has been shown that oxytocin and lysine-vasopressin exhibit a difference in intrinsic uterotonic activity which is referable to the difference in the amino acid residue present at position 3 (Walter et al., 1968, 1969).

A number of neurohypophyseal hormone analogs with replacements in position 4 were found to retain moderate to substantial levels of hormonal activity (Berde and Boissonnas, 1968; Pickering, 1970); others, notably a group of 4-threonine analogs of oxytocin have two- to three-fold higher "oxytocic" (uterotonic, avian depressor, milk-ejecting) potency and one-third to one-half of the antidiuretic and pressor potencies when compared to oxytocin itself (Manning et al., 1970). Still another 4-substituted analog, namely, [4-leucine]-oxytocin (Hruby et al., 1969; Chan and du Vigneaud, 1970); manifests inhibitory activity: it antagonizes the antidiuretic response to arginine-vasopressin (ADH) and thus is an anti-antidiuretic agent. This analog and its closely related congener [4-leucine]-mesotocin were independently found to inhibit competitively the hydro-osmotic action of ADH, but not theophylline, in the toad bladder (Chui and Sawyer, 1970; Eggena et al., 1970)—indicating that the antagonism takes place at the receptor level. In addition, the analog was found to have a potent natriuretic–diuretic effect in rats—during water diuresis as well as during antidiuresis—without changing the glomerular filtration rate (GFR). A related analog, [2,4-diisoleucine]-oxytocin (Hruby et al., 1970), also exhibits a potent natriuretic–diuretic effect without altering GFR but does not antagonize the antidiuretic action of ADH. The latter analog in high doses elicits a small increase in blood pressure in the rat, but [4-leucine]-oxytocin produces a transient decrease in blood pressure under the same conditions. Thus these analogs have made possible a series of studies which provided additional evidence for the specificity of antidiuretic, natriuretic, and vasopressor receptors for neurohypophyseal hormones. In addition, analogs such as [4-leucine]-oxytocin could prove to be useful drugs for treatment of clinical conditions associated with excess production of antidiuretic hormone.

[4-Asparagine, 8-lysine]-vasopressin (Zaoral, 1965; Berde and Boissonnas, 1968) has substantial antidiuretic activity, but the slope of its log dose–response curve differs from, and therefore crosses, that of the parent hormone—the analog being about eight times more potent than lysine-vasopressin at dose levels of 10^{-9} mg, three times as potent at dose levels of 10^{-8} mg, but less potent than lysine-vasopressin at higher dosage. It is perhaps not surprising to find that position 4, a site of evolutionary change in the neurohypophyseal hormones, should possess the potential for generating qualitatively "new" biological activities and at the same time exhibit considerable tolerance for structural variation.

Analogs with structural changes in position 5, in sharp contrast to analogs with similar or identical changes in position 4, exhibit striking reductions in hormonal activities, as exemplified by comparison of the potencies in several biological assay systems of (1) 5-asparagine and 4-glutamine analogs of oxytocin (Jaquenoud and Boissonnas, 1962a,b), (2) 5-valine and 4-valine analogs of oxytocin (Walter and Schwartz, 1966; du Vigneaud et al., 1966), (3) 4-α-aminobutyric acid and 5-alanine analogs of oxytocin (du Vigneaud et al., 1964), (5) 5-serine and 4-serine analogs of lysine-vasopressin (Berde and Boissonnas, 1968), etc. In a dose–response analysis of analogs with deletions (replacement by hydrogen) of the free amino function in position 2 or the carboxamide functions in positions 4, 5, or 9 (Chan and Kelley, 1967), it was found that the absence of the carboxamide group of asparagine ([5-alanine]-oxytocin) was associated with such a drastic reduction of biological activity (affinity and intrinsic activity) that it was impossible to define the upper range of the dose–response curve (even with analog doses of 100 to 200 mM)—whereas the comparable deletion of the glutamine carboxamide group ([4-α-aminobutyric acid]-oxytocin) was associated with retention of substantial biological activity (affinity less than that of oxytocin but more than 4 orders of magnitude greater than that of [5-alanine]-oxytocin; intrinsic activity identical to that of oxytocin). The stringent biological requirement for the presence in position 5 of a carboxamide moiety at a specific location (one methylene group removed from the peptide backbone) is a consequence of the function of this group in maintaining the biologically active conformation of oxytocin, namely, the provision of its carboxyl moiety for hydrogen bonding to the peptide NH of leucine in position 8 (Urry and Walter, 1971; Walter et al., 1971; Walter, 1971).

The fact that analogs with replacement of the asparagine residue in position 5, such as [5-valine]-oxytocin (Walter and Schwartz, 1966; Havran *et al.*, 1969), can despite an extreme reduction in affinity still retain the same intrinsic activity as the normal hormone can be rationalized on the assumption that a very small but finite number of the total population of analog molecules presented to the receptor can assume a conformation that makes possible a biologically effective analog–receptor interaction even in the absence of the carboxamide group in position 5.

The carboxamide moiety in position 9 is also of great importance for hormonal activity, its deletion (Branda and du Vigneaud, 1966) or the deletion of the entire glycinamide residue (Jaquenoud and Boissonnas, 1962*b*), yielding analogs with very low potency (affinity) and altered intrinsic activity (Chan and Kelley, 1967) in mammalian assay systems. Deletions of the proline or leucine residues in position 7 or 8, respectively (Jaquenoud and Boissonnas, 1962*b*), or of the entire prolyl-leucylglycinamide "tail" of the hormone (Ressler, 1956), led to analogs with similarly low levels of potency; the single-deletion analogs (7-de-proline, 8-de-leucine, 9-de-glycinamide) have been reported to inhibit the avian depressor and rat pressor activities of oxytocin at an analog/hormone ratio of 10,000:1. The biological requirement for glycinamide is a reflection of its role in the conformational stabilization of an important structural feature of oxytocin, namely, the β turn involving the tripeptide tail of the molecule (Urry and Walter, 1971; Walter *et al.*, 1971).

The role of the disulfide bridge in the antidiuretic–hydro-osmotic action of neurohypophyseal hormones has been a subject of considerable discussion during the past decade. It is sufficient at this time to note that ample evidence has accumulated to indicate that the sulfur centers in the hormone are not essential for hormonal activity (Rudinger and Jost, 1964*b*; Schwartz *et al.*, 1964*a,c*; Schwartz, 1965; Schwartz and Livingston, 1964; Walter *et al.*, 1967; Jost and Rudinger, 1967; Rudinger, 1969; Yamanaka *et al.*, 1970). However, a ring structure is an important requirement, as evidenced by the fact that acyclic analogs have extremely low hormonal activity (Walter *et al.*, 1967; Rudinger, 1969) and by the finding that potency is reduced when the 20-membered ring is enlarged (Jarvis and du Vigneaud, 1964). Thus the importance of the disulfide bridge derives largely, if not entirely, from the structural constraint it imposes on the hormone— which serves to limit the population of biofunctionally incompetent conformers and, accordingly, to assure the availability at the receptors of enough molecules with the "right" topochemical features for effective hormone–receptor interaction. The fact that some acyclic analogs have been found to exhibit low levels of activity has been interpreted as indicating that a very

small portion of these open-chain molecules take up the conformation required for the biologically effective hormone–receptor interaction. This
interpretation is supported by the finding that uterotonic and milk-ejecting
responses to two of these acyclic compounds were antagonized by the neurohypophyseal hormone inhibitor N-carbamyl-O-methyloxytocin, observations which suggest strongly that the "oxytocic" activities of these acyclic
analogs were indeed "hormone-like" rather than nonspecific (Poláček,
cited by Rudinger, 1969). It was also observed that acyclic analogs can act as
partial agonists, a phenomenon that was rationalized on the assumption
that "the greater conformational freedom permits such molecules to occupy
the binding sites on the receptor without necessarily aligning the functionally
required groups (e.g., the phenolic hydroxyl group) in the correct positions.
The occurrence of such 'nonproductive' as well as the proper 'productive'
modes of binding would be expected to result in the appearance of partial
agonism" (Rudinger, 1969).

When mammalian and amphibian systems are compared with respect
to the structural requirements for neurohypophyseal hormonal function, a
number of differences are encountered; e.g., (1) arginine-vasotocin, the
natural amphibian hormone (see above), is the most potent agonist in all *in
vivo* (frog kidney) and *in vitro* (frog bladder, frog skin, toad bladder) amphibian systems but in none of the mammalian systems; (2) structural changes in
arginine-vasotocin at all loci other than position 8 have proven to be highly
detrimental to hormonal activity in amphibian systems (this includes the
deletion of the free amino group in position 1, a manipulation which—in
mammalian systems—enhances antidiuretic and "oxytocic" activities);
(3) the action of neurohypophyseal peptides on the frog kidney (decreased
free water clearance, decreased filtration rate, increased tubular sodium re-
absorption) appears to require an isoleucine residue in position 3 (tocin
ring) for binding of the hormone to the receptor and a basic amino acid
residue in position 8 to render the hormone–receptor complex capable of
biological function (Jard and Morel, 1963); (4) there is a specific requirement
for a glutamine residue at position 4 for the maintenance of hydro-osmotic
and natriuretic potency in the frog bladder and kidney assay systems but
no such requirement in any mammalian system; (5) the oxytocin analog
with the proline residue deleted from position 7 ([7-de-proline]-oxytocin)
exhibits 1/25 of the frog skin natriferic activity of oxytocin, whereas the 8-de-
leucine and 9-de-proline analogs exhibit about twice the natriferic potency
of oxytocin (Morel and Bastide, 1964)—however, all of these "shortened"
analogs retain only rudimentary hormonal activities in mammalian systems.

The foregoing discussion of structure–activity relationships has for the
most part dealt with hormonal activity only as it is expressed in the final
event of the reaction sequence that follows the formation of a biologically

functional hormone–receptor complex. It is important, however, to know the structure–activity relationships that apply to intermediate and early events in this chain of reactions, particularly to the earliest event amenable to observation and analysis. We have recently gained such information in amphibian and mammalian systems in collaboration with Dr. Oscar Hechter and his associates, Drs. Bär and Dousa—with whom we have compared the effects of neurohypophyseal peptides on early events (adenylate cyclase activation, protein kinase stimulation) and late events (hydro-osmotic flux, antidiuresis) in hormone action.

In studies on the toad bladder, we have employed an adenylate cyclase preparation derived from a homogenate of epithelial cells scraped from the bladder mucosa. This cyclase preparation was specifically stimulated by neurohypophyseal hormones and analogs that elicit a hydro-osmotic response in the intact bladder. A striking parallelism was evident with respect to the peptide concentrations required for half-maximal stimulation (affinity parameter) of the cyclase preparation and of the intact system; i.e., the values of the affinity constant, pD_2 (negative logarithm of the peptide concentration which evokes a half-maximal response), for [8-arginine]-vasopressin (AVP), oxytocin (OXY), and [8-lysine]-vasopressin (LVP) were found to follow the same order (AVP > OXY > LVP) in the adenylate cyclase system and in the hydro-osmotic assay system. This parallelism of pD_2 values was found to hold for a series of oxytocin and vasopressin analogs which were also studied in both systems. In addition, a number of analogs which inhibit the hormone-induced hydro-osmotic response of the bladder were also observed to inhibit the hormone-induced stimulation of the bladder cyclase, the relative antagonistic effectiveness of the latter compounds in the intact bladder paralleling that in the adenylate cyclase assay (Bär et al., 1970; Kirchberger et al., 1972).

As noted earlier in this chapter, we have observed a similar parallelism with respect to the affinity parameter (pD_2) determined for neurohypophyseal peptides at the level of adenylate cyclase activation and at the level of the final hormonal effect (antidiuresis) in the mammalian kidney (Dousa et al., 1970a, 1971).

In contrast with the above-noted parallelism of pD_2 values (affinity parameters) measured with respect to "early" and "late" events, the maximal responses evoked by saturating concentrations of neurohypophyseal peptides (intrinsic activity parameters) did not correlate when the results of studies in the cyclase system and the intact system were compared (Bär et al., 1970; Dousa et al., 1970, 1971; Kirchberger et al., 1972). For example, if the maximal hydro-osmotic response to arginine-vasopressin is defined as 100% (intrinsic activity 1.00), the corresponding maximal responses to [2-O-methyltyrosine]-oxytocin (2-O-MeOt) and [1-β-mercaptopropionic acid, 8-alanine]-oxypressin (De-8-Ala-OP) are 88% and 100% (intrinsic activities 0.88 and 1.00), respectively, in the intact system; however, these analogs have only about 30% of the stimulatory capacity of arginine-vasopressin (intrinsic activity 0.30) in the bladder cyclase preparation. Thus the production of $3',5'$-AMP by hormones and analogs at rates in excess of that evoked by 2-O-MeOT or De-8-Ala-OP appears *not* to be translated into the final (hydro-osmotic) response. In fact, most of the neurohypophyseal peptides that we have studied can stimulate adenylate cyclase to a level of activity greatly exceeding that required to elicit a maximal response in the final effector system. These findings support our earlier proposal that the bladder has a "receptor reserve" with respect to neurohypophyseal hormone–induced changes in membrane permeability (Eggena *et al.*, 1969, 1970). These findings also indicate the need for caution in the application of current receptor theory to the analysis of the mechanism of action of any hormone (or other agent) when (1) the target-tissue response involves a chain of sequential events and (2) the experimental data have been derived from observations on a late event or final event in the chain. It should be noted that we do not know yet whether the basic nature of the "receptor reserve" of the toad bladder is anatomical (spare receptors) or functional (e.g., the "reserve" phenomenon may be the consequence of a very broad range over which adenyl cyclase can be activated); the phenomenon could also depend on a mechanism for altering the efficiency of the process which couples hormone–receptor interaction to adenylate cyclase activation or on a mechanism affecting phosphodiesterase activity or on any combination of the above and other factors. Therefore there are no mechanistic implications in the present use of the term "receptor reserve," which is employed only descriptively to indicate that early steps in hormone action—"at" or "close to" the initial hormone–receptor interaction—often manifest much more activity than can be translated into the final physiological effect. Thus the responsive system is provided with a margin of safety in the event (1) of defective function developing at an "early" step in hormone action, or (2) of a decline in hormone production, release, or transport to target cells.

In conclusion, it may be noted again that we have entered a period in which experimental study and analysis of hormonal structure–activity relationships can be extended from the level of primary to that of tertiary structure and focused on the earliest event(s) in hormone action.

ACKNOWLEDGMENTS

Studies cited in this paper have been supported by U.S. Public Health Service Grants AM 10080 and AM 13567 of the National Institutes of Arthritis and Metabolic Diseases, the Life Sciences Foundation, Inc., and the U.S. Atomic Energy Commission.

REFERENCES

Bär, H. P., Hechter, O., Schwartz, I. L., and Walter, R., 1970, *Proc. Natl. Acad. Sci.* **67**:7.

Berankova-Ksandrova, Z., Bisset, G. W., Jost, K., Krejci, I., Pliska, V., Rudinger, J., Rychlik, I., and Sorm, F., 1966, *Brit. J. Pharmacol. Chemotherap.* **26**:615.

Berde, B., and Boissonnas, R. A., 1966, *Synthetic analogues and homologues of the posterior pituitary hormones*, in "The Pituitary Gland," Vol. 3, (G. W. Harris and B. T. Donovan,

Berde, B., Boissonnas, R. A., Huguenin, R. L., and Sturmer, E., 1964, *Experientia* **20**:42. eds.), p. 624, Butterworth, London.

Berde, B., and Boissonnas, R. A., 1968, Basic pharmacological properties of synthetic analogues and homologues of the neurohypophysical hormones, in "Neurohypophysial Hormones and Similar Polypeptides," Vol. 23 of "Handbook of Experimental Pharmacology," (B. Berde, ed.), p. 802, Springer-Verlag, New York.

Berde, B., and Konzett, H., 1960, *Med. Exper.* **2**:317.

Bespalova, Zh. D., Martynov, V. F., and Titov, M. I., 1968, *Zh. Obshch. Khim.* **38**:1684.

Boissonnas, R. A., and Guttman, St., 1960, *Helv. Chim. Acta* **43**:190.

Boissonnas, R. A., Guttman, St., Jaquenous, P. A., and Waller, J. P., 1956, *Helv. Chim. Acta* **39**:1421.

Boissonnas, R. A., Guttman, St., and Konzett, H., 1961, *Experientia* **17**:377.

Branda, L. A., and Du Vigneaud, V., 1966, *J. Med. Chem.* **9**:169.

Celis, M., Taleisnik, S., Schwartz, I. L., and Walter, R., 1971*a*, *Biophys. J.* **11**:98*a*.

Celis, M., Taleisnik, S., and Walter, R., 1971*b*, *Proc. Natl. Acad. Sci.* **68**:1428.

Chan, W. Y., and du Vigneaud, V., 1970, *J. Pharmacol. Exptl. Therap.* **174**:541.

Chan, W. Y., and Kelley, N., 1967, *J. Pharmacol. Exptl. Therap.* **156**:150.

Chimiak, A., and Rudinger, J., 1965, *Coll. Czech. Chem. Commun.* **39**:2592.

Chiu, P. J. S., and Sawyer, W. H., 1970, *Am. J. Physiol.* **218**:838.

Dousa, T., Hechter, O., Walter, R., and Schwartz, I. L., 1970*a*, *Physiologist* **13**:182.

Dousa, T., Hechter, O., Walter, R., and Schwartz, I. L., 1970*b*, *Science* **167**:1134.

Dousa, T., Hechter, O., Schwartz, I. L., and Walter, R., 1971, *Proc. Natl. Acad. Sci.* **68**:1693.

Draper, W. B., 1927, *Am. J. Physiol.* **80**:90.

Dudley, H. W., 1923, *J. Pharmacol.* **21**:103.

du Vigneaud, V., 1956*a*, *Harvey Lectures.* **50**:1.

du Vigneaud, V., 1956*b*, *Science* **123**:967.

du Vigneaud, V., 1964, *Proc. Robert A. Welch Found. Conf. Chem. Res.* **8**:133.

du Vigneaud, V., Ressler, C., Swan, J. M., Roberts, C. W., Katsoyannis, P. G., and Gordon, S., 1953, *J. Am. Chem. Soc.* **75**:4879.

du Vigneaud, V., Denning, G. S., Jr., Draborek, S., and Chan, W. Y., 1964, *J. Biol. Chem.* **238**:472.

du Vigneaud, V., Flouret, G., and Walter, R., 1966, *J. Biol. Chem.* **241**:2093.

Eggena, P., Walter, R., and Schwartz, I. L., 1969, Stimulus-effect relationship in the anti-diuretic action of neurohypophyseal peptides and prostaglandin E_1, in "Proceedings of the Fourth, International Congress of Nephrology, Stockholm," p. 257.

Eggena, P., Schwartz, I. L., and Walter, R., 1970, *J. Gen. Physiol.* **56**:250.

Eisler, K., Rudinger, J., and Sorm, F., 1966, *Coll. Czech. Chem. Commun.* **31**:4563.

Havran, R. T., Schwartz, I. L., and Walter, R., 1969, *J. Am. Chem. Soc.* **91**:1836.

Hechter, O., and Braun, T., 1971, Peptide hormone–receptor interaction, in "Structure–Activity Relationships of Protein and Polypeptide Hormones," Part I (M. Margoulies and F. C. Greenwood, eds.), p. 212, Excerpta Medica International Congress Series No. 241, Exerpta Medica Foundations, Amsterdam.

Hermann, P., and Zaoral, M., 1965, *Coll. Czech. Chem. Commun.* **30**:2817.

Hruby, V. J., Flouret, G., and du Vigneaud, V., 1969, *J. Biol. Chem.* **244**:3890.

Hruby, V. J., du Vigneaud, V., and Chan, W. Y., 1970, *J. Med. Chem.* **13**:185.

Huguenin, R. L., and Boissonnas, R. A., 1966, *Helv. Chim. Acta* **49**:695.

Jaquenoud, P. A., and Boissonnas, R. A., 1961, *Helv. Chim. Acta* **44**:133.

Jaquenoud, P. A., and Boissonnas, R. A., 1962*a*, *Helv. Chim. Acta* **45**:1601.

Jaquenoud, P. A., and Boissonnas, R. A., 1962*b*, *Helv. Chim. Acta* **45**:1462.

Jard, S., and Morel, F., 1963, *Am. J. Physiol.* **204**:222.

Jarvis, D., and du Vigneaud, V., 1964, *Science* **143**:545.

Jost, K., and Rudinger, J., 1967, *Coll. Czech. Chem. Commun.* **32**:1229.

Kamm, O., Aldrich, T. B., Grate, I. W., Rowe, L. W., and Bugbee, E. P., 1928, *J. Am. Chem. Soc.* **50**:573.

Katzoyannis, P. G., 1957, *J. Am. Chem. Soc.* **79**:109.

Katsoyannis, P. G., and du Vigneaud, V., 1958, *J. Biol. Chem.* **233**:1352.

Kimbrough, R. D., Jr., Cash, W. D., Branda, L. A., Chan, W. Y., and du Vigneaud, V., 1963, *J. Biol. Chem.* **238**:1411.

Kirchberger, M., Walter, R., Schwartz, I. L., Bar, H. P., and Hechter, O., 1971, Unpublished data.

Krejci, I., and Polacek, I., 1968, *Europ. J. Pharmacol.* **2**:393.

Livermore, A. H., and du Vigneaud, V., 1949, *J. Biol. Chem.* **180**:365.

Manning, M., Coy, E., and Sawyer, W. H., 1970, *Biochemistry* **9**:3925.

Morel, F., and Bastide, F., 1964, Relationship between the structure of several analogues of oxytocin and their "natriferic" activity *in vitro*, *in* "Oxytocin, Vasopressin and Their Structural Analogues, Proceedings of the Second International Pharmacology Meeting," Vol. 10 (J. Rudinger, ed.), p. 47, Pergamon Press, Oxford.

Nesvadba, A., Honzl, J., and Rudinger, J., 1963, *Coll. Czech. Chem. Commun.* **28**:1691.

Pickering, B. T., 1970, Aspects of the relationship between the chemical structure and biological activity of the neurohypophyseal hormones and their synthetic analogues, *in* "Pharmacology of the Endocrine System and Related Drugs: The Neurohypophysis," Vol. I (H. Heller and B. T. Pickering, eds.), p. 81, Pergamon Press, Oxford.

Rasmussen, H., Schwartz, I. L., Young, R., and Marc-Aurele, J., 1963, *J. Gen. Physiol.* **46**:1171.

Ressler, C., 1956, *Proc. Soc. Exptl. Biol. Med.* **92**:725.

Ressler, C., and du Vigneaud, V., 1957, *J. Am. Chem. Soc.* **79**:4511.

Rudinger, J., 1969, Neurohypophyseal peptides and synthetic analogues: Structure and hormonal action, *in* "Progress in Endocrinology, Proceedings of the Third International Congress of Endocrinology" (C. Gual, ed.), p. 419, Excerpta Medica International Congress Series No. 184, Excerpta Medica Foundation, Amsterdam.

Rudinger, J., and Jost, K., 1964*a*, Synthetic analogues of oxytocin and vasopressin: Structural relations, *in* "Oxytocin, Vasopressin and Their Structural Analogues, Proceedings of the Second International Pharmacology Meeting, Vol. 10 (J. Rudinger, ed.), p. 3, Pergamon Press, Oxford.

Rudinger, J., and Jost, K., 1964*b*, *Experientia* **20**:570.

Rudinger, J., and Krejci, I., 1962, *Experientia* **18**:585.

Rudinger, J., and Krejci, I., 1968, Antagonists of the neurohypophysial hormones, *in* "Neurohypophysial Hormones and Similar Polypeptides," Vol. 23 of "Handbook of Experimental Pharmacology," (B. Berde, ed.), p. 748, Springer-Verlag, New York.

Rudinger, J., Honzl, J., and Zaoral, M., 1956, *Coll. Czech. Chem. Commun.* **21**:770.

Rychlik, I., 1964, Inactivation of oxytocin and vasopressin by tissue enzymes: A basis for the design of analogues, *in* "Oxytocin, Vasopressin and Their Structural Analogues, Proceedings of the Second International Pharmacology Meeting," Vol. 10 (J. Rudinger, ed.), p. 153, Pergamon Press, Oxford.

Sawyer, W. H., 1965, *Endocrinology* **5**:427.

Schlapp, W., 1925, *Quart. J. Exptl. Physiol.* **15**:327.

Schwartz, I. L., 1955, Adiuretin, in "Mechanisms of Hormone Action, A NATO Advanced Study Institute" (P. Karlson, ed.), p. 121, Academic Press, New York.

Schwartz, I. L., 1971, Structure–activity relationships as a tool for the analysis of sequential events in the action of neurohypophyseal hormones on membrane permeability, in "Structure–Activity Relationships of Protein and Polypeptide Hormones," Part I (M. Margoulies and F. C. Greenwood, eds.), p. 31, Excerpta Medica International Congress Series No. 241, Excerpta Medica Foundation, Amsterdam.

Schwartz, I. L., and Hechter, O., 1966, Insulin structure and function: Reflections on the present state of the problem, in "International Symposium on Insulin" (P. G. Katsoyannis and I. L. Schwartz, eds.), Am. J. Med. 40:765.

Schwartz, I. L., and Livingston, L. M., 1964, Cellular and molecular aspects of the antidiuretic action of vasopressins and related peptides, in "Vitamins and Hormones," Vol. 22 (I. G. Wool, ed.), p. 261, Academic Press, New York.

Schwartz, I., L., Rasmussen, H., Marc-Aurele, J., and Christman, D., 1964a, Molecular phenomena related to the mechanism of action of vasopressin on membrane permeability, in "The Biochemical Aspects of Hormone Action" (A. B. Eisenstein, ed.), p. 66, Little, Brown, Boston.

Schwartz, I. L., Rasmussen, H., Livingston, L. M., and Marc-Aurele, J., 1964b, Neurohypophyseal hormone–receptor interactions, in "Oxytocin, Vasopressin and Their Structural Analogues, Proceedings of the Second International Pharmacology Meeting" (J. Rudinger, ed.) p. 125, Pergamon Press, Oxford.

Schwartz, I. L., Rasmussen, H., and Rudinger, J., 1964c, Proc. Natl. Acad. Sci. 52:1044.

Smyth, D. G., 1967, J. Biol. Chem. 242:1579.

Smyth, D. G., 1970, Biochim. Biophys. Acta 200:395.

Studer, R. O., and Cash, W. D., 1963, J. Biol. Chem. 238:657.

Urry, D. W., and Walter, R., 1971, Proc. Natl. Acad. Sci. 68:956.

Van Dyke, H. B., Sawyer, W. H., and Overweg, N. I. A., 1963, Endocrinology 73:637.

Walter, R., 1971, Conformations of oxytocin and lysine vasopressin and their relationships to the biology of neurohypophyseal hormones, in "Structure–Activity Relationships of Protein and Polypeptide Hormones," Part I (M. Margoulies and F. C. Greenwood, eds.), p. 181, Excerpta Medica International Congress Series No. 241, Excerpta Medica Foundation, Amsterdam.

Walter, R., and Schwartz, I. L., 1966, J. Biol. Chem. 241:550.

Walter, R., Rudinger, J., and Schwartz, I. L., 1967, Am. J. Med. 42:653.

Walter, R., Dubois, B. M., and Schwartz, I. L., 1968, Endocrinology 83:979.

Walter, R., Dubois, B. M., Eggena, P., and Schwartz, I. L., 1969, Experientia 25:33.

Walter, R., Schwartz, I. L., Darnell, H., and Urry, D. W., 1971, Proc. Natl. Acad. Sci. 68:1355.

Yamanaka, T., Hase, S., Sakakibara, S., Schwartz, I. L., Dubois, B. M., and Walter, R., 1970, Mol. Pharmacol. 6:474.

Zaoral, M., 1965, Coll. Czech. Chem. Commun. 30:1853.

Zaoral, M., Kolc, J., and Sorm, F., 1967a, Coll. Czech. Chem. Commun. 32:1242.

Zaoral, M., Kolc, J., and Sorm, F., 1967b, Coll. Czech. Chem. Commun. 32:1250.

CHAPTER 15

USE OF POLYMERIC REAGENTS IN THE SYNTHESIS OF LINEAR AND CYCLIC PEPTIDES

A. Patchornik, M. Fridkin, and E. Katchalski

Departments of Organic Chemistry and Biophysics
The Weizmann Institute of Science
Rehovot, Israel

I. INTRODUCTION

New methods of peptide synthesis have yielded impressive successes in recent years in the preparation of many naturally occurring polypeptides. ACTH (Schwyzer and Sieber, 1963), glucagon (Wünsch, 1967), calcitonin (Sieber *et al.*, 1968; Guttman *et al.*, 1968), ferredoxin (Bayer *et al.*, 1968), ribonuclease A (Denkewalter *et al.*, 1969; Gutte and Merrifield, 1969), and human pituitary growth hormone (Li and Yamashiro, 1970) are representative of a large number of polypeptides which have been prepared chemically.

The rapid progress is a result of improvements in the classical methodology—new coupling reagents for the formation of the peptide bond, new protecting groups, highly efficient methods for the isolation and purification of synthetic products—as well as a result of the development of several new approaches to the synthesis of peptides. The new strategies include systematically controlled synthesis with the aid of *N*-carboxyanhydrides (Denkewalter *et al.*, 1966) and "solid-phase synthesis (Merrifield, 1963). These have recently been explored and developed in response to the requirements for speed, simplicity, and efficiency in the synthesis of large polypeptides and proteins. With the new synthetic tools now available to the peptide chemist and the enzymologist, challenging problems such as the total synthesis of an enzyme or determination of the relation between the primary structure of proteins and their biological activity can be attacked.

Solid-phase synthesis by the Merrifield approach has been covered in several articles and reviews (e.g., Merrifield, 1968, 1969; Marglin and Merrifield, 1970; Stewart and Young, 1969). The purpose of this chapter is to review peptide synthesis by the "polymeric reagents" approach.

II. POLYMERIC REAGENTS AND THEIR POTENTIAL USE IN ORGANIC SYNTHESIS

For many years, the synthesis of high molecular weight polymers and the investigation of their physicochemical properties have been the main concern of polymer chemists. In classical organic synthesis, only a limited class of synthetic polymers such as ion exchange resins and polymers possessing catalytic activity were found useful (Calmon and Kressman, 1957; Helfferich, 1962). Little information is available in the literature on reactions in which synthetic polymers are actively involved as chemical reagents, namely, reactions in which the polymer is used to modify other molecules and is changed chemically as a result of the reaction.

In the following, we intend to deal with one class of such polymers, namely, polymeric reagents carrying a transfer reaction according to the following general scheme:

$$\text{(P)} - \text{A}n + \text{B} \rightarrow \text{(P)} - \text{A}(n-1) + \text{A} - \text{B}$$

According to this scheme, the residues A, which are attached to the polymer P via chemically active bonds, are transferred during the chemical reaction to B, with the formation of A—B.

Any polymeric reagents, soluble or insoluble in the common organic or aqueous solvents, should be applicable to the above type of reaction. Since the A residues are as a rule low molecular weight compounds, one might expect that the physical properties of the high molecular weight carrier, P, will determine the properties of the polymeric reagent $\text{(P)} - \text{A}n$. Using polymeric reagents in chemical reactions, one might anticipate the following advantages:

1. Increased yields and improved synthetic procedures as a result of the use of excess of insoluble or soluble reagents which can be removed readily at the end of the reaction by filtration or by differential solubility of reagents and products.
2. Products of high purity obtained as a result of the technical advantages intrinsically involved in the novel synthetic procedures.

3. Continuous synthesis on columns and automatization of synthesis when insoluble polymeric reagents are employed.

4. Selectivity in the chemical reaction due to steric effects; dilution within the polymer chains; and affinity of reactant B for specific groups within the polymer P.

III. POLYMERIC REAGENTS IN PEPTIDE SYNTHESIS

Successful synthesis of relatively high molecular weight polypeptides with a predetermined amino acid sequence requires that the yield in every single step of the synthesis be close to quantitative; that the main reaction products be easily separable from excess of reagents, starting materials, and side-products; that the purification of the products be carried out readily if possible; and that automatization of the synthesis be feasible.

The use of polymers as tools in the preparation of peptides has been introduced in response to some of the above considerations. When insoluble polymers are being employed in peptide synthesis, either as carriers or as reagents, the polymers should possess good mechanical properties such as stability to prolonged shaking and to elevated temperatures and easy filtration; they should swell in the common organic solvents used for peptide synthesis; steric effects should be minimized; and adsorption of soluble reactants by the polymer should be as small as possible.

In addition, a polymeric reagent should carry a relatively large amount of activated N-blocked amino acids and should show stability on storage.

IV. USE OF POLYMERIC ACTIVE ESTERS OF BLOCKED AMINO ACIDS IN PEPTIDE SYNTHESIS

A polymeric reagent of this type can be prepared by the conversion of N-blocked amino acids or peptides into the corresponding insoluble polymeric active ester according to Fig. 1.

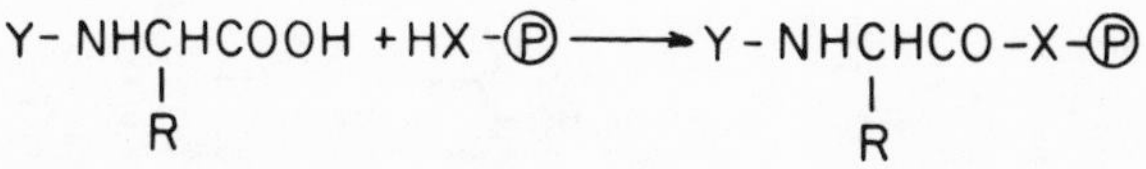

$$\text{Y-NHCHCOOH} + \text{HX-}\textcircled{P} \longrightarrow \text{Y-NHCHCO-X-}\textcircled{P}$$
$$\quad\quad\quad | \qquad\qquad\qquad\qquad\qquad\qquad | $$
$$\quad\quad\quad \text{R} \qquad\qquad\qquad\qquad\qquad\qquad \text{R}$$

Fig. 1. General scheme for preparation of polymeric active esters. Y, amino-terminal protecting group; $\textcircled{P}$, polymeric carrier; X, oxygen or sulfur.

Potentially useful carriers, ℗—XH, should contain functional hydroxy or thio groups to which the carboxylic terminal of the amino acid derivatives can be coupled to yield active ester bonds.

Some of the polymers which were prepared and investigated are shown in Fig. 2. The coupling between the polymer and the carboxylic derivatives is carried out in appropriate organic solvent, using the conventional coupling agents for the synthesis of peptides, such as dicyclohexylcarbodiimide (DCC) (Fridkin *et al.*, 1966, 1972; Laufer *et al.*, 1968; Wieland and Birr, 1966), mixed anhydrides (Patchornik *et al.*, 1967; Wieland and Birr, 1966), and thionyl chloride (Wieland and Birr, 1966). At the end of the reaction, excess of reagents is washed off from the polymer, which is now ready for peptide synthesis. The content of various *N*-blocked amino acids attached to the polymer was determined by the following methods: (1) determination of the increase in weight of the polymer due to the addition of the carboxylic component, (2) determination of the amino acids liberated on exhaustive acid hydrolysis, (3) spectrophotometric determination of the amount of *N*-protected amino acid liberated on treatment with NaOH, (4) elemental analysis (sulfur analysis was of particular use for the *O*-nitrophenylsulfenyl derivative), and (5) reaction of the polymeric reagent with excess of benzyl-amine and back-titration of the latter with perchloric acid in dioxane

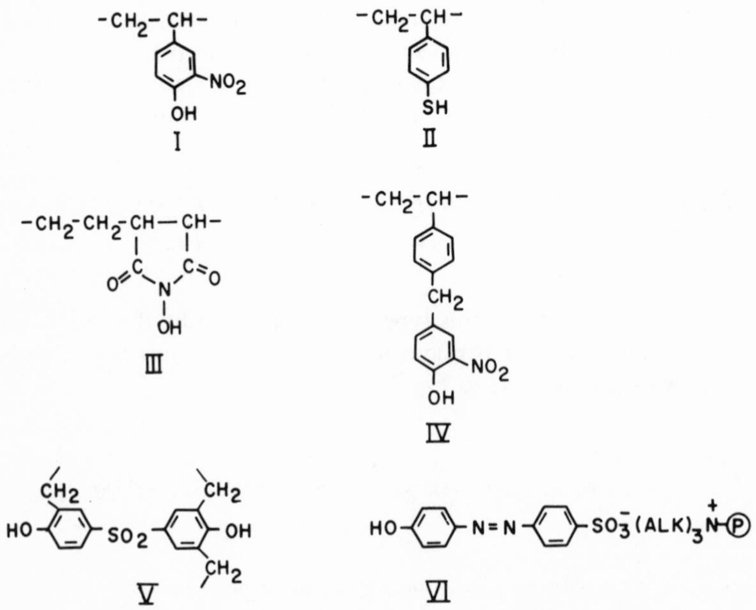

Fig. 2. Polymers used for preparation of polymeric active esters.

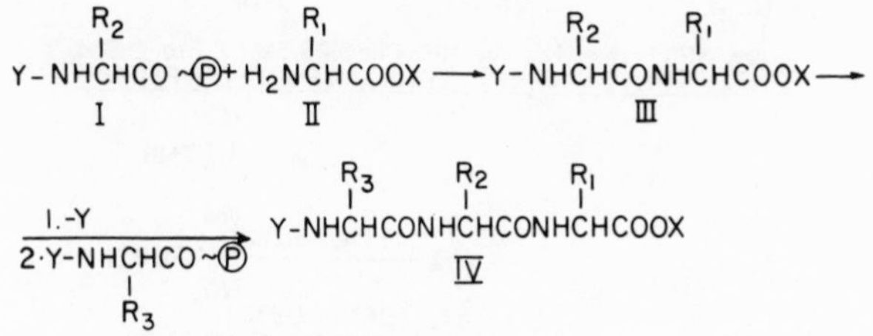

$$Y-\underset{\underset{R_2}{|}}{NHCHCO}\sim\text{\textcircled{P}}+H_2N\underset{\underset{R_1}{|}}{CHCOOX}\longrightarrow Y-\underset{\underset{R_2}{|}}{NHCHCONH}\underset{\underset{R_1}{|}}{CHCOOX}\longrightarrow$$

$$\underset{2\cdot Y-\underset{\underset{R_3}{|}}{NHCHCO}\sim\text{\textcircled{P}}}{\overset{1.-Y}{\rule{3cm}{0.4pt}}}\quad Y-\underset{\underset{R_3}{|}}{NHCHCONH}\underset{\underset{R_2}{|}}{CHCONH}\underset{\underset{R_1}{|}}{CHCOOX}$$

I   II   III   IV

Fig. 3. General scheme for peptide synthesis using polymeric active esters ($\sim$ stands for an active ester bond).

(Kalir *et al.*, in preparation). All five methods yielded similar results within the limits of experimental error.

The use of polymeric active esters in peptide synthesis is illustrated in Fig. 3. Polymeric active ester [*I*] derived from an *N*-blocked amino acid and the appropriate polymer, ℗, is coupled with a desired soluble amino acid ester containing a free α-amino group [*II*] to yield the *N*- and *C*-blocked dipeptide [*III*]. Preferential removal of the *N*-blocking group from the newly formed peptide enables the repetition of the coupling reaction with an insoluble active ester of another *N*-blocked amino acid. Further repetition of this set of reactions leads obviously to the elongation of the peptide chain and formation of a peptide with a predetermined amino acid sequence.

A comparison of this approach for peptide synthesis with that of Merrifield (Merrifield, 1963, 1964) reveals that whereas in solid-phase peptide synthesis it is the peptide which is bound to the insoluble carrier and the activated *N*-blocked amino acid is added in solution, in the method described above a solution of free peptide ester is added to a *N*-blocked amino acid ester attached to an insoluble polymer. Furthermore, purification of the intermediate peptides formed during synthesis can be readily effected in the polymer reagent method, since these peptides are liberated into solution. In the Merrifield synthesis, on the other hand, peptide purification can be carried out only after detachment of the final product from the polymeric carrier.

Polymeric nitrophenol derivatives have been prepared and investigated in several laboratories as potential candidates for peptide synthesis (Fridkin *et al.*, 1966, 1968; Patchornik *et al.*, 1967; Sklyarov *et al.*, 1966; Panse and Laufer, 1970; Kalir *et al.*, in preparation). Cross-linked poly-4-hydroxy-3-nitrostyrene, [PNP] (polymer [*I*], Fig. 2), was used in our laboratory for the preparation of many insoluble active esters derived from *N*-blocked amino acid and *N*-blocked peptides. The polymeric reagents usually contained 1–2 mmoles of activated amino acid per gram polymer. They have been used for the preparation of a variety of peptides. Among the peptides prepared

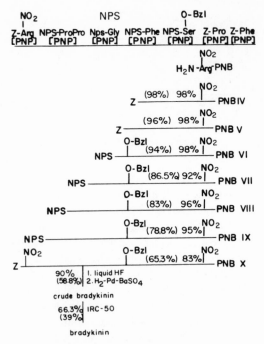

Fig. 4. Synthesis of bradykinin using polymeric active
esters (Fridkin *et al.*, 1968). [PNP], Copoly-4-hydroxy-
3-nitrostyrene–4 %-divinylbenzene.

were oxidized glutathione (90 % overall yield) (Fridkin *et al.*, 1966; Patchornik *et al.*, 1967) and the naturally occurring nonapeptide bradykinin (39 % overall yield) (Fridkin *et al.*, 1968). The synthesis of the latter peptide using the corresponding high molecular weight insoluble active ester is given in Fig. 4.

The peptide which was obtained after purification on IRC-50 ion exchange resin was identical in its physical and chemical properties to authentic bradykinin and possessed full biological activity.

Two main difficulties were encountered during the course of preparation of peptides with the aid of [PNP]-active ester (Fridkin *et al.*, 1971).

1. The mechanical properties of the polymeric carrier were not completely satisfactory. The particles partially disintegrated during the course of synthesis, especially in dimethylformamide, and the products were sometimes contaminated with superfine particles of the polymer.
2. The acylation of amines with the polymeric active esters is strongly dependent on steric factors.

In order to overcome these difficulties, a new type of polymeric carrier was prepared. Copolystyrene–2%-divinylbenzene in bead form was used as the starting material because of its excellent mechanical properties. The preparation of one of the carriers used is shown in Fig. 5 (Kalir et al., in preparation).

Several low molecular weight peptides were prepared, in high yields and with a high degree of purity, with the aid of the new polymer synthesized. Its use in the synthesis of high molecular weight peptides is under investigation. The results obtained so far are encouraging and indicate that the abovementioned difficulties have been overcome to a considerable extent.

Sklyarov et al. (1966) described briefly the use of [PNP] and of poly-(2-hydroxy-5-nitro-*m*-phenylenedimethylene) (obtained by copolymerization of *p*-nitrophenol, phenol, and 1,3-dichloroacetone in dioxane) in the preparation of few dipeptides. Panse and Laufer (1970) described the synthesis of a 4-hydroxy-3-nitrobenzoic acid derivative of polystyrene and its use in the preparation of a few short peptides. Williams (1972) described a new route for the preparation of [PNP] and has investigated some aspects of racemization during peptide synthesis using [PNP]-active esters (Lauren and Williams, 1972).

Cross-linked *p,p'-dihydroxydiphenylsulfone* (polymer [*V*], Fig. 2) was prepared by Wieland and Birr (1966) by condensing *p,p'*-dihydroxydiphenylsulfone with formaldehyde. The insoluble active esters derived from this polymer were used in the synthesis of several di- and tripeptides in 56–71% yields. In another polymeric reagent which was prepared by Wieland and Birr (1967), the *N*-protected amino acid active ester bearing free sulfonic functional group was attached via ionic bonds to an anion exchanger, Dowex 1 × 2. The ionically bound active ester of *N*-3,5-dimethoxybenzyl-oxycarbonyl-L-phenylalanine and 4-hydroxybenzeazo-4-sulfonic acid was allowed to react with L-alanyl-glycine methyl ester in formamide. The corresponding tripeptide ester was obtained in 69% yield based on the free amine. The attractive feature in this variant of the polymeric active ester

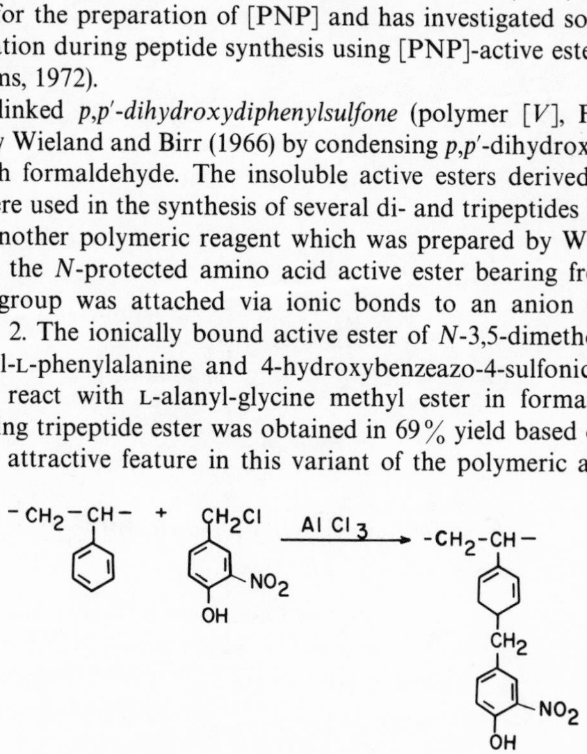

Fig. 5. Preparation of (4-hydroxy-3-nitro)benzylated polystyrene (Kalir et al., in preparation).

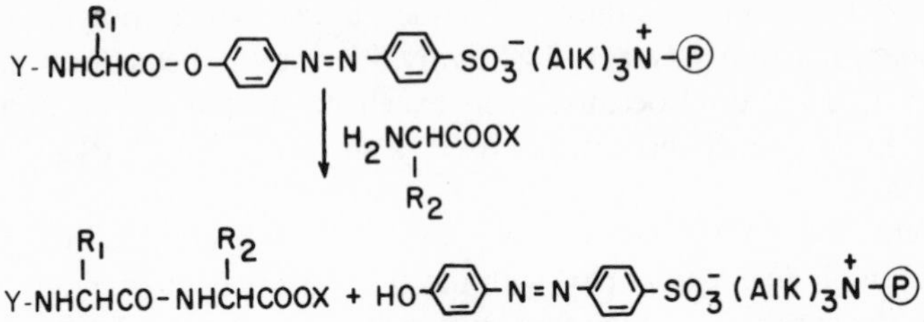

$$Y- \underset{\underset{R_1}{|}}{NHCHCO} - O - \langle \rangle - N=N - \langle \rangle - SO_3^- (AIK)_3 \overset{+}{N} - \textcircled{P}$$

$$\downarrow \underset{\underset{R_2}{|}}{H_2NCHCOOX}$$

$$Y-\underset{\underset{R_1}{|}}{NHCHCO} - \underset{\underset{R_2}{|}}{NHCHCOOX} + HO - \langle \rangle - N=N - \langle \rangle - SO_3^- (AIK)_3 \overset{+}{N} - \textcircled{P}$$

Fig. 6. Synthesis of peptide using insoluble ionically bound active ester
(Wieland and Birr, 1967).

approach (shown in Fig. 6) is the relative ease with which an active ester can be bound reversibly to the polymeric support.

Another interesting modification of the polymeric reagent peptide synthesis was developed by Flanigan and Marshall (1970) and by Marshall and Liener (1970). A 4-methylthiophenol derivative of polystyrene (see Fig. 7) was used as a support for the synthesis of peptides similarly to the Merrifield

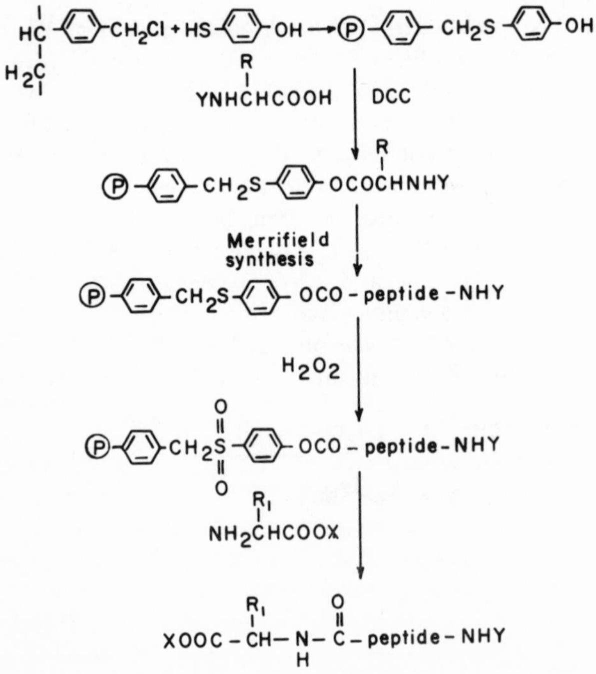

Fig. 7. Synthesis of peptide using "dual function support" (Marshall and Liener, 1970).

technique. However, in this case it was possible to activate the terminal carboxyl of the bound peptide by oxidation of the sulfur of the polymer derivative with hydrogen peroxide to the corresponding sulfone.

Only a limited number of peptides, in rather low yield, have been synthesized so far by this technique. It is worth mentioning that the amino acids cysteine, methionine, and tryptophan can be oxidized easily by hydrogen peroxide. Thus modification of the technique is essential, for the synthesis of peptides containing these amino acids at points other than the C-terminal is required.

A few other polymeric phenol-active esters have been reported in the literature: poly(4-thiostyrene) (Sklyarov et al., 1966; Patchornik et al., 1967), nitrated novolak (Patchornik et al., 1967), poly(5-vinyl-8-hydroxyquinaldine) (Manecke and Haake, 1968), but very little information is included in these reports.

Peptide synthesis by means of insoluble active esters derived from N-5(polystyryl-4-methoxycarbonyl)-imino-4-oximino-1,3-dimethyl-2-pyrazoline and N-blocked amino acids was recently reported by Guarneri et al. (1972). Several dipeptides as well as the N-terminal hen egg-white lysozyme tetrapeptide, N^α, N^ε-Di-Z-Lys-Val-Phe-Gly-OEt, was carried out in high yield (80% overall yield for the lysozyme tetrapeptide).

Because of the well-known reactivity of the N-hydroxysuccinimide esters of N-blocked amino acids (Anderson et al., 1964), high activity of the corresponding polymeric esters was predicted. Since such polymeric esters are aliphatic, less steric hindrance during the synthesis of peptides containing bulky amino acid derivatives was expected (Fridkin et al., 1971). The use of active esters derived from N-blocked amino acids and poly(ethylene-co-N-hydroxymaleimide) [PHMI] (polymer [III], Fig. 2) was developed independently in three laboratories (Patchornik et al., 1967; Laufer et al., 1968; Wildi and Johnson, 1968). Laufer et al. (1968) prepared the copolymer by condensing poly(ethylene-co-maleic anhydride) with hydroxylamine. The copolymer was cross-linked by exposure to high-energy electron irradiation. The t-Boc-amino acid active esters of the linear and cross-linked [PHMI] obtained were employed in the synthesis of 17 different tri- to octapeptides in good yields. Cross-linked poly(ethylene-co-hydroxymaleimide) was prepared in our laboratory by reacting the corresponding polymeric anhydride with hydroxylamine and cross-linking with hexamethylenediamine hydrazine, spermidine, or spermine (Fridkin et al., 1972). Polymeric active esters derived from the cross-linked PHMI and N-t-Boc-amino acids were successfully used in the synthesis of a heptapeptide, corresponding the residue 159–165 of bovine carboxypeptidase A. The stepwise synthesis is summarized in Fig. 8. The high yield obtained in the coupling reactions, the purity of products, and the relatively fast reaction rates in comparison with

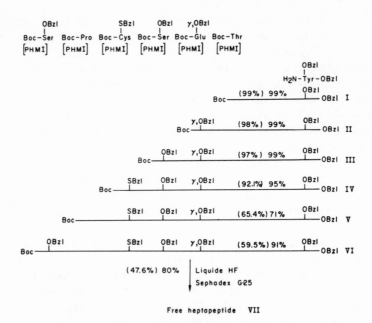

Fig. 8. Synthesis of heptapeptide, corresponding to residues 159–165 of bovine carboxypeptidase A, using polymeric active esters (Fridkin *et al.*, 1972). [PHMI], Cross-linked poly(ethylene-*co*-hykroxymaleimide).

the PNP-active esters suggest that PHMI-active esters might be very useful reagents in peptide synthesis.

Further experiments with insoluble PHMI-amino acid active esters have shown that macromolecules such as poly-ε-benzloxycarbonyl-L-lysine (average molecular weight 10,000) and insulin can be alanylated quantitatively by *t*-Boc-alanine-PHMI in DMF (Fridkin *et al.*, 1972). These results indicate that acyl-PHMI derivatives might be of use in the modification of synthetic and native macromolecules, provided that the latter are soluble in the medium in which the acylation reaction is carried out.

A new polymeric *N*-hydroxysuccinimide derivative (see Fig. 9) was prepared recently in our laboratory using copolystyrene–2%-divinylbenzene, in bead form, as starting material (Kalif *et al.*, in preparation). The product shows the good mechanical properties of the parent polystyrene. It has been used successfully for the preparation of various dipeptides, and we are currently applying it to the synthesis of longer peptides.

V. POLYMER CONDENSING AGENTS

The use of insoluble reagents for the activation of amino acids has not been limited to polyactive esters. Wolman *et al.* (1967) used the *in-*

soluble polyhexamethylene carbodiimide

$$-(CH_2)_6-N{=}C{=}N-[(CH_2)_6-N{=}C{=}N]_n-(CH_2)_6-$$

in peptide synthesis. For example, N,S-di-Z-L-cysteine and glycine benzyl ester were coupled with the aid of an eightfold excess of the polymeric reagent in methylene chloride suspension to give a 93% yield of N,S-di-Z-L-cysteinyl-glycine benzyl ester.

A series of *polymeric carbodiimides* of aromatic or aliphatic nature (see Fig. 10) were developed in our laboratory (Fridkin *et al.*, 1971) and were found to be of use as coupling agents in peptide synthesis.

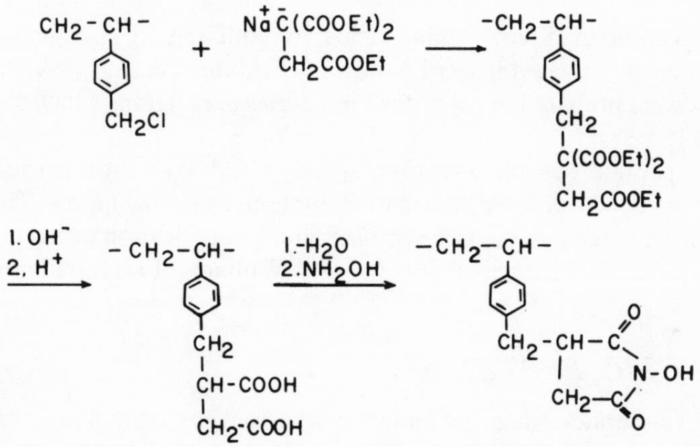

Fig. 9. Preparation of N-hydroxysuccinimide derivative of polystyrene (Kalir *et al.*, in preparation).

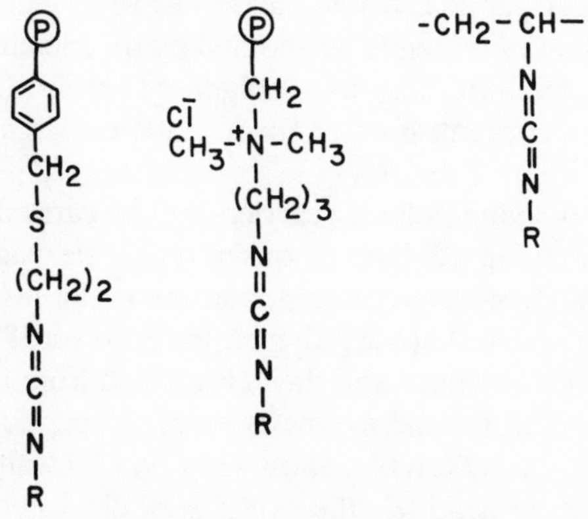

Fig. 10. Polymeric carbodiimides (Fridkin *et al.*, 1971).

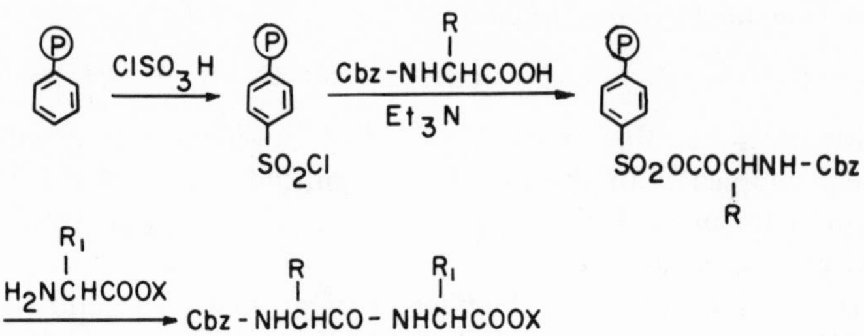

Fig. 11. Peptide synthesis using chlorosulfonated polystyrene as a coupling agent (Fridkin et al., 1971).

Activation of N-acyl amino acids by coupling to a *sulfonated polystyrene* was investigated in our laboratory (Fridkin et al., 1971). A number of peptides were prepared in good yield in a series of reactions which are shown in Fig. 11.

The peptide reagent N-ethoxycarbonyl-2-ethoxy-1,2-dihydroquinoline (EEDQ) was incorporated into an insoluble polymeric support. The use of the resulting reagent in peptide synthesis and its application in an automated process have been described (Brown and Williams, 1971; Williams et al., 1972).

VI. CYCLIC PEPTIDES

Cyclic peptides are generally synthesized by reacting a carboxyl-activated linear peptide with its own amino group (Schwyzer and Sieber, 1957; Wieland and Ohly, 1957; Vogler et al., 1960). The cyclization reaction is usually carried out in high dilution to minimize intermolecular condensation. Nevertheless, yields are usually low, and the products of condensation contain linear peptides which have to be removed. Formation of the linear peptides might in principle be avoided by the attachment of the linear peptide to be cyclized to an insoluble polymeric carrier. If such attachment leads to local low concentration, only intramolecular cyclization should result.

Cyclization involving polymer carriers can be carried out in two ways, which are depicted in Fig. 12. In reactions of type I, cleavage from the carrier occurs simultaneously with cyclization, and the cyclic product is recovered from solution after removal of the polymer. In reactions of type II, the bound linear molecule is first cyclized and then cleaved off from the polymer.

A general scheme for the synthesis of cyclic peptides by reaction of type I, using [PNP]-active esters is shown in Fig. 13. Using this method and the appropriate linear peptides, the cyclic peptides listed in Table I were prepared in our laboratory (Fridkin et al., 1965; Patchornik et al., 1967).

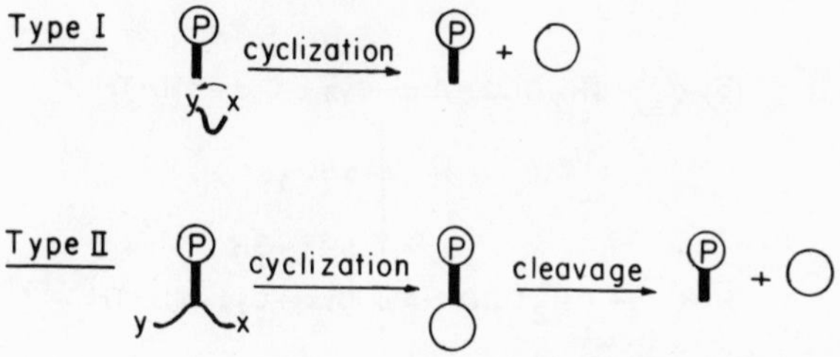

Type I

Type II

Fig. 12. General scheme for cyclization on polymeric support. ⓟ, The polymeric carrier, X and Y, activated terminals of a linear peptide.

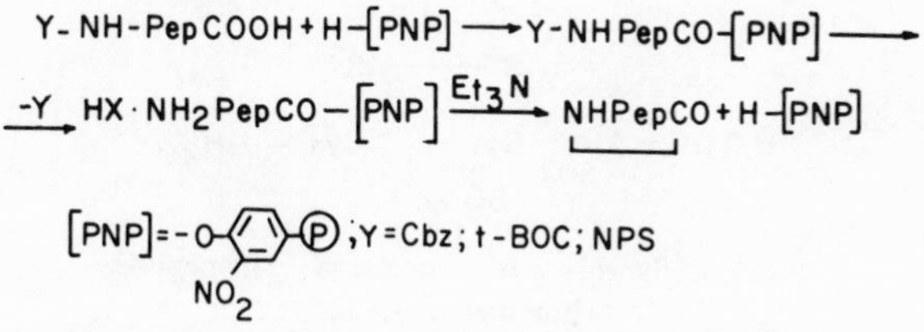

$$Y-NH-PepCOOH + H-[PNP] \longrightarrow Y-NHPepCO-[PNP] \longrightarrow$$

$$\xrightarrow{-Y} HX \cdot NH_2PepCO-[PNP] \xrightarrow{Et_3N} NHepCO + H-[PNP]$$

$$[PNP] = -O-\langle \_ \rangle-ⓟ \; ; Y = Cbz; t-BOC; NPS$$
$$NO_2$$

Fig. 13. Synthesis of cyclic peptides using polymeric active esters (Fridkin *et al.*, 1965).

Table I. Cyclic Compounds Prepared on Polymeric Support

Cyclic compound	Yield (%)
Gly—Gly	75
Ala—Ala	80
Phe—Gly	70
Ala—Gly	77
Ala—Ala—Ala—Ala	65
Ala—Gly—Ala—Ala	56
HN—(CH₂)₅—CO	80
NH—(CH₂)₆—CO	65

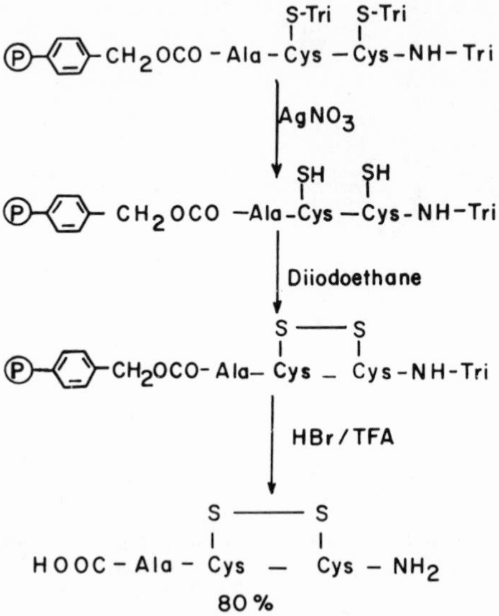

$$
\begin{array}{c}
\quad\quad\quad \overset{\text{S-Tri}}{\underset{|}{}} \quad \overset{\text{S-Tri}}{\underset{|}{}} \\
\text{(P)}\!-\!\langle\ \rangle\!-\!CH_2OCO-Ala-Cys-Cys-NH-Tri
\end{array}
$$

$$\big\downarrow AgNO_3$$

$$
\begin{array}{c}
\quad\quad\quad\quad \overset{\text{SH}}{\underset{|}{}} \quad \overset{\text{SH}}{\underset{|}{}} \\
\text{(P)}\!-\!\langle\ \rangle\!-\!CH_2OCO-Ala-Cys-Cys-NH-Tri
\end{array}
$$

$$\big\downarrow \text{Diiodoethane}$$

$$
\begin{array}{c}
\quad\quad\quad\quad\quad S\!-\!-\!-\!S \\
\quad\quad\quad\quad\quad |\quad\quad | \\
\text{(P)}\!-\!\langle\ \rangle\!-\!CH_2OCO-Ala-\ Cys\ \_\ Cys-NH-Tri
\end{array}
$$

$$\big\downarrow \text{HBr / TFA}$$

$$
\begin{array}{c}
S\!-\!-\!-\!S \\
|\quad\quad\quad | \\
HOOC-Ala-Cys\ -\ Cys-NH_2 \\
80\%
\end{array}
$$

Fig. 14. Synthesis of a cyclic $S{-}S$-tripeptide
(Bondi et al., 1968)

By following a similar approach, (type I) Flanigan and Marshall (1970) synthesized di-, tetra-, and hexacyclopeptides in fair yields using a methyl-thiophenol derivative of polystyrene as the polymeric carrier. An example of cyclization of type II is the synthesis of the cystine-containing tripeptide

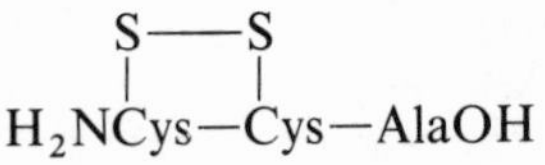

$$
\begin{array}{c}
S\!-\!-\!S \\
|\quad\ | \\
H_2NCys-Cys-AlaOH
\end{array}
$$

The peptide N,S-ditrityl-L-cysteinyl-S-trityl-L-cysteinyl-L-alanine (see Fig. 14) was synthesized on a Merrifield polymer. After removal of the S-trityl groups by silver nitrate, the free thiol groups were oxidized by diiodoethane, and the cyclic cystine peptide was then released from the polymer by treatment with HBr in TFA. Purification on Dowex-50 ion exchanger yielded the desired peptide in 80 % overall yield (Bondi et al., 1968).

The biologically active peptide oxytocin was also prepared with the aid of a polymeric carrier. The linear S-blocked peptide was synthesized in a stepwise manner on a polyphenol support. The thiol protecting groups were removed, and the free SH groups were oxidized by passing air through a suspension of the polymeric-bound peptide. Upon ammonolysis, the oxytocin obtained was released into solution and subsequently purified (Inukai *et al.*, 1968).

Skylarov and Shashkova (1969) have synthesized a glycine analogue of gramicidin S using a benzyl chloroformate derivative of polystyrene as an insoluble carrier. A diblocked linear tetrapeptide was attached to the polymer via a free δ-amino group of an ornithine residue. The tetrapeptide was then elongated in a stepwise manner at the *N*-terminal end to yield an insoluble decapeptide derivative. Upon removal of the *C*-terminal protecting group and esterification of the free carboxyl with *N*-hydroxysuccinimide, an insoluble decapeptide active ester was obtained. The *N*-protecting group was removed, and the free amino group was allowed to react with the activated carbonyl terminal to give the cyclic decaptide bound to the polymer. The cyclic peptide thus formed was split from the polymer by the action of HBr in acetic acid and purified on ion exchangers. The yield was 20% based on the amount of the linear tetrapeptide which was attached to the polymer. The described method might be of use in synthesis of cyclic peptides containing difunctional amino acids.

VII. CONCLUDING REMARKS

The use and applicability of polymeric reagents in the synthesis of peptides has been demonstrated in this chapter. It is pertinent to note, however, that the appropriate polymeric reagents might be of considerable use also in other classical organic reactions. Their use might simplify the synthetic procedure, lead to high yields of the desired products, and increase specificity. As a matter of fact, the few examples already available in the literature illustrate the potential of the polymeric reagent approach in organic synthesis.

Polymers containing quinone residues have been used to oxidize various compounds such as tetraline, cysteine, and ascorbic acid (Manecke *et al.*, 1959); polymers containing free sulfhydryl groups were found to be useful reducing agents of disulfide bonds (Overberger and Lebovitz, 1956; Gorecki and Patchornik, 1972); polymers containing percarboxylic groups were used to hydroxylate compounds with olefinic double bonds to form α-glycols (Helfferich and Luten, 1964); charge-transfer complexes of polyvinylpyridine with bromine could be used as mild brominating agents (Zabicki *et al.*, in preparation); polymers and copolymers of *N*-halomaleimides revealed considerable specificity in the halogenation of aromatic compounds (Yaroslavsky *et al.*, 1970); polymer carrying nitronium ions showed specificity in the nitration of toluene (Wright *et al.*, 1965); polymeric sulfonyl chlorides were useful as coupling reagents in the preparation of nucleotides (Rubinstein and Patchornik, 1972); an insoluble polymeric carbodiimide was used as a reagent in the Moffat oxidation (Weinshenker and Shen, 1972); polymeric ylides were utilized in the Witting reaction (Camps *et al.*, 1971*b*; Tanimoto *et al.*, 1967; McKinley and Rakshys, 1971); polymer-supported rhodium (I) catalyst was used for selective catalytic reduction of olefins (Grubbs and Kroll, 1971).

Since chemically reactive groups can be widely separated from one another while attached to a polymer backbone, one can prevent interaction between the reactive moieties which are attached to the polymer. This might facilitate intramolecular interactions, or specific intermolecular reactions which can be carried out with the corresponding monomeric reactive compounds only with considerable difficulty. The prevention of intermolecular condensation and enhancement of intramolecular condensation on a polymer carrier have been illustrated herein by the synthesis of cyclic peptides using immobilized linear peptide active esters. A similar technique was used to synthesize cyclic ketones (Crowley and Rapoport, 1970). Stable polycarbanions were utilized in the direct monoacylation of esters (Patchornik and Kraus, 1970), for the directed mixed ester condensation of two acids bound to a common polymer backbone (Kraus and Patchornik, 1971*a*), and for the alkylation of esters (Kraus and Patchornik, 1971*b*; Camps *et al.*, 1971*a*).

REFERENCES

Anderson, G. W., Zimmerman, J. E., and Callahan, F. M., 1964, The use of esters of *N*-hydroxysuccinimide in peptide synthesis, *J. Am. Chem. Soc.* **86**:1839.

Bayer, E., Yung, G., and Hagenmaier, H., 1968, Untersuchungen zur Totalsynthese des Ferredoxins-I, *Tetrahedron* **24**:4853.

Bondi, E., Fridkin, M., and Patchornik, A., 1968, The use of polymeric carriers in the synthesis of cyclic peptides containing S—S bonds, *Israel J. Chem.* **6**:22p.

Brown, J., and Williams, R. E., 1971, A regenerable, solid-phase coupling reagent for use in peptide synthesis, *Can. J. Chem.* **49**:3765.

Calmon, C., and Kressman, R. R. E., 1957, "Ion Exchangers in Organic and Biochemistry," Interscience, New York.

Camps, F., Castells, J., Ferrando, M. J., and Font, J., 1971*a*, Organic synthesis with functionalized polymers. I. preparation of polymeric substrates and alkylation of esters, *Tetrahedron Letters* **20**:1713.

Camps, F., Castells, J., Font, J., and Vela, F., 1971*b*, Organic synthesis with functionalized polymers. II. Wittig reaction with polystyryl-*p*-diphenylphosphoranes, *Tetrahedron Letters* **20**:1715.

Crowley, J. I., and Rapoport, H., 1970, Cyclization *via* solid-phase synthesis. Unidirectional Dieckmann produces from solid-phase and benzyl triethylcarbinyl pimelates, *J. Am. Chem. Soc.* **92**:6363.

Denkewalter, R. G., Schwan, H., Strachan, R. G., Beesley, T. E., Veber, D. F., and Hirschmann, R., 1966, The controlled synthesis of peptides in aqueous medium. I. The use of α-amino acid *N*-carboxyanhydrides, *J. Am. Chem. Soc.* **88**:3163.

Denkewalter, R. G., Veber, D. F., Holly, F. W., and Hirschman, R., 1969, Studies of the total synthesis of an enzyme. I. Objective and strategy, *J. Am. Chem. Soc.* **91**:502.

Flanigan, E., and Marshall, G. R., 1970, Synthesis of cyclic peptides on dual function supports, *Tetrahedron Letters* **27**:2403.

Fridkin, M., Patchornik, A., and Katchalski, E., 1965, A synthesis of cyclic peptides utilizing high molecular weight carriers, *J. Am. Chem. Soc.* **87**:4646.

Fridkin, M., Patchornik, A., and Katchalski, E., 1966, Use of polymers as chemical reagents. I. Preparation of peptides, *J. Am. Chem. Soc.* **88**: 3164.

Fridkin, M., Patchornik, A., and Katchalski, E., 1968, Use of polymers as chemical reagents. II. Synthesis of bradykinin, *J. Am. Chem. Soc.* **90**:2953.

Fridkin, M., Patchornik, A., and Katchalski, E., 1971, Some experiments on the synthesis of peptides with the use of polymeric reagents, *in* "Peptides," p. 164, North-Holland, Amsterdam.

Fridkin, M., Patchornik, A., and Katchalski, E., 1972, Peptide synthesis by means of *tert*-butyloxycarbonylamino acid derivatives of poly (ethylene-*co*-*N*-hydroxymaleimide), *Biochemistry* **11**:466.

Gorecki, M., and Patchornik, A., 1973, Polymer bound dihydrolipoic acid. A new insoluble reducing agent for disulfides, *Biochim. Biophys. Acta* **303**:36.

Grubbs, R. H., and Kroll, L. C., 1971, Catalytic reduction of olefins with a polymer-supported rhodium (I) catalyst, *J. Am. Chem. Soc.* **93**:3062.

Guarneri, M., Ferroni, R., Giori, P., and Bennasi, C. A., 1972, Peptide synthesis through 4-nitroso-5-amino-pyrazole insoluble active esters, *in* "Chemistry and Biology of Peptides," p. 17–1, Ann Arbor Science Publishers, Ann Arbor, Mich.

Gutte, B., and Merrifield, R. B., 1969, The total synthesis of an enzyme with ribonuclease A activity, *J. Am. Chem. Soc.* **91**:501.

Guttman, St., Pless, J., Sandrin, E., Jaquenoud, P. A., Bossert, H., and Willems, H., 1968, Syntheses des Tyreocalcitonins, *Helv. Chim. Acta* **51**:1155.

Helfferich, F., 1962, *in* "Ion Exchange," Chap. 11, McGraw-Hill, New York.

Helfferich, F., and Luten, D. B., Jr., 1964, Oxygen-transfer resins, a new type of oxidation-reduction polymers, *J. Appl. Polymer Sci.* **8**:2899.

Inukai, N., Nakano, K., and Murakami, M., 1968, The peptide synthesis. II. Use of the phenol resin for the peptide synthesis, *Bull. Chem. Soc. Japan* **41**:182.

Kalir, R., Fridkin, M., and Patchornik, A., Peptide synthesis by means of improved nitrophenol and *N*-hydroxysuccinimide polymers, Manuscript in preparation.

Kraus, M. A., and Patchornik, A., 1971*a*, The direct mixed ester condensation of two acids bound to a common polymer backbone, *J. Am. Chem. Soc.* **93**:7325.

Kraus, M. A., and Patchornik, A., 1971*b*, Reactive species mutually isolated on insoluble polymeric carriers. II. The alkylation of esters, *Israel, J. Chem.* **9**:269.

Laufer, D. A., Chapman, T. M., Marlborough, D. I., Vaidya, V. M., and Blout, E. R., 1968, A reagent for peptide synthesis. Copoly (ethylene-*N*-hydroxymaleimide), *J. Am. Chem. Soc.* **90**:2696.

Lauren, D. R., and Williams, R. E., 1972, Racemization during use of polymeric supports in peptide synthesis, *Tetrahedron Letters* **26**:2665.

Li, C. H., and Yamashiro, D., 1970, The synthesis of a protein possessing growth promoting and lactogenic activities, *J. Am. Chem. Soc.* **92**:7608.

Manecke, G., and Haake, E., 1968, Peptide synthesis with poly-(5-vinyl-8-hydroxy-chinaldin), *Naturwissenschaften* **55**:343.

Manecke, G., Bahr, Ch., and Reich, Ch., 1959, Anwendung von Redoxharzen, *Angew. Chem.* **71**:646.

Marglin, A., and Merrifield, R. B., 1970, Chemical synthesis of peptides and proteins, *Ann. Rev. Biochem.* **39**:841.

Marshall, D. L., and Liener, I. E., 1970, A modified support for solid-phase peptide synthesis which permits the synthesis of protected peptide fragments, *J. Org. Chem.* **35**:867.

McKinley, S. V., and Rakshys, J. W., Jr., 1971, Ylid resin reagents: The preparation and application of insoluble polymeric ylides, in "American Chemical Society Meeting, Washington D.C.," Abst. 27.

Merrifield, R. B., 1963, Solid phase peptide synthesis. I. The synthesis of a tetrapeptide, J. Am. Chem. Soc. 85:2149.

Merrifield, R. B., 1964, Solid-phase peptide synthesis. III. An improved synthesis of bradykinin, Biochemistry 3:1385.

Merrifield, R. B., 1968, The automatic synthesis of proteins, Sci. Am. 218:56.

Merrifield, R. B., 1969, Solid-phase peptide synthesis, Advan. Enzymol. 32:221.

Overberger, C. G., and Lebovitz, A., 1956, Water soluble copolymers containing sulfhydryl groups, J. Am. Chem. Soc. 78:4792.

Panse, G. P., and Laufer, D. A., 1970, A new polymeric reagent for peptide synthesis, Presented at the Second American Peptide Symposium, Cleveland, Ohio.

Patchornik, A., and Kraus, M. A., 1970, Reactive species mutually isolated on polymeric carriers, I. The directed monoacylation of esters, J. Am. Chem. Soc. 92:7587.

Patchornik, A., Fridkin, M., and Katchalski, E., 1967, Synthesis of linear and cyclic peptides with the aid of insoluble active esters of amino acids and peptides, in "Peptides," p. 91, North-Holland, Amsterdam.

Rubinstein, M., and Patchornik, A., 1972, Polymers as chemical reagents. The use of poly(3,5 diethylstyrene) sulfonyl chloride for the synthesis of internucleotide bonds, Tetrahedron Letters 28:2881.

Schwyzer, R., and Sieber, P., 1957, Die Synthese von Gramicidin S, Helv. Chim. Acta 40:624.

Schwyzer, R., and Sieber, P., 1963, Total synthesis of adrenocorticotrophic hormone, Nature 199:172.

Sieber, P., Brugger, M., Kamber, B., Riniker, B., and Rittel, W., 1968, Die Synthese von Calcitonium M, Helv. Chim. Acta 51:2057.

Sklyarov, L. Yu., and Shadhkova, M., 1969, Synthesis of the glycine analog of gramicidin S on a polymeric carrier, Zh. Obshch. Khim. 39:2778.

Sklyarov, L. Yu, Gurbunov, V. I., and Shchukina, L. A., 1966, Synthesis of peptides with the aid of activating resins, Zh. Obshch. Khim. 36:2220.

Stewart, J. M., and Young, J. D., 1969, "Solid Phase Peptide Synthesis," Freeman, San Francisco.

Tanimogo, S., Horikawa, J., and Oda, R., 1967, Formation of S-ylide polymer from poly(p-vinylbenzyl methyl sulfide), Kogyo Kagaku Zasshi 70(7):1269.

Vogler, V., Studer, R. O., Lergier, W., and Lanz, P., 1960, Synthesen in der Polymyxin-reine, Helv. Chim. Acta 43:1751.

Weinshenker, N. M., and Shen, C. M., 1972, Polymeric reagents. II. Preparation of ketones and aldehydes utilizing an insoluble carbodiimides, Tetrahedron Letters 32:3285.

Wieland, Th., and Birr, Ch., 1966, Syntheses of peptides by means activated units bound to resins, Angew. Chem. Internat. Ed. 5:310.

Wieland, Th., and Birr, Ch., 1967, Synthesis of peptides by means of activated units, ionically bound to resins, Chimia 21:581.

Wieland, Th., and Ohly, K. W., 1957. Peptide cyclization with carbodiimide, Liebigs Ann. Chem. 605:179.

Wildi, B. S., and Johnson, J. H., 1968, Polymeric carrier resins for peptide synthesis, in "155th National Meeting of the American Chemical Society," Abst. A-8.

Williams, R. E., 1972, Polymeric supports in peptide synthesis: An alternate preparation of crosslinked 4-hydroxy-3-nitropolystyrene, J. Polymer Sci. A-1 10:2123.

Williams, R. E., Brown, J., and Lauren, D. R., 1972, Regenerable polymeric reagents for large scale automatic peptide synthesis, Polymer Preprints 13(2):823.

Wolman, Y., Kivity, S., and Frankel, M., 1967, The use of polyhexamethylene-carbodiimide, an insoluble condensing agent, in peptide synthesis, *Chem. Commun.* **1967**:629.

Wright, O. L., Teipel, J., and Thoennes, D., 1965, The nitration of toluene by means of nitric acid and an ion-exchange resin, *J. Org. Chem.* **30**:1301.

Wünsch, E., 1967, The total synthesis of the pancreatic-hormone glucagon, *Z. Naturforsch.* **22b**:1269.

Yaroslavsky, C., Patchornik, A., and Katchalski, E., 1970, Unusual brominations with *N*-bromopolymaleimide, *Tetrahedron Letters* **42**:3629.

Zabicki, J., Oren, I., and Katchalski, E., Use of charge-transfer complexes of polyvinyl pyridine and bromine in the bromination of aromatic and olefinic compounds, Manuscript in preparation.

CHAPTER 16

SOLID-PHASE PEPTIDE SYNTHESIS

R. B. Merrifield

Rockefeller University
New York, N.Y.

I. INTRODUCTION*

It is a privilege to be able to contribute to this volume in which Professor Zervas' friends, students, and colleagues have joined together to honor him and his remarkable contributions to peptide chemistry. Although I did not know him until after his days in the Bergmann Laboratory at the Rockefeller Institute, I can claim the distinction of now working in those very same rooms that they occupied back in the mid-1930s. Like all peptide chemists, I am greatly indebted to Professor Zervas, having depended so heavily on the carbobenzoxy group and on the various modified urethan protecting groups which have been direct extensions of the revolutionary advance that Bergmann and Zervas made.

Just 10 years ago, a new idea for the synthesis of peptides was introduced, called "solid-phase" peptide synthesis. I would like to take this opportunity to discuss the philosophy underlying this unconventional approach, to show how the technique has changed during the intervening period, to point to some of the variations on the theme, and to touch briefly on some of the things that have been achieved through application of the solid-phase principle.

*Abbreviations used in this chapter: Boc, t-butyloxycarbonyl; Z, carbobenzoxy (= benzyloxy-carbonyl); Bpoc, 2-(4-biphenylyl)-2-propyloxycarbonyl; TFA, trifluoroacetic acid; DCC, dicyclohexylcarbodiimide; HOSu, N-hydroxysuccinimide; Et_3N, triethylamine; Bzl, benzyl; NBS, N-bromosuccinimide; NEPIS, N-ethyl-5-phenylisoxazolium-3'-sulfonate; DMF, dimethylformamide; pMZ, a-methoxybenzyloxycarbonyl; S-DVB, styrene–divinylbenzene copolymer. Other abbreviations as recommended by the IUPAC-IUB, *J. Biol. Chem.*, 247:977 (1972).

II. THE SOLID-PHASE IDEA

A. General Considerations

During a period in which I was involved in the synthesis of a series of small peptides by the classical methods, it became more and more clear to me that we needed better methods and that if we intended to extend our horizons to the synthesis of larger peptides we surely must have more efficient techniques. It is evident from the other chapters in this book and from the recent literature that improvements in the classical techniques have been forthcoming and that better methods are now at hand, but I was convinced that a completely new approach to the problem was essential, and the solid-phase idea emerged from these considerations (Merrifield, 1962, 1963).

By attaching the first amino acid of the peptide chain to an insoluble solid support and growing the chain in a stepwise manner, it seemed probable that several basic difficulties could be overcome. The purification of the peptide intermediates could be accomplished by simple rapid-washing procedures, the losses of product would be minimized because of the complete insolubility of the anchored peptide chain, and, ironically, the problem of insolubility of intermediate reactants in the usual sense would be avoided. The overall process would be accelerated, and there would be an opportunity to mechanize and automate the entire synthesis. Some of these advantages were obtained at the outset, some were slowly achieved after more detailed studies, and some have not yet been fully realized (Merrifield, 1969; Stewart and Young, 1969; Marshall and Merrifield, 1971). The success of this approach depends on three things: (1) rapid reactions, (2) complete reactions, and (3) absence of side-reactions. The reactions can certainly be very fast, and in most instances they come close to completion, although it is unlikely that the last molecule ever reacts. A number of side-reactions have been recognized and eliminated or minimized, but some others remain as special problems.

The original idea involved a two-phase system in which a soluble amino acid reagent reacted with the growing peptide chain while it was held in the solid phase on an insoluble support, and was later separated from it by filtration. The name "solid-phase peptide synthesis" was derived from that situation. Basically, however, the requirement is simply for a peptide support which is physically separable from the reactants, and in principle other separation techniques and other two-phase systems are possible: solid–solid, liquid–liquid, or even those involving the vapor phase.

The general requirements for a solid support are insolubility, physical stability, chemical stability, and, most important, ability to provide conditions in which the synthetic reactions can occur rapidly and quantitatively.

Polystyrene-based resin beads containing low concentrations (1 or 2%) of the cross-linking agent divinylbenzene meet the specifications rather well. They are insoluble, they are physically stable to shaking and filtering, and they swell in organic solvents to give a large open matrix in which the coupling reactions can occur. By radioautography at the level of the light microscope, the accessibility of the interior of the beads appears to be quite uniform (Merrifield and Littau, 1968), and a relatively large space is available to the growing peptide chains. However, on a molecular level it is probable that there is heterogeneity and that not all sites are equally accessible. Steric and diffusion problems are possible sources of difficulty that could lead to incomplete reactions, to shortened peptide chains, or to deletion peptides during synthesis.

Several efforts to overcome these presumed problems have been made. An early one used a soluble polystyrene to support homogeneous coupling reactions, followed by a change in solvent to precipitate the product for separation from reactants by filtration (Shemyakin et al., 1965; Green and Garson, 1969). With such soluble macromolecular supports, it may be possible to effect separation by ultrafiltration without prior precipitation of the polymer. It may also be possible to use a liquid–liquid system containing two polymeric solutes (Albertson, 1960) to achieve the separation. Problems of inaccessibility of internal sites might also be avoided by limiting the reactions to the surface of the solid support, and several efforts have been made in that direction. It is important to point out that for a significant effect to be achieved the reactions must really be limited to the surface and not simply to a relatively thin layer of polymeric support, because the latter will still be very thick by molecular dimensions and the diffusion and steric problems will not have been eliminated.

The main efforts in this direction have involved using (1) highly cross-linked suspension copolymers of styrene and divinylbenzene that do not swell and are limited in their reactions to sites near the surface (Merrifield, 1963), (2) macroporous resins which have rigid matrices due to high cross-linking but large internal surfaces (Merrifield, 1965, 1969), (3) glass beads coated with styrene polymers (Horvath et al., 1967; Bayer et al., 1970b), (4) teflon particles radiation-grafted with polystyrene films (Tregear et al., 1967), and (5) glass beads with chains chemically bonded to the silicic acid surface (Bayer et al., 1970b). The use of a column to hold the coated beads has provided a convenient way to handle the larger quantities of solid support required to produce a given quantity of peptide (Scott et al., 1971).

The polymeric supports allow an essentially unlimited variety of synthetic reactions to be devised and carried out so that the range of the chemistry for the attachment and removal of the peptide chain is not seriously restricted by the polymeric nature of the system. The chemistry is strongly dependent, however, on the particular strategy selected for the synthesis.

The solid-phase approach to the synthesis of a peptide or protein can be divided into at least six substrategies, each of which has its own advantages and disadvantages and requires its own set of chemical reactions and conditions. The synthesis can be either a continuous, stepwise process or a fragment condensation process. The stepwise technique can proceed either by attachment of the C-terminal amino acid to the resin and the addition of activated amino acids to the amino end of the growing peptide chain, or by attachment of the N-terminal amino acid to the resin and addition of subsequent amino acids to the carboxyl end of the activated peptide chain, or by attachment at a side-chain and extension in either direction. A solid-phase–fragment scheme can be designed in three ways. In each case, a small peptide is first synthesized on a solid support and is removed under conditions which give a protected fragment that can be purified and characterized. The pure fragment can then be activated and coupled to a peptide chain that is anchored to a solid support, or the fragment can be reattached to a solid support and then be extended by further stepwise or fragment couplings, or two of the pure protected fragments can be coupled in solution to give a larger peptide. These general schemes are illustrated in Figs. 1–6. Other schemes which make use of solid supports for reagents or for isolating and purifying intermediates will not be included here. The purpose of this chapter is to indicate the various chemical reactions that have been applied to these schemes, with special regard for the combinations of protecting groups that are needed in order to have a compatible set of reactions.

B. Stepwise Approaches

1. Extension of the Chain from the Carboxyl Terminus. The original, and by far the most frequently used, solid-phase approach is the one in which the carboxyl-terminal amino acid of the peptide is bonded to the solid support as outlined in Fig. 1. The initial hope was that the reactions could be developed and refined to the point where the stepwise extension of the peptide chain could be continued without interruption to give the final polypeptide sequence. If the operations were essentially the same for all amino acids and if the cycle time were short, this seemed to offer the best chance for the synthesis of long polypeptides or proteins of defined sequence. The problems, of course, are numerous and must be examined one at a time, and then in combination.

The first synthesis that was at all successful made use of the carbobenzoxy group for α-amino protection. A simple benzyl ester bond to the polystyrene–divinylbenzene support was used to attach the first amino acid. It was not stable enough, however, to allow repeated removal of the Z group, and it was

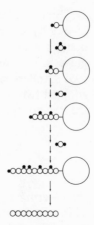

Fig. 1. Stepwise extension of the chain from the carboxyl terminus. The large circle represents the solid support, the small open circles are the amino acid residues, and the filled circles are protecting groups or activating groups.

necessary to either nitrate or brominate the resin to improve the stability and increase the differential. Since the nitrobenzyl ester was then too stable for acidolytic cleavage of the finished peptide from the resin, it was necessary to turn to saponification to remove the peptide:

$$C_6H_5CH_2OCONHCHR^1CONHCHR^2COOCH_2C_6H_3(NO_2)-\textcircled{R} \xrightarrow{\underset{HOAc}{10\% \text{ HBr}}}$$

$$\xrightarrow{NaOH} H_2NCHR^1CONHCHR^2COOH \qquad [1]$$

The symbol $\textcircled{R}$ is used to represent the styrene–divinylbenzene resin. A sufficient degree of selectivity to acidolysis was obtained with the bromo derivative to allow a small peptide to be made in that way (Merrifield, 1963):

$$C_6H_5CH_2OCONHCHR^1CONHCHR^2COOCH_2C_6H_3(Br)-\textcircled{R} \xrightarrow{\underset{HOAc}{10\% \text{ HBr}}}$$

$$H_3\overset{+}{N}CHR^1CONHCHR^2COOCH_2C_6H_3(Br)-\textcircled{R} \qquad [2]$$

$$H_3\overset{+}{N}CHR^1CONHCHR^2COOCH_2C_6H_3(Br)-\textcircled{R} \xrightarrow{\underset{HOAc}{30\% \text{ HBr}}}$$

$$H_3\overset{+}{N}CHR^1CONHCHR^2COOH \qquad [3]$$

Clearly, very much better selectivity was necessary if a multistep synthesis were to be undertaken.

Since it is essential for the α-protecting group to be completely removed ($>99.9\%$) at each step with only small losses ($<1\%$) of the anchoring bond, it is obvious that the rate constants must differ by 10^3 or more. As a move in that direction, the new t-butyloxycarbonyl(Boc) group (Carpino, 1957; Anderson and McGregor, 1957; McKay and Albertson, 1957) was adopted (Merrifield, 1964). Under conditions where we are confident that the Boc group is removed essentially quantitatively

$$(CH_3)_3COCONHCHRCOOH \xrightarrow[CH_2Cl_2]{50\% \, TFA}$$

$$(CH_3)_2C{=}CH_2 + CO_2 + H_3\overset{+}{N}CHRCOOH \qquad [4]$$

the C-terminal benzyl ester is usually lost to the extent of only 1 or 2%. This has, for example, allowed the stepwise synthesis of ribonuclease A (124 amino acids) to be carried out with an overall retention of 17% of the original peptide chains.

Sakakibara *et al.* (1965) replaced the Boc group with the amyloxy-carbonyl (Aoc) group, and Weygand and Ragnarsson (1966) used p-methoxycarbobenzoxy amino acids, which have nearly the same stability to acid as the Boc derivatives. In addition, these derivatives were thought to have the advantages of better crystallinity and higher solubility in the usual solvents. For peptides composed of simple amino acids, therefore, the protection problem was solved, but for trifunctional amino acids the problems have been more difficult.

While it is acceptable to have a small loss of peptide chain from the solid support at each step, it is not acceptable to have similar losses of side-chain protection. When the entire chain is lost one simply suffers a corresponding loss in overall yield, but when side-chain functional groups are exposed there is a probability of side-reactions occurring which will give rise to impurities in the product that are difficult to separate. Thus amino and hydroxy groups will be acylated during coupling steps

$$\begin{array}{c} HO{-}CH_2 \\ | \\ H_2NCHCOO{-}\textcircled{R} \end{array} \xrightarrow[DCC]{BocNHCHRCOOH} \begin{array}{c} BocNHCHRCOOCH_2 \\ | \\ BocNHCHRCONHCHCOO{-}\textcircled{R} \end{array} [5]$$

giving rise to the growth of side-chains, and carboxyl groups can become activated by reagents such as DCC and enter into acylation reactions or undergo intramolecular rearrangement to give a permanently blocked acylurea derivative:

$$\begin{array}{c} CH_2COOH \\ | \\ {-}NHCHCOO{-}\textcircled{R} \end{array} \xrightarrow{DCC} \begin{array}{c} CH_2CON(C_6H_{11})CONH(C_6H_{11}) \\ | \\ {-}NHCHCOO{-}\textcircled{R} \end{array} [6]$$

Problems of this kind have been encountered in the past, particularly with lysine. It has been shown by Yaron and Schlossman (1968) and by Grahl–Nielsen and Tritsch (1969) that the N^ε-carbobenzoxy group on lysine is not entirely stable under the conditions required for the removal of the N^α-Boc group

$$\text{Boc}-\text{Lys(Z)}-\text{O}-\textcircled{R} \xrightarrow[\text{HOAc}]{N\ \text{HCl}} \text{Lys(Z)}-\text{O}-\textcircled{R} + \text{Lys}-\text{O}-\textcircled{R} \qquad [7]$$

leading to the introduction and growth of side-chains at the lysine residues during subsequent stages of the synthesis.

$$\begin{array}{c} \text{NH}_2 \\ | \\ (\text{CH}_2)_4 \\ | \\ \text{H}_2\text{NCHCOO} - \textcircled{R} \end{array} \xrightarrow[\text{DCC}]{\text{BocNHCHRCOOH}} \begin{array}{c} \text{BocNHCHRCONH} \\ | \\ (\text{CH}_2)_4 \\ | \\ \text{BocNHCHRCONHCHCOO} - \textcircled{R} \end{array} \qquad [8]$$

It has also been reported (Nitecki and Goodman, 1970) that the benzyl ester of glutamic acid is not sufficiently stable, although work in this laboratory and elsewhere (F. H. C. Stewart, 1967) has indicated good stability. Ser(Bzl) and Thr(Bzl) are quite stable, but Tyr(Bzl) and Cys(MeOBzl) may be slightly too labile.

Several good studies on the mechanism and rate of cleavage of protecting groups have been made, for example, Blaha and Rudinger (1965), Homer et al. (1965), Losse et al. (1968), and Schnabel et al. (1971). From the data discussed in these papers, it can be seen that a rather wide choice is available, and significant improvements over the earlier solid-phase protecting group combination are now possible. One approach to the side-chain problem is to go to more stable side-chain protection and another is to select more acid-sensitive α-amino-protecting groups.

When Zervas developed his o-nitrophenyl sulfenyl protecting group (Zervas et al., 1963), it was examined for use in solid-phase synthesis (Najjar and Merrifield, 1966; Kessler and Iselin, 1966) and found to be quite suitable. The group could be removed under such mild acidic conditions (0.15 N HCl in acetic acid–chloroform) that essentially no loss of benzyl esters was observed. It was also removable with thioacetamide and other nucleophiles:

$$\text{C}_6\text{H}_4(\text{NO}_2)\text{SNHCHRCOOH} \xrightarrow{\text{CH}_3\text{CSNH}_2} \text{C}_6\text{H}_4(\text{NO}_2)\text{SSC(CH}_3)\text{=NH} +$$

$$\text{H}_2\text{NCHRCOOH} \qquad [9]$$

In a very valuable study, Sieber and Iselin (1968) explored a number of modified urethan protecting groups with acid sensitivities relative to Boc ranging from 1 to 66,000. They recommended biphenylisopropyloxycarbonyl (Bpoc) derivatives (ration 3000:1) as especially suitable for use in conjunction with Boc and t-butyl derivatives and, of course, with any more stable group:

$$C_6H_5-C_6H_4C(CH_3)_2OCONHCHRCOOH \xrightarrow[CH_2Cl_2]{0.5\% \text{ TFA}}$$

$$C_6H_5-C_6H_4C(CH_3)=CH_2 + CO_2 + H_2NCHRCOOH \qquad [10]$$

The Bpoc group has now been applied to solid-phase syntheses (Sieber and Iselin, 1968; Wang and Merrifield, 1969) and provides more than enough differential between the amino group and the α-benzyl ester anchoring bond or the side-chain benzyl derivatives. The selectivity with side-chain Boc or t-butyl derivatives $vs.$ Bpoc is similar to the selectivity in the Bzl $vs.$ Boc situation and may become marginal with very long peptide chains.

The very acid-sensitive amino acid derivatives are not without their problems. They cannot be stored as free acids for extended periods unless extreme precautions are taken, and they are therefore usually kept as salts of dicyclohexylamine or other organic bases. This in turn requires an extra step in the synthesis to release the free acid for coupling. There is, in addition, the possibility that the protecting group can be lost during the coupling and allow the introduction of two residues instead of only one,

$$\text{Bpoc}-\text{Ala} + \text{H}-\text{Leu}-\text{O}-\textcircled{R} \xrightarrow{\text{DCC}} \text{Bpoc}-\text{Ala}-\text{Leu}-\textcircled{R} \rightarrow$$

$$\text{H}-\text{Ala}-\text{Leu}-\text{O}-\textcircled{R} \longrightarrow \text{Bpoc}-\text{Ala}-\text{Ala}-\text{Leu}-\text{O}-\textcircled{R} \quad [11]$$

although this has not actually been observed to occur.

Several attempts have been made to construct side-chain derivatives for lysine with increased acid stability for use in multistep solid-phase syntheses in combination with N^α-Boc protection. Sakakibara et al. (1970) found N^ε-diisopropylmethyloxycarbonyl-lysine (Dipmoc-Lys) to be especially suitable. It was essential stable in N HCl–acetic acid at 20°C for 24 hr, whereas the carbobenzoxy protecting group was removed from N^ε-Z-lysine to the extent of 20% in the same time, and N^ε-4-Cl-Z-lysine was 8% cleaved under these conditions. On the other hand, the Dipmoc group was completely removed from Z-Pro-Lys(Dipmoc)-Gly-NH$_2$ by HF in 60 min at 20°C. Several other urethan derivatives of lysine were examined by Noda et al. (1969). They found Lys (4-CN-Z) and Lys(4-NO$_2$-Z) to be resistant to HF but the 4-chloro and 3-chloro derivatives to be quantitatively removed in 2–3 hr. The cleavage in 30% HBr-HOAc was faster, and even Lys(4-NO$_2$-Z) was deprotected in 8 hr. In the 4 N HCl–dioxane used for α-Boc deprotection, Lys(Z) was removed to the extent of 5–10% in 12 hr, whereas Lys(4-Cl-Z) went 2–4% and Lys(3-Cl-Z) only 1% under these conditions. Thus the stability was improved severalfold over carbobenzoxy, and the differential between the side-chain protection and N^α-Boc was increased to much more satisfactory levels. Recent work by Yamashiro and Li (1972) has shown that

Lys(2-Br-Z) and Lys(4-Br-Z) are also considerably more stable to the deprotection conditions of solid-phase peptide synthesis than is Lys(Z) and are more suitable for multistep synthesis by this method. The relative rates in 50% TFA-CH$_2$Cl$_2$ were approximately 1:15:60.

Quantitative evaluation of the stabilities of some new lysine derivatives has been carried out in our laboratory recently (Erickson and Merrifield, 1972). The apparent first-order rate constants for the deprotection of seven N^ε-carbobenzoxy-lysine derivatives with 50% trifluoracetic acid in methylene chloride at 20°C were measured. The stabilities relative to Lys(Z) are shown in Table I. The 2-chloro, 2,4-dichloro, and 3,4-dichloro derivatives are

Table I. Relative Stability of N^ε-Lysine Derivatives

N^ε substituent	Relative stability,[a] to 50% TFA-CH$_2$Cl$_2$, 20°C
Z	1
4-Cl-Z	3
2-Cl-Z	60
2,4-diCl-Z	88
3,4-diCl-Z	170
2,6-diCl-Z	790
3-Cl-Z	1000

[a]The N^ε-lysine derivative and free lysine were quantitatively measured on the amino acid analyzer.

considered to have the desired stabilities to acid for use in conjunction with α-Boc protection. The ratio of α to ε deprotection rates is greater than 10^5, which means that when 99.99% of the Boc group is removed less than 0.01% of the carbobenzoxy derivative will have been lost. The new Boc-Lys(2,4-Cl$_2$-Z) reagent was tested in the system of Yaron and Schlossman (1968) and shown to completely prevent the formation of branched-chain polylysine products.

Efforts to devise more stable tyrosine-protecting groups have been made by Yamashiro et al. (1972) and by Erickson and Merrifield (1972). Li's group found Tyr(3-Br-Bzl) and Tyr(2,6-Cl$_2$-Bzl) to be approximately 50 times more stable to 50% TFA than Tyr(Bzl) itself. However, about 10% of a side-reaction product was observed on removal of these protecting groups with HF, even in the presence of excess anisole. Erickson and Merrifield (1972) also found the relative rates of cleavages of Tyr(Bzl) and Tyr(2,6-Cl$_2$-Bzl) to be 50:1 and showed the product from O-2,6-dichlorobenzyl-tyrosine to be 3(2,6-dichlorobenzyl)-tyrosine, which was formed by an intramolecular rearrangement from the ether. In contrast, Tyr(Z), which was 25 times more stable than Tyr(Bzyl), gave tyrosine as the sole product after HF deprotection.

Two of the remaining amino acids that have been problems in solid-phase synthesis are histidine and arginine. The introduction of N^{im}-dinitrophenyl-histidine (Shaltiel, 1967) and N^{im}-tosyl-histidine (Fujii and Sakakibara, 1969) has greatly improved the situation for two reasons. These groups are much more readily and conveniently removed than the traditional N^{im}-benzyl group, either by thiolysis for DNP, or by HF for tosyl, and they greatly minimize the racemization problem discovered by Windridge and Jorgensen (1971). Similarly, Arg(Tos) has been an improvement over Arg(NO_2). It has good stability during the synthesis but is readily removed by HF with less danger of byproduct formation. There may be some merit in protecting the amides of Asn and Gln as the 2,4-dimethoxybenzyl (Pietta *et al.*, 1971) or 4,4'-dimethoxybenzydryl (Mbh) (König and Geiger, 1970*b*) derivatives, although the problem has not been fully examined.

At the moment, the best protecting groups for use in conjunction with α-Boc in a normal stepwise solid-phase synthesis appear to be Ser(Bzl), Thr(Bzl), Tyr(2,6-Cl_2-Bzl), Cys(MeOBzl), Met(O), Asn(Mbh), Gln(Mbh), Asp(OBzl), Glu(OBzl), His(Tos), and Lys(2,4-Cl_2Z).

2. Extension of the Chain from the Amino Terminus. As can be seen from Fig. 2, extension of the chain from the amino terminus begins by anchoring the amino group of the *N*-terminal amino acid of the proposed peptide chain to a solid support followed by the extension of the chain through condensation of the carboxyl group to the next amino acid. Since this is the direction in which the chain grows during protein synthesis *in vivo*, it might appear that a scheme designed along the same lines would give the most effective chemical synthesis. However, there are several reasons why this does not appear to be the case.

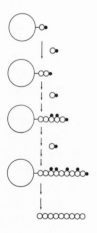

Fig. 2. Stepwise extension of the chain from the amino terminus.

All chemical methods for the activation of an amino acid for the coupling reaction involve activation of the carboxyl group. Therefore, when a peptide chain is to be extended at its carboxyl end by addition of another amino acid it is the carboxyl of the *peptide* which must be activated, not that of the *amino acid*. This has three important consequences in the present context. It means that the activation step must be very efficient; otherwise, part of the growing chain will not be activated and will not couple at that point. It also means that all of the activated chain must then couple to the incoming amino acid derivative but that it cannot be used in excess to force the reaction to completion. The monomer unit can be in excess but not the activated component. It also means that any side-reaction which the activated component may undergo will correspondingly lower the yield of the desired peptide and at the same time introduce a problem of separation of the resulting byproduct. Finally, activation of the peptide carboxy will greatly limit the possibilities for coupling reactions because of the increased chances of racemization. It is well documented that urethan-protected *amino acids* are resistant to racemization during the activation and coupling reactions, whereas the same reactions with protected *peptides* will lead to considerable amounts of racemic products. Under certain conditions, for example, with the classical azide method (Curtius, 1902) or the newer dicyclohexylcarbodiimide–hydroxysuccimide or dicyclohexylcarbodiimide–hydroxybenzotriazole methods (König and Geiger, 1970a) this effect can be minimized, but some potential danger always seems to be present.

The early work of Letsinger and Kornet (1963) made use of this approach by preparing a chloroformate derivative of a polystyrene–divinylbenzene popcorn polymer and coupling leucine ethyl ester to it to give a substituted carbobenzoxy derivative:

$$\text{(R)}-C_6H_4CH_2OCOCl + H_2NCHRCOOC_2H_5 \rightarrow$$
$$\text{(R)}-C_6H_4CH_2OCONHCHRCOOC_2H_5 \qquad [12]$$

Saponification of the ester, activation as the mixed anhydride, and coupling with the glycine ethyl ester gave the first peptide bond. In principle, then, this set of reactions could have been used to extend the chain to give longer peptides, although problems with yields and optical purity could be expected.

In an effort to further test this approach with the aid of better protecting groups and coupling reactions, Felix and Merrifield (1970) studied the use of the azide method. Styrene–divinylbenzene beads were used, and the first amino acid was again attached as a substituted *N*-benzylocarbonyl derivative, but this time the carboxyl was protected as Boc-hydrazide, which could be deprotected and converted to the azide with *N*-butyl nitrite (Honzl and

Rudinger, 1961)

$$\text{(R)}-C_6H_4CH_2OCONHCHRCONHNHCOOC(CH_3)_3 \xrightarrow{\text{TFA}} \xrightarrow{C_4H_9ONO}$$

$$\text{(R)}-C_6H_4CH_2OCONHCHRCON_3 \qquad\qquad [13]$$

and coupled to another aminoacyl-Boc-hydrazide. Conditions were devised
which gave moderate yields of an optically pure model tetrapeptide. This
was not considered to be a replacement for the usual stepwise solid-phase
method, but it was thought to be useful for application in certain special
problems. The work also showed that the Curtius rearrangement did not
occur under these azide coupling conditions.

At the present time, it is not possible to mimic effectively the biological
scheme for protein synthesis in the laboratory. It is far too complex, and the
extreme selectivity and specificity of the reactions cannot be duplicated.

3. Attachment of the Peptide Through a Side-Chain. It should be possible to make use of the side-chain of any of the trifunctional amino acids
for the anchoring site in a solid-phase synthesis. With the side-chain covalently bound, the peptide chain could be extended in either direction: by
addition of an activated amino acid at the amino terminus or by activation
of the peptide carboxyl and extension at the carboxyl end (Fig. 3).

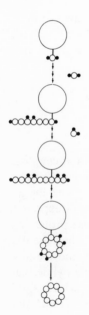

Fig. 3. Attachment of the chain through a side-chain and stepwise
extension at either end.

Several of the early experiments using this principle and some of the potential approaches have already been discussed (Marshall and Merrifield, 1971). The first successful scheme of this type was by Sklyarov and Shashkova (1969), who made use of the δ-amino group of ornithine for the synthesis of a glycine analogue of gramicidin S:

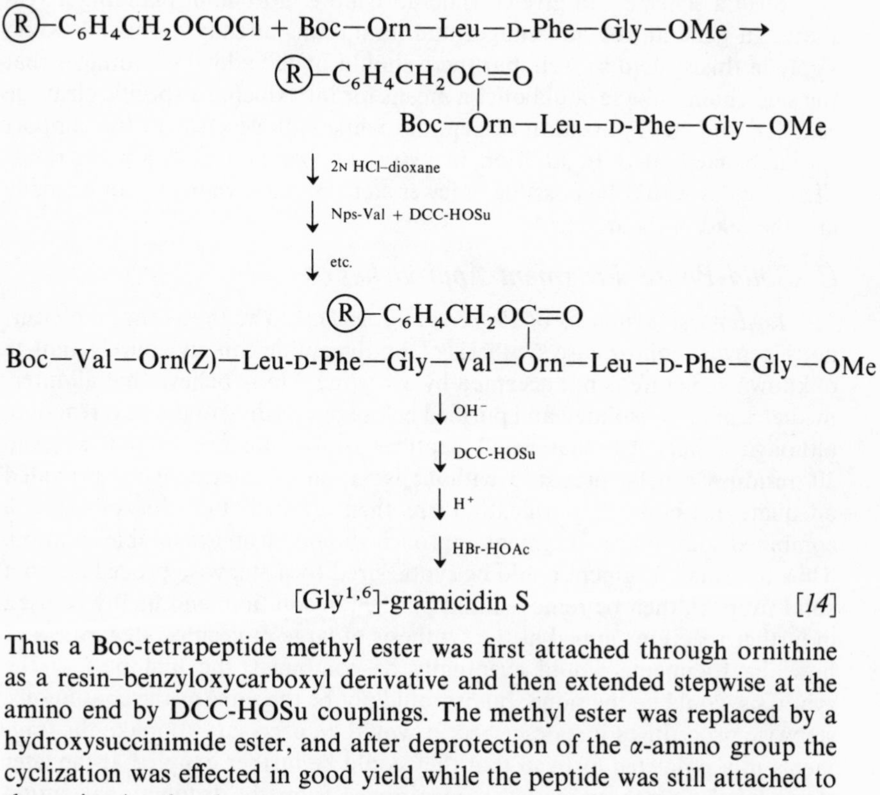

$$\text{(R)}-C_6H_4CH_2OCOCl + Boc-Orn-Leu-D\text{-Phe}-Gly-OMe \rightarrow$$

$$\text{(R)}-C_6H_4CH_2OC=O$$
$$|$$
$$Boc-Orn-Leu-D\text{-Phe}-Gly-OMe$$

$\downarrow$ 2N HCl–dioxane

$\downarrow$ Nps-Val + DCC-HOSu

$\downarrow$ etc.

$$\text{(R)}-C_6H_4CH_2OC=O$$
$$|$$
$$Boc-Val-Orn(Z)-Leu-D\text{-Phe}-Gly-Val-Orn-Leu-D\text{-Phe}-Gly-OMe$$

$\downarrow$ OH⁻

$\downarrow$ DCC-HOSu

$\downarrow$ H⁺

$\downarrow$ HBr-HOAc

$$[Gly^{1,6}]\text{-gramicidin S} \qquad [14]$$

Thus a Boc-tetrapeptide methyl ester was first attached through ornithine as a resin–benzyloxycarboxyl derivative and then extended stepwise at the amino end by DCC-HOSu couplings. The methyl ester was replaced by a hydroxysuccinimide ester, and after deprotection of the α-amino group the cyclization was effected in good yield while the peptide was still attached to the resin. A related approach was employed by Meienhofer and Trzeciak (1971) for a synthesis of [8-lysine]-vasopressin, as shown in equation [15]:

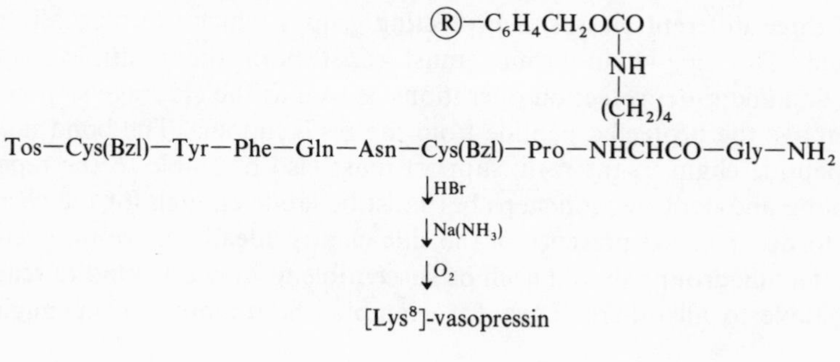

$$\text{(R)}-C_6H_4CH_2OCO$$
$$|$$
$$NH$$
$$|$$
$$(CH_2)_4$$
$$|$$
$$Tos-Cys(Bzl)-Tyr-Phe-Gln-Asn-Cys(Bzl)-Pro-NHCHCO-Gly-NH_2$$

$\downarrow$ HBr

$\downarrow$ Na(NH₃)

$\downarrow$ O₂

$$[Lys^8]\text{-vasopressin} \qquad [15]$$

The attachment of aspartic acid or glutamic acid as a benzyl ester could also serve in a side-chain scheme for extension from either end. Krumdieck and Baugh (1969) used an α-ester linkage in their synthesis of poly-γ-glutamyl folic acid. Anchoring through the sulfhydryl of cysteine also appears to be feasible and to offer certain advantages.

Such a scheme will give considerably more latitude in planning a synthesis. In general, the advantages and limitations already discussed would apply in this system as well, but there should be the added advantages that the side-chain linkage could offer a means for introducing a specific cleavage reaction and the cyclization of peptides while still attached to the support should be facilitated. In addition, in a structure–activity study a wider range of analogues would be possible in fewer steps because changes can be made at either end of the molecule.

C. Solid-Phase–Fragment Approaches

1. Stepwise Synthesis of Protected Fragments. The thesis that a continuous stepwise solid-phase synthesis of a polypeptide can give pure products of known structure is not accepted by everyone. Many believe that all intermediates must be isolated and purified before proceeding to the next reaction, although others feel that small peptides containing five to ten, or even 20, residues can be prepared without isolation of intermediates provided adequate methods of purification are then applied. For this purpose, a combined solid-phase–fragment approach should be an acceptable solution. Thus the small fragment could be synthesized by a stepwise procedure on a solid support, then be removed and purified in solution, and finally be used in further reactions aimed at the synthesis of large molecules. Regardless of how the fragments would eventually be combined, the first part of the synthesis would be the same; but it would not be the same as the continuous, stepwise procedures just described. It would be necessary to make the fragments in a *protected* form so that they could be further coupled at the later stages of the synthesis without interference from the trifunctional amino acids. For that reason, a new set of conditions and reactions must be devised.

The requirements for a general protected-fragment synthesis are for at least three different classes of protecting groups which are mutually compatible. The side-chain groups must resist both the multiple stepwise coupling and α-deprotection operations as well as the cleavage step needed to remove the protected peptide from the resin support. The bond holding the peptide chain to the resin support must also be stable to the repeated coupling and deprotection steps but must be labile enough for the cleavage step to occur in the presence of the side-chains. Ideally, these three classes of protecting groups should each be susceptible to only one kind of reaction and stable to all others. Thus, for example, the α-amino group might be sensitive to acid, the carboxyl attachment to base, and the side-chains to hydrogenation. Alternatively, a group susceptible to oxidation or reduction or photolytic cleavage might be introduced, or a safety-catch principle might be applied. To date, no such general and highly selective scheme has been devised which is completely free of complications.

The first step in that direction was by Marshall and Merrifield in 1965, when they made the angiotensin II 2–8 heptapeptide on the usual S-DVB solid support, cleaved it in HBr-TFA, and then coupled it with α- or β-benzyl Boc-aspartate to give the two octapeptide angiotensin II isomers:

$$H-Arg(NO_2)-Val-Try-Ile-His(Bzl)-Pro-Phe-OH \xrightarrow{\text{Z-Asp(OBzl)-OH}}$$

$$\xrightarrow{\text{H}_2\text{-Pd}} \text{angiotensin II} \qquad\qquad [16]$$

This specialized technique was elaborated on by Anfinsen et al. (1967), but it does not give the general procedure that is needed. The same can be said for the use of hydrazinolysis (Ohno and Anfinsen, 1967) to give protected hydrazide fragments which can then be coupled as azides in further fragment condensations, since hydrazinolysis is limited to peptides not containing side-chain esters and may also be dangerous in the case of arginine-containing peptides.

A different kind of selectively cleavable attachment to the resin support was devised by Weygand (1968) and independently by Mizoguchi et al. (1969). They have synthesized phenacyl esters which can be cleaved by hydrolysis, ammonolysis, or thiophenolate:

$$ClCH_2COCl + C_6H_5-\textcircled{R} \xrightarrow{\text{AlCl}_3} ClCH_2COC_6H_4-\textcircled{R} \rightarrow$$

$$BocNHCHRCOOCH_2COC_6H-\textcircled{R} \xrightarrow{\text{C}_6\text{H}_5\text{S}^-} BocNHCHRCOO- \quad [17]$$

The latter, when used in conjunction with various acid-sensitive protecting groups, provides a very useful route to protected peptide intermediates.

The "safety-catch" principle has been applied in several instances. It depends on having a carboxyl attachment to the resin support which is stable to the synthetic reactions required to assemble the protected peptide chain but which can be selectively activated at the proper time by another kind of reaction. The activated derivatives can then be removed from the resin in a protected form suitable for further fragment condensations. D. L. Marshall and Liener (1970) and Flanigan and G. R. Marshall (1970) have employed a phenolic sulfide resin which could be activated by oxidation with hydrogen peroxide or m-chloroperbenzoic acid to the sulfone. This product is now an active ester that can be cleaved from the resin by saponification, aminolysis, or hydrazinolysis:

$$\text{BocNHCHR}^1\text{CONHCHR}^2\text{COOC}_6\text{H}_4\text{SCH}_2\text{C}_6\text{H}_4-\text{(R)}$$

$$\downarrow \text{H}_2\text{O}_2\text{-HOAc}$$

$$\text{BocNHCHR}^1\text{CONHCHR}^2\text{COOC}_6\text{H}_4\text{SO}_2\text{CH}_2\text{C}_6\text{H}_4-\text{(R)}$$

$$\downarrow \text{H}_2\text{NCHR}^3\text{COOCH}_2\text{C}_6\text{H}_5$$

$$\text{BocNHCHR}^1\text{CONHCHR}^2\text{CONHCHR}^3\text{COOCH}_2\text{C}_6\text{H}_5 \qquad [18]$$

The generality of this support is limited by the oxidation conditions, which preclude its use in conjunction with peptides containing methionine, cysteine, or tryptophan. However, in favorable circumstances these residues might be introduced at the aminolysis step. The method of Wieland *et al.* (1970)

$$\text{BocNHCHR}^1\text{CONHNH}_2 + \text{ClCH}_2-\text{(R)}$$

$$\downarrow$$

$$\text{BocNHCHR}^1\text{CONHNHCH}_2-\text{(R)}$$

$$\downarrow \text{H}^+$$

$$\downarrow \text{ZNHCHR}^2\text{COOH} + \text{DCC}$$

$$\text{ZNHCHR}^2\text{CONHCHR}^1\text{CONHNHCH}_2-\text{(R)}$$

$$\mid \text{NBS}$$

$$\text{ZNHCHR}^2\text{CONHCHR}^1\text{CON}{=}\text{NCH}_2-\text{(R)}$$

$$\downarrow \text{H}_2\text{NCH}_2\text{C}_6\text{H}_5$$

$$\text{ZNHCHR}^2\text{CONHCHR}^1\text{CONHCH}_2\text{C}_6\text{H}_5 + \text{N}_2 + \text{CH}_3-\text{(R)}_{[19]}$$

could also be useful for many peptides but is similarly limited because of an oxidative step in which a protected peptide hydrazide resin is activated before cleavage.

The new safety-catch scheme of Kenner *et al.* (1971) avoids an oxidation step and the accompanying problems and appears to have many of the desirable features required of such a method. The acylsulfonamide bond is completely stable to the trifluoroacetic acid or HBr–acetic acid used for N^α-deprotection, and because of its acidic character the sulfonamide was not further acylated during the chain elongation steps with amino acid active esters in the presence of 2,6-lutidine. The acylsulfonamide bond was labilized by methylation under very mild conditions with diazomethane in ether–acetone solvent:

$$BocNHCHR^1COOC_6H_4(NO_2) + H_2NCHR^2CONHSO_2C_6H_4 - Ⓡ$$

$$\downarrow \text{2,6-lutidine}$$

$$BocNHCHR^1CONHCHR^2CONHSO_2C_6H_4 - Ⓡ$$

$$\downarrow \text{CH}_2\text{N}_2$$

$$\begin{array}{c} CH_3 \\ | \\ BocNHCHR^1CONHCHR^2CONSO_2C_6H_4 - Ⓡ \end{array}$$

$$\downarrow \text{NH}_2\text{NH}_2$$

$$BocNHCHR^1CONHCHR^2CONHNH_2 \qquad [20]$$

The unchanged sulfonamide was then susceptible to cleavage by NaOH, NH_3, or NH_2NH_2. Unfortunately, as the authors point out, this scheme also suffers to some extent from racemization of the C-terminal residue, and $\alpha \rightarrow \beta$ rearrangement of the asypartyl–glycine sequence was observed. This emphasized the difficulties that have been faced in all the efforts to find fully compatible conditions for solid-phase synthesis.

Several schemes for the synthesis of protected peptide fragments have been based on three different degrees of sensitivity to acid. Although this does not meet the ideal of having each class of protecting group susceptible to a different kind of reaction, it has turned out in practice to be one of the best techniques at the moment.

The method devised by Wang and Merrifield (1969) makes use of the very acid-labile 2-(4-biphenyl)-2-propyloxycarbonyl (Bpoc) group (Sieber and Iselin, 1968) for α-amino protection. Side-chain protection is with relatively stable benzyl or tosyl derivatives, and the carboxyl attachment to the resin support is through a t-alkyl ester or t-alkyloxycarbonylhydrazide derivative of intermediate stability to acid:

$$\underset{0.5\%\,TFA}{\underset{CH_3}{\overset{CH_3}{C_6H_5-C_6H_4\underset{|}{C}-O-\overset{O}{\overset{\|}{C}}-NHCHCONHCHRCONHNHCO}}}\overset{CH_2-O\overset{\diagup}{}CH_2C_6H_5}{}\underset{50\%\,TFA}{\overset{\overset{HF}{\diagup}}{\overset{O}{\overset{\|}{-C}}\underset{CH_3}{\overset{CH_3}{\underset{|}{C}}-CH_2CH_2C_6H_4-Ⓡ}}} \qquad [21]$$

In this way, the stepwise removal of the Bpoc group can be accomplished with 0.5 % trifluoroacetic acid in methylene chloride with essentially no loss of side-chain protection and with only minimal losses from the resin. The final cleavage from the resin to give either the carboxyl or hydrazide derivative can be achieved with 50 % trifluoroacetic acid without appreciable losses of side-chain groups. If the final protected fragment requires a blocked amino group, the last coupling step can be made with a more stable derivative such as a carbobenzoxyamino acid. The hydrazide procedure was of quite general utility, while the t-alkyl alcohol–resin was limited in its usefulness because of the difficulty in carrying out the esterification step without accompanying racemization. Wang (1972) has now devised a new resin–ester attachment which overcomes this problem. It involves the synthesis of the p-alkoxybenzyl alcohol derivative of copolystyrene–divinylbenzene. This alcohol is readily esterified without racemization and is susceptible to cleavage in 50% trifluoroacetic acid:

$$HOCH_2C_6H_4OCH_2C_6H_4-\textcircled{R}\xrightarrow{\text{Bpoc-Ala}}$$

$$Bpoc-Ala-O\underset{\underset{50\%\,TFA}{\uparrow}}{-}CH_2C_6H_4OCH_2C_6H_4-\textcircled{R}$$

$$[22]$$

Other resin attachments of intermediate stability to acid have also been made which can be incorporated into similar schemes. Wildi and Johnson (1968) prepared a polymeric hydroxysuccinimidomethyl support, and Southard et al. (1969) prepared a benzhydryl alcohol derivative of a crosslinked polystyrene resin. Both were converted to the alkylhalides and esterified with N^α-protected amino acids. The 1-benzoylisopropenyl (Bip) group was found to be compatible with the benzhydryl ester system

$$C_6H_5COCH=C(CH_3)\underset{\underset{0.4\,N\,HCl}{\nearrow}}{-}NHCHRCOO\underset{\underset{50\%\,TFA}{\searrow}}{-}CH(C_6H_5)C_6H_4-\textcircled{R}$$

$$[23]$$

and was applied to the synthesis of Leu-Ala-Gly-Val. Although not yet tested, it is possible that the Bpoc and Bip groups, or the o-nitrophenylsulfenyl protecting group of Zervas et al. (1963), could be used interchangeably as the acid-labile group in conjunction with any of the anchoring bonds just described.

These are some of the ways in which protected fragments have been synthesized by solid-phase procedures. Several of the fragments have crystallized directly, while some have required purification by other procedures. When the pure fragment is in hand, it may then be incorporated into larger peptides or proteins by at least three general approaches.

2. Fragment Coupling in Solution. The most direct solid-phase–fragment procedure (shown in Fig. 4) is to conduct the fragment coupling in solution by conventional methods. If the protected carboxyl-component fragment is in the form of the hydrazide, it is converted to the azide for

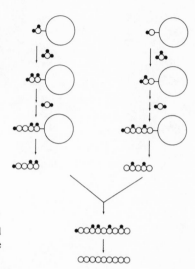

Fig. 4. Fragment coupling in solution of protected peptides synthesized by the stepwise solid-phase method.

coupling. If it has been obtained as the free acid, it must be coupled by another suitable nonracemizing procedure. As with any fragment procedure, the racemization problem can be minimized by designing the synthesis with C-terminal glycine or proline residues. The following examples will serve to illustrate this approach, although there has been no good test in which a large polypeptide has been prepared in this way.

Ohno *et al.* (1969) have prepared the protected tetrapeptide H-Lys(TFA)-Leu-Asn-Ile-OH by a solid-phase synthesis and have coupled it in solution with Boc-Gly(OBzl)-OSu in the presence of N-methyl morpholine to give the crystalline protected pentapeptide. The latter was then coupled by the DCC-HOSu method with a C-terminal nonapeptide ester, which had been prepared by stepwise synthesis in solution to give the protected tetra-decapeptide Boc-Glu(OBzl)-Lys(TFA)-Leu-Asn-Ile-Trp-Ser(Bzl)-Glu(OBzl)-Asn-Asp(OBzl)-Ala-Asp(OBzl)-Ser(Bzl)-Gly-OBzl, corresponding to residues 135–148 of the staphylococcus nuclease. The product was purified by silica gel chromatography.

The use of an azide coupling of protected fragments in solution was demonstrated by Wang and Merrifield (1971). Crystalline Z-Asp(OBzl)-Arg(NO$_2$)-Val-Tyr(Bzl)-NHNH$_2$ was obtained by acid cleavage of the corresponding *t*-alkyloxycarboxylhydrazide–resin. It was converted to the azide with isoamylnitrite and coupled in DMF with H-Val-His(Bzl)-Pro-Phe-OH which had been synthesized on the *t*-alkyl alcohol–resin. Conversely, a protected peptide made on the *t*-alkyl alcohol–resin could be used as the carboxyl component of a fragment condensation. Thus Z-Lys(Z)-Phe-Phe-

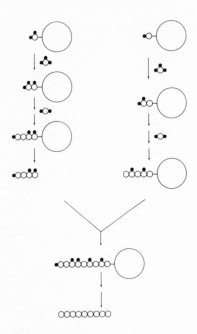

Fig. 5. Fragment coupling to a peptide–resin.

Gly-OC(CH$_3$)$_2$-CH$_2$-CH$_2$-C$_6$H$_4$-resin was synthesized stepwise from Bpoc-amino acids and cleaved in 50% TFA-CH$_2$Cl$_2$. The crystalline protected tetrapeptide was activated with DCC and coupled in solution with H-Leu-Met-OCH$_3$. The hexapeptide was first converted to the amide and then deprotected with TFA at 80° to give the hexapeptide amide analog of eledoisin, H-Lys-Phe-Phe-Gly-Leu-Met-NH$_2$. On a molar basis, it was as active as eledoisin itself in lowering the blood pressure of the rat.

3. Fragment Coupling to a Peptide–Resin. It has been possible in several instances to couple small peptides to a resin-bound peptide chain (Fig. 5). This approach has the important advantage that the peptide chain will grow by several residues at a time, and the physical and chemical properties of the reactants and products will therefore differ from one another by more than if only a single amino acid residue had been added. This allows a better purification to be achieved. Probably the first example was by Weygand and Ragnarsson (1966), who carefully pointed out the above advantages. They coupled pMZ-Leu-Phe to Val-resin by the DCC-HOSu method to give pMZ-Leu-Phe-Val-resin, and later Boc-Pro-Pro-Phe-Phe-OSu was coupled to Ala-Ala-OCH$_2$-CO-resin to give the pure hexapeptide after cleavage and deprotection (Weygand, 1968). Although these examples do not contain trifunctional amino acids, it appears that the scheme would lend itself to such applications. Rothe and Dunkel (1971) prepared a series of nylon oliogomers, $H-[NH(CH_2)_5CO]_n-OH$, where n ranged from 3 to 25. To increase the differences in properties of the members of their series, they coupled ε-aminocaproic acid dimers or tetramers instead of the single amino acid.

This method was also examined by Omenn and Anfinsen (1968). They coupled a soluble protected di-, tri-, or tetrapeptide to the amino terminus of another peptide anchored to a solid support. To minimize racemization, either the azide, or DCC-HOSu, or N-ethyl-5-phenylisoxazolium-3′-sulfonate (NEPIS) coupling methods were applied. Thus Boc-Glu(OBzl)-Lys(Z)-Lys(Z)-Ser(Bzl)-OH was coupled in fourfold excess with H-Leu-Pro-resin with NEPIS in DMF. The product was cleaved in HBr-TFA and obtained in 90 % yield in an excellent state of purity with no racemization. The same peptide in equivalent quality was obtained by DCC-HOSu coupling for 2 hr at room temperature. Coupling Boc-Leu-Ala-Tyr to $H-[Lys(TFA)]_5-$resin by NEPIS was satisfactory, but an azide coupling was incomplete, giving about 30% unreacted pentalysine derivative. Likewise, staphylococcal nuclease 32–39 octapeptide gave only 20–30% yield when coupled to a nonapeptide–resin.

The most successful application of this fragment approach to a high molecular weight polypeptide was made by Sakakibara et al. (1968). They first coupled Aoc-Pro-Pro to Gly-resin with DCC and then extended the chain in a stepwise manner by coupling Aoc-Pro-Pro-Gly. Products containing 10 or 20 tripeptide units were obtained. They were essentially monodisperse and in dilute solution formed triple-helical structures which resembled natural collagen in regard to their sharp transition temperatures. The $H-(Pro-Pro-Gly)_{10}-OH$ was obtained in crystalline form.

4. Reattachment of a Protected Fragment to a Solid Support and Extension of the Chain in the Solid Phase. Yajima et al. (1970) have attached Boc-Lys(Z)-Pro-Val-Gly-OH to a bromomethylated S-DVB polymer and, following α-deprotection, have coupled $Z-Glu(OBzl)-His-Phe-Arg(NO_2)-$Trp-Gly-OH by means of DCC. The free peptide was obtained after cleavage and deprotection with HF and was purified by column chromatography on CM-cellulose. A soluble byproduct, presumably the hexapeptidylurea, was isolated from the resin filtrates after the condensation. It was pointed out that product and byproduct behaved very similarly on the ion exchange columns and that normal fragment condensation in solution would have given a mixture difficult to separate. The work demonstrated the value of this form of fragment synthesis and showed that it is possible to attach peptides of moderate size to resin supports and to couple rather large and complex fragments to peptide–resins. This approach is illustrated in Fig. 5.

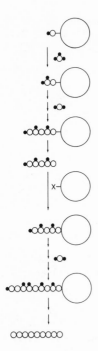

Fig. 6. Reattachment of a protected fragment to a solid support followed by stepwise extension of the chain in the solid phase.

A further permutation of the ways to combine solid-phase and fragment synthesis was suggested by Marshall and Merrifield (1971). It should be possible to synthesize protected fragments on solid supports as already described, to remove them for purification, to reattach them to resin supports, and then to extend them stepwise (Fig. 6). After suitable numbers of residues (perhaps five or ten) have been added, the new peptide could be removed for purification and the chain could then be extended further by a repetition of the cycle. This procedure would have the advantage of rapid stepwise synthesis coupled with a provision for a conventional purification step at frequent intervals. Since it does not require the coupling together of fragments, it avoids the low yields that often accompany such condensations. On the other hand, it requires the attachment of large peptides to the solid support, and although this has not been extensively tested it is likely to be rate limiting. Yajima (personal communication) observed that the efficiency of the attachment decreased as the size of the peptide increased. To carry this idea very far would require new ways of linking the peptide to the support and probably new supports as well.

III. SUMMARY AND CONCLUSIONS

A careful consideration of the substrategies of solid-phase peptide synthesis and the several variations of protecting group combinations that have been examined leads me to conclude that there are three general procedures which are to be preferred at this time. For special purposes, however, each of the techniques has certain advantages. For a problem involving the synthesis of a large polypeptide or small protein where it is absolutely essential to know the identity of all intermediates and to know that the product is free of all closely related peptides, a solid-phase–fragment approach is safest. At the moment, the most practical variant appears to be the coupling of purified protected fragments in solution. The added security of this approach must be weighed against the lower yields, increased time and manpower requirements, and special solubility problems that can be expected.

For problems in which the possibility of some quantity of closely related impurities can be tolerated, the continuous stepwise approach is to be preferred. In this case, the great saving of time and effort for the synthesis and increased yield must be weighed against the added time required for purification and characterization of the product. The stepwise synthesis from the carboxyl terminus is clearly to be preferred over extension of the chain from the amino end. Whether the combination of α-amino- and side-chain-protecting groups should be shifted toward more stable ones or more labile ones is not well established, since each has its advantages and its limitations. It is certain, however, that the selectivity of removal must be quite great. A difference of 10^4 or more in deprotection rates is required if a satisfactory synthesis of a small protein is to be expected.

The definite improvements in conditions for solid-phase synthesis which have occurred during the last 10 years encourage me to believe that similar developments and refinements can also be expected in the future. If they are forthcoming, then I think we have not nearly reached the limits of the peptides that can be synthesized or of the problems that can be attacked by the synthetic approach.

REFERENCES

Albertson, P. A., 1960, "Partition of Particles and Macromolecules," Wiley, New York.

Anderson, G. W., and McGregor, A. C., 1957, *t*-Butyloxycarbonyl amino acids and their use in peptide synthesis, *J. Am. Chem. Soc.* **79**:6180.

Anfinsen, C. B., Ontjes, D., Ohno, M., Corley, M., and Eastlake, A., 1967, The synthesis of protected peptide fragments of a staphylococcal nuclease, *Proc. Natl. Acad. Sci.* **58**:1806.

Bayer, E., Eckstein, H., Hagele, K., Koenig, W. A., Bruning, W., Hagenmeier, H., and Parr, W., 1970a, Failure sequences in the solid phase synthesis of polypeptides, *J. Am. Chem. Soc.* **92**: 1735.

Bayer, E., Jung, G., Halasz, I., and Sebastian, I., 1970b, A new support for polypeptide synthesis in columns, *Tetrahedron Letters* **1970**:4503.

Bergmann, M., and Zervas, L., 1932, Concerning a general method peptide synthesis, *Berichte* **65**:1192.

Blaha, K., and Rudinger, J., 1965, Rates of fission of some substituted benzyloxycarbonyglycines and two heterocyclic analogs with hydrogen bromide in acetic acid, *Coll. Czech. Chem. Commun.* **30**:585.

Boissonnas, R. A., Guttmann, S., and Jacquenoud, P. A., 1960, Synthesis of L-arginyl-L-prolyl-L-prolyl-glycyl-L-phenylalanyl-L-seryl-L-propyl-L-phenylalanyl-L-arginine, a nonapeptide showing the properties of bradykinin, *Helv. Chim. Acta* **43**:1349.

Carpino, L. A., 1957, Oxidative reactions of hydrazines. II. Isophthalmides. New protective groups on nitrogen, *J. Am. Chem. Soc.* **79**:98.

Chillemi, F., and Merrifield, R. B., 1969, Use of N^{im}-dinitrophenylhistidine in the solid phase synthesis of the tricosapeptide 124–146 of human hemoglobin β-chain, *Biochemistry* **8**:4344.

Curtius, T., 1902, Synthetic studies with hippuric acid azide, *Berichte* **35**:3226.

Erickson, B. W., and Merrifield, R. B., 1972, Improved protecting groups for solid phase peptide synthesis, in "Chemistry and Biology of Peptides" (J. Meienhofer, ed.), pp. 191–195, Ann. Arbor Science, Michigan.

Felix, A. M., and Merrifield, R. B., 1970, Azide solid phase peptide synthesis, *J. Am. Chem. Soc.* **92**:1385.

Flanigan, E., and Marshall, G. R., 1970, Synthesis of cyclic peptides on dual function supports, *Tetrahedron Letters* **1970**:2403.

Fujii, T., and Sakakibara, S., 1969, Solid phase synthesis of isoleucine-5-angiotensin II to demonstrate the use of N^{im}-tosyl histidine, *Bull. Chem. Soc. Japan* **43**:3954.

Grahl-Nielsen, O., and Tritsch, G. L., 1969, Synthesis of oliogomeric L-lysine peptides by the solid phase method, *Biochemistry* **8**:187.

Green, B., and Garson, L. R., 1969, Peptide synthesis by the soluble polymer technique, *J. Chem. Soc. (Lond.)* **1969**:401.

Homer, R. B., Moodie, R. B., and Rydon, H. N., 1965, The kinetics and mechanism of the removal of the N-benzyloxycarbonyl group from N-benzyloxycarbonylglycine ethyl ester and related compounds in acetic acid containing hydrobromic acid or sulfuric acid, *J. Chem. Soc.* **1965**:4403.

Honzl, J., and Rudinger, J., 1961, Nitrosyl chloride and butyl nitrite as reagents in peptide synthesis by the azide method; suppression of amide formation, *Coll. Czech. Chem. Commun.* **26**:2333.

Horvath, C. G., Preiss, B. A., and Lipsky, S. R., 1967, Fast liquid chromatography: An investigation of operating parameters and the separation of nucleotides on pellicular ion exchangers, *Anal. Chem.* **39**:1422.

Kenner, G. W., McDermott, J. R., and Sheppard, R. C., 1971, The safety catch principle in solid phase peptide synthesis, *Chem. Commun.* **1971**:636.

Kessler, W., and Iselin, B., 1966, Selective splitting of substituted phenylsulfenyl-protecting groups in peptide synthesis, *Helv. Chim. Acta* **49**:1330.

Kobayashi, Y., Sakai, R., Kakiuchi, K., and Isemura, T., 1970, Physicochemical analysis of poly-(L-prolyl-L-prolylglycine) with defined molecular weight. Temperature dependence of molecular weight in aqueous solution, *Biopolymers* **9**:415.

König, W., and Geiger, R., 1970a, A new method for the synthesis of peptides: Activation of the carboxyl group with dicyclohexylcarbodiimide using 1-hydroxybenzotriazoles as additives, *Chem. Ber.* **103**:788.

König, W., and Geiger, R., 1970b, A new amide-protecting group, *Chem. Ber.* **103**:2041.

Krumdieck, C. L., and Baugh, C. M., 1969, The solid phase synthesis of polyglutamates of folic acid, *Biochemistry* **8**:1568.

Letsinger, R. L., and Kornet, M. J., 1963, Popcorn polymer as a support in multistep syntheses, *J. Am. Chem. Soc.* **85**:3045.

Lin M. C., Gutte, B., Caldi, D. G., Moore, S., and Merrifield, R. B., 1972, Reactivation of des (119–124) ribonuclease A by mixture with synthetic COOH-terminal peptides; the role of phenylalanine-120, *J. Biol. Chem.* **247**:4768.

Losse, G., Zeidler, D., and Grieshaber, T., 1968, Kinetics of the acid splitting of *N*- and *C*-protecting groups of peptides, *Ann. Chem.* **715**:196.

Marshall, D. L., and Liener, I. E., 1970, A modified support for solid phase peptide synthesis which permits the synthesis of protected peptide fragments, *J. Org. Chem.* **35**:867.

Marshall, G. R., and Merrifield, R. B., 1965, Synthesis of angiotensins by the solid phase method, *Biochemistry* **4**:2394.

Marshall, G. R., and Merrifield, R. B., 1971, Solid phase synthesis: The use of solid supports and insoluble reagents in peptide synthesis, *in* "Biochemical Aspects of Reactions on Solid Supports" (G. Stark, ed.), pp. 111–169, Academic Press, New York.

Mazur, R. H., and Plume, G., 1968, Synthesis of bradykinin, *Experientia* **24**:661.

McKay, F. C., and Albertson, N. F., 1957, New amine-masking groups for peptide synthesis, *J. Am. Chem. Soc.* **79**:4686.

Meienhofer, J., and Trzeciak, A., 1971, Solid phase synthesis with attachment of peptide resin through an amino acid side chain: [8-Lysine]-vasopressin, *Proc. Natl. Acad. Sci.* **68**:1006.

Merrifield, R. B., 1962, Peptide synthesis on a solid polymer, *Fed. Proc.* **21**:412.

Merrifield, R. B., 1963, Solid phase peptide synthesis. I. The synthesis of a tetrapeptide, *J. Am. Chem. Soc.* **85**:2149.

Merrifield, R. B., 1965, Solid phase peptide synthesis, *Endeavour* **23**:3.

Merrifield, R. B., 1969, Solid phase peptide synthesis, *Advan. Enzymol.* **32**:221.

Merrifield, R. B., and Littau, V., 1968, Solid phase peptide synthesis. The distribution of peptide chains on the solid support, *in* "Peptides" (E. Brices, ed.), pp. 179–182, North-Holland, Amsterdam.

Mizoguchi, T., Shigezane, K., and Tokamura, N., 1969, Bromacylpolystyrene, a new type of polymer support for solid phase peptide synthesis, *Chem. Pharm. Bull.* **17**:411.

Najjar, V. A., and Merrifield, R. B., 1966, The use of the *o*-nitrophenylsulfenyl group in the synthesis of the octadecapeptide bradykininylbradykinin, *Biochemistry* **5**:3765.

Nitecki, D. E., and Goodman, J. W., 1970, γ-Glutamyl peptides, *in* "Peptides" (B. Weinstein and S. Lande, eds.), pp. 435–450, Marcel Dekker, New York.

Noda, K., Terada, S., and Izumiya, N., 1969, Modified benzyloxycarbonyl groups for protection of ε-amino group of lysine, *Bull. Chem. Soc. Japan* **43**:1883.

Ohno, M., and Anfinsen, C. B., 1967, Removal of protected peptides by hydrazinolysis, *J. Am. Chem. Soc.* **89**:5994.

Ohno, M., Eastlake, A., Ontjes, D. A., and Anfinsen, C. B., 1969, Synthesis of the fully protected carboxyl-terminal tetradecapeptide sequence of staphylococcal nuclease, *J. Am. Chem. Soc.* **91**:6842.

Omenn, G. S., and Anfinsen, C. B., 1968, Solid phase peptide coupling, *J. Am. Chem. Soc.* **90**: 6571.

Pietta, P. G., Covallo, P., and Marshall, G. R., 1971, 2,4-Dimethoxybenzyl as a protecting group for glutamine and asparagine in peptide synthesis, *J. Org. Chem.* **36**:3966.

Rothe, M., and Dunkel, W., 1967, Synthesis of monodisperse oligomers of ε-amino caproic acid up to a degree of polymerization of 25 by the Merrifield method, *J. Polymer Sci. B (Polymer Letters)* **5**:589.

Sakakibara, S., Shin, M., Fujino, M., Shimonishi, Y., Inoue, S., and Inukai, N., 1965, *t*-Amyloxycarbonyl as a new protecting group in peptide synthesis, *Bull. Chem. Soc. Japan*, **38**:1522.

Sakakibara, S., Shimonishi, Y., Okada, M., and Kishida, Y., 1967, Removal of protective groups by anhydrous hydrogen fluoride, *in* "Peptides" (H. C. Beyerman, A. van de Linde, and W. Maassen van den Brink, eds.), pp. 44–49, North-Holland, Amsterdam.

Sakakibara, S., Kishida, Y., Kiruchi, Y., Sakai, R., and Kakiuchi, K., 1968, Synthesis of poly-(L-prolyl-L-prolylglycyl) of defined molecular weight, *Bull. Chem. Soc. Japan* **41**:273.

Sakakibara, S., Fukuda, T., Kishida, Y., and Honda, I., 1970, A new protective group suitable for masking specific amino groups during peptide synthesis, *Bull. Chem. Soc. Japan* **43**:3322.

Schnabel, E., Klostermeyer, H., and Berndt, H., 1971, The selective acidolytic splitting of the *tert*-butyloxycarbonyl group, *Ann. Chem.* **749**:90.

Schwyzer, R., 1959, Total synthesis of B-corticotropin, *Angew. Chem.* **71**:742.

Scott, R. P. W., Chan, K. K., Kucera, P., and Zolty, S., 1971, The use of resin coated glass beads in the form of a packed bed for the solid phase synthesis of peptides, *J. Chromatog. Sci.* **9**:577.

Shaltiel, S., 1967, Thiolysis of some dinitrophenyl derivatives of amino acids, *Biochem. Biophys. Res. Commun.* **29**:178.

Shaltiel, S., and Fridkin, M., 1970, Thiolysis of dinitrophenylimidazoles and its use during synthesis of histidine peptides, *Biochemistry* **9**:5122.

Shemyakin, M. M., Ovchinnikov, Yu A., Kiryuskin, A. A., and Kozhevnikova, I. V., 1965, Synthesis of peptides in solution on a polymeric support, *Tetrahedron Letters* **27**:2323.

Sieber, P., and Iselin, B., 1968, Peptide synthesis with the use of the α-(*p*-diphenyl)-isopropyloxy-carbonyl (Dpoc)-amino protecting group, *Helv. Chim. Acta* **51**:622.

Sklyarov, L. Y., and Shaskova, I. V., 1969, Synthesis of the glycine analog of gramicidin S on a polymeric carrier, *J. Gen. Chem. U.S.S.R.* **39**:2714.

Southard, G. L., Brooke, G. S., and Petee, J. M., 1969, A mild method of solid phase peptide synthesis employing an enamine nitrogen protecting group and a benzhydryl resin as a solid support, *Tetrahedron Letters* **1969**:3505.

Stewart, F. H. C., 1967, Comparative acidic cleavage experiments with methyl substituted benzyl esters of amino acids, *Aust. J. Chem.* **20**:2243.

Stewart, J. M., and Young, J. D., 1969, "Solid Phase Peptide Synthesis," Freeman, San Francisco.

Stewart, J. M., Knight, M., Paiva, A. C. M., and Paiva, T., 1972, *in* "Peptides" (S. Lande, ed.), pp. 59–64, Gordon and Breach, New York.

Tregear, G. E., Catt, K., and Niall, H. D., 1967, Applications of synthetic polymers to protein chemistry, *Proc. Roy. Aust. Chem. Inst.* **1967**:345.

Visser, S., Roeloffs, J., Kerling, K. E. T., and Havinga, E., 1968, Synthesis of protected hexa- and octa-peptide hydrazides related to the *N*-terminal part of bovine pancreatic ribonuclease A using the solid phase method, *Rec. Trav. Chim.* **87**:559.

Visser, S., Raap, J., Kerling, K. E. T., and Havinga, E., 1970, Combination of solid phase and classical methods of peptide synthesis in the preparation of 1–13 [13-leucine] bovine ribonuclease, *Rec. Trav. Chim.* **89**:865.

Wang, S. S., 1972, Solid phase synthesis of protected peptide fragments, *in* "Chemistry and Biology of Peptides" (J. Meienhofer, ed.), pp. 179–182, Ann Arbor Science, Michigan.

Wang, S. S., and Merrifield, R. B., 1969, Preparation of a *t*-alkyloxycarbonylhydrazide resin and its application to solid phase peptide synthesis, *J. Am. Chem. Soc.* **91**:6488.

Wang, S. S., and Merrifield, R. B., 1971, *t*-Alkyloxycarbonylhydrazide resin and *t*-alkyl alcohol–resin for preparation of protected peptide fragments, *in* "Peptides" (E. Scoffone, ed.), pp. 74–83, North-Holland, Amsterdam.

Wang, S. S., and Merrifield, R. B., 1972, A solid phase–fragment approach to peptide synthesis, *Internat. J. Peptide and Protein Res.* **4**:309.

Weygand, F., 1968, Bromacetyl polystyrene as supporting material for peptide synthesis on a solid phase, *in* "Peptides" (E. Bricas, ed.), p. 183, North-Holland, Amsterdam.

Weygand, F., and Ragnarsson, U., 1966, Peptide synthesis on a solid phase, *Z. Naturforsch.* **21b:**1141.

Weygand, E., Hoffman, D., and Wünsch, 1966, Peptide synthesis with dicyclohexycarbodiimide on addition of *N*-hydroxysucciminide, *Z. Naturforsch.* **21b:**426.

Wieland, T., Lewalter, J., and Birr, C., 1970, Subsequent activation of carboxylic derivatives by oxidation or elimination of water and its use for solid phase synthesis of peptides and cyclization of peptides, *Ann. Chem.* **740:**31.

Wildi, B. S., and Johnson, J. H., 1968, Polymeric carrier resins for peptide synthesis, *in* "155th National Meeting of the American Chemical Society," Abst. A-8.

Windridge, G. C., and Jorgensen, E. E., 1971, Racemization in the solid phase synthesis of histidine-containing peptides, *Intra-Sci. Chem. Rep.* **5:**375.

Yajima, H., Kawatani, H., and Watanobe, H., 1970, Synthesis of the decapeptide ACTH-(5–14) as an example of peptide-fragment condensation on the polymer support, *Chem. Pharm. Bull.* **18:**1333.

Yamashiro, D. H., Noble, R. L., and Li, C. H., 1972, New protecting groups in solid phase synthesis, *in* "Chemistry and Biology of Peptides" (J. Meienhofer, ed.), pp. 197–202, Ann Arbor Science, Michigan.

Yaron, A., and Schlossman, F., 1968, Preparation and immunologic properties of stereospecific α-dinitrophenylnonalysines, *Biochemistry* **7:**2673.

Zervas, L., Borovas, D., and Gazis, E., 1963, Tritylsulfenyl and *o*-nitrophenylsulfenyl groups as *N*-protecting groups, *J. Am. Chem. Soc.* **85:**3660.

Zimmerman, J. E., and Anderson, G. W., 1967, The effect of active ester components on racemization in the synthesis of peptides by the dicyclohexylcarbodiimide method, *J. Am. Chem. Soc.* **89:**7151.

CHAPTER 17

MONITORING IN
SOLID-PHASE PEPTIDE SYNTHESIS

J. Hirt, E. W. B. de Leer, and H. C. Beyerman

Laboratory of Organic Chemistry
Technische Hogeschool
Delft, The Netherlands

I. INTRODUCTION

After its invention by R. B. Merrifield (1962), solid-phase peptide synthesis has passed through a considerable evolution (Merrifield, 1969; Stewart and Young, 1969; Losse and Neubert, 1970*a*; Marshall and Merrifield, 1971); it evolved from the synthesis of a tetrapeptide (Merrifield, 1963) to the synthesis of a ribonuclease preparation (124 amino acid residues, AA) with 78% enzymatic activity (Gutte and Merrifield, 1971).

The most important condition for a synthesis by this method to be successful is that all the reactions must proceed quantitatively, since no intermediate purifications can take place. In case of incomplete couplings, *failure sequences* arise, the structure of which greatly resembles that of the desired product, in consequence of which the purification is made very difficult, if not impossible. In view of these difficulties, some authors (Geiger, 1971; Wünsch, 1971) consider the synthesis of a purifiable peptide with 12 or 15 amino acids to be the extreme that can be achieved by this method. The relative statistical distribution of the peptide mixture that is formed can be calculated as a function of the constant coupling yield (Baas *et al.*, 1971; Bayer *et al.*, 1970). This distribution is shown in Fig. 1 for the synthesis of human growth hormone (190 AA; calculated, however, for 188 residues according to the first sequence proposal).

As appears from Fig. 1, a coupling yield of 99.0% in each step gives a product containing only 15 mole % of the correct sequence side by side with

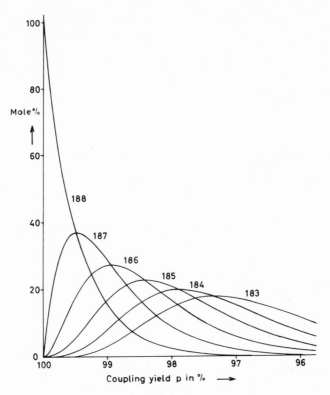

Fig. 1. Relative statistical distribution of peptides after 187 coupling
steps, as a function of a constant coupling yield p in each step. The
curves marked 187, 186, etc., represent the total mole percentage of all
possible peptides containing 187, 186, etc., residues, (Boas *et al.*, 1971).

29 mole % of all possible failure sequences of 187 amino acids. Since no
unambiguous product can be isolated from so complex a mixture by means
of present-day purification techniques, such a synthesis must be considered a
failure.

The problem of the failure sequences was clearly demonstrated for the
first time by Bayer *et al.* (1970) with the synthesis of [Leu-Ala]$_6$. After partial
acid hydrolysis, the dipeptides formed were esterified and trifluoroacetylated,
and separated with the aid of gas chromatography. In this way, it was possible to demonstrate the presence of Leu-Leu and Ala-Ala, which must have
been formed as a result of incomplete coupling reactions. In the case of the
synthesis of [Pro^3, Ile^5]-angiotensin II (Khosla *et al.*, 1968), it was even
possible to isolate a failure sequence which had been formed as a result of an
incomplete coupling of Boc-Arg(NO$_2$) to Pro.

An incomplete removal of the N-protecting group also leads to failure sequences or, if the blocking is permanent, to a *truncated sequence.*

These incomplete reactions and side-reactions inherent in the amino acid to be coupled or in the coupling method make it essential to monitor all the reactions during a solid-phase synthesis. However, it must be noted that even when all reactions go to completion, it is possible that failure sequences are formed by interchain aminolysis, which results in *chain-doubling* (Beyerman *et al.,* 1972*b*).

By applying methods of analysis which can be automated, it would also become possible to really automate the mechanized solid-phase method (Merrifield, 1965; Merrifield *et al.,* 1966), i.e., to monitor it by feedback control.

II. MONITORING OF THE COUPLING REACTION

As is evident from the preceding, a quantitative conversion of the free amine on the polymer in the coupling step is essential in solid-phase peptide synthesis. Incomplete coupling reactions caused by, for example, too short a reaction time or sterical hindrance lead to failure sequences, which usually cannot be separated or even distinguished from the desired sequence by analytical means.

If the coupling is not complete at one time under the normal conditions, it must be repeated or a larger excess of the acylating agent should be used. This has already been demonstrated by Weygand and Obermeier (1968). (See also Corley *et al.,* 1972.)

When the last few percentages of amino groups are acylated very slowly, a permanent blocking (causing a truncated sequence) can be applied. For this reaction, terminating agents such as acetic anhydride (Merrifield, 1963), N-acetylimidazole (Markley and Dorman, 1970), isopropenyl formate (van Melick and Wolters, 1972), 3-nitrophthalic anhydride (Wieland *et al.,* 1969*b*), or 3-sulfopropionic anhydride (Wissmann and Geiger, 1970) have been used.

It is essential for the prevention of failure sequences that all amino groups on the polymer have been acylated; thus the best method for monitoring is the determination, after the coupling, of the remaining free amine on the polymer. If no free amine can be demonstrated, this does not necessarily mean that the desired coupling reaction is quantitative, because a side-reaction might have caused a truncated sequence.

Methods in which the decrease of the acylating agent or the increase of the leaving group of the acylating agent liberated during the reaction is measured continuously or discontinuously are too inaccurate to indicate a quantitative coupling. These methods may be important when studying the coupling kinetics, e.g., as a function of the position in the peptide chain or the conformation of the peptide chain, the penultimate amino acid, or the side-chain-protecting group.

A. Qualitative Methods

During the synthesis of a small peptide, the intermediate product may be split off from the polymer by hydrobromic acid in trifluoroacetic acid and may then be examined, e.g., by electrophoresis (Jorgensen et al., 1969) and thin layer chromatography (Gisin et al., 1969; Losse and Neubert, 1970b; Losse and Ulbrich, 1971; Dorman and Markley, 1971). Other methods are removal by alcoholysis (Beyerman et al., 1968b; Beyerman, 1972; Beyerman et al., 1971a) and hydrazinolysis (Sieber and Iselin, 1968) and examination with thin layer chromatography.

Miura et al. (1968) and Gisin et al. (1969) ascertained by means of infrared spectroscopy of the peptide-on-polymer (v C=O absorption of the secundary amide) that a dipeptide or tetradepsipeptide, respectively, had been formed. Felix and Merrifield (1970) examined in this way the rate of reaction of polymer-L-leucine azide and polymer-S-p-methoxybenzyl-L-cysteine with L-alanine Boc-hydrazide by following the disappearance of the azide band at 2240 cm^{-1}.

Bayer et al. (1968a) in their synthesis of S,S'-dibenzyl-[Ser^4]-oxytocin monitored the coupling steps up to the pentapeptide by mass spectrometry after acidic removal of the peptide from the polymer, esterification, and trifluoroacetylation; they did not detect any failure sequences. Neither was an error found when the S,S'-dibenzyl-[Ser^4]-oxytocin was subjected to partial acid hydrolysis. The resulting peptides were separated by gas chromatography (of the N-trifluoracetyl-O-methyl esters) and by Sephadex partition chromatography; they were identified by mass spectrometry (Bayer et al., 1968a; Bayer and Koenig, 1969; Bayer, 1969, 1970).

Losse and Klengel (1971) described a method (Fig. 2) in which after the coupling a resin sample is treated with a solution of 2-hydroxy-5-(phenylazo)-benzaldehyde or 2-hydroxy-5-(p-nitrophenylazo)-benzaldehyde in ethanol. The aldehyde and the remaining free amine will form an aldimine (Schiff base), which leads to an orange-red resin after washing out of the excess of the reagent.

It appears from a comparative study of Losse and Ulbrich (1971) that this method, carried out with 2-hydroxy-5-(p-nitrophenylazo)-benzaldehyde, gives the same result as the bromocresyl purple test (Beyerman and Hindriks, 1971), but an examination by thin layer chromatography of the product (dipeptide) after acidic removal from the resin is more sensitive.

Disadvantages of this method are that reaction time is long (5 hr), with colored resins it is difficult to apply, and a wrong conclusion may be drawn if steric hindrance interferes with the reaction. Because the Schiff base forma-

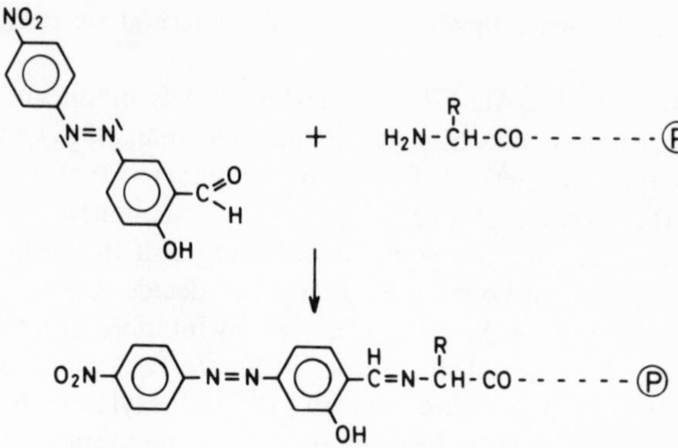

Fig. 2

tion is an equilibrium, the result is dependent on the aldehyde concentration (Fankhauser *et al.*, 1971).

Zhukova *et al.* (1970) dansylated a sample of peptide-on-polymer after removal of the *N*-protecting group from the coupled amino acid. On hydrolysis, they tried to detect the amino acid which had not been acylated in the coupling step, in the presence of the amino acid added, by means of thin layer chromatography. This is a time-consuming procedure, in which the sensitivity depends on the stability of the dansylated amino acid during the hydrolysis.

The bromocresyl purple test of Beyerman and Hindriks (1971), found independently by Loffet (1968), proved to be satisfactory according to a comparative examination by Losse and Ulbrich (1971). The last authors used dimethylformamide as the solvent, which is preferable because of increased sensitivity. This test was compared with quantitative methods by means of 2-hydroxy-1-naphthaldehyde (Esko *et al.*, 1968), pyridine hydrochloride (Dorman, 1969), and pyridine hydrobromide (Dorman procedure, modified by Losse and Ulbrich), and with qualitative methods by means of 2-hydroxy-5-(*p*-nitrophenylazo)-benzaldehyde (Losse and Klengel, 1971) and thin layer chromatography after removal of the peptide from the polymer. In our laboratory (Hirt, 1971), bromocresyl purple in ethanol as solvent proved to be less sensitive than ninhydrin (Kaiser *et al.*, 1970). It is a disadvantage that side-products, probably related to dicyclohexylurea, cause a disturbing greenish coloration of the resin.

With ninhydrin reagents, a coloration of the resin and supernatant can be observed in the case of an incomplete coupling reaction when a sample of peptide-on-polymer is heated at 100°C for 5 min. The color is purple unless the unacylated amino acid is proline or β-benzyl aspartate; in that case, a brown or reddish-brown coloration occurs. This method has been described by Kaiser *et al.* (1970).

The reaction of Boc-Ala-OH and H-Phe-P was monitored quantitatively by the pyridine hydrochloride method (Dorman, 1969) and qualitatively by the ninhydrin method. From this, it appeared that 5 μmoles/g of free amine on the resin can still be detected by the qualitative method.

This fast method also has some disadvantages. If the resin is colored during the synthesis, it is sometimes difficult to decide whether the test is negative or slightly positive. Steric hindrance may interfere with the reaction.

Before the test is carried out, the resin has to be washed thoroughly, because a mixture of Boc-amino acid and dicyclohexylcarbodiimide (also *N*-cyclohexyl-*N'*-β-(*N*-methylmorpholino)-ethyl-carbodiimide *p*-toluene sulfonate) gives a positive result (Hirt, 1971). Boc-amino acid, the carbodiimides, dicyclohexylurea, the *N*-acylurea rearrangement product of Boc-Ile and DCC, Z-, Nps-, of Pht-amino acids or Boc-dipeptides and DCC, Boc-amino acids, and *N,N'*-carbonyldiimidazole do not show this phenomenon.

Application of this method has been reported in the literature by Rivaille *et al.* (1971), Potts *et al.* (1971), Tregear *et al.* (1971), Fankhauser *et al.* (1971), Rivier *et al.* (1972), Sievertsson *et al.* (1972), and Beyerman *et al.* (1972a).

Chou *et al.* (1971) have modified this method to a (semi-)quantitative one; this was applied by Bayer *et al.* (1971).

B. Quantitative Methods

1. Colorimetric Methods. Amino acid analysis, after direct hydrolysis of the peptide-on-polymer or of the peptide after removal from the resin, has been applied frequently as a check for the coupling reaction (Merrifield, 1964; Najjar and Merrifield, 1966; Rudinger and Gut, 1967; Richards *et al.*, 1967; Beyerman *et al.*, 1967; Gisin *et al.*, 1969; Jollès and Jollès, 1968; Gut and Rudinger, 1968; Ney and Polzhofer, 1968; Beyerman *et al.*, 1968a; Wieland *et al.*, 1969a; Westall and Robinson, 1970; Kaiser *et al.*, 1970; Sano *et al.*, 1971; Lübke, 1971; Polzhofer, 1972).

This method is time-consuming and too inaccurate to indicate a quantitative coupling. The data are more difficult to interpret if the relevant amino acid occurs already in the peptide chain.

However, amino acid analysis may be useful for the monitoring of a synthesis if diagnostic amino acids (Beacham *et al.*, 1971) occur throughout the peptide chain, as, e.g., in acyl carrier protein (Hancock *et al.*, 1971).

For special purposes, the separation of peptides, after removal from the resin by HBr/TFA or HF, has been developed. The peptides are separated on an ion exchange resin and determined by means of ninhydrin (Noda *et al.*,

1970; Mitchell and Roeske, 1970; Sano *et al.*, 1971; Dorman *et al.*, 1971; Beyerman *et al.*, 1972*b*).

The first attempts to follow a coupling reaction in solid-phase peptide synthesis were carried out by means of colorimetric determinations.

Bodanszky and Sheehan (1964, 1966), Hörnle (1967), and Bodanszky and Bath (1969) measured the absorption of the active ester at 270 nm and of the phenol at 314 nm in the filtrate of a *p*-nitrophenyl ester coupling.

Continuous coupling experiments, in which liquid samples were taken, were carried out by Rudinger and Gut (1967), Dutta and Morley (1971), and Lübke (1971).

Rudinger and Gut (1967) measured the appearance of the quinoline, liberated during the reaction of *t*-butyloxycarbonyl-leucine 5-chloro-8-hydroxyquinoline ester with glycine–resin in dimethylformamide, at 330 nm (61% conversion after 4 hr and 72% after 20 hr).

Dutta and Morley (1971) measured the absorption of 2-pyridyl esters at 265 nm and of 2-hydroxypyridine at 305 nm, of *p*-nitrophenyl esters at 275 nm and of *p*-nitrophenol at 305 nm. It was shown that the 2-pyridyl esters were more reactive than the corresponding *p*-nitrophenyl esters in methylene chloride as the solvent, but this situation was reversed in dimethylformamide.

Lübke (1971) measured the absorption of the liberated *p*-nitrophenol during the reaction of Z-Leu-ONp and alanyl–resin. With a 4:1 molar ratio, the conversion was about 70% after 6 hr and 77% after 24 hr, while the reaction of the active ester with leucylalanylamide in a 2:1 molar ratio was complete within 6 hr.

By the use of a flow-through apparatus, Gut and Rudinger (1968) followed continuously the reaction of a Nps-amino acid *N*-hydroxysuccinimide ester with an aminoacyl polymer by measuring the increase in the transmittance at 380 nm. The removal of the Nps group could also be examined in this way. It is a disadvantage that the "colored" compound has to be present in a dilute form, in contrast to the usual synthesis conditions.

The colorimetric methods mentioned above have limited application. One does get an idea as to the kinetics of the relevant coupling but not an answer to the question whether the reaction has proceeded quantitatively.

Esko *et al.* (1968) determined the residual free amino groups after the coupling by condensation with 2-hydroxy-1-naphthaldehyde to an aldimine (Schiff base), which, after removal of the excess reagents, was decomposed by means of an excess of benzylamine (Fig. 3). The absorption of this soluble aldimine was measured at 420 nm, from which the amount of free amino groups could be calculated. The reaction of the resin sample with 2-hydroxy-1-naphthaldehyde takes 12 hr when it is performed in absolute ethanol (Esko *et al.*, 1968) or absolute ethanol–methylene chloride 1:1 (v/v) (Esko

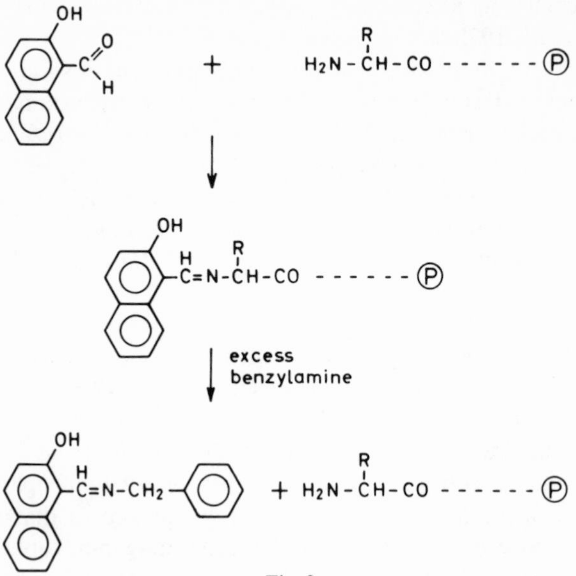

Fig. 3

and Karlsson, 1970), but this reaction time can be reduced when methylene chloride (1.5–2.0 hr; Yamashiro *et al.*, 1972) or dioxane (1 hr; Corley *et al.*, 1972) is used as solvent. By means of this method, fundamental investigations were carried out, such as the following of coupling reactions (Esko *et al.*, 1968; Karlsson *et al.*, 1970b) and deprotection (Karlsson *et al.*, 1970a), catalysis in active ester couplings (Ragnarsson *et al.*, 1970), competition experiments (Ragnarsson *et al.*, 1971), and study of Boc-amino acids in connection with yield and racemization (Esko and Karlsson, 1970).

Losse and Ulbrich (1971) in a comparative study showed that a small number of the remaining amino groups are poorly accessible for the aldehyde, while protonation according to the Dorman procedure (Dorman, 1969) is still possible.

From a study of Fankhauser *et al.* (1971), it appears that the result depends on the 2-hydroxy-1-naphthaldehyde concentration. It also appears that with an increasing chain length the quantity of the free amino groups after the deprotection determined by this method decreases considerably faster than is possible according to other data.

Li and Yamashiro (1970) applied this method during their synthesis of the sequence formerly ascribed to human growth hormone.

The qualitative ninhydrin method of Kaiser *et al.* (1970) was worked out by Chou *et al.* (1971) into a quantitative method. A known amount of resin was heated with ninhydrin reagent, and the absorption of the supernatant was measured after dilution at 570 nm in a relation to a ninhydrin blank. The number of micromoles was then determined via a glycine color standard.

The quantitative ninhydrin method was applied by Bayer *et al.* (1971) for the determination of the remaining free amino groups after the coupling reaction. The method does not give absolute values, because in that case a correction factor dependent on the relevant amino acid and its place in the peptide chain would be needed.

Zhukova *et al.* (1970) treated a sample of the resin after the coupling with a deprotection agent and then blocked the amino groups by means of dinitrofluorobenzene. After hydrolysis and thin layer chromatographic separation, the Dnp-amino acids were extracted and determined spectrophotometrically. These values were corrected for destruction during hydrolysis by carrying out a parallel experiment with known amounts of pure Dnp-amino acids. By means of this method, the authors ascertained that double couplings gave better results than one coupling with the same overall reaction time.

Recently, a fast method for the determination of free amino groups on an insoluble polymer was described by Gisin (1972). The polymer was treated with 0.1 N picric acid in methylene chloride. After removal of the excess of the reagent, the picric acid was removed by means of a solution of diisopropylethylamine or 0.1 M pyridine hydrochloride in methylene chloride and was determined spectrophotometrically. To illustrate this method, Boc-Pro-OH and Boc-Gly-OH, respectively, were coupled to a hydroxymethyl resin by means of N,N'-carbonyldiimidazole. The amino acid content of samples which were periodically taken was determined by the picric acid method after deprotection and by amino acid analysis. For contents of over 100 μmoles/g, the results agreed within a range of 1–8 %. The reproducibility was ±2 %.

This method was used to monitor the loss of dipeptide from the polymer, caused by carboxylic acid–catalyzed intramolecular aminolysis, producing the corresponding dioxopiperazine (Gisin and Merrifield, 1972). Application of the method for determination of the remaining free amino groups after the coupling reaction has not yet been published.

Garden and Tometsko (1972) presented a method in which an accurately weighed sample of dansylated resin is subjected to hydrazinolysis and the quantity of dansylpeptide hydrazide is measured spectrofluorometrically. It is a highly sensitive method but time consuming. No application in monitoring solid-phase synthesis has been described.

2. Volumetric Methods. A monitoring method carried out on the whole resin batch is the procedure according to Dorman (1969). The free amino groups are converted into the hydrochloride by means of pyridine hydrochloride. After washing out of the excess reagent, the hydrochloric acid is removed by triethylamine. In this filtrate and those of the subsequent washings, the chloride is determined potentiometrically by titration with silver nitrate. The whole procedure takes $1-1\frac{1}{2}$ hr and gives reproducible results.

A number of applications have been reported (Markley and Dorman, 1970; Kaiser et al., 1970; Hagenmaier, 1970; Baba et al., 1971; Dorman and Markley, 1971; Schaich and Schneider, 1971; Jakubke and Baumert, 1971; Bayer et al., 1972). The synthesis of LH-FSH-releasing hormone by Beyerman et al. (1972a) was monitored by this method (see Table I).

Loffet and Dremier (1971) used tribenzylamine hydroiodide instead of pyridine hydrochloride and determined iodine, after oxidation of iodide, spectrophotometrically.

G. Losse and Ulbrich (1971) reported that better results are obtained when pyridine hydrobromide is used and bromide is determined selectively with respect to chloride. The main reason was that on treatment of a chloromethylated resin with triethylamine in dimethylformamide a small amount of chloride was found in the filtrate (possibly owing to new cross-linking of the chloromethylated resin).

The pyridine hydrobromide method was applied by A. Losse (1971).

With a view to automation, Brunfeldt et al. (1969, 1971, 1972a,b) developed a monitoring procedure in which the number of free amino groups, both after the deblocking and after the coupling, are determined by titration with a solution of perchloric acid in acetic acid. The whole resin batch is then suspended in a 1:1 (v/v) mixture of acetic acid and methylene chloride.

Other titratable groups, such as in N^{im}-benzylhistidine, are also determined. Titration of methionine-containing peptides did not lead to sulfon formation. Acid-labile protecting groups, such as the Bpoc group, are probably not applicable when this monitoring method is used. The duration of the whole titration procedure is about 2 hr.

In the method described by Mehlis et al. (1971), the remaining free amino groups are acetylated after the coupling reaction, and the N-protecting group is removed quantitatively. The chloride value divided by the chloride value of the deprotection before the coupling (both corrected for a blank value) gives the conversion. By means of this method, the reaction progress of Boc-Phe-OTcp and H-Ala-P (90% conversion in 30 min) and Boc-Ile-OTcp and H-Phe-P (76% conversion in 960 min) was followed.

3. *Radioactive Methods*. Richards et al. (1967) synthesized [D-Ala-L-(^{14}C)-Ala]$_5$-ε-Dnp-Lys. After attachment of Boc-D-Ala an amino acid

Table I. Determination of the Free Amino Groups During the Synthesis of the Blocked Sequence of the LH-FSH-Releasing Hormone

Amino acid:	1 (Pyro)Glu	2 His(Tos)	3 Trp	4 Ser(Bzl)	5 Tyr(Bzl)	6 Gly	7 Leu	8 Arg(NO$_2$)	9 Pro	10 Gly-P
NH$_2$ value after removal of the Boc group (mmoles)[a]	—	—	1.07	1.14	1.19	1.20	1.20	1.24	1.31	1.41
1st coupling (2 eq)										
Reaction time (hr)	24		1½	1½	1½	1	1	1½	1	1
NH$_2$ value in mmoles[a,f]		—[b]	0.319	0.015	0.018	0.027	0.010	0.010	0.082	0.025
NH$_2$ value in %			30	1.3	1.5	2.2	0.8	0.8	6.3	1.8
2nd coupling (1 eq)										
Reaction time (hr)	24		2½	1	1½	¾	1	1	1½	1
NH$_2$ value in mmoles[a,f]		—[d]	0.195	0.009[d]	0.080	0.018[d]	0.010	0.008[d]	0.073	0.014[c]
NH$_2$ value in %			18	0.8	0.7	1.5	0.8	0.6	5.6	1.0
3rd coupling (2 eq)										
Reaction time (hr)			20							
NH$_2$ value in mmoles[a,f]			0.229[c,h]							
NH$_2$ value in %			21							
Acetylation[e]										
NH$_2$ value in mmoles[a,f]			—[c]						0.082[g]	
NH$_2$ value in %									6.3	

[a] Determined potentiometrically with 0.2 N and 0.01 N silver nitrate, respectively.
[b] Ninhydrin test positive.
[c] Ninhydrin test weakly positive.
[d] Ninhydrin test negative.
[e] Acetylation with N-acetylimidazole (10 eq) in dimethylformamide for 30 min.
[f] Corrected for blank value.
[g] We do not believe that this figure designate., the correct NH$_2$ value; probably an appreciable increase of the blank value. The new value thus becomes 0.082 + 0.015 (old blank value).
[h] After the coupling of Boc-His(Tos), the Dorman process is no longer of any use on account of the high blank value. Moreover, the pyridine-HCl partially splits off the tosyl group.

analysis was carried out, and after coupling of Boc-L-(^{14}C)-Ala the radioactivity of the supernatant was measured, both after alkaline hydrolysis.

Krumdieck and Baugh (1969) in their synthesis of polyglutamates of
folic acid used N-(2-^{14}C-butyloxycarbonyl)-L-glutamic acid α-benzyl ester.
The yield of the coupling was determined by measuring the nonincorporated
^{14}C after the mixed anhydride coupling. This yield was found to be consistently between 95 and 100%.

Beyerman *et al.* (1971*b*) performed experiments in which ^{14}C-Boc-
amino acids were used for monitoring the coupling reaction in dimethylformamide and ethyl acetate as solvents. The progress of the reaction was
followed by taking liquid samples and measuring their radioactivity with a
liquid scintillation counter. An example is shown in Fig. 4.

Hörnle and Geising (1971) and Geising and Hörnle (1971) made use of
Edman degradation to determine free amino groups on the resin (Fig. 5).
They used ^{14}C-phenylisothiocyanate, so that the quantity of ^{14}C-phenyl-
thiohydantoin formed could be measured by means of a liquid scintillation
counter. By this method, the reaction progress was followed (Geising and
Hörnle, 1971) of Boc-Ile-OH/DCC with two undecapeptides with valine

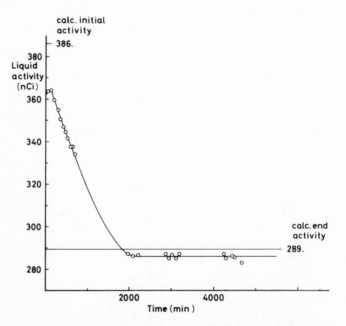

Fig. 4. Coupling of Boc-Gly to Val-P (esterified by the sodium carbonate method) in dimethylformamide. The total radioactivity in the
liquid (nCi) as a function of time (min), (Beyerman *et al.*, 1971*b*).

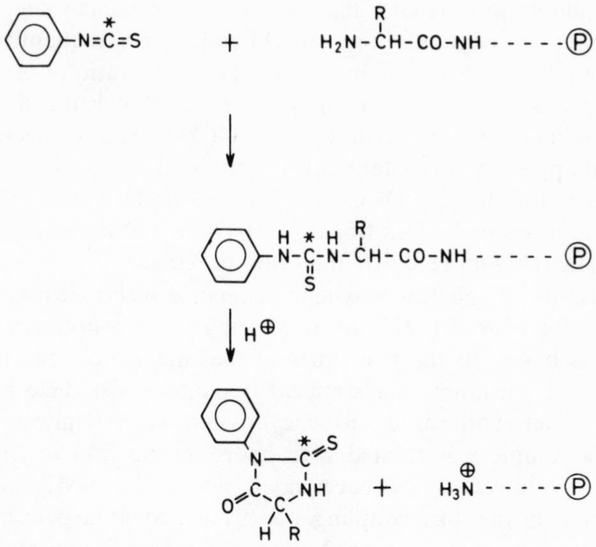

Fig. 5

as *N*-terminal amino acid. It appeared that the coupling to the undeca-peptide, which contained only aliphatic amino acids, proceeded more slowly. The Ile-Val formation was examined as a function of the chain length. From Ile-Val to [Ile-Val-Leu-Ala]$_5$, the reaction rate decreased considerably.

Fankhauser *et al.* (1971) acetylated, after the coupling, a sample of peptide-on-polymer with ^{14}C-acetylimidazole or ^{14}C-acetic anhydride. After washing and drying, a weighed part of the resin was burned and the $^{14}CO_2$ was measured by means of a liquid scintillation counter.

The method was applied to the synthesis of hexaleucine on a hydroxy-methyl resin and compared with the 2-hydroxy-1-naphthaldehyde method of Esko *et al.* (1968). After all the couplings, the ninhydrin test (Kaiser *et al.*, 1970) was negative, while small quantities of free amine could be detected by the Esko method. The results of the Esko method, however, appeared to be dependent on the aldehyde concentration. The values measured by the acetylation methods were ten to 30 times higher than those determined by the Esko method. The results of the two acetylation methods were not consonant with each other. Diacetylation and other side-reactions are probably the cause of the difficulties.

4. Mass Spectrometry and ^{19}F Nuclear Magnetic Resonance Spectros-copy. Weygand and Obermeier (1968) employed a combination of Edman degradation and mass spectrometric dilution analysis to determine the remaining amino groups after the coupling. After Edman degradation, the resulting phenylthiohydantoin was mixed with a known quantity of the corresponding p-fluorophenylthiohydantoin. From the ratio of the intensity of the mass peaks, the unknown quantity could be calculated. From their experiments (couplings to H-Ile-Leu-Val-OCH$_2$-CO-copolystyrenedivinylbenzene), it appeared, for instance, that there is no real difference in reaction rate between Boc-Ile-OH/DCC and Z(OMe)-Ile-OH/DCC, that Boc-Ile-OSu reacts more slowly than Boc-Ile-OH/DCC, and that Boc-Leu-Ile-OSu reacts considerably more slowly than Boc-Ile-OSu.

The use of ^{19}F nuclear magnetic resonance spectroscopy was recently reported by Bayer et al. (1972) for monitoring of the coupling reaction. This technique is based on the phenomenon that the signals of trifluoroacetyl groups of different amino acids appear in characteristic field positions. As an example, they synthesized L-isoleucyl-L-alanyl-L- valylglycine. After each coupling, a sample was treated with hydrobromic acid in trifluoroacetic acid, the peptide was trifluoroacetylated, and a ^{19}F NMR spectrum was taken. Even after the third coupling, free glycine could be determined. It is a disadvantage that much resin has to be sacrificed for the monitoring.

III. MONITORING OF THE DEPROTECTION STEP

An important step in solid-phase peptide synthesis is the selective removal of the N-terminal protecting group. An incomplete removal of this group causes a failure sequence, or even a truncated sequence when the peptide chain remains permanently blocked (Fig. 6).

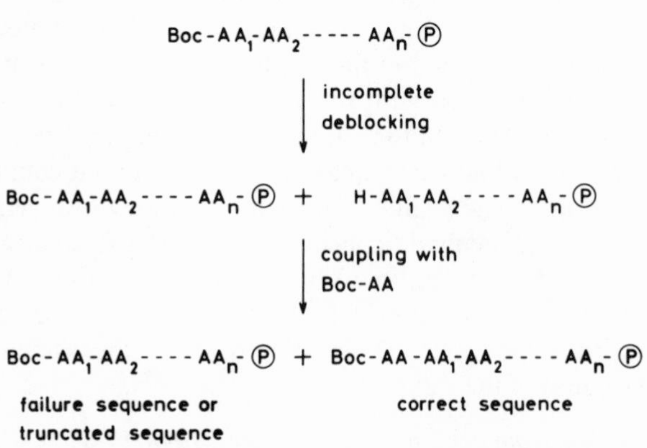

Fig. 6. Formation of failure or truncated sequences as a result of incomplete deblocking.

As the N-protecting group in solid-phase peptide synthesis, the t-butyloxycarbonyl (Boc) group is most widely used. This group can be split off with 1 N HCl/HOAc, as proposed originally by Merrifield (1964). Later, other deblocking reagents, such as 4 N HCl/dioxane (Stewart and Woolley, 1965), trifluoroacetic acid in methylene chloride (1:1) (Gutte and Merrifield, 1969), and 5.35 N HCl in dimethylsulfoxide and methylene chloride (1:1) (Chou et al., 1971) were introduced. These reagents were introduced especially to swell the resin during the deblocking step, by which the penetration of the acid into the resin was reported to be facilitated. However, one should take into consideration a dilution of these deblocking reagents by the solvent left in the resin from the previous washing, for which reason a prewashing with the deblocking reagent is recommended. Recently, Loffet and Dremier (1971) recommended a 10% solution of mercaptoethane sulfonic acid in acetic acid as a deblocking agent; this has the special advantage that tryptophane is stable under these conditions.

When the Boc group was introduced in solid-phase synthesis, Merrifield (1964) investigated the splitting-off of this group with 1 N HCl/HOAc as the deblocking agent. The reaction was monitored by treating samples of Boc-Arg(NO$_2$)-resin during a certain time with 1 N HCl/HOAc. After the excess deblocking agent had been washed out, the hydrochloride formed was neutralized with triethylamine in dimethylformamide and the chloride was determined in the filtrate by the Volhard procedure. It appeared that the Boc group was completely removed in 15 min, since the amount of chloride was then in excellent agreement with the amino acid content of the resin as found by a quantitative amino acid analysis.

A great many investigators have used this method for the monitoring of the removal of the Boc group (see, for instance, Bayer et al., 1968b; Klostermeyer, 1968; Okuda and Zahn, 1969; Brunfeldt and Halstrøm, 1970; Hagenmaier, 1970; Hammerström et al., 1970; Polzhofer and Ney, 1970; Beyerman et al., 1972a). An incomplete deblocking was never observed in these cases; however, in most cases the total chloride content decreased with the progress of the synthesis (see, for instance, Hagenmaier, 1970; Beyerman et al., 1972a).

This phenomenon was explained by partial scission of the peptide—polymer bond during the deblocking step. When 1 N HCl/HOAc is used, this loss of peptide is about 1% in each deblocking step, while the loss is about 2–3% when trifluoroacetic acid–methylene chloride (1:1) is used.

Brunfeldt and Christensen (1972) found in each step a decrease of 2–2.5% of the total number of free amino groups of an Ala-resin when they repeated the titrations with 0.05 N HClO$_4$ in HOAc. This decrease appeared to be caused by an irreversible blocking of amino groups by a nonidentified impurity in the methylene chloride (May and Baker) used.

Another commercial methylene chloride (Merck, *pro analysi*) did not give this side-reaction.

Polzhofer and Ney (1970) noticed a sharp decrease in the total number of free amino groups after the coupling of Glu(OBzl); we presume this was caused by pyroglutamic acid formation. Takashima *et al.* (1968) suggested the formation of pyroglutamic acid from glutamine to be the cause of a truncated sequence in the synthesis of oxytocin. An investigation by Beyerman *et al.* (1971c) into the formation of pyroglutamic acid from glutamine in solid-phase synthesis under acidic conditions revealed that especially acetic acid was productive of pyroglutamic acid formation. Strongly acidic reagents caused hardly any pyroglutamic acid formation (Table II). Since acetic acid should not be used after the removal of the Boc group from glutamine, the method of choice for this deblocking should be use of 4 N HCl-dioxane or trifluoroacetic acid–methylene chloride (1:1).

Table II. Formation of Pyroglutamyl Peptide During the Cleavage of the Boc Group from Boc-Gln-Gly-P

Acid solvent[a]	Pyroglutamyl peptide (%) formed[b] after		
	$\frac{1}{2}$ hr	1 hr	70 hr
4 N HCl in dioxane	—	—	1
1 N HCl in AcOH	—	—	1
CF_3COOH	—	—	4
CF_3COOH-CH_2Cl_2, 1:1	—	—	11
CF_3COOH-CH_2Cl_2, 1:5	—	trace	17
AcOH-H_2O^c, 1:1			45
AcOH[c]	1–2	2–3	60

[a]All ratios by volume.
[b]At 23°C; percentage yield estimated after triethylamine-catalyzed ethanolysis by comparison on thin layer chromatogram with equivalent amounts of pGlu-Gly-OEt.
[c]After previous removal of the Boc group with trifluoroacetic acid for 15 min at 23°C.

A very elegant control for the complete deblocking of the Boc groups was worked out by Krumdieck and Baugh (1969). They synthesized polyglutamates of folic acid and used for this synthesis Boc-Glu-OBzl marked with ^{14}C in the Boc group. The removal of the Boc group therefore could be checked by measuring the radioactivity left in the resin after the deblocking step. They found that with a 20% solution of trifluoroacetic acid in methylene chloride the Boc group was completely removed in 30 min.

Geising and Hörnle (1971) investigated the deblocking step by means of their radioactive phenylisothiocyanate procedure. They found that with 1 N HCl/HOAc, the Boc group was completely removed from peptides with N-terminal Leu in about 20 min, while for peptides with N-terminal Ile the time required was about 30 min. They also found that during the deblocking step a certain amount of peptide appeared to be lost from the resin, which occurred mainly in the dipeptide stage. This might also be caused by dioxopiperazine formation during the neutralization step (Lunkenheimer and Zahn, 1970; Gisin and Merrifield, 1972; Rothe and Mazánek, 1972; Brunfeldt et al., 1972a,b) or interchain aminolysis (chain doubling) (Beyerman et al., 1972b).

In fact, incomplete deblocking in solid-phase synthesis was reported in only a limited number of cases.

Dorman and Markley (1971) met with difficulties in the removal of the Boc group from a Boc-Ile-resin with 1 N HCl/HOAc (30 min) or with 3 N HCl/dioxane (40 min), followed by a treatment with 1 N HCl/HOAc (30 min.). Complete deblocking was achieved by two successive treatments with 1 N HCl/HOAc for 30 min. Steric hindrance of the amino acid side-chain as well as resin phenomena may play a role in this, since no difficulties were encountered during the deblocking of the next two amino acids, Boc-Leu and Boc-Ala.

A number of incomplete deblockings were reported in the solid-phase synthesis of two encephalitogenic peptides,

Ala-Arg-Thr-Thr-His-Tyr-Gly-Ser-Leu-Pro-Gln-Lys-Gly-resin   [1]

Ser-Tyr-Ser-Met-Glu-His-Phe-Arg-Trp-Gly-Lys-Pro-Val-resin   [2]

(Chou et al., 1971; Shapira et al., 1971). The deblocking reaction was followed with a quantitative ninhydrin reaction; it was thus found that after Ser and His in [1] and after Trp and His in [2], the Boc group was incompletely removed with trifluoroacetic acid–methylene chloride (1:1), 5.6 N HCl/dioxane, or a 1:1 mixture of these two reagents. A new reagent, 5.35 M HCl in dimethylsulfoxide–methylene chloride (1:1), gave good results in some cases, while good results were also obtained when the resin was prewashed with 1.5 M urea in dimethylformamide. The unsatisfactory experience of Shapira et al. (1971) in the deblocking reaction induced them to state that "the major cause of the formation of peptides with undesired sequences during the Merrifield synthesis is inadequate deblocking rather than incomplete coupling." However, Stewart (1971), who repeated one of these syntheses on a 1 % cross-linked resin, did not meet with any difficulty in the deblocking reactions.

A detailed investigation into the effectiveness of various deblocking reagents for the Boc group was carried out by Karlsson et al. (1972a). The total number of free amino groups after the deblocking reaction was determined with the 2-hydroxy-1-naphthaldehyde method of Esko et al. (1968). In all cases investigated, trifluoroacetic acid–methylene chloride (1:1) gave a complete removal of the Boc group in 10–30 min, while 3.8 N HCl/dioxane turned out to be insufficient. 1 N HCl/HOAc was effective only when the resin was swelled in methylene chloride before the deblocking step. When this pretreatment was omitted, the Boc group was split off only 55% from a Boc-Ala-resin after 30 min as compared to 92% from a Boc-Arg(NO_2)-resin after 60 min. [Merrifield (1964) found a complete removal of the Boc group from a Boc-Arg(NO_2)-resin in 15 min with the same reagent.]

Bayer et al. (1972) found an incomplete deblocking of the Boc group from Thr(Bzl) in the penultimate step of the synthesis of Boc-Val-Thr(Bzl)-Val-Leu-Thr(Bzl)-Ala-Leu-Gly-P. After cleavage of the peptide from the solid support by transesterification (Beyerman et al., 1968b) and partial purification by gel filtration, a sample was deprotected and then trifluoroacetylated. The ^{19}F NMR spectrum showed it to be a mixture of two components in a ratio of 65% octapeptide and 35% heptapeptide. This incomplete deblocking was also confirmed by the chloride titration method.

In our laboratory, incomplete deblocking was observed in a few cases.

In our solid-phase synthesis of luteinizing hormone–follicle stimulating hormone releasing hormone (Beyerman et al., 1972a), PGA-His-Trp-Ser-Tyr-Gly-Leu-Arg-Pro-Gly-NH_2, it appeared from amino acid analysis that pyroglutamic acid had been attached about 50% to His(Tos)-peptide-on-resin, while the ninhydrin test was negative after two couplings. Treatment of this peptide-on-resin with trifluoroacetic acid–methylene chloride (1:1) gave a positive ninhydrin test. After another coupling, the pyroglutamic acid incorporation rose to about 85% with respect to histidine. In a second synthesis, the Boc group of Boc-His(Tos)-peptide-on-resin was completely removed when this reaction was repeated (once for 5 min, once for 25 min, and once for 30 min).

In the synthesis of the C-terminal sequence of human calcitonin (Hindriks, 1972; Hirt, 1971), Boc-Pro-Gln-Thr(Ztf)-Ala-Ile-Gly-Val-Gly-Ala-Pro-P, incomplete deblocking was found after the coupling of Boc-Gln; this observation could be duplicated. The Boc group was split off 80–90% with trifluoroacetic acid–methylene chloride (1:1) in 30 min. Repetition of the deblocking with the same reagent for 30 min brought this reaction to completion.

It appears from the literature as well as from our own experience that the removal of the Boc group presents some problems in a number of cases, although the extreme difficulties met with by Chou et al. (1971) seem to have been the exception up to now. However, this necessitates the monitoring of the removal of the Boc group in a new synthesis, since otherwise considerable numbers of failure sequences may be formed. For best results, it is advisable to use trifluoroacetic acid–methylene chloride (1:1) or possibly (1:3) and to repeat the deblocking step routinely. Double deblocking has already been employed by Hancock *et al.* (1971) in the synthesis of an acyl carrier protein analog.

IV. DISCUSSION

It is quite understandable that investigators who want to quickly produce, with the aid of a synthesizer, a peptide which can be purified and identified unambiguously tend to follow the principle "don't worry, but hurry." We think, however, that it will be wise to monitor, in any case for a first synthesis of a sequence, at least the couplings qualitatively, e.g., by the ninhydrin method. A new and possibly more sensitive method might be the fluorescamine test (Weigele *et al.*, 1972a,b).

Quantitative monitoring of both the coupling and deprotection by one method is in practice not sufficiently accurate. The only method for determining the remaining Boc groups after deblocking is the radioactive method of Krumdieck and Baugh (1969). This method, however, involves practical difficulties for routine use, so that an indirect method (determination of free amino groups) must still be applied.

If hydrogen chloride is used as deprotecting agent, the amount of chloride equivalent to the free amine can be directly determined in the filtrate resulting from the triethylamine treatment. If a mixture of trifluoroacetic acid and methylene chloride is used, the trifluoroacetate from the peptide-on-polymer must first be neutralized, after which the free amino groups can be determined. It is also possible after washing out of the excess of trifluoroacetic acid with methylene chloride to directly carry out the Dorman method (Hirt, 1971).

For the quantitative determination of free amino groups on the resin, there are three nondestructive methods. Only these methods can be applied to the whole batch involved in the synthesis. The nondestructive methods are the pyridine hydrochloride method of Dorman (1969), the perchloric acid titration of Brunfeldt *et al.* (1969), and the picric acid method of Gisin (1972). It is a common drawback of the three methods that they are not universally applicable, since basic centers in the peptide chain will be involved in the determination. The perchloric acid titration has already been automated (Brunfeldt *et al.*, 1972a,b).

The reported destructive methods have the disadvantage that they are time consuming. It is an advantage that the methods are performed on a sample, because monitoring methods applied on the whole resin batch can cause lowering of the yield of the synthesized peptide (Hancock *et al.*, 1973).

In our hands, the pyridine hydrochloride method gave reproducible values, but it was sometimes difficult to interpret the absolute values. In most cases, the ninhydrin method was used at the same time, if there was uncertainty whether the coupling was complete. An unsolved problem is the blank value, which is occasionally increased in the Dorman procedure (Hirt, 1971; Beyerman *et al.*, 1972a; see Table I).

This survey covers the literature up to the middle of 1972.

REFERENCES

Baas, J. M. A., Beyerman, H. C., van de Graaf, B., and de Leer, E. W. B., 1971, The relative statistical distribution of peptides in stepwise synthesis, *in* "Peptides—1969" (E. Scoffone, ed.), p. 173, North-Holland, Amsterdam.

Baba, T., Sugiyama, H., and Seto, S., 1971, Solid phase synthesis of the active center octapeptide of *Candida krusei* cytochrome C apoprotein, *Sci. Rep. Res. Inst. Tokoku Univ. Ser. A* **23**:85.

Bayer, E., 1969, Neuere Ergebnisse der Peptidesynthese am Festkörper, *Chem. Labor Betr.* **20**:193.

Bayer, E., 1970, New results in the solid phase method for the synthesis of peptides, *in* "Peptides: Chemistry and Biochemistry" (B. Weinstein and S. Lande, eds.), p. 99, Marcel Dekker, New York.

Bayer, E., and Koenig, W. A., 1969, Sequence analysis of polypeptides by chromatography and mass spectrometry, *J. Chromatog. Sci.* **7**:95.

Bayer, E., Hagenmaier, H., Jung, G., and König, W., 1968a, Solid phase synthesis of oxytocin and apoferredoxin, *in* "Peptides—1968" (E. Bricas, ed.), p. 162, North-Holland, Amsterdam.

Bayer, E., Jung, G., and Hagenmaier, H., 1968b, Untersuchungen zur Totalsynthese des Ferredoxins-I; Synthese der Aminosäuresequenz von *C. pasteurianum* Ferredoxin, *Tetrahedron* **24**:4853.

Bayer, E., Eckstein, H., Hägele, K., König, W. A., Brüning, W., Hagenmaier, H., and Parr, W., 1970, Failure sequences in the solid phase synthesis of polypeptides, *J. Am. Chem. Soc.* **92**:1735.

Bayer, E., Hagenmaier, H., Jung, G., Parr, W., Eckstein, H., Hunziker, P., and Sievers, R. E., 1971, The problem of failure sequences in the solid phase synthesis of peptides, *in* "Peptides —1969" (E. Scoffone, ed.), p. 65, North-Holland, Amsterdam.

Bayer, E., Hunziker, P., Mutter, M., Sievers, R. E., and Uhmann, R., 1972, ^{19}F Nuclear magnetic resonance for the control of peptide synthesis, *J. Am. Chem. Soc.* **94**:265.

Beacham, J., Dupuis, G., Finn, F. M., Storey, H. T., Yanaihara, C., Yanaihara, N., and Hoffmann, K., 1971, Studies on polypeptides. XLIX. Fragment condensations with peptide derivatives related to the primary structure of ribonuclease T, *J. Am. Chem. Soc.* **93**:5526.

Beyerman, H. C., 1972, Some fundamental experiments in solid-phase peptide synthesis, *in* "Progress in Peptide Research," Vol. II (S. Lande, ed.), p. 25, Gordon and Breach, New York.

Beyerman, H. C., and Hindriks, H., 1971, Determination of the completeness of coupling in solid phase peptide synthesis by a colour indicator, *in* "Peptides—1969" (E. Scoffone, ed.), p. 145, North-Holland, Amsterdam.

Beyerman, H. C., Boers-Boonekamp, C. A. M., van Zoest, W. J., and van den Berg, D., 1967, Active esters and bifunctional catalysts in solid-phase peptide synthesis, *in* "Peptides" (H. C. Beyerman, A. van de Linde, and W. Maassen van den Brink, eds.), p. 117, North-Holland, Amsterdam.

Beyerman, H. C., Boers-Boonekamp, C. A. M., and Maassen van den Brink-Zimmermannová, H., 1968a, Synthesis of oxytocinoic acid and oxytocin on a solid support *via* triazolides from active esters, *Rec. Trav. Chim. Pays-Bas* **87**:257.

Beyerman, H. C., Hindriks, H., and de Leer, E. W. B., 1968b, Alcoholysis of Merrifield-type peptide–polymer bonds, *Chem. Commun.* **1968**:1668. '

Beyerman, H. C., Hindriks, H., Hirt, J., and de Leer, E. W. B., 1971a, Alcoholysis of the peptide–polymer bond in solid-phase synthesis, *in* "Peptides—1969" (E. Scoffone, ed.), p. 87, North-Holland, Amsterdam.

Beyerman, H. C., van der Kamp, P. R. M., de Leer, E. W. B., Maassen van den Brink, W., Parmentier, J. H., and Westerling, J., 1971b, Monitoring of solid-phase peptide synthesis by making use of N-(2-^{14}C-Boc)-amino acids and determination of the peptide content with ^{35}S-sulfuric acid, *in* "Proceedings of the Eleventh European Peptide Symposium, Vienna, Austria" (H. Nesvadba, ed.) (1973), p. 138, Amsterdam.

Beyerman, H. C., Lie, T. S., and van Veldhuizen, C. J., 1971c, On the formation of pyroglutamyl-peptides in solid-phase peptide synthesis, "Proceedings of the Eleventh European Peptide Symposium, Vienna, Austria" (H. Nesvadba, ed.) (1973), p. 162, Amsterdam.

Beyerman, H. C., Hindriks, H., Hirt, J., de Leer, E. W. B., and van der Wiele, A., 1972a, A synthesis of the decapeptide sequence proposed for the LH- and FSH-releasing homone, *Rec. Trav. Chim. Pays-Bas* **91**:1239.

Beyerman, H. C., de Leer, E. W. B., and van Vossen, W., 1972b, Inter-chain aminolysis, a novel side-reaction in solid-phase peptide synthesis, *Chem. Commun.* **1972**:929.

Bodanszky, M., and Bath, R. J., 1969, Active esters and resins in peptide synthesis: The role of steric hindrance, *Chem. Commun.* **1969**:1259.

Bodanszky, M., and Sheehan, J. T., 1964, Active esters and resins in peptide synthesis, *Chem. Ind. (Lond.)* **1964**:1423.

Bodanszky, M., and Sheehan, J. T., 1966, Active esters and resins in peptide synthesis, *Chem. Ind. (Lond.)*, **1966**:1597.

Brunfeldt, K., and Christensen, T., 1972, Process monitoring in solid phase peptide synthesis, amino group blocking effect of impure methylenechloride, *FEBS Letters* **19**:345.

Brunfeldt, K., and Halstrøm, J., 1970, Tritylation of a partially protected pentapeptide synthesized by the Merrifield solid-phase method, *Acta Chem. Scand.* **24**:3013.

Brunfeldt, K., Roepstorff, P., and Thomsen, J., 1969, Process control in the solid phase peptide synthesis by titration of free amino groups, *Acta Chem. Scand.* **23**:2906.

Brunfeldt, K., Roepstorff, P., and Thomson, J., 1971, Process control in automated peptide synthesis, *in* "Peptides—1969" (E. Scoffone, ed.), p. 148, North-Holland, Amsterdam.

Brunfeldt, K., Christensen, T., and Villemoes, P., 1972a, Automatic monitoring of solid phase synthesis of a decapeptide, *FEBS Letters* **22**:238.

Brunfeldt, K., Bucher, D., Christensen, T., Roepstorff, P., Rubin, I., Schou, O., and Villemoes, P., 1972b, Automated monitoring of solid phase peptide synthesis by perchloric acid titration, *in* "Chemistry and Biology of Peptides" (J. Meienhofer, ed.), p. 183, Ann Arbor Science, Ann Arbor.

Chou, F. C.-H., Chawla, R. K., Kibler, R. F., and Shapira, R., 1971, Incomplete deblocking as a cause of failure sequence in solid phase peptide synthesis, *J. Am. Chem. Soc.* **93**:267.

Corley, L., Sachs, D. H., and Anfinsen, C. B., 1972, Rapid solid-phase synthesis of bradykinin, *Biochem. Biophys. Res. Commun.* **47**:1353.

Dorman, L. C., 1969, A non-destructive method for the determination of completeness of coupling reactions in solid phase peptide synthesis, *Tetrahedron Letters* **1969**:2319.

Dorman, L. C., and Markley, L. D., 1971, Solid phase synthesis and antibacterial activity of *N*-terminal sequences of melittin, *J. Med. Chem.* **14**:5.

Dorman, L. C., Markley, L. D., and Mapes, D. A., 1971, A model system for studying solid phase peptide reactions, *Anal. Biochem.* **39**:492.

Dutta, A. S., and Morley, J. S., 1971, Polypeptides. Part XII. The preparation of 2-pyridyl esters and their use in peptide synthesis, *J. Chem. Soc.* (C) **1971**:2896.

Esko, K., and Karlsson, S., 1970, Effects of Boc-amino acid adsorption with respect to yield and racemisation in the Merrifield method, *Acta Chem. Scand.* **24**:1415.

Esko, K., Karlsson, S., and Porath, J., 1968, A method for determining free amino groups in polymers with particular reference to the Merrifield synthesis, *Acta Chem. Scand.* **22**:3342.

Fankhauser, P., Brenner, M., Schilling, B., and Fries, P., 1971, Reaktionen an Trägerharzen: Zur quantitativen Bestimmung des Totalumsatzes, *in* "Proceedings of the Eleventh European Peptide Symposium, Vienna, Austria" (H. Nesbadva, ed.) (1973), p. 153.

Felix, A. M., and Merrifield, R. B., 1970, Azide solid phase peptide synthesis, *J. Am. Chem. Soc.* **92**:1385.

Garden, J., II, and Tometsko, A. M., 1972, Fluorometric method for quantitative determination of free amine groups in peptide-containing Merrifield resins, *Anal. Biochem.* **46**:216.

Geiger, R., 1971, Die Syntheses physiologisch wirksamer Peptide, *Angew. Chem.* **83**:155.

Geising, W., and Hörnle, S., 1971, Bestimmung freier Aminogruppen mittels ^{14}C-markiertem Phenylisothiocyanat bei der Peptidsynthese in fester Phase, *in* "Proceedings of the Eleventh European Peptide Symposium, Vienna, Austria" (H. Nesvadba, ed.) (1973), p. 146.

Gisin, B. F., 1972, The monitoring of reactions in solid-phase peptide synthesis with picric acid, *Anal. Chim. Acta* **58**:248.

Gisin, B. F., and Merrifield, R. B., 1972, Carboxyl-catalyzed intramolecular aminolysis. A side reaction in solid-phase peptide synthesis, *J. Am. Chem. Soc.* **94**:3102.

Gisin, B. F., Merrifield, R. B., and Tosteson, D. C., 1969, Solid-phase synthesis of the cyclo-dodecadepsipeptide valinomycin, *J. Am. Chem. Soc.* **91**:2691.

Gut, V., and Rudinger, J., 1968, Rate measurements in solid phase peptide synthesis, *in* "Peptides—1968" (E. Bricas, ed.), p. 185, North-Holland, Amsterdam.

Gutte, B., and Merrifield, R. B., 1969, The total synthesis of an enzyme with ribonuclease A activity, *J. Am. Chem. Soc.* **91**:501.

Gutte, B., and Merrifield, R. B., 1971, The synthesis of ribonuclease A, *J. Biol. Chem.* **246**:1922.

Hagenmaier, H., 1970, The influence of the chain length on the coupling reaction in solid phase peptide synthesis, *Tetrahedron Letters* **1970**:283.

Hammerström, K., Lunkenheimer, W., and Zahn, H., 1970, Peptides 78, Merrifield synthesis of the insulin sequence B 1–8 using various thiol protecting groups, *Makromol. Chem.* **133**:41.

Hancock, W. S., Prescott, D. J., Nulty, W. L., Weintraub, J., Vagelos, P. R., and Marshall, G. R., 1971, The synthesis of a protein with acyl carrier protein activity, *J. Am. Chem. Soc.* **93**:1799.

Hancock, W. S., Prescott, D. J., Vagelos, P. R., and Marshall, G. R., 1973, Solvation of the polymer matrix. Source of truncated and deletion sequences in solid phase synthesis, *J. Am. Chem. Soc.* **38**:774.

Hindriks, H., 1972, Thesis (this laboratory), Delft, Vaste-fase peptide synthesen.

Hirt, J., 1971, unpublished results (this laboratory).

Hörnle, S., 1967, Synthese der Rinder-Insulin-A-Kette nach der Merrifield-Methode unter ausschliesslicher Verwendung von *tert.*-Butyloxycarbonyl-aminosäure-*p*-nitro-phenylestern, *Hoppe-Seyler's Z. Physiol. Chem.* **348**:1355.

Hörnle, S., and Geising, W., 1971, Determination of free amino groups by the use of ^{14}C-labelled isothiocyanate during solid phase peptide synthesis, *Hoppe-Seyler's Z. Physiol. Chem.* **352**:5.

Jakubke, H. D., and Baumert, A., 1971, Vergleich einiger für die Festphasen-Peptidsynthese eingesetzten Kupplungsmethoden am Beispiel eines Modellpeptids, *in* "Proceedings of the Eleventh European Peptide Symposium, Vienna, Austria" (H. Nesvadba, ed.) (1973), p. 132, Amsterdam.

Jollès, P., and Jollès, J., 1968, Synthèse par la methode de Merrifield d'un octapeptide faisant partie du lysozyme de blanc d'oeuf de Poule (enchaînement Cys^{64}-----Gly^{71}), *Helv. Chim. Acta* **51**:980.

Jorgensen, E. C., Windridge, G. C., Patton, W., and Lee, T. C., 1969, Angiotensin II analogs. I. Synthesis and biological evaluation of Gly^1,Gly^2,Ile^5-angiotensin II, Ac-Gly^1,Gly^2,Ile^5-angiotensin II, and Gly^1,Gly^2,Ile^5,His(Bzl)^6-angiotensin II, *J. Med. Chem.* **12**:733.

Kaiser, E., Colescott, R. L., Bossinger, C. D., and Cook, P. I., 1970, Color test for detection of free terminal amino groups in the solid-phase synthesis of peptides, *Anal. Biochem.* **34**:595.

Karlsson, S., Lindeberg, G., Porath, J., and Ragnarsson, U., 1970*a*, Removal of *t*-butyloxycarbonyl groups in solid phase peptide synthesis, *Acta Chem. Scand.* **24**:1010.

Karlsson, S., Lindeberg, G., and Ragnarsson, U., 1970*b*, Reactivity of *p*-nitrophenyl esters in solid phase peptide synthesis, *Acta Chem. Scand.* **24**:337.

Khosla, M. C., Chaturvedi, N. C., Smeby, R. R., and Bumpus, F. M., 1968, Synthesis of 1-isoleucine-, 3-proline-and 5-alanine-angiotensins. II., *Biochemistry* **7**:3417.

Klostermeyer, H., 1968, Synthese von Gramicidin S mit Hilfe der Merrifield-Methode, *Chem. Ber.* **101**:2823.

Krumdieck, C. L., and Baugh, C. M., 1969, The solid-phase synthesis of polyglutamates of folid acid, *Biochemistry* **8**:1568.

Li, C. H., and Yamashiro, D., 1970, The synthesis of a protein possessing growth-promoting and lactogenic activities, *J. Am. Chem. Soc.* **92**:7608.

Loffet, A., 1968, personal communication.

Loffet, A., and Dremier, C., 1971, A new reagent for the cleavage of the tertiary butyloxycarbonyl protecting group, *Experientia* **27**:1003.

Losse, A., 1971, Einfluss der Harz-Struktur auf den Peptid-Bindungsschritt der Merrifield-Synthese, *Tetrahedron Letters* **1971**:4989.

Losse, G., and Klengel, H., 1971, Synthese des Depsipeptides Valinomycin nach der Festphasenmethode, *Tetrahedron* **27**:1423.

Losse, G., and Neubert, K., 1970*a*, Peptidesynthese an hochpolymeren Verbindungen, *Z. Chem.* **10**:48.

Losse, G., and Neubert, K., 1970*b*, Synthese von Gramicidin S durch neue Varianten der Festphasensynthese, *Tetrahedron Letters* **1970**:1267.

Losse, G., and Ulbrich, R., 1971, Kontrolle des Umsatzgrades bei der Merrifield-Synthese, *Z. Chem.* **11**:346.

Lübke, K., 1971, Solid-phase synthesis of Arg^8-vasopressin, *in* "Peptides—1969" (E. Scoffone, ed.), p. 154, North-Holland, Amsterdam.

Lunkenheimer, W., and Zahn, H., 1970, Peptide, LXXIX, Merrifield-Synthese symmetrischer Cystinpeptide, *Justus Liebigs Ann. Chem.* **740**:1.

Markley, L. D., and Dorman, L. C., 1970, A comparative study of terminating agents for the use in solid-phase peptide synthesis, *Tetrahedron Letters* **1970**:1787.

Marshall, G. R., and Merrifield, R. B., 1971, Solid phase synthesis: The use of solid supports and insoluble reagents in peptide synthesis, *in* "Biochemical Aspects of Reactions on Solid Supports" (G. R. Stark, ed.), p. 111, Academic Press, New York and London.

Mehlis, B., Fischer, W., and Niedrich, H., 1971, A method for the determination of conversion in the Merrifield peptide synthesis, *in* "Peptides—1969" (E. Scoffone, ed.), p. 146, North-Holland, Amsterdam.

Merrifield, R. B., 1962, Peptide synthesis on a solid polymer, *Fed. Proc., Fed. Am. Soc. Exptl. Biol.* **21**:412.

Merrifield, R. B., 1963, Solid phase peptide synthesis. I. The synthesis of a tetrapeptide, *J. Am. Chem. Soc.* **85**:2149.

Merrifield, R. B., 1964, Solid-phase peptide synthesis. III. An improved synthesis of brady-kinin, *Biochemistry* **3**:1385.

Merrifield, R. B., 1965, Automated synthesis of peptides, *Science* **150**:178.

Merrifield, R. B., 1969, Solid-phase peptide synthesis, *Advan. Enzymol. Rel. Areas Mol. Biol.* **32**:221.

Merrifield, R. B., Stewart, J. M., and Jernberg, N., 1966, Instrument for automated synthesis of peptides, *Anal. Chem.* **38**:1905.

Mitchell, A. R., and Roeske, R. W., 1970, Amino acid insertion in solid-phase peptide synthesis, *J. Org. Chem.* **35**:1171.

Miura, Y., Toyama, M., and Seto, S., 1968, Solid phase synthesis of a nonapeptide of melittin, *Sci. Rep. Res. Inst. Tohoku Univ. Ser. A* **20**:41.

Najjar, V. A., and Merrifield, R. B., 1966, Solid phase peptide synthesis. VI. The use of the *o*-nitrophenylsulfenyl group in the synthesis of the octapeptide bradykininylbradykinin, *Biochemistry* **5**:3765.

Ney, K. H., and Polzhofer, K. P., 1968, Synthese von Casein-Peptiden nach der Merrifield-Methode, *Tetrahedron* **24**:6619.

Noda, K., Terada, S., Mitsuyasu, N., Waki, M., Kato, T., and Izumiya, N., 1970, Snake venom peptides. I. Solid phase synthesis of a hexapeptide fragment of cobrotoxin, *Mem. Fac. Sci. Kyushu Univ. Ser. C* **7**:189.

Okuda, T., and Zahn, H., 1969, Synthese der B-Ketten des Human- sowie des Rinderinsulins nach dem Festkörper-Verfahren von Merrifield, *Makromol. Chem.* **121**:87.

Polzhofer, K. P., 1972, Synthese eines Labempfindlichen Pentadecapeptide aus Kuh-κ-Casein, *Tetrahedron* **28**:855.

Polzhofer, K. P., and Ney, K. H., 1970, Synthese von Casein-Peptiden nach der Merrifield-Methode, *Tetrahedron* **26**:3221.

Potts, J. T., Jr., Tregear, G. W., Keutmann, H. T., Niall, H. D., Sauer, R., Deftos, L. J., Dawson, B. F., Hogan, M. L., and Aurbach, G. D., 1971, Synthesis of a biologically active *N*-terminal tetratriacontapeptide of parathyroid hormone, *Proc. Natl. Acad. Sci.* **68**:63.

Ragnarsson, U., Lindeberg, G., and Karlsson, S., 1970, Note on the use of active esters in combination with 1,2,4-triazole in solid phase peptide synthesis, *Acta Chem. Scand.* **24**:3079.

Ragnarsson, U., Karlsson S., and Sandberg, B., 1971, Studies on the coupling step in solid phase peptide synthesis. Some preliminary results from competition experiments, *Acta Chem. Scand.* **25**:1487.

Richards, F. F., Sloane, R. W., Jr., and Haber, E., 1967, The synthesis and antigenic properties of a macromolecular peptide of defined sequence bearing the dinitrophenol hapten, *Biochemistry* **6**:476.

Rivaille, P., Robinson, A., Kamen, M., and Milhaud, G., 1971, Synthèse en phase solide de l'hormone de libération de l'hormone lutéotrophique (LH-RH), *Helv. Chim. Acta* **54**:2772.

Rivier, J., Monahan, M., Vale, W., Grant, G., Amoss, M., Blackwell, R., Guillemin, R., and Burgus, R., 1972, Solid phase peptide synthesis on a benzhydrylamine resin of LRF (luteinizing hormone releasing factor) and analogues including antagonists, *Chimia* **26**:300.

Rothe, M., and Mazánek, J., 1972, Nebenreaktionen bei der Festphasen-Peptidesynthese als Folge der Bildung von Cyclopeptiden, *Angew. Chem.* **84**:290.

Rudinger, J., and Gut, V., 1967, Discussion, in "Peptides" (H. C. Beyerman, A. van de Linde, and W. Maassen van den Brink, eds.), p. 89, North-Holland, Amsterdam.

Sano, S., Tokunaga, R., and Kun, K. A., 1971, Solid phase method for peptide synthesis using macroreticular copolymers, *Biochim. Biophys. Acta* **244**:201.

Schaich, E., and Schneider, F., 1971, Experiments on the synthesis of a partial sequence of a streptococcal protease by the Merrifield method, *Hopp-Seyler's Z. Physiol. Chem.* **352**:4.

Shapira, R., Chou, F. C.-H., McKneally, S., Urban, E., and Kibler, R. F., 1971, Biological activity and synthesis of an encephalitogenic determinant, *Science* **173**:736.

Sieber, P., and Iselin, B., 1968, Peptidsynthesen unter Verwendung der 2-(p-Diphenyl)-iso-propyloxycarbonyl (Dpoc)-Aminoschutzgruppe, *Helv. Chim. Acta* **51**:622.

Sievertsson, H., Chang, J.-K., von Klaudy, A., Bogentoft, C., Curry, B., Folkers, K., and Bowers, C., 1972, Two syntheses of the luteinizing hormone releasing hormone of the hypothalamus, *J. Med. Chem.* **15**:222.

Stewart, J. M., 1971, personal communication.

Stewart, J. M., and Woolley, D. W., 1965, All-D-bradykinin and the problem of peptide anti-metabolites, *Nature (Lond.)* **206**:619.

Stewart, J. M., and Young, J. D., 1969, "Solid Phase Peptide Synthesis," Freeman, San Francisco.

Takashima, H., du Vigneaud, V., and Merrifield, R. B., 1968, The synthesis of deamino-oxytocin by the solid phase method, *J. Am. Chem. Soc.* **90**:1323.

Tregear, G. W., Niall, H. D., Potts, J. T., Jr., Leeman, S. E., and Chang, M. M., 1971, Synthesis of substance P, *Nature New Biol.* **232**:87.

Van Melick, J. E. W., and Wolters, E. T. M., 1972, Selective formylation of amino groups under neutral conditions, *Synth. Commun.* **2**:83.

Weigele, M., Blount, J. F., Tengi, J. P., Czajkowski, R. C., and Leimgruber, W., 1972a, The fluorogenic ninhydrin reaction, structure of the fluorescent principle, *J. A. Chem. Soc.* **94**:4052.

Weigele, M., DeBernardo, S. L., Tengi, J. P., and Leimgruber, W., 1972b, A novel reagent for the fluorometric assay of primary amines, *J. Am. Chem. Soc.* **94**:5927.

Westall, F. C., and Robinson, A. B., 1970, Solvent modification in Merrifield solid-phase peptide synthesis, *J. Org. Chem.* **35**:2842.

Weygand, F., and Obermeier, R., 1968, Messung des Umsatzes bei der Festkörper-Peptid-synthese mit Hilfe einer massenspektrometrischen Verdünnungsanalyse, *Z. Naturforsch.* **B23**:1390.

Wieland, T., Birr, C., and Flor, F., 1969a, Synthese von Antamanid mit der Merrifield-Technik, *Justus Liebigs Ann. Chem.* **727**:130.

Wieland, T., Birr, C., and Wissenbach, H., 1969b, 3-Nitrophthalsäureanhydrid als Blockierungs-reagens bei der Polypeptidsynthese nach Merrifield, *Angew. Chem.* **81**:782.

Wissmann, H., and Geiger, R., 1970, Abtrennung von Fehlsequenzen in der Festphasen-Peptidsynthese mit dem Anhydrid der 3-Sulfopropionsäure, *Angew. Chem.* **82**:937.

Wünsch, E., 1971, Synthese von Peptid-Naturstoffen, *Angew. Chem.* **83**:773.

Yamashiro, D., Blake, J., and Li, C. H., 1972, The use of N^α,N^{im}-bis(tert-butyloxycarbonyl)-histidine and N^α-2-(p-biphenylyl)isopropyloxycarbonyl-N^{im}-tert-butyloxycarbonylhisti-dine in the solid-phase synthesis of histidine-containing peptides, *J. Am. Chem. Soc.* **94**:2855.

Zhukova, G. F., Ravdel', G. A., and Shchukina, L. A., 1970, Synthesis of eledoisin analogs on a polymer, *J. Gen. Chem. (U.S.S.R.)* **40**:2750.

CHAPTER 18

NEED FOR SOLID-PHASE
THINKING IN SOLID-PHASE SYNTHESIS

P. Fankhauser and M. Brenner

Institut für Organische Chemie
Universität Basel
Basel, Switzerland

Age-old wisdom of alchemists:
Corpora non agunt nisi fluida.

I. INTRODUCTION

The invention of the solid-phase method looked like an ingenious trick to overcome some of the unpleasant features of the classical methods. As we know today, the ingenuity of the trick remains, but only a large investment of heavy real effort will eventually, if ever, work it into a real progress over the classical approach.

Departure from solution chemistry was felt to solve three major classical problems:

1. Nonquantitative coupling and deblocking yields.
2. Separation of homogeneous peptidic material from excess coupling and excess deblocking reagents.
3. Decreasing solubility of the growing peptide chain.

Such feeling expressed itself in the very term "solid-phase" synthesis, somehow implying chemical and physical inertness of the "solid" support as well as of the growing peptide chain. Euphoria seized biochemists and others: disappearance of the solubility problem and of the isolation problem meant disappearance of the yield problem, because quantitative yield now looked merely like a matter of simple mass action!

Actually, the journey away from solution chemistry did not lead to solid-phase dreamland. Instead, it ended in confrontation with a phase neither liquid nor solid, but something in between and appropriately designated as a gel. Consequently, the solubility problem persists, although in a disguised form, and the reactivity problem has even gained in weight, because of new sources of failure due to noninertness of both support and peptide chain and because of error accumulation due to lack of error elimination from permanently fixed material. On the other hand, the separation problem with respect to excess reagent and soluble byproducts is found to be largely solved. This latter fact together with the solid-phase method's adaptability to automation is so important that the unsatisfactory situation with respect to product homogeneity and its control now has become a challenge to peptide chemistry. In that context, a redefinition of the very problems to be dealt with seems helpful. Such redefinition must start from the recognition of the true physical state of the peptide–resin ensemble submitted to chemical reaction or to analysis, which of course is but another type of chemical reaction, and it must take into account the chemical reactivity of the peptide bond and its potential role under coupling, deblocking, and washing conditions. In the same context, a redefinition of what is meant by a peptide—when synthesized by a solid-phase procedure—will probably become inevitable. For the time being and for reasons discussed later, the term "solid-phase peptide" will be used in order to distinguish a solid-phase product from its analogue as prepared by classical methods. A solid-phase peptide still attached to the resin, in other words, any stage of the outgrowing peptide chain, will be designated by the term "resinopeptide."

As indicated by the title, this chapter does not pretend to be a comprehensive review. It is rather an essay, based on experience and thought, with occasional substantiation by sporadic reference to the literature.

II. THE NONSOLID STATE OF THE SOLID PHASE

A. General Remarks

The solid phase, denoted henceforth by the letters "SP", is chemically made up of varying amounts of resinopeptide and of a matrix, in these days most frequently a styrene–divinylbenzene copolymer. For SP synthesis and analysis, it is suspended in a solvent, and a good solvent is invariably a powerful swelling agent for the SP matrix. It is tempting to explain this important fact in terms of facilitated mass transport, facilitation being merely due to easier permeation in the less densely packed swollen state. Mass transport is, however, only a necessary and not a sufficient condition for securing chemical reaction. There is the additional requirement of solvation.

Potentially reactive sites remain unreactive when kept in a nonsolvated state. Full appreciation of the role of swelling therefore involves recognition of the general resemblance between swollen and dissolved states. Swelling is indeed a process of incipient dissolution of segments between cross-links within the polymer network. The chemical nature of the network and its depth-to-surface ratio are in this context of secondary importance. With respect to the need for solvation prior to chemical reaction, there is no essential difference between grafts on bead, popcorn, pellicular, brush, or even soluble polymers. In brief, success of SP reactions depends much less on reagent transport than on appropriate solvation of the loci envisaged for reaction.

It may at this point be useful to recall the mechanics of the swelling process and its driving force: as more and more solvent is absorbed by the polymer, its network is progressively expanded. The chains are forced to assume more elongated, less probable configurations, with an accompanying loss of configurational entropy. Opposing, there is an entropy increase due to more mixing of solvent with polymer. Equilibrium between the solvent–polymer system and the pure solvent will be attained when the net entropy change compensates for the solvent–polymer-dependent heat of interaction, thus making the free energy of swelling disappear (cf. Flory, 1950; Flory and Rehner, 1943).

If the heat of mixing is solvent–polymer dependent, it must change, and swelling must change, too, when hydrogen atoms of a hydrocarbon matrix are substituted by nonhydrocarbon residues. Depending on the particular matrix–graft–solvent system, the matrix and the graft may at a certain ratio become incompatible with each other, with the result of a certain extent of demixing, and at some other ratio the matrix–graft ensemble may even become incompatible with itself, with the result of eventual mechanical disintegration. The general circumstances are quite similar to the well-known case of mutually incompatible aqueous polymer solutions. Upon mixing together, these separate into two layers, behaving as two phases in equilibrium. An example is 5% (w/w) dextran ($\overline{M_w} = 460,000$, $\overline{M_n} = 180,000$) and 4% (w/w) polyethylene glycol ($\overline{M_n} = 6000\text{–}7500$). Let us quote from Albertsson (1970):

As a first approximation the gain in entropy of mixing is related to the *number* of molecules involved in the mixing process. The entropy of mixing per mole is therefore, again, as a first approximation, the same for small and large molecules. The interaction energy between molecules, however, increases with the size of the molecule since it is proportional to the number of sement–segment contacts. For very large molecules, the interaction energy per mole will therefore dominate over the entropy of mixing per mole. Hence, it is the type of interaction between the macromolecules that determines the result of their mixing. If the type of interaction is repulsive or, expressed in a different way, if the attraction between like molecules is larger than between unlike molecules, the system will have its energetically most favorable state when the two polymers are separated. The result of mixing of two such polymers is "incompatibility"; one phase contains predominantly the one polymer, and the other phase the other polymer.

A semisolid matrix–graft–solvent system tends, like Albertsson's system, toward its lowest free energy state, and, within the limits imposed by the network, its way to acquire that state is qualitatively the same. Resino-peptides, unless very large, cannot of course be compared with macro-molecules as regards their surface area. But the difference in the number of available interaction sites can easily be compensated for by the intimate resinopeptide–matrix contact as compared to the very loose contact between dissolved macromolecules and by a higher degree of incompatibility between resinopeptide and matrix.

Given for the matrix the gel state, with a mobile rather than a rigid network, and given the translational effects of thermal motion, the idea of resinopeptides mechanically separated from each other by obstructing polymer material seems inadequate. We should not exclude contact and even association between resinopeptides. Under the influence of attractive self-interaction and repulsive interaction with the matrix, repartition might well lead all the way to internal precipitation of resinopeptides, with concomitant loss of accessibility and reactivity of their end groups. But—to come now to the main point—an unfavorable repartition may in principle be counter-acted by an appropriate choice of solvent! Let us remember that according to theory the solvent plays a decisive role in the type of equilibrium state acquired by the system. Let us not forget, on the other hand, that the system changes its swelling properties with each additional amino acid residue and that such change may be slight or very abrupt or something in between. Systematic control of the swelling capacity at all stages of a SP synthesis would seem to be essential in order to avoid risks, quite apart from the general interest which swelling does deserve. A simple technique from our laboratory, using the polarization microscope, is to be published shortly.

Even without occurrence of precipitation and occlusion, the two-phase quasi-liquid character of the resinopeptide–matrix ensemble must duly affect the course of chemical events within SP particles. Reaction rates being rarely diffusion controlled in peptide chemistry, except perhaps in the case of large fragment couplings, impairment of diffusion within the gel might be of minor importance. But the existence of virtual phase boundaries calls for liquid–liquid partition and Donnan-type phenomena. While we cannot at the present time be certain about their quantitative significance in the particle's interior, the effect of a boundary between the phase representing the swollen particle and the phase representing the surrounding liquid is measurable: there are net differences in solute concentrations within and outside the SP particle. Inversely, such difference in chemical potential constitutes the very proof for the existence of a two-phase system as postulated. Inasmuch as net reaction rates are governed by concentration, unfavorable reagent partition may well be a reason for observed adverse SP rate effects. At any event, there are no data at present to correlate concentration levels within and outside the SP particles. This unsatisfactory state of affairs must be borne in mind whenever conclusions are to be drawn from concentration data referring to supernatants of SPs.

B. Swelling

Data in Table I demonstrate swelling of polystyrene–2%-DVB as a function of an increasing load by a polar substituent ($-CH_2OH$) in various solvents. The response reversal to CH_2Cl_2 and DMF, respectively, at a content of about 1 mmole of $-CH_2OH$ per gram dry weight of hydroxy-

Table I. Swelling of Hydroxymethylated Polystyrene–2%-DVB as a Function of Hydroxymethyl Content and the Solvent

mmoles of $-CH_2OH/g$ resin	mmoles of $-CH_2Cl/g$ original resin[a]	Volume swelling[b] (ml solvent/g hydroxymethyl resin)			Molar ratio of solvent uptake	
		CH_2Cl_2	Dioxane	DMF	CH_2Cl_2/dioxane	CH_2Cl_2/DMF
0.75	1.2	2.15	2.15	1.67	1.33	1.54
1.0	1.4	2.00	2.23	1.89	1.19	1.26
1.8	2.5	1.43	2.11	2.08	0.90	0.82
1.8[c]	2.5	2.05	2.22	1.92	1.23	1.28

[a]The difference between $-CH_2Cl$ and $-CH_2OH$ content originates from partial ether formation between $-CH_2Cl$ groups and methylcellosolve during conversion to acetate in methylcellosolve followed by alkaline hydrolysis in dioxane–water.
[b]As found by a centrifugation method (cf. Pepper et al., 1952) at $3000 \times g$. Standard deviation about 1%, depending on solvent volatility.
[c]About 50% of $-CH_2OH$ still acetylated.

methylresin should be noted. Contrary to popular opinion (Marshall and Merrifield, 1971), swelling properties apparently change long before resino-peptides contribute substantially to the mass of the SP. And contrary to a widely accepted view, CH_2Cl_2 is not under all circumstances—i.e., regardless of resinopeptide load and nature—the agent of choice for the swelling of polystyrene–DVB supports. This latter conclusion is further substantiated by the data collected in Fig. 1 referring to comparative swelling experiments with various resino-BOC-oligoleucines and resino-oligoleucine hydro-chlorides in various solvents as a function of number, n, of Leu residues; in this context, the degree of swelling in the case $n = 0$, representing the

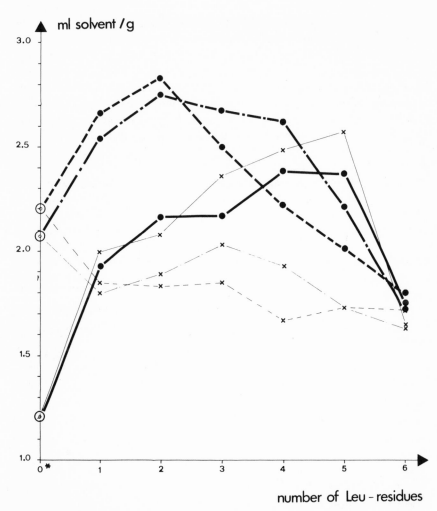

Fig. 1. Swelling of resino-Boc-oligoleucines and resino-oligoleucine hydrochlorides (0.64 mmole/g polystyrene–2%-DVB) as a function of number of leucine residues and of solvent. Solvent uptake in ml/g unsubstituted polystyrene–2%-DVB. ●, Boc-oligoleucines; ×, oligoleucine hydrochlorides; – – – –, CH_2Cl_2; – · – · –, dioxane; ——, N,N-dimethyl-formamide; *, unsubstituted polystyrene–2%-DVB. Standard deviation 0.03 ml/g.

unsubstituted corresponding polystyrene, should be noted. Table II offers supplementary information about polystyrene swelling as a function of the degree of cross-linking, again in various solvents.

The response to solvent polarity is regular up to a certain number of leucine residues, depending on the particular solvent, and shrinking occurs

Table II. Volume Swelling of Styrene–DVB Copolymers
as a Function of Cross-Link Density and Solvent[a]

Percent DVB	CH_2Cl_2	Dioxane	DMF
1	2.85	2.70	1.60
2	2.20	2.05	1.22
3	1.80	1.65	1.00
4	1.47	1.37	0.80
8	0.90	0.87	0.55
12	0.68	0.64	0.34

[a]ml solvent/g resin, by centrifugation (cf. Pepper et al., 1952).

when the number of residues increases. Such results do call for an interpretation in the sense of increasing solvation in the first stages of chain growth, followed first by self-solvation (association) and possibly by precipitation at later stages. Evidently, we find ourselves confronted with a disconcerting reincarnation of the solubility problem. Insolubility in solution chemistry is matched by shrinkage in SP chemistry!

Without fully excluding other mechanisms such as unfavorable reagent partition, several literature reports on steadily diminishing conversion rates (coupling yield) should, on the basis of Fig. 1, probably be assessed in terms of a gradual reduction of the resinopeptide solvation. Corroborating evidence (Geising and Hörnle, 1971) comes from comparative couplings of Ile to N-terminal Val of an aliphatic and of a mixed resino-undecapeptide, the coupling rate being smaller in the first instance. Also, the coupling rate of Val to [Ile-Val-Leu-Ala]$_n$-resin diminishes with increasing n.

If solvation in a given case tends to decrease as a result of chain length of the resinopeptide, and also as a result of increasing residue lipophilicity, it may in another case tend to increase upon resinopeptide prolongation, as a result of changing conformation due, for instance, to a changing lipophilicity/hydrophobicity ratio. Indeed, conversion rate in the course of Merrifield's angiotensinyl-bradykinin synthesis (Merrifield, 1967) fell to 20% in the stage 11 coupling (His). Advancing the synthesis under identical coupling and deblocking conditions did not raise the conversion rate before stage 15 (Arg), when the yield attained 35%. On the other hand, deblocking in 4N HCl/dioxane instead of 1N HCl/CH_3COOH improved conversion to 65% for His and to 85% for Arg. Very interestingly, no such irregularity was observed with respect to these same steps in the same sequence when the bradykinin moiety of the resinopeptide was missing (Marshall and Merrifield, 1965). A final improvement with reported conversion rates of 100% was achieved by reducing the cross-linking in the polystyrene support from 2 to 1%. This latter effect, however, must not be overemphasized. We should certainly not overlook the possible influence of cross-link density, but its importance is probably secondary. As a matter of fact, with deblocking in 1 N HCl/CH$_3$COOH, Merrifield found conversion deficiency almost identical when the DVB cross-linking was 1 or 2%. Thus the important difference demonstrated by Merrifield's series of experiments is to be seen between effects of glacial acetic acid and dioxane, respectively. The latter medium appears in general superior with respect to the resultant reactivity of resinopeptidyl-polystyrenes, and it is highly probable that the inferiority of glacial acetic acid is not to be explained (Marshall and Merrifield, 1971) on the grounds of a once assumed occasional incidence of irreversible amino-terminal acetylation (Merrifield, 1967).

Other examples of milieu-dependent SP reactions are Westall and Robinson's (1970) observations of bad coupling or good coupling, respectively, of Glu onto Gly-Glu-Ser-Pro-Phe-Gly-Lys-resin in CH$_2$Cl$_2$/DCCI or in urea-DMF/DCCI, and cases described by Chou *et al.* (1971), where deblocking necessitated mixtures of common deblocking media. Again, like in Merrifield's angiotensinyl-bradykinin example, deblocking difficulties reported by the Chou group were not encountered with corresponding partial sequences.

Partition has been measured between a sulfonated styrene–DVB copolymer and surrounding acetone–water or dioxane–water mixtures (Davies and Owen, 1956). Water is preferentially absorbed, and partition coefficients change by one order as a function of cross-link density and milieu composition. We found (unpublished) inversed preference, i.e., for the organic component, with a chloromethylated styrene–DVB copolymer exposed to 50% (v/v) dioxane/1 N aqueous NaOH. With 2% cross-linked material loaded with 2 mmoles of chloromethyl per gram of dry polymer, the concentration within and outside the resin was with respect to NaOH 0.36 and 0.54 N, with respect to water 55 and 46% (v/v).

Donnan effects are reported for systems comprising the acidic ion exchange resin, Dowex-50, and concentrated HCl (Baumann and Gibbon, 1947). More spectacular, the sulfonated dye Chlorazol Sky Blue FF in water is not accepted at all by a swelling acid ion exchanger. By measuring the concentration increase of the dye, Pepper *et al.* (1952) could quantitatively assay for water uptake.

In forced deswelling of peptidylresins by heating *in vacuo*, a current practice in preparing samples for analysis, the drying process leads to considerable alteration in the architecture of SP particles. There is a good chance of increasing the exposure of functional sites and/or of adsorbed material. For example, Boc-hexaleucyl-polystyrene dried at 80°C and 0.001 mm Hg and soaked with CH$_2$Cl$_2$ for 5 min contained the same amount of solvent as a sample conditioned with CH$_2$Cl$_2$ after foreign solvent elimination by mere washing procedures. In DMF instead of CH_2Cl_2, solvent uptake after 5 min soaking amounted to only 70% of the equilibrium value attained after 30 min, the latter being identical to the equilibrium value observed on an exclusively solvent-treated sample. Nevertheless, intermediary drying must have had an unfavorable effect on the resolvation process and on the resulting reactivity: in parallel 4 N HCl/dioxane deblocking (2 hr) of a resolvated and a solvated sample of BOC-[leucyl]$_6$-polystyrene, there appeared per gram of polystyrene 350 and 650 μequivalents, respectively, of titrable amino groups, the latter value being consistent with titration values as determined in all the preceding stages ($n = 1$ to 5) in the synthesis of the resinohexapeptide.

With respect to adsorption, an astonishing example is furnished by retention of dicyclohexylurea (or some precursor thereof) upon BOC-aminoacylation of hydroxymethyl-polystyrene by means of DCCI. Extensive washing with CH_2Cl_2, CH_2Cl_2–dioxane (1:1), dioxane, dioxane–ethanol (1:1), ethanol, ethanol–water (1:1), ethanol, ethanol–dioxane (1:1), dioxane, dioxane–methylene chloride (1:1), and methylene chloride gave a product containing an amount of nitrogen approximately equivalent to expectation in the event of complete esterification. However, drying *in vacuo* at 80°C followed by a repetition of the washing cycle reduced the nitrogen content to 3–40% of its original value, depending on the sample. The first CH_2Cl_2 washings even deposited dicyclohexylurea crystals!

C. Internal Mobility

A corollary to the nonsolid state of the SP would be internal mobility. Pertaining evidence is seen in a paper of Lunkenheimer and Zahn (1970) on the oxidation of resinocysteinepeptides. Resinocysteinylglycine and resinocysteinylvaline, respectively, with an approximate peptide load of 0.5 mmole per gram of dry resin, were almost completely converted into the corresponding resinocystinepeptides. Assuming in the swollen state a volume of 3–4 ml per gram of dry resin, a peptide load of 0.5, 1, and 3 mmoles per gram of dry resin and a regular (NaCl-type) lattice for the average distribution of peptide attachment sites, the distance between sites would be about 220, 180, and 130 Å. Even if allowance is made for hypothetical free bond rotation within the polyvinyl part of the resin structure, and for fully extended peptide chains, the latter could not in a rigid lattice meet to an extent corresponding to the observed degree of oxidation.

D. Incompatibility and Disintegration

Partial mechanical SP disintegration is often responsible for heavy material losses during SP synthesis. A pertinent experiment has unequivocally demonstrated that the main reason for such disintegration is a physicochemical one. Partial abrasion by friction and bead breaking by mechanical impact make a comparatively unimportant contribution. In the course of the authors' hexaleucine synthesis on a commonly used styrene–2%-DVB bead polymer, the peptide load corresponding to 650 µequivalents per gram of dry resin, SP particle disintegration started at the tetrapeptide state. Under the microscope, a certain fraction of the beads appeared broken. This fraction greatly increased in stages 5 and 6. Sieve analysis of the hexaleucylresin revealed 8% by weight of material below 400 mesh, consisting almost exclusively of bead fragments, and about 25% of broken beads in the 200–400 mesh portion. A control with everything, including Dorman titration and washings, identical to conditions in the resinohexaleucine preparation—except coupling, which was simulated five times by omission of Boc-Leu addition to the medium—left the SP particles practically unchanged. Bead fragments were extremely rare (0.1–0.2%); less than 1% by weight of the total material was below 400 mesh, and even that portion consisted largely of unbroken beads. With resinopeptides from mixed amino acids, incompatibility with the matrix may arise not at all or at later stages: Sano and Kurihara (1969) report sudden particle disintegration after the first 50 stages of their synthesis of horse cytochrome c.

E. Permeation

Within a resting viscous liquid material, transport is, apart from convection, a pure diffusion phenomenon, its direction depending only on concentration gradients. Within a gel, and more so within a semisolid matrix, there may be mechanical barriers to transport in one direction, leaving some sort of holes for transport in other directions. Such impeded diffusion we call "permeation." Its occurrence would seem to become probable with increasing size of transported molecules. For fast reactions within a semisolid matrix, permeation involving large reagent molecules may become rate determining. However, before an observed SP effect on rate is interpreted in terms of hindered transport, the situation demands very careful examination. For example, nitrophenylbenzoates in solution aminolyze about five times faster than corresponding esters of nitrophenols and carboxyl-bearing swollen polystyrene–DVB copolymers. Comparing now kinetics of aminolysis by 2-aminoethanol and n-tetradecylamine, respectively, of nitrophenol fixed to carboxylated bead and popcorn polymers, the first-order rate constants (min^{-1}) were not significantly different, e.g., 0.12/0.11 for aminoethanol and 0.15/0.18 for tetradecylamine. But then in studying imidazol-catalyzed alcoholysis of fixed dinitrophenylesters by benzyl alcohol, Letsinger and Jerina (1967) found N-methylimidazol equally and poly-N-vinyl-imidazol eight times more effective in the case of a 0.2%-DVB popcorn support as compared to the case of a 2%-DVB bead polymer. Hence, the above-mentioned rate depression by a factor of 5 when passing from dissolved to resin-fixed nitrophenyl ester must be regarded as a solvent effect, or a partition effect, but not as evidence for hindered transport.

True permeation effects on rate may involve more than just a loss of time. Remembering the general strategy of peptide synthesis, we become aware how heavily this depends on the selectivity of the available reactions. Selectivity being a matter of rate ratios, hindered diffusion of a particular reagent is liable to affect product quality. Familiar examples of relative pathway shifting are found in the peptide chemists' dealing with the racemization problem.

III. THE SOLID-PHASE PEPTIDE

A. General Remarks

It is undeniable that peptides produced by SP synthesis have become instrumental in biochemical and pharmacological research. Emotional objections to this irreversible development should give way to a rational analysis of the facts to be faced and to be dealt with in work with SP peptides. The goal we should now strive for is indeed a fixation of limits with respect to the usefulness of a given product for a given purpose. The following discussion is an attempt in this direction.

First, the notion of failure sequences needs an extension of its original definition. This term shall henceforth denote all sequences [too short (= underdone), too long (= overflow), wrong composition] not corresponding to a given synthesis program. Now, how much weight is to be given to failure sequences in a SP peptide? The answer depends above all on the mole fraction of each of the undesired molecular species present.

When error probabilities are small and constant throughout the course of the synthesis, especially of large peptides, most of the very numerous undesired species are formed in insignificant molar quantities. They merely contribute to the so-called background material, even if their combined mass amounts to a considerable fraction of the total. Purification procedures will in general remove only that portion of the background which sufficiently differs from the correct sequence in molecular size and physical properties. That same portion may or may not in part be endowed with a specific property in common with the correct sequence, and therefore should indeed be removed. The then remaining portion will still contain background, but must be used as such, because otherwise SP synthesis loses its interest, which is simplicity.

This latter peptide mixture will henceforth by definition be denoted as a "fractionated solid-phase peptide," symbolized by "SPP."

In the absence of nonstatistical errors, the mole fraction X of the correct molecular sequence in the SPP is greater than the mole fraction of any other molecular species, even though the sum of the many mole fractions of failure sequences with, for example, one missing amino acid may be greater than X. Further, X tends to decrease with increasing size of the peptide, because then the SPP will become a broader section out of a flatter sequence distribution.

In the case of a nonstatistical error incidence—which is more probable in view of the ever-changing interplay between the matrix and the growing resinopeptide—the mole fraction of the program-conformable sequence may not significantly exceed *all* other mole fractions. The solid-phase product will thus in general be separable into clearly distinct portions, every one being still contaminated by background material, and only a single one among them will approximately correspond to what has just been defined as the SPP. There is, however, a practical difficulty: we are never certain about the efficiency of the available separation methods. With respect to the portion considered to represent the SPP, we have therefore no guarantee that *all* mole fractions other than X are in fact insignificant. An apparent SPP with mole fraction X of the correct sequence, and a comparable mole fraction Y of a failure sequence, is a doubtful object for further investigation. If the respective failure sequence has properties responding to the system applied in testing the correct sequence, its presence with mole fraction $Y \approx X$ will falsify the result of the test. How are we going to know whether a SPP is a good $(X > Y)$ or a bad $(X \approx Y)$ preparation?

The answer to this question must come from a recognition of error types, an insight into error sources, and a safe analytical technique for continuous error detection. The reason is quite clear. Having no possibility to check the "purity," i.e., the condition $X > Y$, of a SPP, we can at best—choosing at every stage appropriate experimental conditions—exclude nonrandom failures and must at least—at every stage on the basis of appropriate analyses—exclude their unnoticed occurrence. A SPP is probably then a good preparation when its analytical flowsheet marks no gross deviation from statistical expectation.

We can of course not exclude that a whole class of failure sequences, e.g., all those missing one amino acid or all those with failures in an inactive region of a programmed biologically active chain, are endowed with activity, and it is clear that in a given SP the sum Y' of mole fractions belonging to that class can exceed the mole fraction X. If the activities belonging to Y' and X are qualitatively identical, the SPP will look better than it is from a purely chemical point of view. There is no harm in this event as long as the user's mind does remain critical!

B. Error Types and Sources

Differentiation seems necessary between trivial errors such as incomplete reaction, either in deblocking or coupling, and nontrivial errors caused by side-reactions such as transpeptidation and backbone acylation followed by aminoacyl insertion or chain shortening. Instances belonging to the first error class* are encountered as a consequence of occasionally or constantly reduced activity, which itself is a consequence of the physical state of the SP. The occurrence of side-reactions (second error class) is either an inherent property of the reacting system, that is, of the particular chemistry involved, or else a property imparted to the reacting system, for example, in an attempt to overcome reactivity reduction by severing reaction conditions.

This chapter does not deal with the type of chemistry actually available in SP synthesis. Suffice it to say that it is far from being appropriate for the preparation of pure peptides, because no reaction whatsoever is sufficiently selective to be satisfactory for application in an error-accumulating system. Only the effect on chemistry of the nonideal behavior of the SP is pertinent to the subject of this chapter.

1. Underdone Sequences. As already discussed, incompatibility between resinopeptides and matrix may produce inhomogeneities in the SP particle, the result being local or general closing off, i.e., inaccessibility of resin-fixed potential reaction sites, or else local or general lock-out of reagents. There is of course uncertainty about the respective mechanisms on the molecular level, but we should like to interpret "closing off" in a mechanical sense and "lock-out" in a thermodynamic sense. An unambiguous differentiation between so-defined phenomena will in most cases demand a systematic investigation.

Development of local inhomogeneity on incompatibility grounds is probably accentuated by preexisting inhomogeneity within the polymer structure, such as uneven distribution of cross-link density (cf. Patterson, 1971). Much might be learned about preexisting and developing inhomogeneities from a wide and systematic application of Merrifield's radioautogram technique (Merrifield and Littau, 1968), eventually supplemented by an extension of the resolving power by an electron microscopic method.

Reactivity reductions being liable to vary in space and time, their statistical consequence is product inhomogeneity, i.e., the occurrence *inter alia* of underdone sequences—"underdone" meaning shorter because of residue omission or because of chain truncation.

*A statistical evaluation of accumulation of class 1 errors has been presented by Baas *et al.* (1971) and by Bayer *et al* (1970).

a. Temporarily Interrupted Chain Growth—Omission: Omission has been discussed and defined by Bayer *et al.* (1970) as a result of temporary nonreactivity of potential reaction sites. It is thus due to temporary local conditions, predominance of which is not necessarily restricted to single, nonconsecutive reaction stages (cf. Merrifield's angiotensinyl-bradykinin example discussed in Section IIB).

b. Permanently Interrupted Chain Growth—Truncation: Truncation may occur freely, i.e., on individual resinopeptide chains, by fortuitous yet permanent closing off. Events of this type will in a constant environment yield truncates of all possible molecular weights, with no one in significant molar excess over the others.

Changing the environment of reactive sites during resinopeptide growth may cause a monotonous change in the extent of truncation. Such expectation, and also the expectation of milieu and reagent dependence of truncation, is indeed met by observations: in the course of preparation of our hexaleucylresin, the number of titrable amino groups (protonation by HCl/dioxane, removal of excess HCl by dioxane and $CHCl_3$, and elution of chloride by $CHCl_3/NEt_3$) per gram of dry resin remained constant throughout stages 1–6; the number of amino groups reactive to ^{14}C-acetylimidazol in CH_2Cl_2, however, decreased in every stage by about 4% of the preceding value (Fankhauser *et al.*, 1973a).

Truncation may also occur as a coordinated event involving whole chain populations situated in areas of deswelling and shrinkage, or otherwise particular. This type of event will lead to one or only a few species of truncates which may be isolable from contaminating background material and from the accompanying complete peptide. The amino acid residue situated next to the matrix is naturally more than others liable to become involved in incompatibility situations. As regards pertinent examples, attention may be drawn to the numerous cases of *C*-terminal amino acids encountered in crude SP peptide preparations. A pertinent contribution comes also from kinetic work of Scotchler *et al.* (1970) on the effect of 6 N aqueous HCl in acetic and propionic acids on resin-fixed Boc-amino acids. Initial rates being largely identical, the final extent of amino acid removal from the resin (50 and 75%) was solvent dependent. On the assumption of primary liberation and consequent protonation of amino groups, it is tempting to explain the noncompletion of hydrolysis on the basis of Donnan equilibria responsible for low HCl concentration at nonhydrolyzed ester sites. This would then amount to an example of a lock-out rather than a closing-off mechanism.

Monotonously repeated chain lengthening as in the resinohexaleucine case must not necessarily exert a monotonously increasing effect on the reactivity of the terminal amino group: in the SP synthesis $[Ala]_{10}$-ε-DNP-Lys,

Richards *et al.* (1967) found among all detectable byproducts only two, ε-DNP-Lys and [Ala]$_5$-ε-DNP-Lys, to be present in a significant amount. Abrupt reactivity differences as revealed by the appearance of the hexapeptide must probably be attributed to some sort of defined phase change, e.g., a sudden occurrence or disappearance at a certain stage of some ordered structure.

    c. Chain Breaking—Pseudo-omission: Chain breaking is not specifically related to—but perhaps sometimes favored by—the character of the solid phase, except when linked to transpeptidation (see below) and therefore is not a subject to be discussed in the present context. Mechanistically, it is one of the possible reactions following backbone acylation, diacylimides being easily solvolyzed, or aminolyzed (including "activated" transpeptidation), or intramolecularly degraded with oxazolone formation (cf. Halstrøm *et al.*, 1971).

    2. Overflow Sequences. The origin of excess residues in a sequence is always the result of chemical side-reactions. Most discussed is transpeptidation between neighboring (or associated?) resinopeptides. It is a kinetic consequence of small average distance between reactive centers and may insofar be favored in certain states of the SP. Allowed average distances calculated on the hypothesis of regular arrangement within the SP are rather meaningless for the SP in the swollen state. A second possibility is back-bone acylation by excess acylating reagents and consequent aminoacyl-insertion at the same point. This process does occur (Fankhauser *et al.*, 1973*b*), together with chain breaking, but the respective findings will not be discussed here, since they do not specifically depend on the state of the solid phase.

IV. SOLID-PHASE ANALYSIS

A. General Remarks

    Classical peptide synthesis is made up of repetitive cycles of unit operations. Each cycle and each unit of the whole process are at least in principle based on a known quantity of a pure intermediate. With respect to information, these intermediates represent well-defined and equivalent levels. Analytically, we deal with separation and characterization and with mass relations based on simple balance readings.

    SP synthesis is made up of the same cycles and units, but now these latter represent a series of consecutive steps. The task of analysis now is characterization of every single step by furnishing information on the effected amount of change. The goal is a maximum of knowledge about the state of the completed series. Since data on every step are based on data on the preceding step, because these latter define the respective starting situation, we cannot afford errors. Otherwise, relative errors in consecutive steps being additive, our information falls rapidly to an insignificant level.

A routine analytical method is considered to be satisfactory if the standard deviation amounts to about 1 % of the difference between blank and sample readings. As a rule, this condition is met only within a certain range of sample size. Outside, especially below a certain limit, relative errors increase. But the order of magnitude of the absolute error remains rather constant, depending largely on the reproducibility of the blank reading. Since in SP analysis we are always and exclusively interested in differences, a small absolute error, although relatively large, at a lower sample level (for example, residual amino groups after coupling) may practically vanish as compared to a large absolute error, although relatively small, at the upper sample level (for example, total amino groups after deblocking). A simple, rapid, and cheap method with small absolute error at the upper level and not very sensitive at the lower one is preferable to a sophisticated method with a larger absolute error at the upper level and a much smaller absolute error at the lower level. The reader is invited in this context to compare Dorman's titration of amino groups in its original $Py \cdot HCl/DMF/NEt_3$ version (Dorman, 1969) with Esko's 2-hydroxy-1-naphthaldehyde procedure (Esko *et al.*, 1968; Karlsson *et al.*, 1970). It is quite true that standard deviations of the order of $1\% + 1\% = 2\%$ and more are too large for an efficient step control in SP synthesis, where step yields are supposed to approach the 100 % limit. But the answer to the problem cannot be found in improvement on a single one of the two sample levels under consideration. If it is agreed that satisfaction on both sides is perhaps beyond practical possibility, the answer must come from an approach which avoids the concept of lower and upper sample levels altogether. Instead of sometimes determining what is there after a step (e.g., deblocking), and sometimes what is left after a step (e.g., coupling), we should concentrate our full attention on the latter task. That one is more rewarding, because absolute errors in trace analysis are small by definition and because too large relative errors are often susceptible to being dealt with by due sophistication. Our conclusion, then, is to propose the invention of blocking groups which are directly accessible to quantitative analysis, thus opening the way to a direct estimation after deblocking of residual blocking groups, as an analog to direct estimation after coupling of residual amino groups. The same reasoning applies evidently to any other type of group analysis to be used in improving SP synthesis control. In composition analysis, for example, when estimating relative and absolute amino acid content of the resinopeptide, we are always by the very nature of the problem at the upper sample level and therefore have to accept large absolute errors, even in the optimal case of standard deviations around 2 %.

B. Stoichiometry

At first sight, the problem of how to relate findings at stage i to those at stage $i + 1$ seems trivial. Given the fact that SP processes are neither quantitative nor devoid of side-reactions, there is always a starting mass M_0 of polymeric support, and there are at stages i and $i + 1$ masses M_i and M_{i+1} of resinopeptide polymer SP_i and SP_{i+1}, respectively. We can at these same stages always try to estimate the molar amounts x_i and x_{i+1} of resinopeptide present per gram of dry SP_i and SP_{i+1}, for example, by measuring N-terminal amino groups in a moist sample and then drying to constant weight. The validity of that estimation depends entirely on the reliability of the applied method and often is grossly overestimated. Whenever possible, the result on the sample should be checked by a concomitant analysis of the bulk of the SP, except for the drying step, which might be detrimental. Besides, this sort of duplication yields estimates of M_i and M_{i+1} in the dry state, the latter being essential with respect to material balances.

1. A Definition of Yield. In the hypothetical case of a quantitative conversion* when going from SP_i to SP_{i+1} by addition of the mass $M'_{Ri} = M'_i \cdot x'_i \cdot R_i$, R_i being the residue weight of the residue added in this step, we find on the basis of conservation of mole numbers

$$M'_i \cdot x'_i = M'_{i+1} \cdot x'_{i+1} \tag{1}$$

and by elimination of masses M

$$\frac{1}{R_i}\left(\frac{1}{x'_{i+1}} - \frac{1}{x'_i}\right) = 1 \tag{2}$$

which is the relation between x'_{i+1} and x'_i when the yield to stage $i + 1$ is 100%.

In the real case of partial conversion, side-reactions being absent (!), the mass added to M_i is now $M_{Ri} = M_i \cdot y_i \cdot R_i$, with $y_i < x_i$, where x_i is the estimated molar amount of reactive sites present per gram of dry SP_i, and y_i the molar amount of reactive sites per gram of dry SP_i that actually do react. The total mass of SP_{i+1} now is $M_{i+1} = M_i + M_{Ri} < M'_{i+1}$, and the estimated molar amount of reactive sites per gram of dry SP_{i+1} now is $x_{i+1} > x'_{i+1}$. Instead of equations [1] and [2], we now have [3, 3a] and [4]:

$$M_i \cdot x_i = M_{i+1} \cdot x_{i+1} \tag{3}$$

$$M_0 \cdot x_0 = M_n \cdot x_n \tag{3a}$$

*The following symbols for mass and molar quantities referring to such hypothetical processes are marked with prime signs in order to avoid confusion with the corresponding quantities referring to real cases, used later in the text.

expressing conservation of mole numbers. This conservation should be experimentally verified, because equality of total molar amounts of resinopeptide in SP_i and SP_{i+1} as demonstrated on the bulk of the material is a necessary (although insufficient) check with respect to the postulated absence of side-reactions. Rearranging [3], we obtain

$$\frac{1}{R_i}\left(\frac{1}{x_{i+1}} - \frac{1}{x_i}\right) = \frac{y_i}{x_i} \qquad [4]$$

expressing the actual yield of the reaction between stages i and $i + 1$— It should be noted that x_i and x_{i+1} must come from sample analysis if drying of the bulk of the SPs is to be avoided. Multiplying both sides of all equations [4], with $i = 1, 2, 3, \ldots, \ldots n$, by R_i and summing up, we get

$$\frac{1}{x_n} - \frac{1}{x_0} = \sum_{i=0}^{i=n-1} R_i \frac{y_i}{x_i} \qquad [5]$$

and we find for the average yield

$$\overline{\left(\frac{y_i}{x_i}\right)} = \frac{1/x_n - 1/x_0}{\sum_{i=0}^{i=n-1} R_i} \qquad [6]$$

When mass losses of all kinds are excluded, the mass $\Delta M = M_n - M_0$ of the resinopeptide on the SP_n is given by

$$\Delta M = N\left(\frac{1}{x_n} - \frac{1}{x_0}\right) \qquad [7]$$

N denoting the number $M_n \cdot x_n = M_0 \cdot x_0 = M_i \cdot x_i$ of moles of resinopeptide in the SP at any stage i.

Equations [4–6] still hold in case of mere abrasion loss of the SP, which does not affect its composition, while [3] and [7] need a correction factor. An analytical control of abrasion loss is essential but tedious, and therefore the SP process should be conducted in such a way as to exclude abrasion as much as possible. An observed loss of mass, especially a sudden one, is then due to other reasons, and it indicates that the process is getting out of control. Such development will in general also become evident from yields calculated on the basis of [4]. On the other hand, correct mass relations and reasonable data on yield do not exclude trouble.

2. *Hazards.* By comparison of potential synthesis errors according to their respective effects on M_i and x_i, it is readily seen that simultaneous effects from different sources may compensate each other: there is in the specific case of amino group titration after deblocking

1. An error group 1, with M_i too large and x_i too small, comprising aminoacyl insertion, formation of SP fixed quaternary nitrogen, and solute adsorption, e.g., dicyclohexylurea.
2. An error group 2, with M_i and x_i unchanged, comprising trans-peptidation.
3. An error group 3, with M_i too small and x_i too large, comprising truncation, deblocking of side-chain protection, chain breaking, delayed chain initiation, and loss of support material by SP disintegration.
4. An error group 4, with M_i too small and x_i correct or uncertain, comprising abrasion loss of SP and loss of resinopeptide by SP disintegration.

Theoretically, errors may thus accumulate without becoming at once evident. Practically, however, there is a good chance that this does not occur, provided that more than just one analytical method is used and provided that the choice of analytical methods is such as to yield interrelated results amenable to comparison within a framing pattern (see Section IVD). In view of the heavy burden brought about by performance of numerous analyses and in view of the inevitable loss of material implied by repeated sampling, there is an urgent need for the development of techniques amenable to automation and applicable to the bulk of the SP.

C. A Glance at Practice

Having outlined principles for an analytical approach, we now proceed first to a brief exposition of the situation with respect to lower-level sample analysis, then to a warning, and finally to an example of conflicting pieces of evidence.

Equations [4–6] are only useful if the results on the x values are reliable. Actually, they suffer definitely from lack of references for blank determinations and from lack of standards for procedure testing. The consequences of nonavailability of SP preparations of exactly known composition and reactivity are self-evident.

The uncertainty with respect to blanks weighs particularly heavily in view of the present trend to reduce resinopeptide load of the SP to 100–500 μmoles/g, in order to minimize side-reactions. Blanks for Dorman titrations, to give an example, may change in the course of a synthesis due to changing adsorption of pyridine, formation of quaternary nitrogen, reappearance of temporarily blocked amino groups, hydrolysis of imidazole urethane (formed from CDI and free hydroxymethyl groups; unpublished results), and unnoticed side-reactions such as those of chloroform and methylene chloride with triethylamine, *inter alia* mediated by the SP (unpublished results). To cite another example, we were unable to find a satisfactory blank for quantitative acetylation experiments: polystyrene–DVB copolymer and its chloromethyl derivative when used as references consumed 1.5 μmoles/g and 4 μmoles/g, respectively, of ^{14}C-acetylimidazole (Fankhauser et al., 1971).

Just as means are missing to define a blank, they are missing with respect to a test of the extent of SP response to a given reagent. We can never tell right away whether or not a result is falsified, e.g., by closing off or partition phenomena. What we should do, therefore, is to submit the result to a coherence test, as discussed in Section IVD.

With such uncertainties becoming apparent, it is indeed no wonder that on one hand people may feel tempted to forget analysis altogether, and that on the other hand standardization of repeatable SP procedures—analytical and synthetic—is at present virtually impossible. This state of affairs must above all cast serious doubt on any generalization with respect to claims of superiority of a given method over another one. As wanted as a procedure test and the like (Bayer et al., 1970; Dorman et al., 1971) may be, each outcome risks remaining an isolated event difficult to interpret as long as we have no better control of the parameters which define the state of the SP.

As an illustration of an unclear result produced by a recommended method, reference is made to recently published data on a failure sequence test. Two samples of an SP sequential dodecapeptide $[AB]_6$, the first one prepared with intervening acetylation after each coupling in order to block unreacted amino groups, and the second one prepared without intervening acetylation, were subjected to partial hydrolysis and yielded dipeptide fractions containing 0.1 and 4%, respectively, of peptides [AA] or [BB]. The conclusion seemed obvious. In the first sample, omission chains were ruled out and the difference in [AA] or [BB] from the first and from the second samples looked like a measure of the omission incidence to be attributed to the respective coupling method. Indeed, in case of a statistical distribution of omission errors, the two samples should have contained about 99 and 65% of correct sequential product, and in the case of a fully selective distribution the corresponding numbers would have been 99.8 and 90%. Now the amount of titrable amino groups per gram of dry SP should in the first sample have been 10–35% less than in the second one, because of truncation by acetyl groups of all omission chains. By experiment, however, no such difference was found. The explanation probably is heavy loss of resinopeptide from the first sample by acetylating degradation (cf. Halstrøm et al., 1971). But whatever the reason for the discrepancy may be, the interesting result is not amenable to unbiased interpretation.

Opposing pieces of information must not be considered as a nuisance. They help to avoid premature conclusions, and, being a hint to overlooked events, they may even provide clues for discovery of the latter and encourage further investigation, e.g., a look at the obtaining mass situation. Mass control does not seem to be popular in SP synthesis. In the rare cases where weight changes have been reported, there was no accordance with other analytical data (Bayer *et al.*, 1968; Klostermayer, 1968). This may be one reason for a widespread lack of interest, which we consider not to be in accord with the art.

D. A Suggestion: The Coherence Test

Uncertainties in present SP analysis are overwhelming. But uncertainties can always be dealt with to some extent by collecting different types of data and correlating them through construction of the underlying pattern. A familiar example is correlation of a number of scattered points by fitting a curve. Present-day habits in SP analysis do rarely allow such averaging of analytical information. Instead of providing a single set of data about a single quantity, the analytical approach should indeed provide data sets on as many quantities as possible, all of them being interrelated by the quality of the analyzed material. The information coming from each set must then fit into a common pattern consistent with expectation. Data not passing such a coherence test should be ruled out. Converging information, on the other hand, is a most valuable indication of validity of a hypothesis. We have indeed no other means than such patterns for knowing whether a certain result is significant—within the limits of the pertaining experimental conditions—or not interpretable at all.

V. CONCLUSIONS

Habits acquired in dealing with solution chemistry are not necessarily adequate when applied to solid-phase synthesis and analysis. Comprehension of problems as well as of results must be duly readjusted. There is an urgent need for solid-phase thinking, which means adaptation of thought, imagination, and inventiveness to the implications of virtual nonexistence of a solid phase and to the implications of multiphase chemistry.

REFERENCES

Albertsson, P.-A., 1970, Partition of cell particles and macromolecules in polymer two-phase systems, *Advan. Protein Chem.* **24**:309.

Baas, J. M. A., Beyermann, H. C., van de Graaf, B., and de Leer, E. W. B., 1971, The relative statistical distribution of peptides in stepwise synthesis, *in* "Peptides—1969" (E. Scoffone, ed.), pp. 173–176, North-Holland, Amsterdam.

Baumann, W. C., and Eichhorn, J., 1947, Fundamental properties of a synthetic cation exchange resin, *J. Am. Chem. Soc.*. **69**:2830.

Bayer, E., Jung, G., and Hagenmaier, H., 1968, Untersuchungen zur Totalsynthese des Ferredoxins-I, *Tetrahedron* **24**:4853.

Bayer, E., Eckstein, H., Hägele, K., König, W. A., Brüning, W., Hagenmaier, H., and Parr, W.,
1970, Failure sequences in the solid phase synthesis of polypeptides, *J. Am. Chem. Soc.*
92:1735.

Chou, F. C. H., Chowla, R. K., Kibler, R. F., and Shapira, R., 1971, Incomplete deblocking as
a cause of failure sequence in solid-phase peptide synthesis, *J. Am. Chem. Soc.* **93**:267.

Davies, C. W., and Owen, B. D. R., 1956, The behavior of ion-exchange resins with mixed solvents, *J. Chem. Soc.* **1956**:1676.

Dorman, L. C., 1969, A non-destructive method for the determination of completeness of
coupling reactions in solid phase synthesis, *Tetrahedron Letters* **1969**:2319.

Dorman, L. C., Markley, L. D., and Mapes, D. A., 1971, A model system for studying solid
phase peptide reactions, *Anal. Biochem.* **39**:492.

Esko, K., and Karlsson, S., 1970, Effects of Boc-amino acid adsorption with respect to yield
and racemisation in the Merrifield method, *Acta Chem. Scand.* **24**:1415.

Esko, K., Karlsson, S., and Porath, J., 1968, A method for determining free amino groups in
polymers with particular reference to the Merrifield synthesis, *Acta Chem. Scand.* **22**:3342.

Fankhauser, P., Schilling, B., Fries, P., and Brenner, M., 1973a, Reaktionen an Trägerharzen:
Zur quantitativen Bestimmung des Totalumsatzes, *in* "Peptides—1971" (H. Nesvadba, ed.),
pp. 153–161, North-Holland, Amsterdam.

Fankhauser, P., Schilling, B., and Brenner, M., 1973b, Nachweis von Backbone-Acylierung und
Aminoacyleinlagerung unter Solid Phase Synthesis-Bedingungen, *in* "Peptides—1972"
(H. Hanson and H. D. Jakubke, eds.), pp. 162–169, North-Holland, Amsterdam.

Flory, P. J., 1950, Statistical mechanics of swelling of network structures, *J. Chem. Phys.*
18:108.

Flory, P. J., and Rehner, J., Jr., 1943, Statistical mechanics of cross-linked polymer networks.
I. Rubberlike elasticity, *J. Chem. Phys.* **11**:512; II. Swelling, *J. Chem. Phys.* **11**:521.

Geising, W., and Hörnle, S., 1973, Bestimmung freier Aminogruppen mittels ^{14}C-markiertem
Phenylisothiocyanat bei der Peptidsynthese in fester Phase, *in* "Peptides—1971" (H.
Nesvadba, ed.), pp. 146–152, North-Holland, Amsterdam.

Halstrom, J., Brunfeldt, K., and Kovacs, K., 1971, Acidolytic cleavage of peptide bonds during
acetylation, *Experientia* **27**:17.

Karlsson, S., Lindeberg, G., Porath, J., and Ragnarsson, U., 1970, Removal of *t*-butyloxycar-
bonyl groups in solid phase peptide synthesis, *Acta Chem. Scand.* **24**:1010.

Klostermayer, H., 1968, Synthese von Gramicidin S mit Hilfe der Merrifield-Methode, *Chem.
Ber.* **101**:2823.

Letsinger, R. L., and Jerina, D. M., 1967, Reactivity of ester groups on insoluble polymer supports, *J. Polymer Sci. Part A-1* **5**:1977.

Lunkenheimer, W., and Zahn, H., 1970, Merrifield-Synthese symmetrischer Cystinpeptide,
Anal. Chem. **740**:1.

Marshall, G. R., and Merrifield, R. B., 1965, Synthesis of angiotensins by the solid phase method,
Biochemistry **4**:2394.

Marshall, G. R., and Merrifield, R. B., 1971, Solid phase synthesis: The use of solid supports
and insoluble reagents in peptide synthesis, Effects of physical properties of polymeric
supports, *in* "Biochemical Aspects of Reactions on Solid Supports" (G. Stark, ed.),
pp. 148–154, Academic Press, New York.

Merrifield, R. B., 1967, New approaches to the chemical synthesis of peptides, *Rec. Progr.
Hormone Res.* **23**:451.

Merrifield, R. B., and Littau, V., 1968, Solid phase peptide synthesis: The distribution of peptide chains on the solid support, *in* "Peptides—1968" (E. Bricas, ed.), pp. 179–182, North-Holland, Amsterdam.

Patterson, J. H., 1971, Preparation of cross-linked polystyrenes and their derivatives for use as solid supports or insoluble reagents, *in* "Biochemical Aspects of Reactions on Solid Supports" (G. Stark, ed.), pp. 189–213, Academic Press, New York.

Pepper, K. W., Reichenberg, D., and Hale, D. K., 1952, Properties of ion-exchange resins in relation to their structure. Part IV. Swelling and shrinkage of sulphonated polystyrenes of different cross-linking, *J. Chem. Soc.* **1952**:3129.

Richards, F. F., Sloane, R. V., Jr., and Haber, E., 1967, The synthesis and antigenic properties of a macromolecular peptide of defined sequence bearing the dinitrophenyl hapten, *Biochemistry* **6**:476.

Sano, S., and Kurihara, M., 1969, Synthesis of an analogue of horse heart cytochrome *c* by the solid phase method, *Hoppe-Seyler's Z. Physiol. Chem.* **350**:1183.

Scotchler, J., Lozier, R., and Robinson, A. B., 1970, Cleavage of single amino acid residues from Merrifield resin with hydrogen chloride and hydrogen fluoride, *J. Org. Chem.* **35**:3151.

Westall, F. C., and Robinson, A. B., 1970, Solvent modification in Merrifield solid-phase peptide synthesis, *J. Org. Chem.* **35**:2842.

INDEX